Marine Ecology: Processes, Systems, and Impacts

Marine Ecology

Processes, systems, and impacts

Michel J. Kaiser
Martin J. Attrill
Simon Jennings
David N. Thomas
David K. A. Barnes
Andrew S. Brierley
Nicholas V. C. Polunin
David G. Raffaelli
Peter J. le B. Williams

OXFORD
UNIVERSITY PRESS

OXFORD
UNIVERSITY PRESS

Great Clarendon Street, Oxford OX2 6DP

Oxford University Press is a department of the University of Oxford.
It furthers the University's objective of excellence in research, scholarship,
and education by publishing worldwide in

Oxford New York

Auckland Cape Town Dar es Salaam Hong Kong Karachi
Kuala Lumpur Madrid Melbourne Mexico City Nairobi
New Delhi Shanghai Taipei Toronto

With offices in

Argentina Austria Brazil Chile Czech Republic France Greece
Guatemala Hungary Italy Japan Poland Portugal Singapore
South Korea Switzerland Thailand Turkey Ukraine Vietnam

Oxford is a registered trade mark of Oxford University Press
in the UK and in certain other countries

Published in the United States
by Oxford University Press Inc., New York

First published 2005

British Library Cataloguing in Publication Data
Data available

Library of Congress Cataloging-in-Publication Data

Marine ecology: processes, systems, and impacts / Michel J. Kaiser...
[et al.].
p. cm.
Includes bibliographical references and index.
ISBN–13: 978–0–19–924975–6
ISBN–10: 0–19–924975–X
1. Marine ecology. I. Kaiser, Michel J.
QH541.5.S3M2567 2005
577.7—dc22

2005015178

Typeset by Newgen Imaging Systems (P) Ltd., Chennai, India
Printed in Great Britain
on acid-free paper by
Ashford Colour Press Ltd., Gosport, Hants.

ISBN 0-19-924975-X (Pbk.) 978–0–19–924975–6

1 3 5 7 9 10 8 6 4 2

'It is likely that much present day published science depends on "fact" which has not been sufficiently checked. James Elroy Flecker, in his play Hassan, wrote: "Men who think themselves wise believe nothing until the proof. Men who are wise believe anything until the disproof." Perhaps in this complicated world, one should steer a careful path between Flecker's two extremes.'

G. E. (Tony) Fogg (1919–2005)

PREFACE

Marine Ecology: an Introduction

Approximately 2.2 billion people live within 100 km of a coastline. This figure is set to double by 2025. While the average population density along coastlines is currently 80 people km^{-2}, this increases to up to 1000 km^{-2} in countries such as Egypt and Bangladesh. Many of these coastal inhabitants depend directly upon marine resources for their subsistence or income. The world's oceans provide a wealth of goods and services and are used as repositories for our waste products, yield renewable (wind and tidal energy) and non-renewable (oil and gas) forms of energy generation, provide important bulk transportation routes, provide a source of food, yield mined commodities (diamonds), and provide recreational benefits that support important tourism industries. Technological advances have enabled us to utilize areas of the oceans that previously were inaccessible to humans, and many of the natural resources within the marine environment are either fully or over-exploited. Despite this, much of the marine realm has never been viewed by the human eye, indeed it is sobering to think that we probably know more about the surface of the moon than we do about the marine environment of our own planet. Understanding marine ecological processes and systems are currently urgent research priorities if we are to comprehend the ecological effects of human activities that impact upon them and more widely on global systems. This makes marine ecology an exciting and pivotal subject that has matured into an integrated science that encapsulates biological, chemical, and physical processes from the microscopic to the global scale.

● Nearly 39% of the world's population live close to the coastline.

● The world's oceans are heavily exploited for mineral and biological resources, yet much of the ocean remains unexplored.

The Evolution of Marine Ecology

The development of marine ecology can be charted through three major eras. Early naturalists worked in an age when seafarers' stories of sea monsters abounded and authors such as Jules Verne romanticized exploration of the deep and the battle with the leviathan. The observations of early naturalists such as Darwin were mostly restricted to the shoreline, while scientific sampling of the abyss was performed by lowering crude sampling devices to the distant seabed. This must have been (and remains) an incredibly exciting time as every sample probably contained an organism viewed for the first time by human eyes. Exploration of the oceans continues to be an ongoing task with the discovery of a new phylum in the last decade (Chapter 1). Then in the early to mid twentieth century ecologists such as Petersen and Thorson began to consider ecological rules of assemblage structure for benthic biota living on the seabed. This coincided with the early beginnings of fisheries science and concerted efforts to understand the processes of recruitment and mortality in fish stocks.

● Marine ecology can be considered to fall into three main eras: (1) exploration and description; (2) experimental manipulation; (3) integration and application.

The second era of marine ecology began in the late 1960s and early 1970s, when ecologists such as Connell and Paine undertook their seminal research on the effects of disturbance and competition in ecology, using marine systems as models for their studies (Chapter 5). These early marine studies had a pivotal role in the development of general ecological theory, which then springboarded into the more easily studied terrestrial systems. This theme has developed to the present day, and spawned many manipulative studies of the role of predators and grazers in marine systems and long-term studies of food-web dynamics. Technological advances have enabled us to understand better the processes of primary and microbial production, and the advent of stable isotope analysis has provided a common means to assess the trophic status (i.e. top predator, forage species, detritivore, secondary producer) of species in marine communities from around the world (Chapters 6, 7, and 9).

While we still have a lot to learn about marine ecosystem processes, we have entered a new phase (the third era) in which research has become more urgently focused on the ecological ramifications of an ever-increasing list of human impacts (Chapters 12, 13, and 14). These impacts occur against a background of climate change that is occurring at a rate faster than previously recorded in our time. Activities such as commercial fishing have occurred for hundreds of years, but have now reached such intensive levels that we have witnessed complete changes in ecosystem status. An increasing cocktail of contaminants is pouring into coastal waters, and their pernicious sub-lethal effects cause reduced survival of larval stages and alteration of sexual characteristics in adult organisms. The incidences of deoxygenation in coastal waters have increased in recent years with catastrophic implications for the associated fisheries and aquaculture activities, not to mention degradation of ecosystem functioning, goods and services (Box 1). The latter provide a clue to the direction of marine ecology of the future. While there are still many gaps to be filled, the latest exciting advances are being made by studies that are multidisciplinary combining oceanography, biogeochemistry, and sedimentology together with marine ecology, a trend that is set to continue into the future.

● Human impacts on the marine environment are set against a background of an increasing rate of global climate change.

These are exciting times, but even the multidisciplinary approach is constrained within the boundaries of 'marine science'. This has been the typical approach of scientists, to offer up our findings to the wider

Box 1 Ecosystem goods and services

The concept of ecosystem goods is fairly easy to grasp in tangible economic terms, for example, the value ($US) of fish or other commodities extracted or farmed in the ocean. Perhaps less obvious are the values attributed to the regulating functions such as flood control and coastal defence and the cultural non-material benefits derived from marine ecosystems. Cultural values are often deeply embedded within human society as typified by the blessing of the fleet by the local Roman Catholic Bishop in Provincetown Harbour, New England (inset), where a large proportion of the fishing community can trace their roots to Portugal or the Azores. (Photographs: M.J. Kaiser.)

Provisioning	Regulating	Cultural
Products obtained from ecosystems	Benefits obtained from regulation of ecosystem processes	Non-material benefits obtained from ecosystems
food freshwater fuel biochemicals genetic resources	climate regulation disease control flood control detoxification pollination	spiritual recreational aesthetic inspirational educational communal symbolic

Supporting

Services necessary for the production of all other ecosystem services

Soil formation
Nutrient cycling
Ecological processes/functioning

● Marine ecology is beginning to break out of traditional scientific boundaries and is beginning to interface with economics and the social sciences to understand the wider societal importance of marine biodiversity.

public and governmental bodies at which point we have fulfilled our duty. This approach is rapidly being replaced by the realization that marine ecosystems have tangible value to society, not only in terms of the goods (fish, aggregate, oil) that are yielded, but also in terms of the services that they provide (carbon sequestration, coastal defence, waste repositories). Degradation of biodiversity is thought to reduce the ability of the ecosystems to deliver goods and services, thus biodiversity loss has real economic, societal, and cultural costs for human society.

Most experimental manipulations of biodiversity indicate some relationship between biodiversity and ecosystem processes, however few have drawn out the links between biodiversity change and the output of goods and services. A good example would be the loss of the cultural services that occurs when a pristine flora and fauna is degraded (e.g. loss of whales from Arctic waters deprives Inuit peoples of the cultural aspect of hunting whales, regardless of the practical need to acquire food) (Box 1). Different biodiversity change scenarios (e.g. habitat fragmentation, contamination, over-exploitation) are likely to affect processes to a varying extent, because different kinds of taxa (large or small, primary producers or top predators) are lost under the different scenarios. A good example of the latter is the over-harvesting of bottom-dwelling fish and incidental removal of other seabed animals (**benthos**) on the Scotian Shelf off Canada. This has led to a decoupling of the ecosystem link between water column and seabed processes (**bentho-pelagic coupling**) and ultimately resulted in a system dominated by mid-water (**pelagic**) fishes. In addition, specific goods and services may be underpinned by several different processes.

● While ecosystem goods are reasonably simple concepts to grasp, e.g. the acquisition of food or minerals, the cultural value of biodiversity is a more abstract concept.

Using This Book to Study Marine Ecology

Marine Ecology: Processes, Systems, and Impacts has been written to address the current need to understand the application of marine ecology in a marine environment strongly influenced by human activities. The structure of the book reflects the integrated approach to marine ecology that is necessary to answer many of today's key marine environmental and conservation problems, that form the focus of the last section of the book. The book is divided into 15 chapters arranged in four distinct sections. The opening chapter deals with the processes that affect patterns at a variety of spatial and temporal scales, diversity, community organization, and structuring processes, and provides a palaeoecological perspective on present-day marine systems. As a marine ecologist, it is important to have a grasp of the relevant time-scales that impact upon the systems in which we work. While much of the ecology we encounter deals with either instantaneous or relatively short term (1–3 years in duration) processes, evolutionary ecologists think in terms of thousands to millions of years. Some environments such as the deep sea have remained relatively stable on an evolutionary timescale whereas coastal and shelf habitats have experienced far more frequent changes. This dynamic flux in the near coastal habitat is brought into focus by present-day findings, at sites currently 40 m beneath the sea to the west of Florida in the Gulf of Mexico, of human artefacts and animal bones from the early Holocene with evidence of butcher cuts and other implement shaping.

Processes contains two chapters that address the fundamental global processes of primary and microbial production that fuel marine systems and the ecology of the organisms responsible. These two chapters underpin the following section. Systems then addresses in more detail estuaries, rocky and sandy shores, the pelagic environment, continental shelf seabed, the deep sea, mangroves and seagrass meadows, coral reefs, and polar seas. Impacts tackles some of the most pressing environmental issues relevant to the marine environment that span the systems described beforehand, with chapters on fisheries, aquaculture, disturbance, pollution and climate change, and finally marine conservation. The penultimate chapter also includes a consideration of the experimental approach needed to determine the effects of human impacts and deals with common pitfalls made by students undertaking field and laboratory projects.

Other textbooks have traditionally dealt with key physical and chemical environmental processes as a discrete unit, usually towards the beginning of the book. This section can be overlooked or its importance not appreciated at the time of reading or forgotten by the time it becomes relevant in the text. In *Marine Ecology: Processes, Systems, and Impacts*, we have integrated the environmental processes at key points that are cross-referenced throughout the text, so that their relevance is immediate and the learning process enhanced. Some of these key processes are reiterated from a slightly different perspective or even repeated in a number of chapters so that learning is reinforced. Key words and concepts are highlighted in bold throughout the text to aid learning and revision, and explanatory and supplementary text appears in the marginal column. In addition to the definitions and explanations given in the text, an excellent online glossary of important marine biological terms can be found at the web site of the Marine Life Information Network for Britain and Ireland (MARLIN). See www.marlin.ac.uk for more information. Boxes are used to expand specific points, and offer interesting examples and case studies that illuminate the concepts being introduced. Finally, we have provided a short list of further reading, trying wherever possible to recommend widely accessible literature and have provided a list of websites that will enable you to learn more about specific subjects or disciplines. Full citations for all references given in the chapters can be found at the end of the book.

Marine Ecology is a highly relevant and challenging subject that is fascinating to amateur, student, and professional alike. We hope that students using this textbook will gain a real feeling for the excitement of marine ecology as a subject of interest, a hobby and a potential career.

Online Resource Centre

Marine Ecology: Processes, Systems, and Impacts is supported by an Online Resource Centre, which includes figures from the book available to download and a web link library including all the URLs cited in the book. Visit **www.oxfordtextbooks.co.uk/orc/kaiser/**.

Michel Kaiser

ACKNOWLEDGEMENTS

As with any large project, this book could only be brought to fruition with the help and goodwill of many others.

The authors thank the following for the collaboration and help with the reproduction of illustrative materials: Alice Alldredge, Kirsty Anderson, Peter Auster, Phillip Assay, Gen Broad, S.C. Cary, Patrice Ceisel, Chelsea Instruments, Joey Comiso, Paul Dando, Gerhard Dieckmann, Emily Downes, Sarah Foster, Brett Glencross, Sönnke Grossmann, Julian Gutt, Christian Hamm, Heinz Walz GmbH, Emma Jackson, Joachim Henjes, Keith Hiscock, James Hrynyshyn, Atsushi Ishimatsu, Matt Johnson, Johanna Junback, Paul Kay, Gunter Kirst, Christopher Krembs, Harri Kuosa, Lewis LeVay, Alan Longhurst, Ian Lucas, Jan Michels, Peter Miller, Lubos Polerecky, Sophie McCully, NOAA, Stathys Papadimitriou, João Quaresma, Ivor Rees, Marcus Reckermann, David Roberts, Craig Rose, Royal Society for the Protection of Birds, Bill Sanderson, Sigrid Schiel, Shedd Aquarium, Smithsonian Institute, South Australian Research and Development Institute, Michael Stachowitsch, Amanda Vincent (Project Seahorse), Toru Takita, Paul Tucker, Tuna Boat Owners of South Australia/Australia and Fishing Management Authority, U.S. Fish and Wildlife Service, Richard Woodcock, Jeremy Young. The opening quote is from an article by Professor G.E. Fogg who died in 2005. The authors are grateful to the family of Professor Fogg for permission to reproduce his thoughts.

Michel Kaiser was in receipt of a senior research fellowship from the Woods Hole Oceanographic Institute, Marine Policy Center, that was essential to conclude this book. David Barnes would like to thank Peter Fretwell for help preparing some of the figures and to the Sedgewick Geology Museum, Cambridge University for the loan of the ammonites for photography.

Michel Kaiser dedicates this book to Diane, Holly and Florence. Martin Attrill would like to thank Karen, Erin and Meryl for always being there. David Thomas dedicates his efforts in this book to his parents, Christine and Neville Thomas. Andrew Brierley dedicates his chapter to his parents, and thanks Jennie, Abigail and Laura for tolerating his absences from home necessary for work at sea. David Barnes dedicates this book to Finn and Sorcha. Nic Polunin dedicates this book to Cally, Christopher and Alexander.

OUTLINE CONTENTS

DETAILED CONTENTS

PART TWO Systems

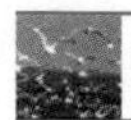

PART THREE **Impacts**

Chapter 1

Patterns in the Marine Environment

CHAPTER SUMMARY

Most space for life on earth is in the vast oceans and seas. Life began in the oceans and remained confined to this environment for hundreds of millions of years. Most major groups (phyla) of animals never left it. The wide expanse of the oceanscape changes dramatically, from undersea mountain ranges to sediment plains and coral reefs to forests of kelp. Patterns of organisms, so obvious at the shore, are also evident from the poles to the tropics, from coasts to ocean centres, from the shallows to the deep abyss, and from millions of years ago to the present day. Patterns occur in species richness, abundance, ancientness, or size, all of which are indicators of powerful changes on the planet surface in time and space. Oceans have widened or been compressed, risen and fallen, heated and cooled and remain dynamic places; most will have changed drastically in just the lifespan of the reader of this book. Examining some of the major patterns in organisms and their biology gives a strong insight into the processes that determine success and evolution of life on earth.

1.1 Introduction

Humans are a land-living species and consequently most familiar with the terrestrial environment, yet the earth is a blue planet and the oceans cover >70% of its surface. Close to the continents this aquatic ecosystem takes the form of shallow seas just a few hundred metres deep known as the **continental shelf**. Most of the oceanic habitat (and 51% of the earth's surface) is nearly 4000 m deep. This three-dimensional environment is populated throughout its depth by many organisms. The steep rises up to the continental shelf contrast with the large sediment-covered basins that span tens of thousands of square kilometres broken only by the undersea mountain ranges of **mid-ocean ridges**. At these locations oceanic crust is formed and gradually moves outwards,

● Our impoverished knowledge of the ocean's inhabitants is emphasized by the fact that it is not just new species that are described, but that entirely new classes or phyla (the highest taxonomic levels of animal types) have been discovered in just the last couple of decades.

pushing continents apart, and eventually disappears down beneath continental crust, forming deep oceanic trenches. Considering its volume, the vast majority (*c*.99%) of the earth's habitat is marine and where most major types of animal (phyla) evolved and continue to live exclusively. Most of the vast water column and seabed (**benthic**) habitat remains unobserved by human eyes. New species are still routinely found in deep-sea samples and even some of the larger animals on earth which live there, megamouth sharks and giant squid, have never been seen alive in their natural habitat.

Standing at the edge of a forest looking out over a prairie, lake, or into the tree canopy, it is easy to see how fragmented the land can be.

Box 1.1 A new phylum

In 1995 an entire new phylum of tiny animals, the Cycliophora, was reported from a discovery two years earlier. As it is only 350 μm in size and superficially resembles individuals of several other phyla of small animals (Gastrotricha, Rotifera, and Entoprocta), it could be considered unsurprising that such animals are still being discovered. Surprisingly, though, the single species (*Symbion pandora*) lives on the mouthparts of *Nephrops norvegicus* – a very common, well studied and widely consumed species, often referred to as scampi. *S. Pandora*, or the 'Pandora', attaches to its host using a sucker and suspension feeds on particles in the water, a small parasitic male (whose sole purpose seems to be for breeding) is also shown attached in the picture below.

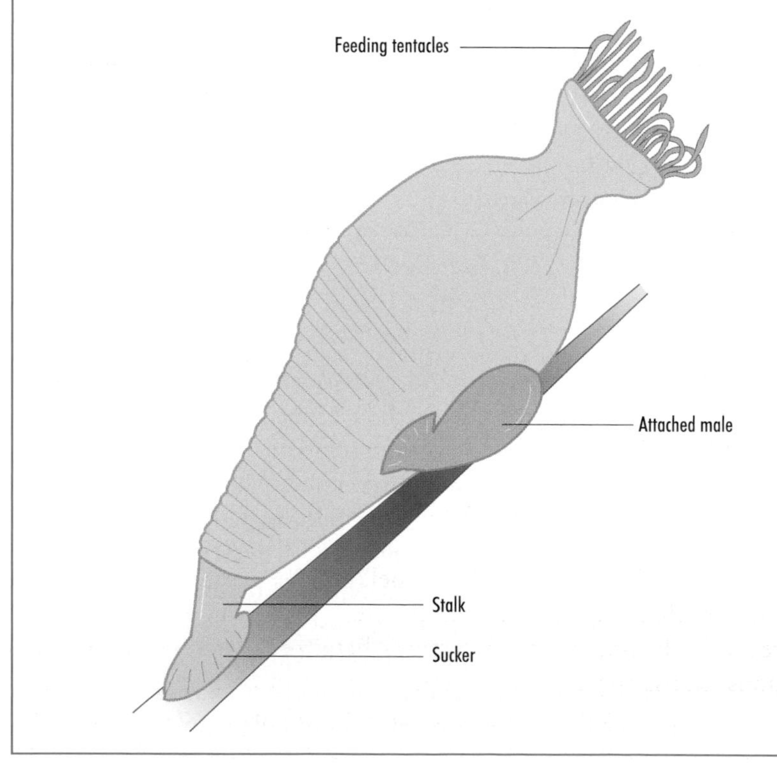

Furthermore even a simple climb up a mountain can reveal altitudinal changes and, though few have experienced it, most people understand that polar regions are deserts compared to the generally species-rich tropics. In contrast, the water column and wide ocean basins might be envisaged as fairly monotonous, uniform ecosystems. However, there are many features that punctuate them abruptly or gradually into many different environments. Changes in time, topography, chemistry, and oceanography allow for the development of patterns across a wide range of scales in time and space, and form the subject of this first chapter and are themes that reoccur throughout this book.

1.1.1 Zonation

Patterns in the marine environment are often beyond our immediate perception; we cannot always see them. Often, patterns are only revealed through sampling and subsequent interpretation. The number and size of samples collected and the type of equipment used will have a profound effect on what is found. Strong differences in opinion exist about even the most basic marine biotic patterns, each defined by evidence from a discrete set of samples. In many respects, patterns in the sea resemble those on land, for example at a large scale along gradients of solar radiation (latitudinal gradients), altitudinal (which in the marine environment is depth, thus bathymetry), from the coast to ocean/continent centres and from young to old areas (e.g. from the mid-Atlantic ridge (new) to the far eastern or western Atlantic sea-floor (oldest)). In the shallowest parts of the sea there are plenty of places where the type or nature of the organisms changes over tens of centimetres or metres, which we term **zonation** (Chapter 5). In warm tropical waters, the type and dominance of corals changes quickly with depth (Chapter 10). In polar seas the abundance and richness of marine life alters equally sharply in response to decreasing physical disturbance by floating icebergs that scour the seabed (Chapter 11). In certain types of environment, such as estuaries (Chapter 4) rapid changes in biological constituents occur along the length of the estuary in response to a suite of changing environmental variables; patterns that are repeated at virtually all latitudes. Zonation is most apparent where the land meets the sea and it is here that it has been studied in most detail (Fig. 1.1a). Zonation is not just driven by tolerance to physical conditions (though this is most important at the high shore level). Connell's (1961) work with barnacles demonstrated that interspecific competition and differential predation pressures strongly influence the location where species survive. In temperate regions across the globe different species and colours of algae and lichens indicate the gradient of immersion on the lower and upper regions of the shore. Shore zonation is equally apparent in some muddy shores in

● In many ways, patterns in the sea resemble those on land with large-scale gradients in solar radiation, altitude (depth), and geological age.

● Zonation is driven at the land/sea interface by a combination of physiological tolerance and competition for space and predation pressure.

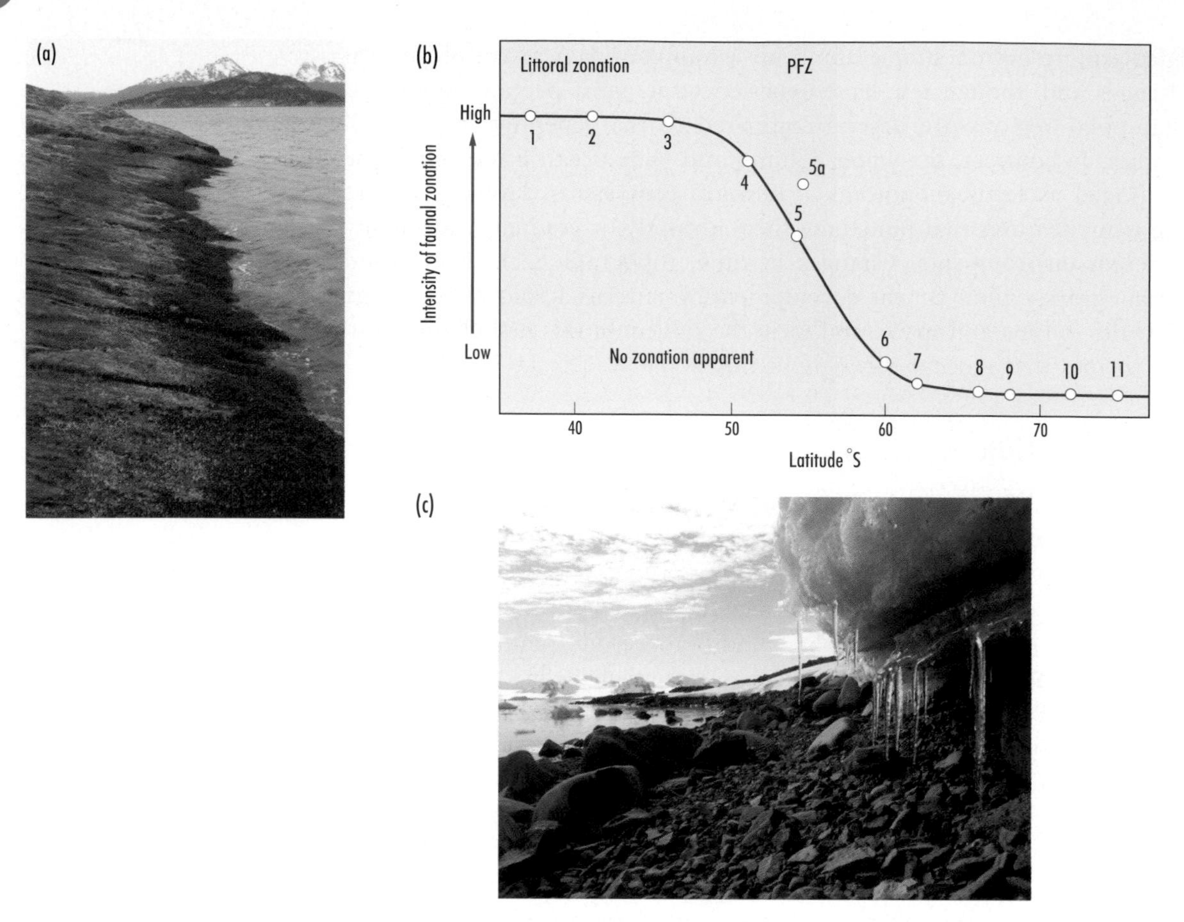

Fig. 1.1 Shore zonation. (a) Bands of barnacles, algae and mussels in Tierra del Fuego (54 °S), southern Argentina. (b) Intensity of zonation in relation to latitude and the position of the Polar Frontal Zone (PFZ). Diagram is a schematic suggesting manner of change of intensity of zonation based on strength of littoral macrobiota patterns at sites at latitudes indicated by 1. Tristan da Cunha, 2. Gough Is., 3. Prince Edward Archipelago, 4. Falkland Islands, 5. South Georgia, 6. Signy Island, 7. South Sandwich Archipelago, 8. Haswell Is., 9. Adelaide Is., 10. Vestfold Hills and 11. McMurdo Sound. 5a is Tierra del Fuego which is actually north of the PFZ. Adapted from Barnes & Brockington (2003). (c) a denuded polar shore (Adelaide Is.).

changing salt marsh vegetation or, at tropical latitudes, in mangrove trees and their associated fauna. Only towards the polar regions does shore zonation become increasingly constrained and ultimately disappear altogether at very high latitudes (Fig. 1.1b). At high polar latitudes (Fig. 1.1c) very few organisms can survive the constant abrasion by floating ice during the summer coupled with encasement in winter ice as the sea surface freezes. Conversely this phenomenon causes zonation in the **sublittoral zone** as the frequency of ice scour (Chapter 11) decreases with depth and distance from the shore.

Box 1.2 Determinants of zonation patterns

Sessile, armoured, cirriped crustaceans (commonly known as barnacles) are very abundant and even the dominant organisms on hard surfaces on temperate rocky shores. Connell (1961) studied the distribution of one species (*Chthamalus stellatus* (below left), which occurs in a distinct vertical zone. Relative to most other marine organisms in the littoral, this zone is high up, so *C. stellatus* individuals have to withstand highly variable temperatures, dessication and longer periods without food. This zone is not, however, of their choosing. *C. stellatus* larvae settle considerably below (and even above) this level on the shore. Other factors, principally competition with other organisms, such as the barnacle *Semibalanus balanoides* (below right), reduce the survival of recruits on the lower shore to near zero. The more heavily armoured plates of *S. balanoides* crush *C. stellatus* individuals when space becomes limiting. The competitors (and many of the predators) of *C. stellatus* find the harsh conditions higher up on the shore too difficult to endure so only at this level can *C. stellatus* prosper.

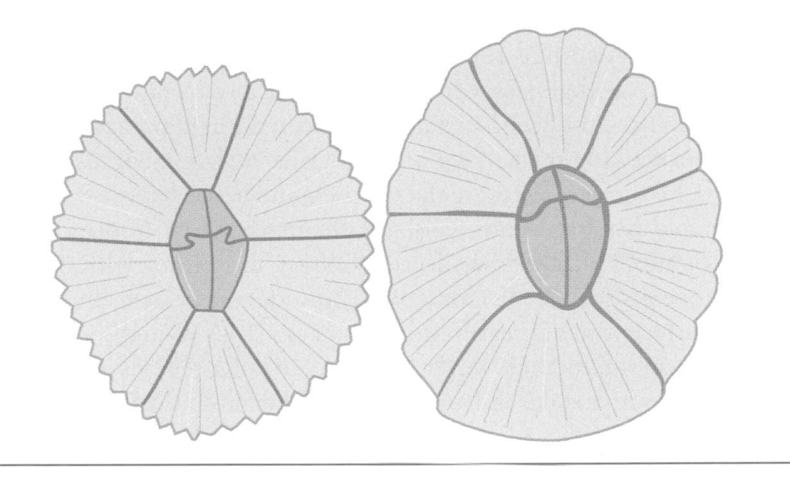

The **littoral** zone has gradients of immersion, exposure, shade, roughness (rugosity), topography, and many others. So it is easy to understand why the littoral zone is species rich. Species specialize in response to different tolerances, exposure, food, and feeding methods (as well as to other factors), so there is potential for many species to co-occur when many environmental factors interact. Examples of this on a rocky shore include those species specialized to living on exposed rock face, in crevices, on the undersides or tops of boulders, overhangs and in pools. But what about variability of conditions away from the seabed, in the water column? In this seemingly uniform environment, many species also occur in close proximity. Hutchinson (1961) referred to this as the 'paradox of the plankton'. The plankton contains many permanent representatives, but also many transient constituents such as the larvae of seabed animals (**benthos**). Larvae of individual species may differ in their time of release, residence time in the plankton, whether they are

● The diversity of environmental conditions found in the littoral zone enhances species richness.

feeding (**plantotrophic**), or have their own yolk sacs (**lecithotrophic**) and behaviour. The water column, like the land and seabed, has many sharply changing physical features that effectively form environmental barriers that constrain the organisms that live there (Chapters 2, 3, and 6).

1.1.2 Oceanography

On a larger scale the water column has a strong pattern of zonation that occurs across its full depth range, similar to that found in the littoral zone. Most importantly in the top few to 200 metres of the water column there is enough light (during the day) for primary producers to photosynthesize, termed the **euphotic zone** (Chapters 2 and 6). As a result, the top 100 to 200 m of open ocean water and 1 to 50 m of coastal water is a very different environment to that found below. Of course, even within the euphotic zone there is a strong gradient of light intensity and wavelength with depth. Beyond a depth of 1000 m (most of the world's ocean volume) the ocean is effectively lightless, with a few small-scale exceptions, such as the bioluminescence produced by bacteria found in the light organs of deep sea biota. Light striking the surface of the marine environment also imparts heat, thus the surface layers are the warmest and hence have lower density than the cold water beneath. Globally the temperature of deep water is relatively uniform at just a few degrees Celsius, in contrast with shallower water temperatures that fluctuate with latitude and season. In the polar regions, surface water is near freezing point at −1.85 °C in the winter, and just positive in summer. However, the water in polar regions is well mixed as dense cold water sinks and wave height and wind speed are greatest at a latitude of 50–60° (Fig. 1.2). Moving away from the polar regions, with a few exceptions such as regions of upwelling, the global ocean is stratified into a warm upper layer, a rapidly cooling zone, and a lower cold zone. The nature of **thermoclines** changes from place to place but in general the thermoclines in the tropics are more than 10 °C warmer than those in temperate regions and they are permanent rather than seasonal. Such stratification is important as organisms require nutrients and minerals as well as light and respiratory gases. Thus away from permanently mixed areas such as polar seas or upwelling zones, the surface layers of the sea also have strong changes in molecules used or produced by organisms, such as nitrates. Locally other gradients, e.g. **haloclines** (salinity) or **pycnoclines** (oxygen) may also stratify water layers. The salinity of seawater is typically about 35 psu. Generally changes in salinity that occur with either latitude or longitude across oceans are small, but are a little greater in the tropics (due to increased evaporation of surface water) and lower close to continents and in the arctic due to the influence of greater fresh water run-off (Chapter 4).

● Thermoclines are a vertical zone of rapid temperature change in the water column.

● psu – practical salinity units, (see 4.1.4.).

There are many features that partition the ocean environment, and the positions of continents, islands, and subsurface marine mountain chains form obvious physical barriers. In the open ocean, differences in water density, currents, and **fronts** provide the conditions that lead to distinct water masses. Although current direction and velocity of water masses and patterns of wind across the globe are complex, there are general large-scale patterns that occur, most obviously in water flow. Most deep (seabed) water is derived from the Southern Ocean. Cooled dense water (termed Antarctic Bottom Water, AABW) sinks in the Southern Ocean and flows away from Antarctica well into the northern hemisphere abyssal regions. There may be several currents that occur at different depths between the deep AABW and surface water. For example, cool northern water flows southwards in the mid region of the water column

● Fronts occur at the boundary between water masses with different physical characteristics.

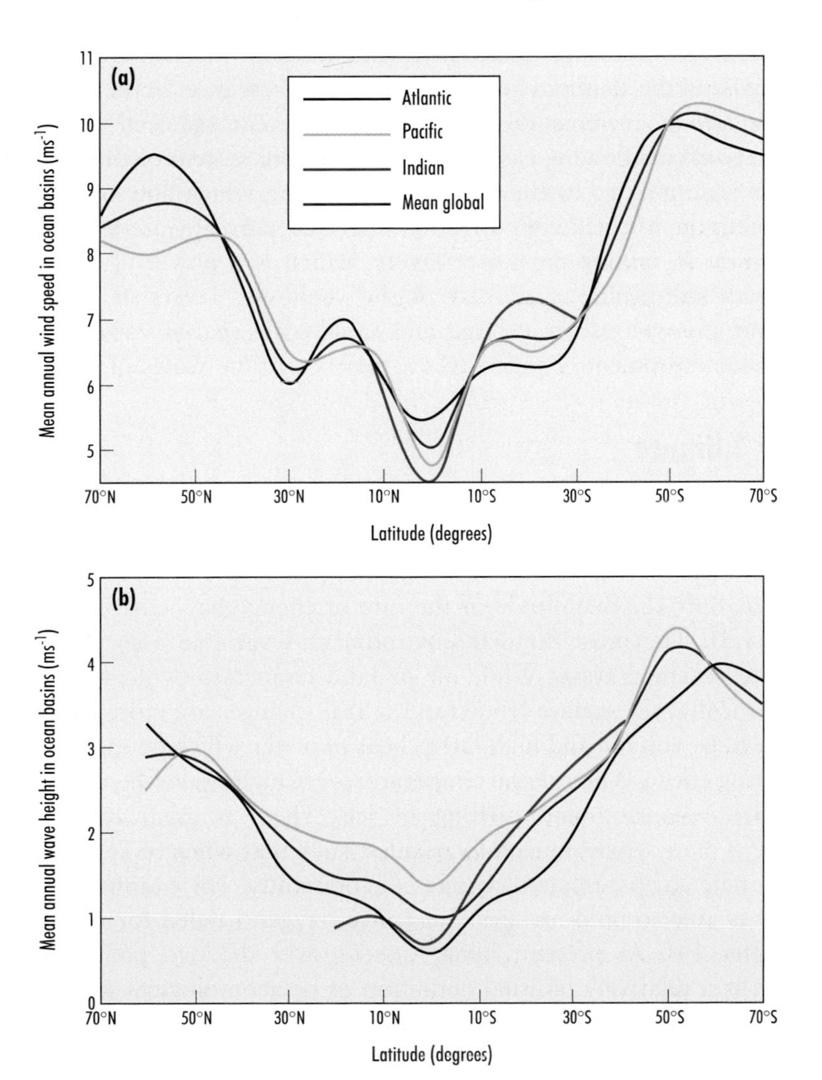

Fig. 1.2 The distribution of mean wind speed and wave action in the marine environment with ocean basin and latitude. Plotted from data from Bentamy et al. 1996.

Table 1.1 Example of changes in the marine environment with time scale.

Time scale	Feature
Hours	High to low tides
Days	Spring to neap tides
Months	Seasonal temperature/salinity/currents
Years	El Niño Southern Oscillation
Decades	Climate warming, ice sheet retreat
Centuries to millennia	Ice ages, Milankovitch cycles, sea level fluctuations, sea floor spreading

in the Atlantic. At the sea surface, the world's oceans are dominated by a series of gyres rotating clockwise in the northern hemisphere and anticlockwise in the southern hemisphere. Where these meet in the equatorial region strong currents flowing westwards occur (as well as smaller countercurrents flowing eastwards). The current system of the Southern Ocean is dominated by the circumpolar current, which flows around the continent in a clockwise direction. Surface currents are particularly important in mixing the water layers, which is a powerful agent that counters stratification, at least in the shallower layers of the water column. As well as these large and small-scale spatial variations, the marine environment changes over a variety of time scales (Table 1.1).

● Surface currents are particularly important in mixing the water layers due to the frictional stress that occurs between two moving bodies of water.

1.1.3 Climate

There is now much concern about climate change and the extent to which human activities are linked to this phenomenon (Chapter 14). The climate of the earth, however, has been in a state of constant change, only the magnitude of the rate of change has varied with time (Fig. 1.3). The most familiar environmental variable associated with climate is temperature. While air or land temperatures fluctuate more dramatically, sea surface temperature (SST) changes are more subtle due to the huge volume and high latent heat of water which gives it a strong buffering effect. When global temperatures are high sea levels rise through thermal expansion and melting of ice. There is great connectivity between many environmental variables, such that when SSTs rise, other environmental parameters change concomitantly. For example warmer water is able to hold less gas and hence oxygen needed for respiration (Chapter 14). At present, icecaps occur over the two polar regions, which is a relatively unusual condition in palaeontological time scales. Ice sheet expansions occur every winter (the geographical extent, but not the volume, of Antarctic ice doubles from summer to winter). On a very

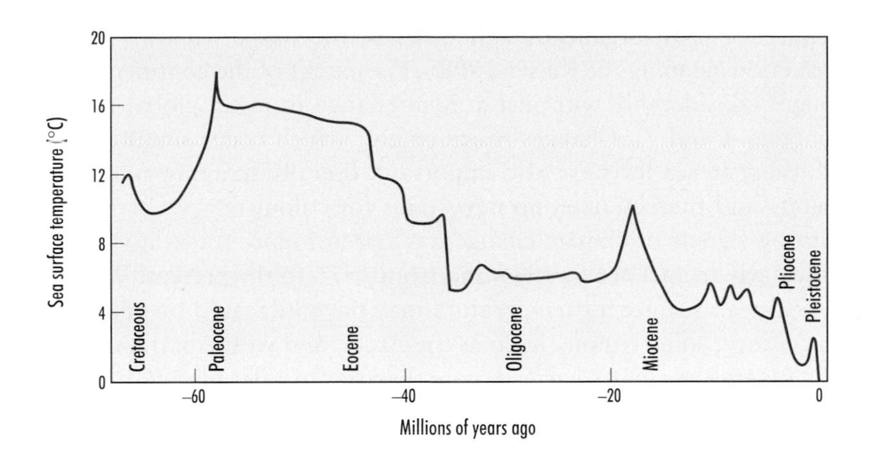

Fig. 1.3 Sea temperature change with time in the Southern Ocean (Adapted from Clarke & Crame 1989).

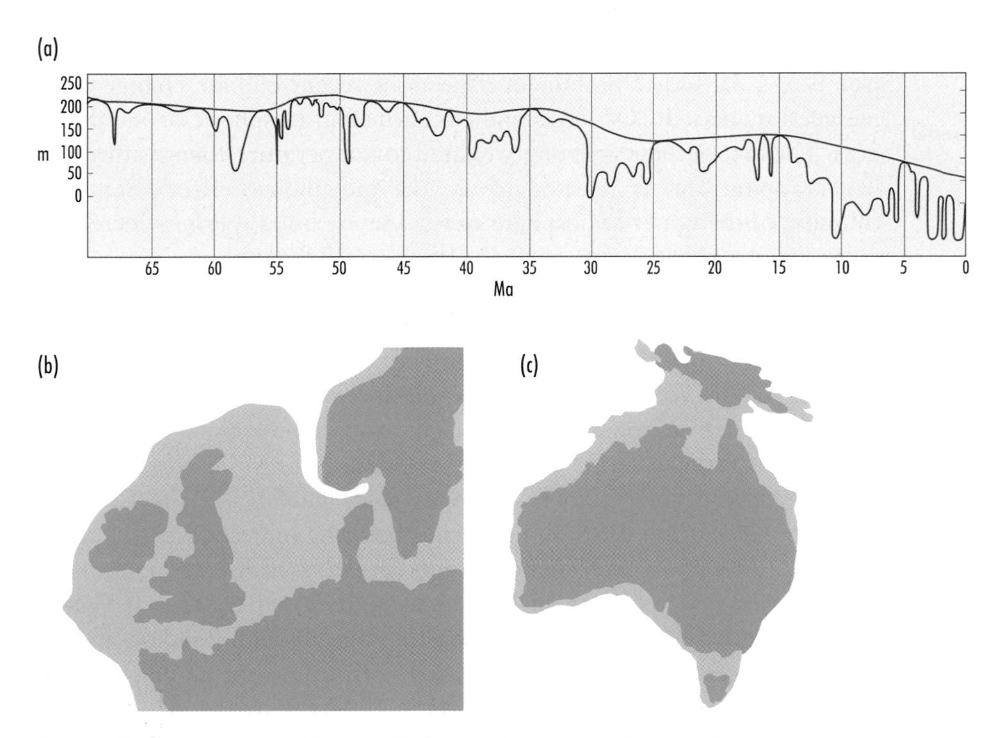

Fig. 1.4 Sea level change with time. Global sea level change from the Mesozoic to present (a) and the coastline of NW Europe (b) and Australasia (c) *c.*18 thousand years ago (in light green) and now (in dark green).

much larger scale, each ice age spreads the icecap to lower latitudes from the poles, then retreats in interglacial periods. The rise and fall of sea level has clear and important implications for the marine habitat (Fig. 1.4): the shallow sloping continental shelf is one of the most important habitats for most organisms with 90% of the world's marine primary production (Jennings & Kaiser 1998). The extent of the continental shelf changes considerably with just a 50 m change in mean global sea level (Chapters 4 and 7). Changes in ice extent, which occur simultaneously with those in sea level are also important, literally scraping away entire habitats and thereby open up new space for colonists.

● The extent of the continental shelf changes considerably with just a 50 m change in mean global sea level.

Strong signals of climate change have been found in the last century, particularly from 1916 to 1945 and from 1976 to the present. While the earth's mean surface air temperature may have increased by only 0.6 °C in a century, some regions, such as the Arctic and west Antarctic air and soil temperatures, have warmed more than this in a decade (Walther et al. 2003). This is not without precedent; there have been drastic and rapid changes in the earth's surface temperature before, even in the sea. These changes occurred in response to massive methane clathrate releases from below the seabed, major volcanic events and after meteorite collisions (such as that near the Yucatan at the end of the Cretaceous period) (see Box 1.3). Other prominent aspects of recent climate change have included increased CO_2 levels and changes in stratospheric ozone thickness. The first of these is strongly related to temperature changes through heat-trapping and is referred to as 'the greenhouse effect'. Seasonal thinning of the upper atmosphere ozone (**ozone holes**) permits increased penetration of the ultraviolet wavelengths of light with their potential damaging effects to organisms. UV penetration of water is limited to just a few metres of the sea surface, so although potentially a serious issue in terrestrial, littoral, and very shallow aquatic habitats it does not affect the majority of the volume of the global ocean environment. Aside from a strong signal of seasonality in the oceans and climate change, there are various other longer-term but cyclical events such as **Milankovitch cycles** (which concern solar activity, such as solar flares) and the **El Niño Southern Oscillation** (ENSO). ENSO has been the subject of considerable scientific discussion after major events in 1972, 1982/1983 and the early 1990s. An ENSO event centres around raised equatorial Pacific sea surface temperatures, raised coastal rainfall and flooding in the western Americas but is also associated with knock-on global climatic anomalies. One example of such an anomaly is the weakening of ocean currents such as the Gulf Stream (which takes warm water to north-western Europe), resulting in colder climatic conditions and changes in rainfall patterns. The phenomenon of ENSO has been recorded for more than a hundred years and discrete events have been traced back nearly 500 years. ENSO and other elements of climate change, as well as physical and chemical oceanography have a strong influence on the magnitude of productivity in the oceans (below). This has consequences for higher predators, such as fish, mammals, and seabirds as well as fisheries (Grantham et al. 2004).

● Past climate changes occurred in response to massive methane clathrate releases from below the seabed, major volcanic events and after meteorite collisions.

● Although the penetration of UV light affects only the upper few metres of seawater, it has the potential to adversely affect fish larvae and other biota.

Box 1.3 Methane clathrate

Methane clathrate is a type of ice that has methane (CH_3) bound within its crystalline structure. Build-ups of this hydrate are probably created by gases moving along geological fault lines where they crystallize on mixing with cold water. Methane clathrate has long been thought to be common in the outer solar system but more recently it has been discovered in considerable quantities below some oceanic sediments. This is important for several reasons: first, it is a major source of fossil fuel (maybe >90% of natural gas reserves) and second, release of overlying pressure on these deposits could result in large scale discharges of methane through the ocean into the atmosphere. Such releases are thought to have occurred rapidly, on various occasions and may have been responsible for sudden climate warming. For example, some of the rapid increases in the temperature of the Southern Ocean (Fig. 1.3) seem likely to be caused by methane clathrate changing from storage in ice crystals to atmospheric gas and then having a profound effect as a greenhouse gas.

1.1.4 Productivity

Unlike on land, true plants generate little of global marine production (except in intertidal saltmarshes, seagrass meadows, and mangrove swamps). In cool shallow coastal waters, macroalgae such as subtidal kelps, or intertidal green, brown and red macroalgae, are highly productive and can grow rapidly (Fig. 1.5). The productivity of the sea mostly depends on microscopic free-living single-celled algae collectively referred to as phytoplankton (see Chapter 2). Although phytoplankton are individual (or aggregations of) single cells they vary in size from large (200–20 μm diameter), diatoms (microplankton) to tiny (2–0.2 μm diameter) bacteria (**picoplankton**) or even smaller cells (see Chapters 2 and 3). In tropical coral reef systems much of the primary productivity is associated with microscopic algae that occur as symbionts within the tissues of animals, such as corals (Chapter 10). Primary productivity in the marine environment varies over several orders of magnitude from grams of carbon $m^{-2} d^{-1}$ to just tens of mg carbon $m^{-2} d^{-1}$. Total annual global phytoplankton primary productivity is patchy in space but is typically highest around continental shelves and lowest in the centre of ocean gyres (Fig. 1.6) (Chapter 2). Secondary production (mainly by zooplankton Fig. 1.5) is linked directly to phytoplankton production though often appears to have a higher biomass because the **turnover** of primary productivity is so rapid.

● Turnover describes the process whereby cells are eaten or die almost as quickly as new ones are produced.

The rate of primary production varies along gradients of light and nutrient availability. As a result productivity varies across the tropics, temperate and polar regions, with depth, with the proximity to the coast and other features such as upwellings. Pelagic productivity is typically more than an order of magnitude lower than for benthic systems. The majority of benthic production is, however, restricted to only a tiny proportion of the global marine environment on the continental shelf (Chapter 7). The lowest levels of total annual productivity are concentrated around the centre of oceanic gyres such as the Pacific Ocean basins north and south of the equator. Seasonally high productivity is

Fig. 1.5 Macro- and micro- phytoplankton and zooplankton. Examples of macro- primary producers are subtidal kelp (a), intertidal red and brown macrophytes (b) and a seagrass meadow (c). Examples of micro- primary producers are a diatom (*Coconeis*) (d), a silico-flagellate (e) and a radiolarian (f). Foraminiferans are protoctist micro-consumers (g). Example zooplankton (h) includes a jellyfish, a polychaete worm, a euphausid crustacean and a ctenophore (comb jelly) and various invertebrate larvae (i). Photographs: (a, b and c) David Barnes, (d and e) Sandra Wilks, British Anarctic Survey, (f) Claire Allen, British Antarctic Survey, (g) Mark Williams, (h) Simon Brockington, (i) David Bowden.

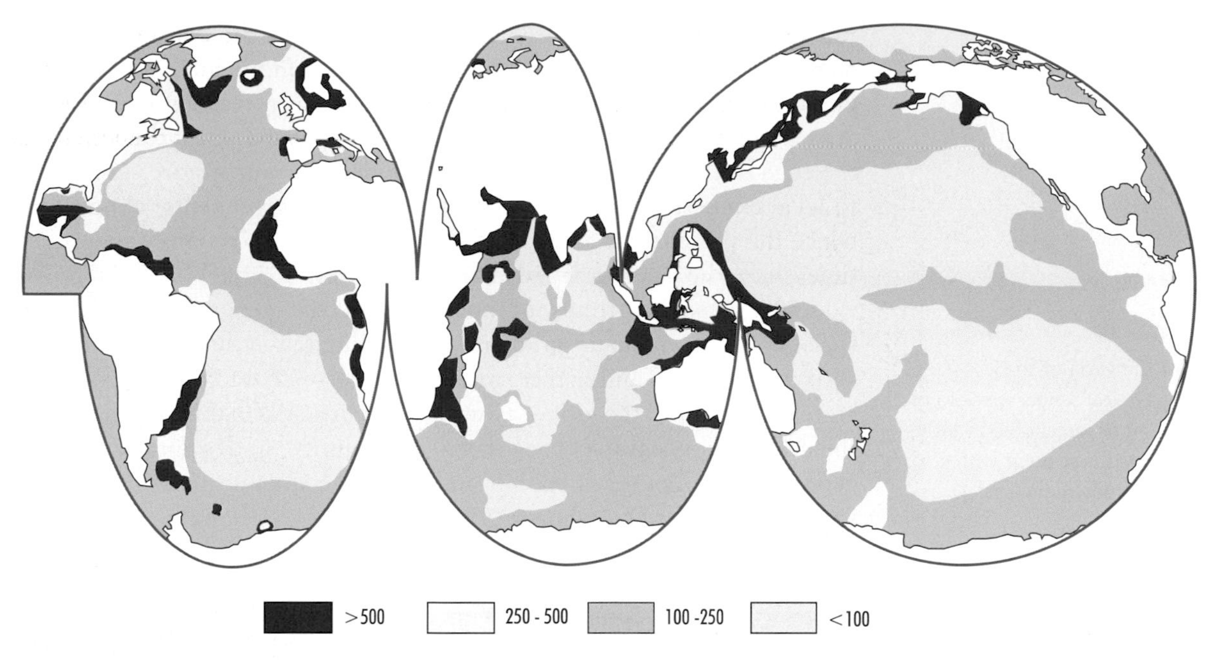

Fig. 1.6 Patterns of annual phytoplankton productivity in the global oceanic environment expressed as g Cm^{-2} year^{-1} (Adapted from Barnes & Hughes 1999).

associated with much of the shallow polar or high temperate coastline and high annual productivity is particularly notable across the tropics of much of the coastal Indo-West Pacific. Patterns of high and low productivity have a largely predictable distribution and are likely to be a major cause in the way other marine organisms are arranged globally. The process of primary production is dealt with in detail in Chapters 2 and 3.

This first chapter examines large-scale patterns in the distribution of marine organisms. The emerging discipline of macro-ecology has grown to quantify, analyse, and explain the patterns of organism groups on large scales (Brown 1995). Biogeography is the science of geographical and historical relationships between and among different organisms (taxa), and is a natural starting point for the examination of large-scale patterns in the marine environment. Later chapters consider patterns and processes that occur in specific systems.

● Patterns of high and low productivity have a largely predictable distribution and are likely to be a major cause in the way other marine organisms are arranged globally.

1.2 Biogeography

Biogeography explores the distribution of species, how groupings of such species form distinct ecosystems and their geographical limits. Individual species (and in some cases higher taxa) each occur across a certain geographic range. Under experimental conditions, many species can endure environmental conditions (e.g. temperature, salinity, sedimentation, and pressure) far in excess of the conditions in which their populations naturally are found. There are many reasons why species are not found across the entire range of their physiological tolerance. At the edge of their geographical range, species are much more susceptible to local extinctions that occur with changing environmental conditions, hence the presence of a species along a gradient may simply reflect the time since the last local extinction event. At a larger scale, there is evidence in some groups of animals that the geographic ranges of species alter along major gradients, and ultimately influence biodiversity (Box 1.4). Species and higher taxa tend to form characteristic groupings by regions around the globe, as well as in particular habitats. The major theme of biogeography has been to identify and characterize the geographic groupings of species and the biogeochemical conditions that make them differ (Longhurst 1998).

● At the edge of their geographical range, species are much more susceptible to local extinctions that occur with changing environmental conditions.

Box 1.4 Variability in latitudinal ranges of North American gastropod mollusc species

Each species occurs across a certain range of latitude. Roy et al. (1998) compared the number of gastropod species with latitude along the Pacific (blue line) and Atlantic (red line) coasts of North America (insert plot). They then investigated whether these species-richness patterns could be explained by species at lower latitude having narrower latitudinal ranges (Rapoport's rule). They found, however, no obvious link and furthermore latitudinal ranges of many tropical gastropod molluscs were higher than those at high latitudes. Roy et al.'s (1998) study was the most serious test of Rapoport's rule and one of the strongest data sets for latitudinal ranges of species for any marine group.

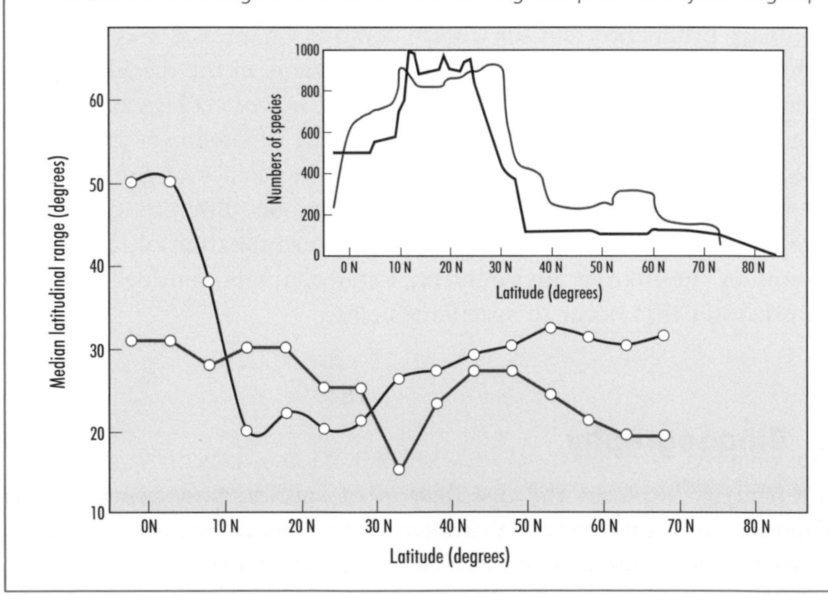

Biogeography has been of interest to both terrestrial and marine biologists for more than a century since early botanists began to match vegetation structures to regional climate types. A succession of pioneering botanists classified groupings of vegetation structures by climate into a series of life zones, **biomes** or **ecoregions** according to environmental characteristics such as precipitation, temperature, and humidity. Similar advances were not easy in the marine environment where mismatches in different aspects of biogeographic knowledge occurred even in the most accessible systems. In the last few decades, underwater scientific instrumentation, advances in marine sampling platforms (ships and autonomous underwater vehicles), and most recently analysis of high quality oceanographic data (Coastal Zone Colour Scanner, CZCS) from the NIMBUS satellite have revolutionized our understanding of ocean surface biogeochemistry. In terms of marine climate, the global marine environment has been considered to consist of three broad domains: polar, temperate, and tropical. In contrast, in terms of their fauna the early zoologists divided the globe into 6 broad realms: Nearctic (North America) and the similar Palearctic (most of Eurasia and Northwest Africa), Neotropical (Central and South America), Ethiopian (Africa), Oriental, and Australian. Boundaries and subdivisions have gone through many subsequent refinements. Although designed around terrestrial biology, theories of island biogeography represent a good introduction to many of the important factors that determine the distributions of organisms.

● Biomes or ecoregions are defined by their environmental characteristics.

1.2.1 Island biogeography

The global terrestrial environment is formed of a series of islands that vary in size, age, isolation, nature, and history. The fauna of islands reflects these factors. The larger, older, and closer to a continent an island is, the greater the number of taxa present. MacArthur and Wilson's (1967) theories of island biogeography explained many of these concepts. Whether marine or terrestrial, animals or plants, the number of species (S) (or higher taxa) increases proportionally to the area of the island (A) according to the **Arrhenius relationship**:

$$\mathrm{Log}\, S = \mathrm{c} + \mathrm{z}(\log A), \text{ where the values of the constants are } \mathrm{c} = c.0.3 \text{ and } \mathrm{z} = c.0.16.$$

Thus on a plot with log scales on both axes, the relationship approximates to a straight line – shown for example by plant numbers on Australian islands or even (approximately) for species on continents. At the other end of spatial scale a number of authors have immersed settlement panels of various sizes into the sea and demonstrated similar increases of species with area of these miniature islands (see e.g. Jackson 1979). There are a number of causes of **species–area relationships**, such as higher numbers of samples (the more samples you take the more species you encounter), more habitats, **immigration** and **extinction** rates, and greater ratios of **speciation** to extinction. Smaller islands are less likely to be encountered by would-be colonizers because they represent a smaller target than a larger land mass. Similarly the more isolated an island, the smaller the pool of organisms that are capable of reaching it. A water or airborne organism would take a long time to reach Bouvet Island in the Southern Ocean, which is thousands of kilometres from the nearest land of any kind. Of course prevailing winds, currents, and chance play a major role in determining which organisms arrive, where and when, which in part explains why there is often much 'noise' in species–area plots of sample data.

● Island size and isolation are important determinants of species richness.

There are all sorts of islands in the marine environment varying from enclosed coastal lagoons, sea lochs, and fjords to seamounts or hydrothermal vents in deep water (Chapter 8). The same principals of colonization and dynamics of fauna apply to these environments. However, species–area relationships tend to be complicated by other factors: for example, Jamaica contains many more species of most organisms than the Falkland Islands even though their land masses are of similar size and distance from their adjacent continental land mass. Jamaica and the Falkland Islands are situated in very different regions (Caribbean cf. Southern Atlantic Ocean), which explains the differences we observe in their species compliment.

1.2.2 Local versus regional patterns

Some regions are richer in taxa than others. Japanese coastal waters are, for example, particularly rich in brachiopods and the Antarctic continental shelves rich in sea spiders (pycnogonids). Overall the Indo-West Pacific region, and in particular around Indonesia, has higher numbers of species of fish, corals, molluscs and probably many other marine animals than elsewhere. On land there are typically more plant species (per unit area) in the neotropics than in other tropical regions, which are in turn more speciose than nearctic or palearctic regions. Thus small areas, or volumes (though these are rarely measured), in species-rich regions will contain more species than similar sized areas in species-poor regions. So in addition to effects such as species–area relationships, there is typically a positive relationship between local and regional diversity. This relationship is useful as it enables prediction of changes at one scale from observations at another. To date, examples of this regional local species relationship principally have been from terrestrial systems (Caley & Schluter 1997). In order to appreciate the relationship of a local

● Generally, there is a strong positive relationship between local and regional diversity, although to date this has been demonstrated best in terrestrial systems.

parameter with a larger scale unit, the unit in question needs to be defined. Although we have already considered a very brief history of the partitioning of the terrestrial and marine realms into smaller areas or units (1.2), it is appropriate to consider this in more depth due to the importance of this issue to biogeography.

1.2.3 Biomes, oceans, and provinces

Using a small set of factors, it is possible to predict a series of community types or 'biomes' on land. Using high-resolution CZCS images of phytoplankton 'greenness' (a proxy for the biomass of a phytoplankton bloom), an analogous series of biomes can be generated for the marine environment. Thus four primary biomes are recognizable from characteristic patterns of phytoplanktonic algal growth, the environmental forcing agents (wind, currents) and fauna at other trophic levels. Longhurst (1998) details these as the (1) Trades, (2) Westerlies, (3) Polar, and (4) Coastal biomes. These are all defined by the agent that determines the depth of the mixed layer of seawater: wind acting over large-scale distances (Trades), local wind speed and light intensity (Westerlies), surface reduced salinity water (Polar), or a complex of processes (Coastal). With the exception of the Trades biome, all include several disjunct water masses. For example, the Polar biome occurs (at high latitude) in each hemisphere and can be split into Antarctic, North Atlantic, and North Pacific secondary associated biomes. As biomes are defined largely on the basis of physical oceanographic conditions and the response of phytoplankton, the exact boundaries are prone to alter with seasonal, ENSO, or other climatic alterations that have even longer cycles. The extent and shape of biomes and their subdivision into provinces is very much dependent on geographic positioning of continents, features of coastlines, and oceanic circulation patterns. As coastlines and even oceanic frontal zones are approximately fractal, a serial subdivision of units could be argued as a valid concept, however, this would require progressively more and more detailed data from which to construct them. The establishment of provinces has followed similar biogeochemical steps to those for the designation of biomes (climatology of mixed layer depth, water transparency, and surface nutrient status), but at a high resolution. The designation of biomes and provinces is very useful for examining many aspects of large-scale patterns in organism ecology and evolution (Fig. 1.7). This is particularly the case for the investigation of the relationship of species-groupings in one area to those in others. One of the most fundamental areas of biogeographic research is the study of the degree of endemism, i.e. how many species (or higher taxa) are unique to a particular area?

● The Trades, Westerlies, Polar, and Coastal biomes are all defined by the environmental factors that determine the depth of mixed layer of the water column.

● The definition of provinces has followed similar biogeochemical steps to those for the designation of biomes (climatology of mixed layer depth, water transparency, and surface nutrient status).

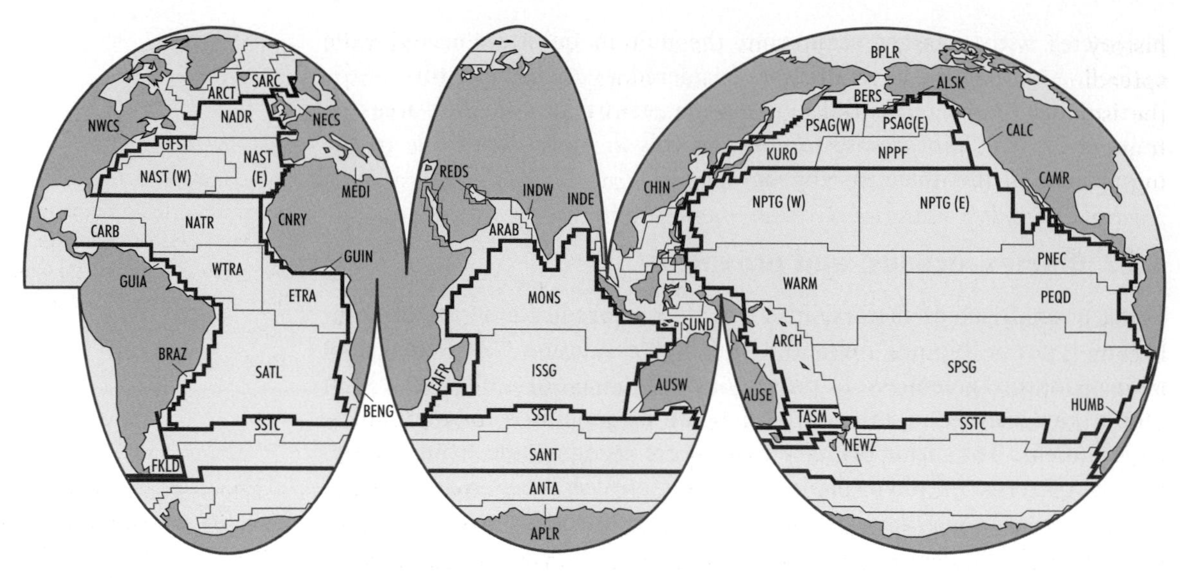

Fig. 1.7 Longhurst's (1998) biogeographic divisions of the marine environment. The oceanic boundaries of biomes (▬) and provinces (—) are shown. The biomes (and provinces within) are Antarctic Polar (ANTA, APLR), Antarctic Westerly Winds (SANT, SSTC), Atlantic Coastal (BENG, BRAZ, CNRY, FKLD, GUIA, GUIN, NECS, NWCS), Atlantic Polar (ARCT, BPLR, SARC), Atlantic Trade Wind (CARB, ETRA, NATR, SATL, WTRA), Atlantic Westerly Wind (GFST, MEDI, NADR, NAST), Indian Ocean Coastal (ARAB, AUSW, EAFR, INDE, INDW, REDS), Indian Ocean Trade Wind (ISSG, MONS), Pacific Coastal (ALSK, AUSE, CALC, CAMR, CHIN, HUMB, NEWZ, SUND), Pacific Polar (BERS), Pacific Trade Wind (ARCH, NPTG, PEQD, PNEC, SPSG, WARM), and Pacific Westerly Winds (KURO, NPPF, PSAG, TASM).

1.2.4 Endemism

At a global scale all species are endemic, at a whole ocean scale many taxa are likely to be endemic but decrease at smaller spatial scales. Typically endemism is used in the context of countries, archipelagos, or islands and seas but also between different sorts of organisms (e.g. while all but one of the marine lizards in the Galapagos Islands are endemic, less than 20% of the fish are endemic). As with most aspects of biogeography and biodiversity, terrestrial data is considerably more advanced compared with that for the marine environment, and the values for many marine invertebrate taxa are simply rough estimates. Table 1.2 shows the similarity in endemism (except in polar archipelagos) of two of the better-known marine and terrestrial taxa at example islands. Care is needed in the interpretation of such data, animals with long lived planktonic larvae (e.g. crabs) are likely to have lower rates of endemism than those with only limited opportunities for dispersal, such as animals that brood their young (e.g. amphipods). Nevertheless, levels and patterns of endemism provide a powerful insight into the evolutionary

● Animals with long-lived planktonic larvae have lower rates of endemism than those with limited dispersal, such as animals that brood their young.

history of areas. Isolation prevents taxa evolving locally and then spreading into other regions, thus the more extreme and longer the isolation the more intense the tendency to endemism. During the fragmentation of the supercontinent of Gondwana, many of the resulting fragments became isolated. Some of the resulting islands, such as Australia, Madagascar, and New Zealand, are famous for their endemic terrestrial vertebrates. Geographically isolated islands such as Hawaii and, on a continental scale, Australia have very high levels of endemism in marine, freshwater, and terrestrial faunas. Another island, Antarctica, became even more oceanographically and climatologically isolated, and a high proportion of the species of marine phyla are endemic (Arntz et al. 1997). Antarctica's marine fauna has few genera and even fewer families that are endemic, which contrasts sharply with patterns in the Australasian and Madagascan vertebrates. Young islands, such as Ascension Island (tropical, mid Atlantic) or regions, such as the Arctic basin, have few endemics. Before comparisons of endemism can be made between locations, considerations of the principles of island biogeography must be remembered. Ascension Island is tiny and young, and even the arctic basin is small in comparison with the Southern Ocean (which covers *c.*10% of the surface of the globe). In one sense, the use of similar-sized areas would be ideal for comparisons of endemism, but what size would be meaningful and would this be consistent from place to place? In the marine environment biomes and provinces are obvious starting points

● Australia and North America lost most of their indigenous animals *c.*100 000 years ago, and Madagascar and New Zealand have lost an even greater proportion in the last 10 000 years.

● Before comparisons of endemism can be made between locations, the principles of island biogeography must be remembered.

Table 1.2 Endemism in marine and terrestrial archipelagos (modified from Myers 1997). Shallow water amphipods and vascular plants are used as example taxa from the two realms. For each locality the island size, total number of species, and percentage of endemics is given. The one polar locality, South Georgia, is shown in italics.

	Land area × 100 km^2	Shallow amphipods		Terrestrial plants	
		No. species	% endemics	No. species	% endemics
New Caledonia	160	172	70	2700	95
New Zealand	2680	113	66	1618	81
Madagascar	5877	314	?	10 000	80
Fiji	183	80	41	1628	50
Japan	3800	300	40	4022	34
Galapagos	80	51	33	701	41
Tristan da Cunha	1.0	24	33	70	29
Society Is.	6.4	32	31	623	44
Bermuda	0.5	51	12	165	9
Ireland	820	263	<1	1210	<1
South Georgia	*36*	*152*	*35*	*25*	*4*

but so are the coastal zones of islands as they are often geographically distinct. The degree of endemism is often a most important variable, particularly in the consideration of conservation issues. However, comparisons between distinct regions are often based on either species richness or diversity per unit area.

1.3 Biodiversity

1.3.1 What is it?

Since the **Convention on Biological Diversity** (CBD) held at the city of Rio de Janeiro, Brazil in 1992, the abbreviated term biodiversity has become an accepted term. Interpretations of what the term means vary, but effectively it refers to the life on the planet, and is used to encompass extinct and living organisms (Box 1.5). There are three main threads of biodiversity: ecological (e.g. biomes, ecosystems, and habitats), organismal (e.g. kingdoms, phyla, and species), and genetic (populations, individuals, and genes). Some of these components, such as genes are unambiguous, while others such as populations or taxonomic ranks are difficult to define with precision. Species is probably the unit of most common reference and species richness one of the most commonly used methods of quantifying diversity. As with any taxonomic rank (or indeed biodiversity), the term 'species' does not have a single definition. An interbreeding group of organisms that is unable to generate young, which are themselves reproductively viable, with any other species is a suitable definition. However, it is somewhat difficult to demonstrate the latter for fossil species! Thus the term **morpho-species** has been widely used (those with observable distinct structural differences, e.g. the colour and shape of a gastropod shell) – but genetic studies have demonstrated that some morpho-species actually comprise many **cryptic species**.

● Biodiversity comprises ecological, organismal, and genetic components.

Even small (0.1 m^2) samples of species assemblages in the marine environment can contain anything from 1 to 10 000 individuals representing from 1 to 150 species. In most cases there will be from 20 to 50

Box 1.5 Biodiversity

This term, a combination of the words 'biological' and 'diversity', essentially refers to 'the variety of life' (see Gaston & Spicer 2004). The CBD used the definition 'the variability among living organisms from all sources including, inter alia, terrestrial, marine and other aquatic ecosystems and the ecological complexes of which they are part; this includes diversity within species, between species and of ecosystems.' Gaston & Spicer (2004) suggested that, amongst the many definitions, this is a good general one if 'living' is altered to 'living and all those that have ever lived' to take into account past forms (the vast majority of life).

species within the sample, but a few species will account for most of the individuals. This means that many different species within a sample are likely to have similar functional roles within the assemblage. This raises the ecological question of 'what is the purpose of so much duplication of similar ecological roles within an assemblage?' Perhaps more intriguing is the notion that some species are **redundant** such that their extirpation from a defined area would have no consequences for local **ecosystem function** (Chapter 5). Hence the prevalence of rare species that share similar functional roles may explain, to an extent, the resilience of marine ecosystems to environmental and anthropogenic influences. As a result, species-poor systems are expected to be those that are most vulnerable to external forcing factors, as the elimination of one or two species may have a proportionately much greater effect on ecosystem functioning.

● Species-poor systems are expected to be those that are most vulnerable to external forcing factors.

The realization that the functional diversity of benthic assemblages may differ considerably from the diversity ascertained from the quantification of separate species (e.g. species richness) has led to the development of new techniques that describe community structure and composition. **Taxonomic diversity** (D) and **taxonomic distinctness** (D^*) are proposed to be more sensitive to variation in environmental stress than traditional diversity indices as they use information derived from the hierarchical taxonomic tree upon which species identities are based (Chapter 14). Taxonomic diversity is defined as the average path length between every pair of individual organisms identified from within a sample. Individuals derived from the same genus have a shorter average path length than individuals within the same family. Taxonomic distinctness is defined as the average path length between every pair of individuals ignoring those between individuals of the same species. These indices circumvent two problems that normally confound detection of more subtle responses to stress by standard diversity indices such as **Shannon-Wiener** (H') and **Margalef's** (d). For example, in stressed environments such as those subjected to severe organic pollution, the dominant species include polychaetes in the *Capitella capitata* complex of species and the genus *Ophryotrocha*. This group contains related species that have a similar trophic role or function. Hence despite the presence of a high number of species from within this group, the weighting given to these species is down-weighted compared with a less perturbed assemblage comprising only a few individuals of these closely related species but greater numbers of taxonomically less related species (Fig. 1.8). In addition, these indices incorporate a multivariate component by retaining the information derived from species identity whereas standard indices give no weighting to species identity such that two assemblages with an entirely different species composition could have the same index of diversity.

● Standard indices of diversity do not incorporate any consideration of species identity, hence two samples with an entirely different complement of species can generate the same value for the diversity index in question.

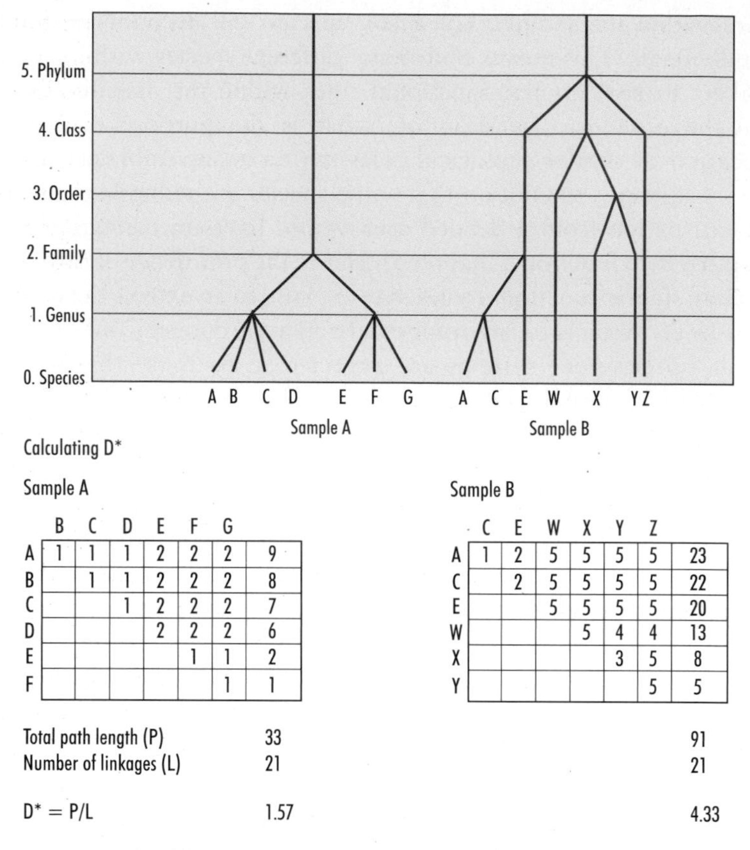

Sample A

	B	C	D	E	F	G	
A	1	1	1	2	2	2	9
B		1	1	2	2	2	8
C			1	2	2	2	7
D				2	2	2	6
E					1	1	2
F						1	1

Sample B

	C	E	W	X	Y	Z	
A	1	2	5	5	5	5	23
C		2	5	5	5	5	22
E			5	5	5	5	20
W				5	4	4	13
X					3	5	8
Y						5	5

	Sample A	Sample B
Total path length (P)	33	91
Number of linkages (L)	21	21
D* = P/L	1.57	4.33

Fig. 1.8 An example of the method of calculating D* for two different samples; A, typical of a polluted site and B, typical of an unpolluted area. The hierarchical linkages are shown with the path length at each level. The species (A, B, C, D, E, F, G, W, X, Y, Z) matrix shows how the path lengths between each species are calculated. D* is the product of the total path length divided by the total number of linkages.

A principal issue with the quantification of biodiversity is the scale at which measurement or sampling is undertaken. There are vast numbers of studies of biodiversity whose scale in time and space varies over many orders of magnitude according to the types of organisms under investigation. For a specific community, there is a standard method for estimating how representative sampling has been. Samples are repeatedly taken, and the number of new species from each successive sample is plotted against the cumulative number of samples. When species numbers no longer increase (the curve reaches an asymptote), the sampling is gauged as sufficient (though this does make a number of assumptions). This is known as a species accumulation curve. The problem is that different scales are appropriate for sampling different organisms, communities and ecosystems. While a 1 m^2 sample would be too large as a unit to assess the richness and abundance of tiny infauna, the same scale would be far too small to assess their wading-bird or fish predators that are far less abundant and much more mobile (Kaiser 2003). So the time or spatial scale to be used to measure biodiversity has to be specifically geared to the ecosystem, community, trophic level, and lifespan of organisms, and even the type of sampling apparatus and protocol (Chapter 14). Given such differences in the method of measurement, comparison of biodiversity values in time and space is not straightforward. Certainly when making or evaluating comparisons that have been made, it is vital to consider the implications and bias due to **scale effects**.

● Sampling needs to be undertaken in proportion to the abundance and mobility of biota, there is no point trying to sample fish communities at a scale of <1 m^2.

1.3.2 Biodiversity through time

By measuring biodiversity of particular types of habitats or particular groups of organisms, it is easy to see that biodiversity is not static in time or space. The number of lower taxonomic units (species, genera, families, etc) has increased many-fold over the approximately 570–600 million years for which we have a fossil record of macro-organisms (Fig. 1.9a). This represents less than 5% of living species past and present. However, we can only estimate the number of species that may have existed at any one time, as for many the fossil record is poor or absent. For every one species for which we have fossil evidence, there may be a thousand whose existence remains unknown. To complicate matters further, the duration of the existence of a species can vary from approximately 1 million to 10 million years. From the limited evidence we have of fossil taxa (Fig. 1.9), the overall increase in numbers of animal families has been uneven, punctuated by a series of sudden decreases. These decreases are referred to as **extinctions**, and the largest occurrences as mass extinctions, which themselves have interesting patterns in time. The rate of speciation also appears to have been variable and major **radiations** have generally occurred 'soon' after the occurrence of extinctions. One theory to explain this sudden radiation is **'barrel filling'**, a term used to refer to the rapid diversification of biota to fill the many empty niches following an extinction event. The pattern shown in Figure 1.9a shows the trend in the number of animal families across time and uses data pooled from all animal families. Such general figures mask the different patterns that occur for each of the component animal groupings. For example, trilobites (Fig. 1.9b) were one of the more abundant and species-rich benthic animal types found in Palaeozoic seas, but it seems that none survived the biggest mass extinction that occurred at the end of the Permian period. The patterns of increase and decrease speciation differed with both the taxonomic level considered and between different specific taxa. For example, in the bryozoan orders Cheilostomatida and Cyclostomatida, contrasts in

● One of the explanations given for major radiation in diversity after mass extinctions is the so-called 'barrel filling' of the resulting empty niches.

● Bryozoans are benthic, colonial, suspension feeders.

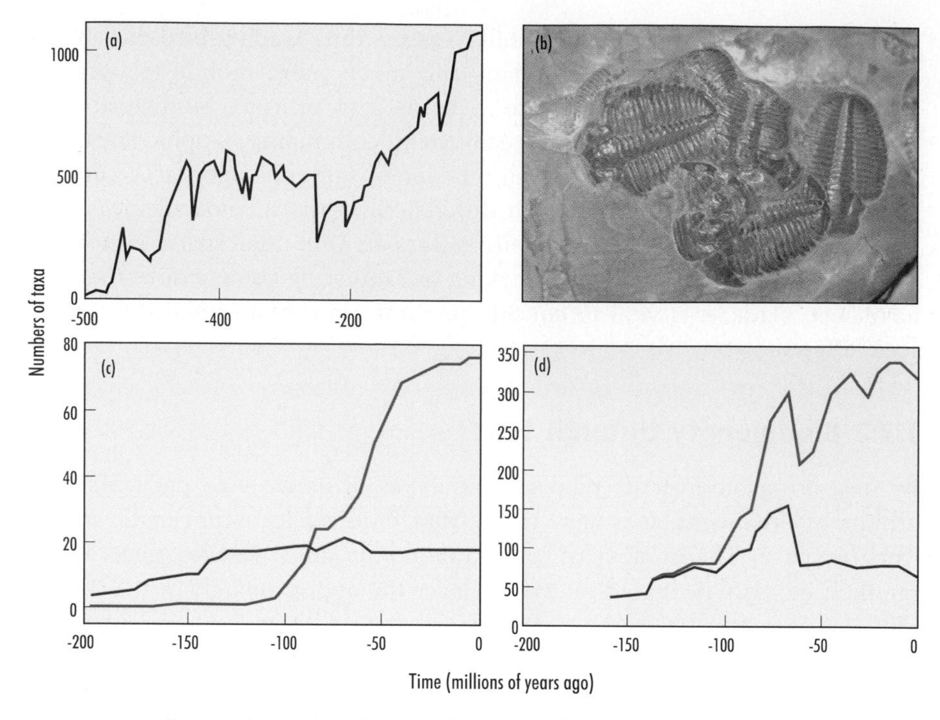

Fig. 1.9 Changes in taxon richness with time (millions of years). (a) Overall numbers of families, (b) fossil trilobites, an extinct phylum, (c) numbers of bryozoan families, and (d) numbers of bryozoan genera. The blue line codes for the order Cheilostomatida and the red line for the order Cyclostomatida. Plot (a) redrawn from Jablonski (1999) and plots (c) and (d) redrawn from Sepkoski et al. (2000). Photograph: David Barnes.

patterns can be seen both among and within representatives of these orders. Furthermore family and genus level patterns contrast strongly, particularly in relation to the **KT mass extinction** 65 million years ago (Fig. 1.9c, d; Box 1.6).

While overall patterns of lower taxonomic levels indicate a substantial increase with time, this may not be the case for the higher taxonomic levels, such as phylum. Stephen J. Gould popularized the work of several scientists that investigated a particularly fossiliferous and ancient outcrop of rock called the **Burgess Shale** (Gould 1989). The fossil organisms revealed in these rocks were not overly speciose but they were very different from each other. Most of the phyla known today are represented in early Cambrian rocks but in addition there were many oddities that could not be assigned easily to existing phyla and resulted in many early misinterpretations of body form and function. The number of taxa present in this early fauna has attracted considerable debate, but there is undoubtedly an almost unprecedented diversity of body types (**disparity**). Whether different organisms are 'lumped' into existing groups or 'split' into new taxa are two extremes of approaches taken by taxonomists. Since the dominance of the enigmatic *Marella* (Box 1.7) in

● Organisms in the Burgess Shale were subject to early misinterpretations of body form and function and continue to generate debate.

Box 1.6 The KT mass extinction

The KT mass extinction refers to the sudden decline of organisms at the boundary of the Cretaceous and Tertiary periods. In the KT mass extinction (K is for *Kreide*, meaning 'chalk' in German, which describes the chalky sediment layer from that time; T is for Tertiary, the next geologic period), all land animals over about 25 kg went extinct, as did many smaller organisms. In total 11% of marine families were lost (which compared with other mass extinctions, except the Permian-Triassic in which 52% of marine families went extinct). The KT mass extinction is famously linked with several events: first, it was approximately at this time that a large meteorite collided with earth in the Caribbean near the Yucatan Peninsula. Second, the remaining dinosaurs perished at this time.

mid-Cambrian rocks there has been a sequential series of dominance by higher taxa, such as trilobites and brachiopods in the Palaeozoic or ammonites and marine reptiles in the Mesozoic. Since the Cretaceous, the cheilostome bryozoans (Fig. 1.9c,d), gastropod molluscs, and teleost fish have dramatically radiated in richness and disparity, a pattern also reflected in the radiation of land mammals. Some taxa, such as the corals and sponges, have been speciose and abundant throughout their fossil record to the present day. Other groups such as the priapulan worms, exhibit remarkably few morphological changes over nearly 600 million years, and remain a minor group in terms of disparity and richness. Based on evidence from only the Cambrian period, or even that from the whole Palaeozoic, it would be difficult to predict which of the higher taxa would be successful today or to predict which phyla might come to dominate in 50–100 million years time. While palaeontologists are primarily interested in the pattern of species through time, for biologists that work with present-day taxa, patterns in space have attracted most attention and debate.

● Based on evidence from only the Cambrian period, or even that from the whole Palaeozoic, it would be difficult to predict which of the higher taxa would be successful today or to predict which phyla might come to dominate at some point in the future.

1.3.3 Biodiversity in space

How many species are alive in the global marine environment today? In total about 2 million species of organism have been described to date from all habitats, of which most have been terrestrial insects or plants. Terrestrial organisms and habitats have been much better sampled, but many believe that they will still dominate species global species richness, even when marine habitats are better studied. Nevertheless, it is important to remember that the seas are much richer in higher taxonomic levels (e.g. Phyla and Classes). At present, providing estimates of the total number of species in the ocean is problematic for the

● Approximately 2 million species have been described to date, but these are mostly terrestrial insects and plants.

Box 1.7 Typical Cambrian fauna

Marella (A) an enigmatic species belonging to an unknown phylum was the most common in Burgess Shale rocks (Cambrian mudstone). Early representatives of extant phyla, such as the priapulan *Ottoia* (B) and polychaetes (C) crawled on these muds together with strange forms like *Hallucigenia* (D), *Wiwaxia* (E) and *Burgessia* (F). Other strange forms were stalked, such as *Dinomischus* (G) or were large swimming predators such as *Anomlacaris* (H). *Anomlacaris* (whose name means strange shrimp) was originally described from just the curled front-most appendages – which were thought to be shrimps. It's circular mouth and main body were also described as separate species until all three fossils were realized as belonging to one larger organism. Other common animals at this time (and well before this period through to present day) were cnidarians such as the jellyfish (I) and the sponges (J).

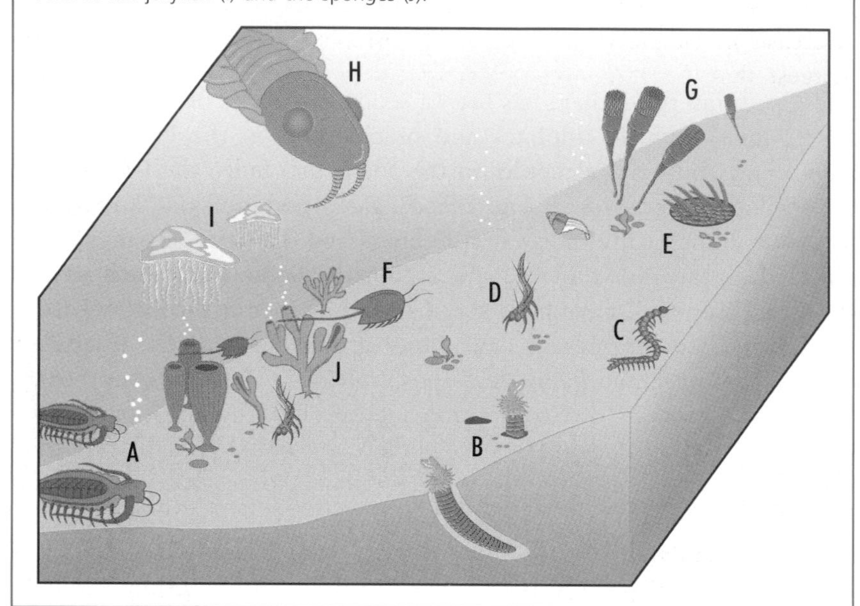

following reasons: most of the oceanic environment is so poorly studied; some taxa, e.g. nematodes, may be very rich in species but there are few taxonomists to describe the species; and some previously well-studied taxa appear to be a complex of many cryptic species, whose identification will require extensive genetic studies. Along a spatial scale gradient, biodiversity is essentially considered at point (α diversity), habitat (β diversity) or regional (γ diversity) levels. Discussions of large-scale patterns of biodiversity almost invariably deal with γ level data. For about two centuries scientists have considered the species richness of organisms to increase from the poles to the tropics. This is supported by robust data sets from the terrestrial environment for plants, insects, birds, reptiles, and mammals. However, considerable debate exists as to the cause of such a **cline** (trend) or whether it is real or partly an artefact of confounding factors (e.g. land area). Data from the marine environment is even patchier, but similar debates rage over the global latitudinal pattern of increasing species richness towards the equator. To date only a few marine data sets cover entire hemispheres or even just multiple oceans. Even the most robust data sets cover only the northern hemisphere and, even then only the American or Atlantic coasts. One of the best-known species–latitude data series is that reported by Roy et al. (1998) for North American gastropods. Gastropod species richness is considerably depressed at high latitudes (Fig. 1.10a) but equatorial values are also significantly lower than those at latitudes from 15 to 30 °N. North Atlantic data for bryozoans and sea-weeds also decrease in richness towards the North Pole, but studies elsewhere suggest that Atlantic taxon richness levels are not representative and

● Clines in species richness from the poles to the tropics could be caused by artefacts such as differences in land area.

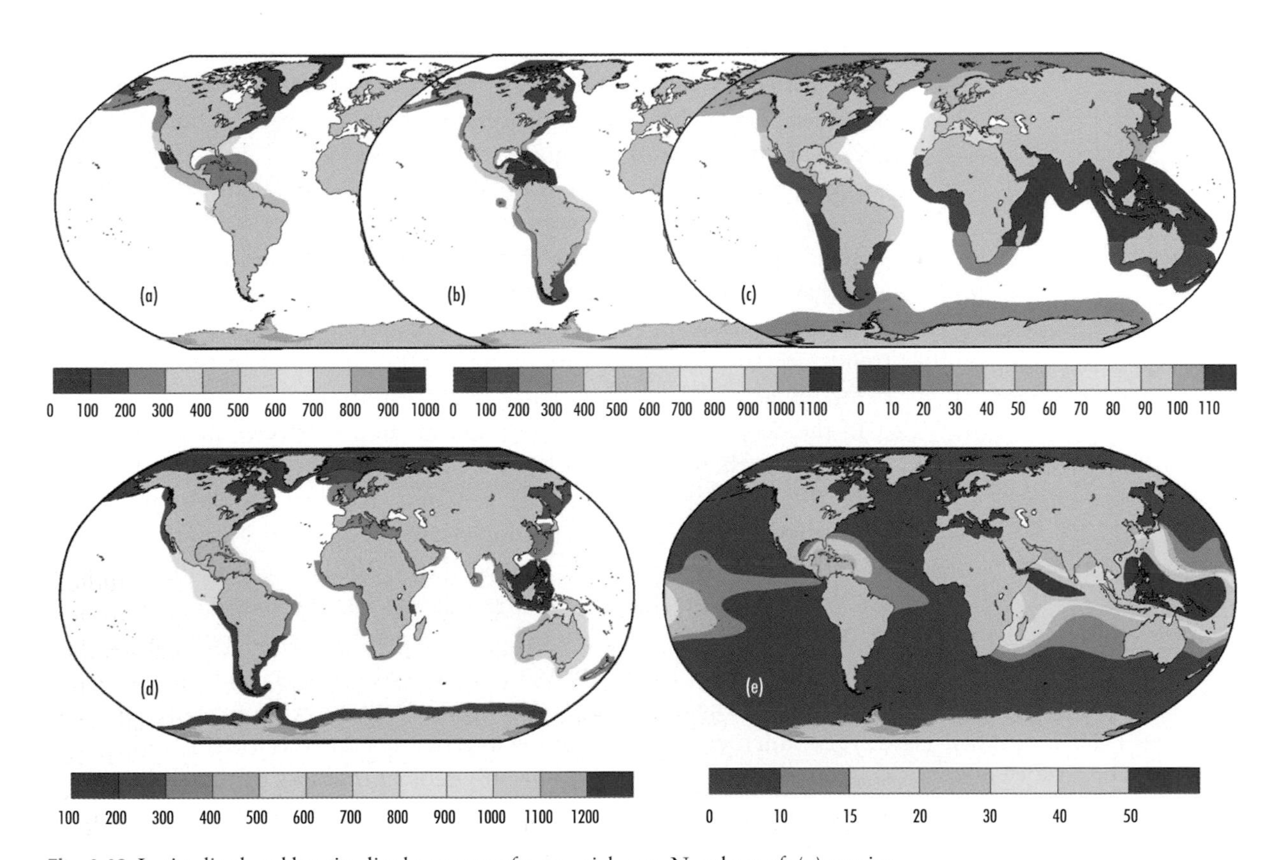

Fig. 1.10 Latitudinal and longitudinal patterns of taxon richness. Numbers of: (a) species of gastropod molluscs, (b) species of decapod crustaceans, (c) species of sabellid polychaetes, (d) species of bivalve molluscs, and (e) genera of hermatypic corals. Plots redrawn from data in Roy et al. (1998), Boschi (2000), Giangrande & Licciano (2004), Crame (2000) and Stehli & Wells (1971) respectively.

thus wide-scale generalizations should not be based on such data alone. Patterns of coastal decapod crustacean (e.g. crabs and lobsters) richness have now been described across the Americas (Fig. 1.10b), and demonstrate a strong decrease in richness towards the Arctic and a weaker decrease towards higher southern latitudes.

● Part of the problem of discerning gradients in diversity is the patchy coverage of major animal and plant groups in the marine environment.

The only truly global data series are those for the bivalves (Fig. 1.10c) and estimates for the reef-building (hermatypic) corals (Fig. 1.10d). Both of these groups show distinct latitudinal gradients, but longitudinal gradients appear equally strong. Basically both patterns radiate out from an **Indo-West Pacific 'hotspot'** (in contrast to terrestrial plants, which are richest in the American tropics). Clarke (1992) suggested that while there was a clear northern cline in species richness, evidence for a similar pattern in the southern hemisphere was at best inconclusive, and remains so today. New data reinforces the existence of the northern trend, but while some studies have supported a southern decrease for decapod crustaceans and bivalve molluscs, other studies demonstrate that gastropod molluscs increase in richness towards the pole. The Southern Ocean has disproportionately high numbers of some marine taxa, notably the Pycnogona, Polychaeta (Fig. 1.5i top right), Brachiopoda, and Bryozoa. Debate on causes of a global pattern in the sea would seem premature given uncertainties surrounding the evidence for consistent patterns in the southern hemisphere. One such explanation for both the latitudinal and longitudinal patterns of richness in the sea centres on sea basin age, which might explain why latitudinal trends are strong towards the young arctic (it is still in the process of being invaded) but weak towards older Antarctica.

● The Southern Ocean has disproportionately high numbers of Pycnogona, Polychaeta, Brachiopoda, and Bryozoa.

In the deep sea, a northern latitudinal cline in species richness is also evident but there have also been suggestions of a decrease in richness from the shallows down into deep waters. Support for this came from early abyssal samples, but sampling elsewhere (e.g. off Norway) has cast doubts on the generality of a trend (Gray 2001). Very shallow faunas are young (major changes occurred at the last age between 14 000 to 10 000 years ago) as sea level changes require constant reinvasions and ice-sheet maxima have repeatedly bulldozed colonists to the shelf edges (Chapter 7). The deep-sea fauna also must be quite young as the flow of oxygen-laden cold water from Antarctica is relatively recent (older than shallow shelf environments, but only tens of millions of years old). Thus the oldest and richest fauna might be expected at shelf edges presently at a depth of about 200 to 300 m. The richness of large (mega)benthos and fish increases towards and away from shelf edge depths. Any bathymetric cline is unlikely to be as sharp or directional as the altitudinal richness clines in the terrestrial environment.

● The oldest and hence richest fauna might be expected to occur at the margin of the continental shelf.

1.3.4 Biodiversity and ecosystem functioning

The species that constitute ecosystems can be assigned to functional types, such as pioneer encrusting suspension feeders (e.g. some polychaetes and bryozoans), competitively dominant encrusting suspension feeders (e.g. some sponges and ascidians), benthic zooplankton feeders (e.g. some anemones), deposit feeders (e.g. echiuran worms), scavengers (e.g. amphipods), mobile carnivores, and others. The concept of biodiversity can also be used in the context of the number of functional groups within a system. Thus the **functional diversity** of an ecosystem is higher if the number of functional groups is higher and/or the number of members of each functional group is higher. Higher functional diversity should mean that ecosystems are more robust to change and environmental stress. This is currently a hot topic of research, which seeks to answer the question 'if the number of species in an ecosystem decreases does this affect its ability to respond to change?' (See also Chapters 5, 7 and 14.)

● One of the key questions in ecology is role of functional diversity in ecosystem resilience.

Some species have been described as 'keystone' to their communities (Chapters 5 and 7). Such species when experimentally removed can have major ecological consequences for the other species within the system (Pace et al. 1999). Few species, even those considered as 'keystone', have been demonstrated to have a pivotal role in their community. Some simple hypotheses have been generated, however, to predict change in ecosystem functionality with species number (i.e. what happens following removal of any random species). Examples of experimental diversity-ecosystem function studies in low richness (estuarine) systems are given in Chapter 4. The redundancy hypothesis predicts that an increasing number of species increases ecosystem functionality proportionally less as the number of species rises to a point when further additions have no net effect (a curve to asymptote). If this is correct, species could be lost from species-rich ecosystems without any loss of functions. This is termed the functional redundancy of species. Support for such an idea comes from the response of many coral reefs to herbivore losses. In the Caribbean, losses of herbivorous, manatees, turtles and fish have occurred but major system deterioration only occurred when the last big herbivore populations (sea urchins) crashed and resulted in overgrowth of coral by algae. Other hypotheses suggest that any species addition or subtraction changes the level of ecosystem functions (**rivet hypothesis**) or that functionality is driven by species interactions rather than the species *per se* (Box 1.8). All of these hypotheses suggest that biodiversity is intricately linked to ecosystem stability. Intrinsic to the idea of ecosystem functionality, ecosystem trophic and food-web structure, are the numbers of individuals of each species present within the system. The latter also exhibit large-scale patterns in time and space.

● In species-rich systems, ecosystem functionality is unlikely to be adversely affected by the loss of one or two species.

Box 1.8 Species richness and ecosystem function: the rivet and redundancy hypotheses

At low levels of species richness, ecosystem functions increase with species richness but at high levels ecosystem functions stop increasing with species number according to the rivet theory. So if 4 species are removed (decrease richness from 32 species A to 28 species B) the rivet hypothesis would suggest measurable decreases in ecosystem functions, whereas the redundancy theory suggests that an experimenter would find no change in ecosystem function.

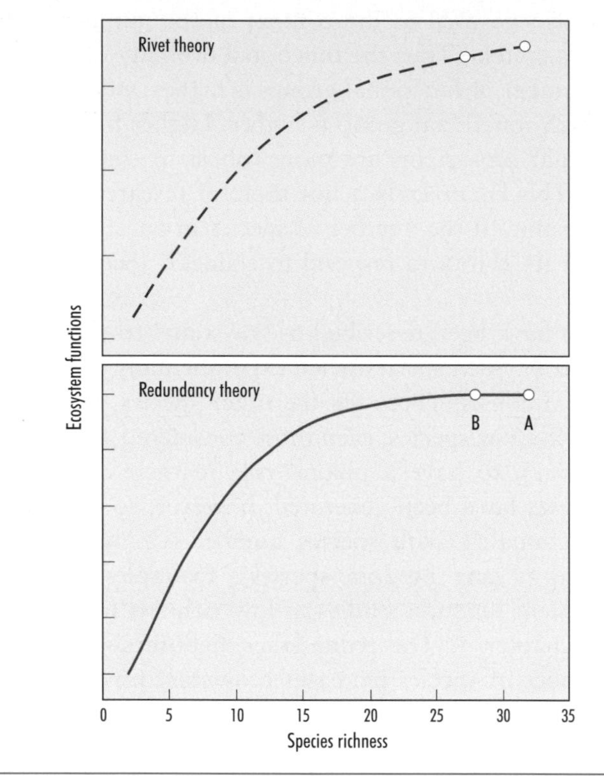

1.4 Abundance and Size

1.4.1 Scale, time, and space

● Low diversity habitats are typified by high abundances of single species, which are themselves highly variable in space and time.

Some animals are very abundant and some are not. Abundant animals are often very small, such that billions would weigh less than a single individual of some of the rare larger ones. Many low diversity habitats are instead characterized by huge numbers of a few species (e.g. *Hydrobia* snails on intertidal mudflats, capitellid worms in marine sediments or sea cucumbers in deep ocean trenches). The abundance of most organisms is very dynamic at certain scales in time and space. Thus it is possible for one species to be super-abundant one year and almost absent the next, while it is not uncommon to encounter thousands of individuals in one sample and hardly any in an adjacent sample. The

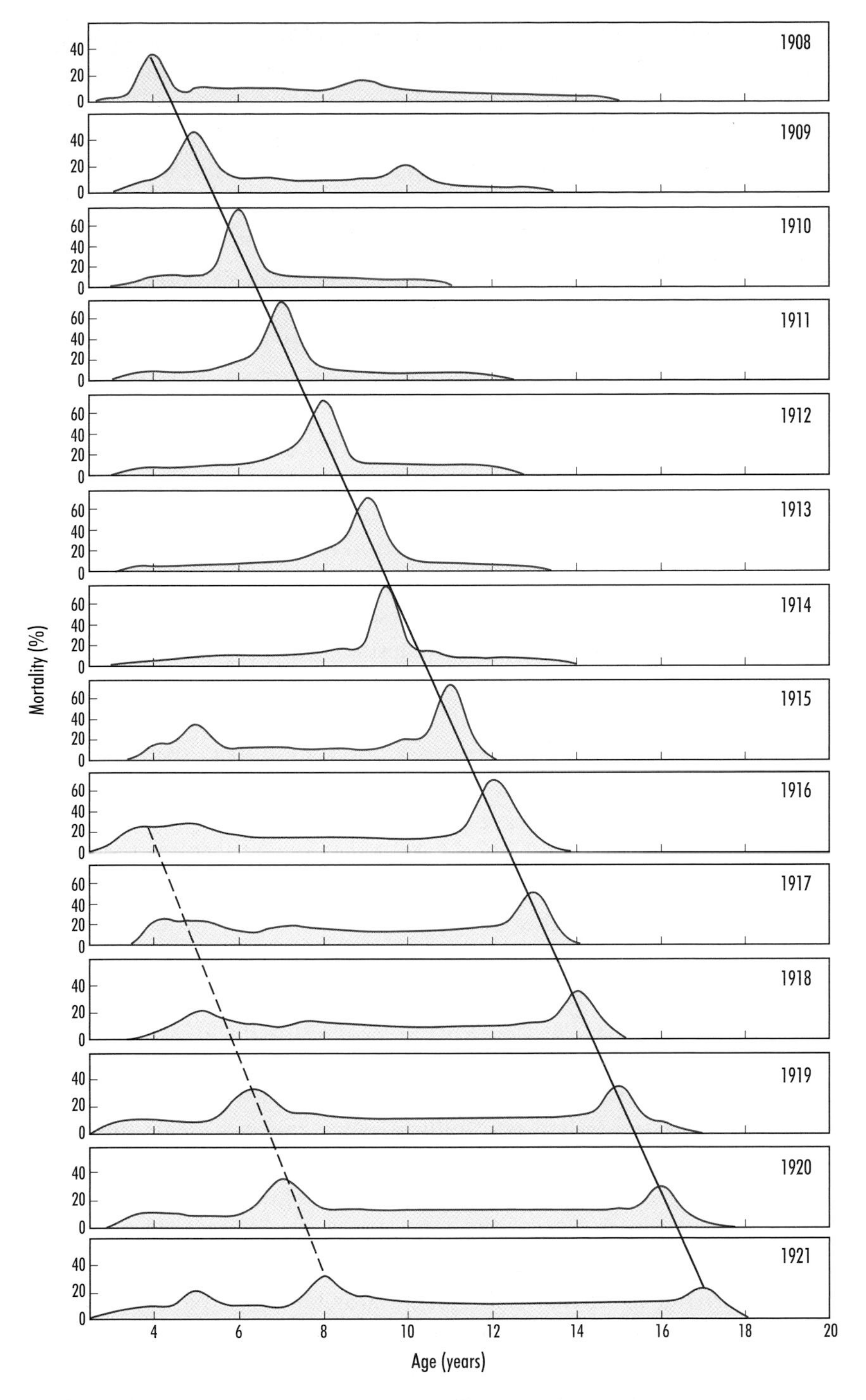

Fig. 1.11 An example of inter-annual variability in recruitment. A major recruitment event of Herring (*Clupea harengus*) in 1904 dominates a Norwegian fishery for more than a decade. Adapted from Cushing (1975).

abundance of a taxon has so far only been considered as an aid to quantifying diversity, but many large-scale patterns in the sea occur not between but within taxa. On land, seasonal changes in the abundance of many species are obvious, and in the sea this is most clearly seen in terms of seasonal blooms of phytoplankton. Annual changes in the abundance of organisms are often due to differential survival of year-classes (**cohorts**) of young that are related to environmental conditions (e.g. temperature), abundance of predators and competitors. There are a number of long-term data sets that demonstrate major inter-annual differences in the abundance of animals across the world's oceans. Long-term data sets of fisheries landings demonstrate the cyclical nature of 'good' and 'bad' year-classes in various fish species such as salmon, cod, and herring (Fig. 1.11). Cyclical patterns have also been observed at interdecadal scales or linked to ENSO events, such as sea stars (Crown-of-thorns *Acanthaster* on coral reefs) or temperate North American sea urchins (*Strongylocentrotus* in macroalgal beds) (Chapters 7 and 10).

- Interdecadal cyclic fluctuations in the abundance of invertebrates such as starfishes and sea urchins have a strong influence on the ecology of the systems in which they occur.

We can gain a much longer-time scale perspective by reference to the fossil record that enables us to compare patterns in the abundance of taxa as well as species richness. Changes in the abundance levels of species form the patterns of mass extinction in the fossil record. Patterns of abundance are often decoupled from patterns of species richness. Two examples that illustrate this are the present-day Antarctic shallow benthic fauna and in the fossil record (Mesozoic to present-day) of cyclostome bryozoans. In terms of the numbers of taxa, many benthic groups of animals found in the Antarctic are dominated by those that brood their young (e.g. sea urchins or gastropod molluscs, see Fig. 1.12). However, the most abundant species in these groups are **broadcast-spawners** (animals that release their gametes directly into the water column). Studying the fossil record, McKinney et al. (1998) found that the proportion of global bryozoan genera and species in the order Cyclostomatida, has steadily decreased relative to the order Cheilostomatida over the last 100 million years (Fig. 1.13a). Prior to the end-Cretaceous mass extinction (KT boundary), the same pattern was evident in abundance, that is cheilostomes became progressively dominant in terms of bryozoan abundance. However, immediately after the KT boundary, the cyclostomes became very abundant again (Fig. 1.13b). This recovery lasted only a few million years, but is a clear illustration that historical abundance and richness can be quite separate. Both cyclostomes and cheilostomes are similar types of organisms (encrusting, benthic suspension feeders), which occur in the same habitats. Competition for resources (space in the case of bryozoans) is thought to be the main driving force that underpins major temporal patterns that are repeated in many co-occurring organism groups.

- Patterns of abundance are often decoupled from patterns of species richness.

1.4.2 Competition

Competition for resources can take many forms (Chapter 5) and can drive large-scale biogeographic and biodiversity patterns. Many consider

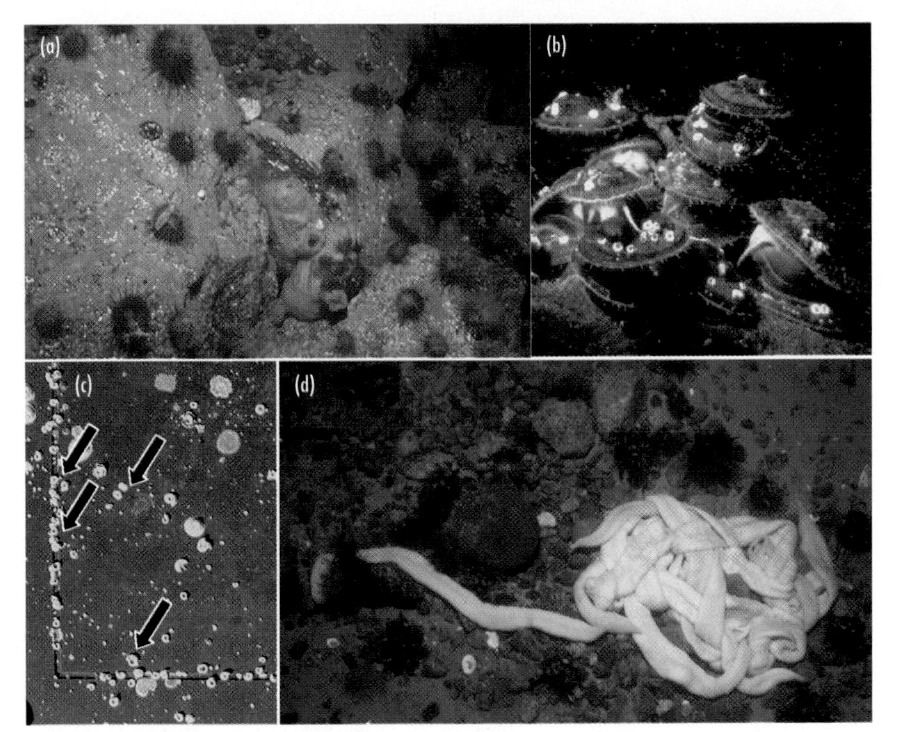

Fig. 1.12 Dominant fauna of Antarctic shallow waters. The echinoid *Sterechinus neumayeri* (a) and the limpet *Nacella concinna* (b) Spirorbid polychaete worms (c) and the nemertean worm, *Parbolasia corrugatus* (d). Photographs: British Antartic Survey.

that one of the fundamental changes that occurred in Cambrian oceans *c*.550+ million years ago was the initiation of competition driven by the rapid increase in the abundance and diversity of marine organisms. Investigation of past evidence for competition is problematical. McKinney (1995) measured the outcomes of spatial interference competition between cyclostome and cheilostome bryozoans preserved in fossils (Fig. 1.13c). Cheilostomes consistently out-competed the cyclostomes in *c*.66% of encounters. This suggests that one of these groups of bryozoans has had a clear, but not increasing, competitive advantage. Both of these organism types are **extant** (alive in present day), hence Barnes and Dick (2000) were able to look for similar patterns across the world, but found no such constancy in the outcome of competition (Fig. 1.13c). There are two possible explanations of this discrepancy: greater variation in competition in modern assemblages or constancy of outcome within the type of environment that preserves well in the fossil record (warm shallow continental shelf). The latter seems more likely and as such makes it difficult to interpret the implication of McKinney's (1995) long-term pattern of competition, as the dominance of cheilostomes would then depend on the proportion of warm shallow shelf compared to other suitable habitats, which are unknown.

● Competition driven by the rapid increase in the abundance and diversity of marine organisms may have exacerbated the rate of diversification in Cambrian oceans.

● Reconstructing the outcome of competition between biota in the fossil record is complicated by potential artefacts such as the likelihood of preservation in particular types of habitat.

Perhaps the most famous example of inferred competition is that between (articulate) brachiopods and bivalve molluscs. Both of these

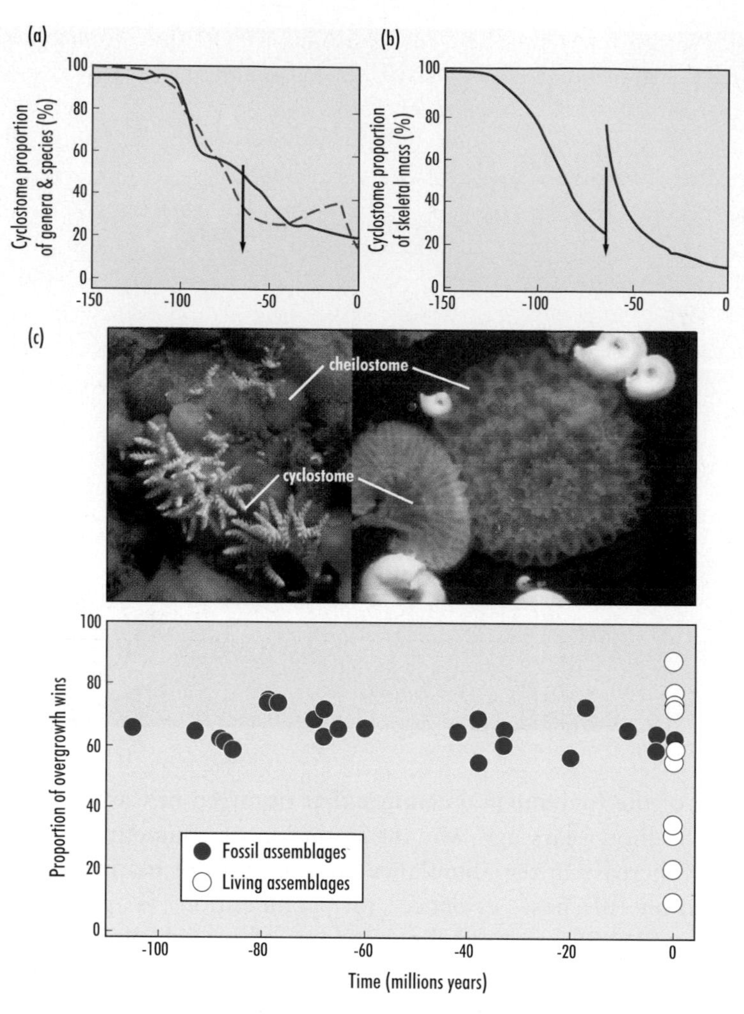

Fig. 1.13 Cyclostome and cheilostome bryozoans: (a) taxonomic richness and (b) relative abundance over the last 140 million years, trend lines plotted from data in McKinney et al. (1998). Arrow shows timing of KT mass extinction, dashed blue line is species number and continuous red line is genus number. Photographs show competition for space between the two groups. McKinney's (1995) data from fossil assemblages (blue circles) suggest cheilostomes win *c.*66% of competitive encounters (c). Competition in living assemblages (white circles) shows much more variability, data from Barnes & Dick (2000). Photographs: David Barnes.

taxa are superficially similar, both have a shell composed of two valves and they are benthic and suspension feeding animals. The brachiopods were initially more abundant and diverse than bivalves, but this trend was reversed following the biggest of the mass extinctions, at the end-Permian (Fig. 1.14a,b). Brachiopods are presently most abundant and speciose in the deep sea and polar regions, two of the few areas where bivalves are not dominant. More than a hundred years ago the match of the two patterns was held-up as a 'classic' example of competition driving the gradual replacement of one group of organisms by another superior competitor. Modern bivalves and brachiopods compete for resources in some environments, notably in New Zealand fjords. In these situations bivalves seem to be generally superior but the outcomes of competition are complicated by predation and other factors. Gould & Calloway (1980) reanalysed the fossil data and found that, except at the end-Permian extinction, there was no negative relationship between the species richness of the two groups (Fig. 1.14c). Furthermore, they actually found a weak positive correlation, so that there were more

● The pattern of occurrence of bivalves and brachiopods was previously viewed as a classic example of the outcome of species that compete for the same resources.

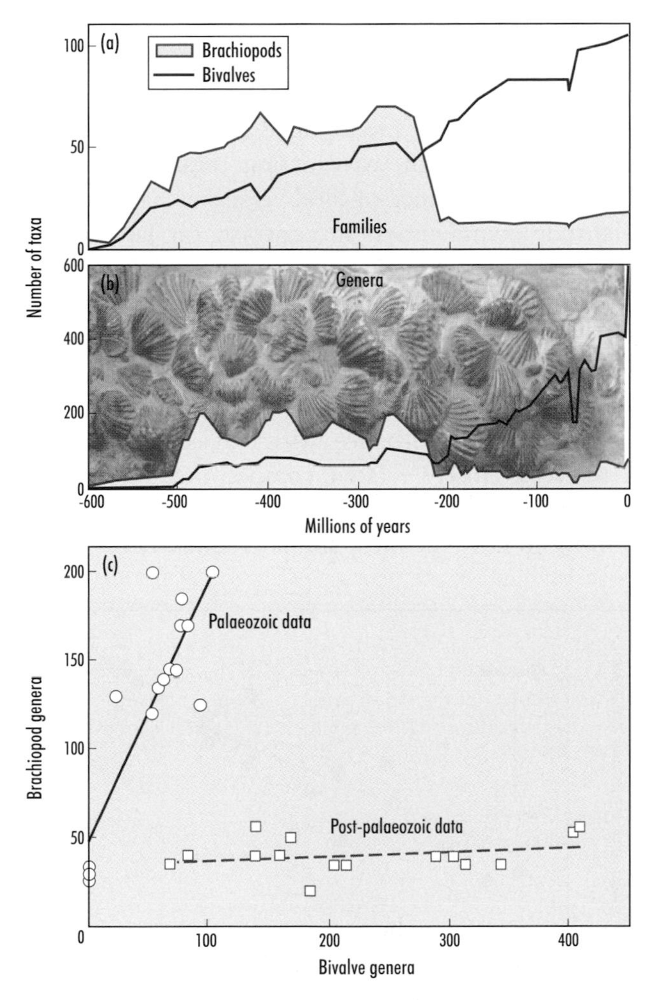

Fig. 1.14 Patterns of, and between, brachiopod and bivalve (mollusc) richness. (a) Numbers of families with time over the last 600 million years, (b) numbers of genera over the same period, and (c) generic richness of brachiopods vs that of bivalves before and after the end-Permian extinction event. Data from Gould & Calloway (1980). The photograph illustrates abundant fossil spiriferid brachiopods.

brachiopod species when there more bivalve species. The positive relationships were, as the authors acknowledged, probably not causal. The possibility that competition explains the patterns of richness in these two co-occurring animals is still debated.

Competition has been investigated in many different regions, habitats, and among and within a wide variety of marine organisms. Relatively few large-scale spatial and taxonomic patterns have emerged. One of the few patterns that has emerged is that of shallow encrusting (bryozoan) communities that are typically much more hierarchical at high latitude (polar seas) (Fig. 1.15). This means that the dominance hierarchy in polar regions is more pronounced: the top competitor wins virtually all encounters with any other species. The next highest ranked competitor wins all encounters except those with the most dominant, and so on down the dominance hierarchy. If it were not for the influence of disturbance, one species would ultimately monopolize all the available space. In this system, disturbance, through wave action (Fig. 1.2) and ice-scour, is important for the maintenance of biodiversity as has been described for some low-latitude communities. In contrast, on Jamaican coral reefs, Jackson (1979) found that in many interactions between species pairs there was no definitive winner. A number of studies have reported similar findings in the competition for space among corals or sponges. In these situations many species seem to be able to exist in close proximity to each other and biodiversity is maintained through biological interactions rather than by physical disturbance. Most investigations of competition have been between closely related competitors, often among species belonging to the same genus. Intuitively this makes sense, as similar species are likely to have the most similar requirements and methods of

● At high latitudes, diversity is maintained by disturbance whereas lower latitudes biological interactions.

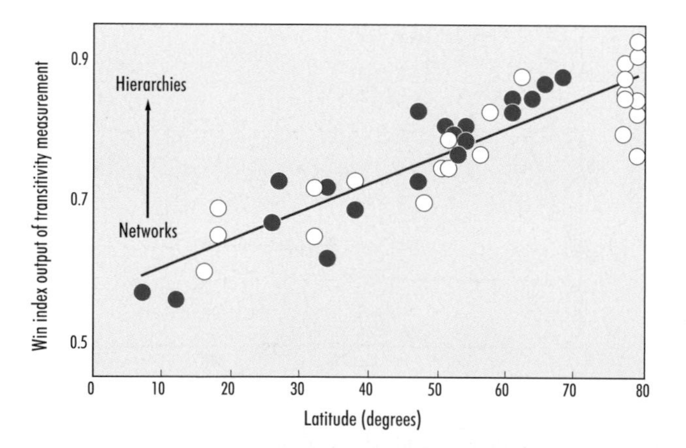

Fig. 1.15 Changes in the structure of marine competition with latitude. Data are how hierachical 'pecking orders' of bryozoan communities are in the northern (○) and southern hemispheres (●), scored using Tanaka & Nandakumar's (1994) transitivity formula. Plot redrawn from Barnes (2002) with additional data.

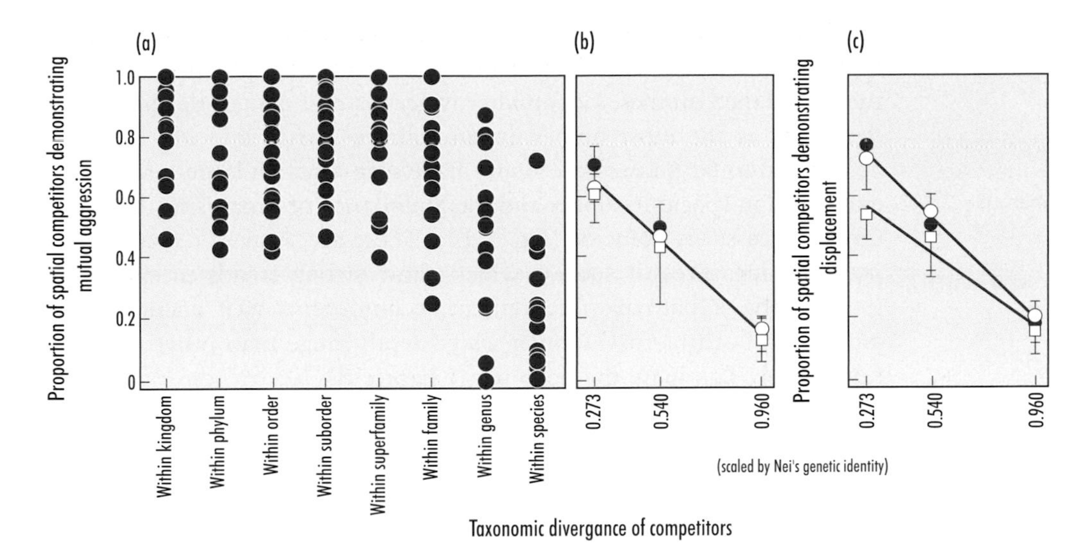

Fig. 1.16 Competition asymmetry with taxon divergence in shallow seas. Proportion of spatial competitors demonstrating mutual aggression decreases with increasing relatedness in temperate bryozoans (red filled circles)(a). Scaled by Nei's genetic index (0.273 = within family, 0.54 = within genus and 0.96 = within species), the relationship seems robust: it also holds true for Antarctic bryozoans (white circles) and temperate ascidians (squares). (b) Plot modified from Barnes (2003).

resources acquisition. In addition, very different species compete for the same resources, for example. humans, seals, and seabirds compete for sand eels (*Ammodytes* spp.). Barnes (2003) examined how the intensity of competition varied with relatedness in two groups of marine organisms (e.g. bryozoans and ascidians) in two regions of matching latitude (north-western Europe and the Antarctic Peninsula). The outcomes of competition were as asymmetrical between organisms from different kingdoms (e.g. algae and ascidians) as they were between animals of different families (Fig. 1.16). This study showed that intraspecific competition (two competitors of the same species) was less likely to result in aggressive competition for space than between congenerics (two species from the same genus), which in turn were less likely than those in the same family but from different genera. Furthermore, the more distantly related the two competitors the more uneven the outcome of competition. Trends in the outcomes of competition, abundance, and richness could be considered as patterns of success. This means that it is possible to measure the success of different taxa.

● The more distantly related the two competitors the more uneven the outcome of competition.

1.4.3 Success

To measure success, we first need a standardized definition of this term. In evolutionary terms, insects are considered a successful taxon because they have radiated to a high level of species richness. Alternatively higher taxonomic levels could be considered (genus, phylum), as could the abundance of taxa. Copepod crustaceans and nematodes have been described as the most numerous animals on earth, and are therefore considered to be 'successful'. As we have seen already, large-scale trends exist both in species richness and the abundance of organisms, but these do not necessarily coincide (Fig. 1.13). There are, however, many more potential measures of success which show strong trends in space and time: Along a bathymetric gradient, echinoderms (sea cucumbers in particular) occur across a much wider depth range than other taxa and numerically dominate the deep sea (Chapter 8).

● Nematodes are considered to be 'successful' because they are among the most numerous animals on earth.

Crustaceans, molluscs and some other phyla occur in marine, freshwater, and terrestrial systems, so an analysis of occupancy of the available habitats in these systems with time could be considered as an analysis of success. A frequently used proxy of success is the evolution of advanced features or complexity. Organism evolution has produced progressively more complex designs with time or the number of evolutionary steps that have occurred (Dulvy et al. 1997). Whether or not there is real directionality in the development of complexity is debated, but most organisms (bacteria and protoctists) remain simple in structure. At this point we should perhaps consider whether more complex organisms are more successful? In many examples this depends on the time scale under consideration. Some of the more speciose and abundant groups in today's seas (e.g. molluscs and crustaceans) were relatively unspecialized in Cambrian seas. Trilobites and brachiopods, in contrast, were more specialized and for considerable periods of evolutionary time were more speciose and abundant (see Fig. 1.10), though not since the palaeozoic.

● A frequently used proxy of success is the evolution of advanced features or complexity.

On smaller (but still large) time scales, increased complexity has made some animal types more abundant and speciose but more prone to extinction on larger time scales. A good example is the evolution of a group of Mesozoic molluscs (ammonites) that resembled the shelled cephalopod *Nautilus* in today's seas. Over tens of millions of years, families and genera of ammonites with complex suture lines in their shells became more speciose and abundant than simpler forms (Fig. 1.17). Despite this, those that survived the mass extinctions were mostly the ammonites with the simpler shell form, which illustrates that selection pressures for evolutionary and ecological success can be very different. Ammonite evolution also illustrated another major trend with time, change in body size.

● Ammonites with complex suture lines in their shells became more speciose and abundant than simpler forms, but it was the latter that survived the mass extinctions.

1.4.4 Size

Body size has a key role in the function, ecology, physiology, and evolution of individual organisms. Contrasts in animal size surround us, many are simply between organism types: all nematode worms are, for

Fig. 1.17 Complexity in suture lines of Jurassic ammonites. Four ammonite species with suture line boundaries coloured to illustrate evolving complexity from simple (left) to progressively more complexly folded ones (right). Photograph: David Barnes.

example, smaller than brachiopods, which are all smaller than marine reptiles. However, contrasting representatives of some taxa span more than four orders of magnitude in size (e.g. for molluscs, while many snails are just a few millimetres in length, giant squid can measure nearly 30 m). Other animal types, such as sponges and colonial forms, do not have definite limits to body size, much like many fungi, lichens, and mosses. There are strong trends in time and space for organism size. In the Precambrian (*c.*570 million years ago), all known animals were fairly small. While in the Mesozoic, we find the largest known ocean 'predator' (*Liopleurodon*). Nevertheless the largest organisms ever to have occurred are found in today's seas (fin (*Balaenoptera physalus*) and blue (*B. musculus*) whales). Amazingly, two of the largest living animals, the megamouth shark (*Megachasma pelagios*) and the giant squid (*Architeuthis dux*), are yet to be seen alive in their natural habitat. Within taxa, many types of animals (mostly those with calcareous skeletons) have tended to get larger with time. It may be that this reflects evolution **away from** small rather than **towards** large size and is an artefact of increased specialization. The increasingly complex ammonites also tended to become larger (Fig. 1.13). For nearly all animals, an increase in volume will change their surface area to volume ratio. This is crucial to functions such as gas exchange and feeding. There has been much discussion (and little generally accepted conclusion) on whether (1) there have been more opportunities for new roles (empty niches) for small or large animals; (2) rates of evolution of body size vary with population size or reproductive mode; and (3) oxygen, temperature or other factors were responsible for the within-fauna size variability seen in modern faunas. For example, amphipods, isopods, pycnogona, and some other taxa in the deep sea and polar regions can achieve very large

● Many types of animals have tended to increase in size with time, this is known as Cope's rule.

● The link to oxygen explains gigantism (e.g. in dragonflies) during the Carboniferous period when atmospheric oxygen levels were higher.

size (sometimes referred to as **polar gigantism**). On careful analysis of amphipod sizes from around the world, Chapelle & Peck (1999) found that size spectra of not only marine, but also freshwater faunas showed a strong correlation to oxygen. They later tested this hypothesis by investigating amphipod size in a high-altitude lake (Lake Titicaca), which confirmed their predictions (they were mainly very small).

Decreases as well as increases in size tend to occur in terrestrial island faunas. Only rarely does no change occur relative to mainland counterparts. On islands, certain animal types (most famously large mammals such as carnivores or mammoths) become dwarfs while others (such as rodents) tend to increase in size. For the sea, we know of far fewer size changes due to isolation or along resource limitation gradients. There is evidence of decreases in size for some radiolaria (Fig. 1.3f), but conversely increases in others. Many estuarine animals seem to be smaller than their fully marine equivalent species (Chapter 4), but any conclusions about either reduced growth or maximum size are usually confounded by their shorter life history. There have been many explanations for why animals should increase in size (such as reduction of predation or increasing feeding efficiency), but there has been little consensus of the importance of dwarfism other than in the formation of colonial modular units. One effective method of escaping surface area to volume constraints on respiratory or feeding surfaces is coloniality. In virtually all cases of colonial taxa (e.g. many ascidians, bryozoans, many corals), the colonial modules are smaller than similar solitary/unitary species, but the colony may be unlimited in size.

● Thus colonies of organisms can increase in size, without any change in the surface area of respiratory or feeding surfaces, because the modules remain fixed in size.

The current interest in global biodiversity means that there is intense interest in the relative extinction rates of small versus large bodied animals. In recent time the extinction of many of the large vertebrates from continents and islands coincided with human inhabitation far less is known about size related extinctions further back in time, most notably in the largest extinction at the end-Permian. Predictions based on energetics alone have suggested that larger mammals have higher extinction probabilities. Various lines of evidence suggest that adaptive breakthroughs in faunas occurred at small body size. This set the scene for rapid increases in size as well as the number of species at the start of the Cambrian and after the various Mesozoic extinctions.

CHAPTER SUMMARY

- The marine environment accounts for 99% of the Earth's living space and it is where nearly all major types of animal (phyla) evolved. Most of these animals remain endemic (unique) to the marine environment. Mass extinctions and radiations of species affected marine organisms most dramatically. As marine organisms tend to be better preserved than those on land, most of our perception of past life comes from marine fossils.

- New evidence, such as satellite imagery, has revealed that the oceans and seas are more compartmentalised than we once thought. Broadly, marine life seems spatially divided into twelve major biogeographic biomes, each with different patterns of productivity and their own specific biotas.
- It has long been recognised that terrestrial organisms exhibit distinct trends in endemism, diversity, abundance, size and other characters with physical environmental characteristics. Intense work on marine organisms during the last couple of decades has revealed some similarities and striking differences with such patterns on land.
- With more extreme and longer isolation of geographic regions, the tendency towards endemism is greater. Continental shelf margins around islands, such as Hawaii or as big as Antarctica, can contain the majority of species found nowhere else. How many depends very much on the scale at which we look for such species.
- Biodiversity is much discussed but often poorly understood. It can be measured using a variety of methods at various taxonomic levels (though usually species). There is very good evidence that richness (the number of taxa) has changed through time and differs with latitude, longitude and other factors.
- Other major aspects of ecology also change drastically along time and space gradients. This chapter discussed the evidence for just a few, including how competition appears to be more hierarchical at the poles than at lower latitudes and how it changes with the taxonomic relatedness of competitors.
- Success can be measured in many ways; some organism types have become successful ecologically (abundance, ubiquity) and evolutionarily (great geological age, many taxa), though not always at the same time. Many others have had some ecological or evolutionary success, whilst some may always have been rare – but was this through design or chance?
- Finally, evidence to date suggests the largest organisms that ever lived are alive today, but the largest land animals died tens of millions of years ago and the largest insects lived even earlier. Gigantism and dwarfism occurs in certain environments even now (deep sea and polar regions). The evolution of body size is complex and depends on many factors.

FURTHER READING

Gaston and Spicer (2004) provide an excellent treatment of biodiversity and current topical issues, while McShea (1996) provides a detailed treatment of trends between metazoan complexity and evolution.

- Gaston, K. J. and J. I. Spicer 2004. *Biodiversity: An introduction* (2nd edition). Blackwell Science, Oxford.
- McShea, D. W. 1996. Metazoan complexity and evolution: Is there a trend? *Evolution* 50: 477–92.

For an introduction to the wider implications of body size, see LaBarbera (1986) for one of the best reviews of this subject.

- LaBarbera, M. 1986. The evolution and ecology of body size. In D.M. Raup and D. Jablonski (eds), *Patterns and Processes in the History of Life*. Springer-Verlag, Berlin, pp. 69–98.

Part 1

Processes

Chapter 2

Primary Production Processes

CHAPTER SUMMARY

Primary production is the starting point of all other life in marine systems. Primary producers in the oceans span many orders of magnitude in size from bacteria less than a micron (μm) in diameter through to 50 m long giant kelps that grow 0.5 m a day. Production is measured using bottled incubations or increasingly from space using satellite-borne ocean colour sensors that detect photosynthetic pigments in surface waters. The conversion of inorganic carbon into biomass, its subsequent sinking to the seabed and sequestration over thousands of years is fundamental for our understanding of the ocean as a potential sink for increasing levels of atmospheric carbon dioxide. This chapter introduces the major factors that control primary production, and how to measure it.

2.1 Introduction

The photosynthetic organisms of the ocean, as on land, are for the most part the fundamental food source on which marine ecosystems are based (Field et al. 1998; Falkowski et al. 2000). In coastal waters, the large stands of seaweeds exposed at low tide, submerged kelp beds or gently wafting meadows of seagrasses that fill coastal lagoons are the obvious plants. These primary producers grow in much the same way as their terrestrial counterparts: assimilating carbon through photosynthesis, and growing by taking up nitrogen, phosphorus, and a host of necessary other minerals and trace substances to generate new biomass.

When considering photosynthesis and the production of new biomass we need to consider both production and loss processes, and both are important for this chapter (Fig. 2.1). In the most simplistic terms, light energy is trapped and used to produce organic matter through photosynthesis, and this organic matter is broken down through respiration to release energy and heat.

● There is a large body of information on primary production processes in aquatic systems, and for this chapter, rather than an extensive list of citations and reference list (as used by other chapters), readers are rather encouraged to use the extended reading list at the end of the chapter. These have been selected for the overviews they give and the synthesis of the primary scientific literature. Citations are used for more specialized points, possibly not so widely covered in the more generalized texts.

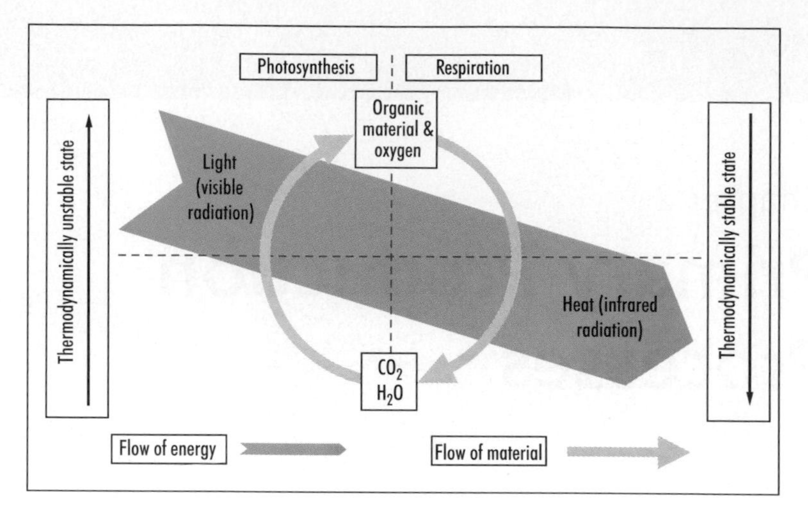

Fig. 2.1 The massive biological cycle. The energy of light, inorganic nutrients, CO_2, water, and salts are converted to a complex mix of organic compounds and oxygen by photosynthetic organisms. Respiration releases CO_2 and energy, at the expense of oxygen and recycles nutrients to the inorganic state. (Image: Peter J. le B. Williams).

● Primary production is the formation of organic matter through the trapping of light energy and assimilation of inorganic elements.

The global scale of this cycle is massive and the annual cycle of production and consumption has been calculated to have the same energy production of about 1.5×10^{14} watts per year, equivalent to the annual output of around 150 000 nuclear power stations. It is also a relatively efficient process, with 40% of the solar radiation absorbed converted into organic material, which is slightly better than the best modern power station.

2.1.1 Marine plants and algae

● Seaweeds, seagrasses, and microscopic algae and bacteria are the primary producers in the oceans.

Looks can be deceptive. While the seagrasses are true flowering plants, the seaweeds are not. Seaweeds are algae, and although they are photosynthetic organisms, in contrast to the terrestrial plants, they are non-flowering and do not have roots, leafy shoots, or sophisticated tissues for transporting water, sugars, and nutrients. Their major evolutionary lineages remain controversial, although recent molecular advances clearly indicate that apart from a few of the green algal species the algae are only remotely linked to land plants.

Seaweeds, or rather macroalgae as they are known, are a diverse group, ranging from mere encrustations on rock surfaces, to giant brown algae such as *Macrocystis prolifera* and *Nerocystis leutkaena*, which reach lengths over 50 m long (Fig. 2.2). The latter species is an annual, and grows taller than a mature oak tree in a single year.

Macroalgae are easily seen by the human observer, but we need the assistance of a microscope to observe the microalgae that generate most

(a) 

(b) 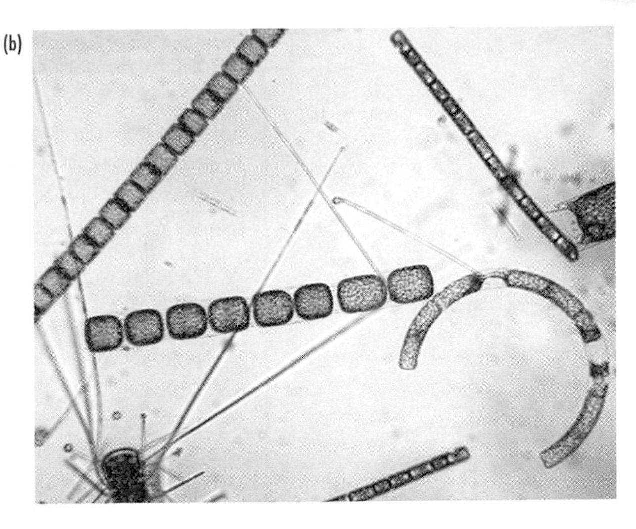

Fig. 2.2 The primary producers of marine systems range from large seaweeds (macroalgae) that grow attached to the sea floor (a), down to microscopic phytoplankton (b). Note the ×10 000 difference in scale between the two images. (Photographs: Ian Lucas/Gerhard Dieckmann).

of the primary production in the oceans (Fig. 2.2). We may not be able to see these individual **phytoplankton** cells, but we can certainly see their effects: clear waters can be turned brown almost overnight, when light, temperature, and nutrient conditions in the water are favourable, such that phytoplankton are induced to grow at such a rapid rate they form an **algal bloom**.

● Phytoplankton can bloom rapidly given enough light and a sufficient supply of inorganic nutrients.

2.1.2 Global ocean primary productivity

Global ocean net primary productivity estimates are numerous and varied (Geider et al. 2001). Most of the variability is largely to do with the methods used to measure ocean primary production, and the different metabolic processes that these methods quantify (Table 2.1). The most recent estimates based on satellite images of phytoplankton biomass in surface waters tend to range between 40 and $50\,\mathrm{Pg\,C\,y^{-1}}$ (note: **P = peta** and 1 Pg is equivalent to 10^{15} g). Recent estimates of terrestrial primary production is estimated to be between 50 and $60\,\mathrm{Pg\,C\,y^{-1}}$, which combined with the oceanic primary production gives a total primary production on Earth of *c.*$10^{16}\,\mathrm{g\,C\,y^{-1}}$ (Table 2.1).

● Marine and terrestrial primary production are roughly the same at approximately $50\,\mathrm{Pg\,C\,y^{-1}}$.

We can measure the distribution of phytoplankton from satellites in space (Fig. 2.3). Like all photosynthesizing organisms the algae contain chlorophyll. With modern-day satellites (see 2.13 below), it is possible to estimate the chlorophyll concentrations in the surface waters of the world's oceans and therefore monitor the growth of the phytoplankton.

Box 2.1 Comparison of terrestrial and aquatic primary production

The amount of bacterial, algal or plant biomass (**primary producers**) built up over time through the process of photosynthesis is generally referred to as **primary production**. This is normally expressed as the amount of carbon fixed by photosynthesis, per unit area of space or volume, per unit of time. Most estimates are expressed as net primary production, which takes into account the costs of respiration as well.

Net primary production
= Total photosynthetic carbon assimilation – respiration carbon losses

The production per square metre of seaweeds and seagrasses is equal to, or in many cases greater than, that of terrestrial plant based systems. For instance *Laminaria* spp. dominated communities have annual productivity rates of approximately 2 kg carbon $m^{-2}y^{-1}$, and the macroalgal sea palm *Postelsia* has been estimated to produce up to 14 kg carbon $m^{-2}y^{-1}$. These estimates contrast with those for mature rainforests and intensive alfalfa crop production (1 to 2 kg carbon $m^{-2}y^{-1}$), and temperate tree plantations or grasslands and prairies which are generally less than 1 kg carbon $m^{-2}y^{-1}$.

● Macroalgae can produce up to 14 kg carbon $m^{-2}y^{-1}$.

Macroalgae can be very productive, and *Laminaria* spp. dominated communities have annual productivity rates of approximately 2 kg carbon $m^{-2}y^{-1}$. (Photograph: David Roberts.)

● Although phytoplankton productivity is much less than that of macroalgae per unit area, on a global scale total phytoplankton productivity is far greater than that contributed by macroalgae.

Coastal phytoplankton annual production is also generally less than 1 kg carbon $m^{-2}y^{-1}$. One study estimated the annual productivity of all the seaweeds in a 1 m wide strip of shoreline, 360 m long, was c.600 kg carbon y^{-1}. An equivalent productivity by the phytoplankton in a 1 m wide strip of the adjacent seawater would require a stretch of water extending out 3.4 km from the shore. From figures like these it would be easy to think that the seaweed production is more important than that of phytoplankton production to the overall productivity of the oceans. However, compared to the vast area of the globe covered by the oceans (80%) in which phytoplankton grow, the seaweed covered strips of coastlines are rather small, and at best can be viewed as sites of intense localized production.

2.1.3 The phytoplankton

The microalgae vary considerably in size ranging from about 2 μm in diameter to over 200 μm. They are very varied in form, some of the most elaborate being the silicate (glass) encased diatoms that have beguiled

Table 2.1 Comparison of annual primary production between marine and terrestrial systems. It must be stressed that at best these values are good estimates, but they do allow a comparison of the magnitudes of primary production from various components of the biosphere.

Domain	Global Annual primary production (*Pg C y^{-1})
Marine	
Temperate westerly winds	16.3
Tropical & subtropical trade winds	13.0
Coastal waters	10.7
Polar	6.4
Marshes/estuaries/macrophytes	1.2
Coral reefs	0.7
Terrestrial	
Tropical rainforests	17.8
Savannas	16.8
Cultivation	8.0
Mixed broadleaf & needleleaf	3.1
Needleleaf evergreen forest	3.1
Perennial grasslands	2.4
Broadleaf deciduous forests	1.5
Needleleaf deciduous forest	1.4
Broadleaf shrubs with bare soil	1.0
Tundra	0.8
Desert	0.5

*P = *peta*, and 1 Pg is equivalent to 10^{15} g.

naturalists since the first microscope lenses became available (Box 2.2). However, these **microphytoplankton** are not the only photosynthetic organisms to be found in the phytoplankton. Since the 1980s a host of much smaller **prokaryotic** photosynthetic organisms have been shown to be an important component of the phytoplankton. These include **cyanobacteria** such as those from the genus *Synechococcus*, cells about 1 μm in diameter, that are found in all waters except the polar oceans. Pelagic **prochlorophytes** in the genus *Prochlorococcus* (cells of 0.7 μm diameter) were discovered in the late 1980s. These very small **picoplankton** are thought to be found in most waters around the globe, and contribute a high percentage of the total primary production of open waters (Fig. 2.4). However, we are only just beginning to understand their role in global primary production, and in many oceanographic studies these tiny organisms remain overlooked (Binder et al. 1996; Karl 2002; Fuhrman & Capone 2001; Scanlan & West 2002).

● Femtoplankton – 0.02 to 0.2 μm;
Picoplankton – 0.2 to 2.0 μm;
Nanoplankton – 2.0 to 20 μm;
Microplankton – 20 to 200 μm;
Mesoplankton – 0.2 to 200 mm.

● Picoplankton, 0.2 to 2 μm in diameter, are important contributors to plankton primary production.

Fig. 2.3 The advent of satellite-borne ocean colour sensors enables scientists to look at the global distributions of phytoplankton in surface waters. This image shows the SeaWiFS average chlorophyll concentration collected from January 1997 to July 2005. (Image: SeaWiFS Project, NASA/Goddard Space Flight Center and ORBIMAGE).

This chapter describes production processes in marine systems with a focus on photosynthetic organisms and the constraints acting on this process, while in Chapter 3 we explore the role of non-photosynthetic organisms. The growth of all photosynthetic organisms is restricted primarily by the supply of light, carbon dioxide, oxygen, and the main macro-nutrients (phosphate, silicate, and nitrate). However, even when all these factors are adequate the lack of trace elements can be enough to restrict growth (see Box 2.7).

There are many other mechanisms for utilizing energy sources other than light, as well as different sources of carbon for the generation of new biomass. It is only proper that at least some mention is given to the diversity of metabolism, especially since this variety is fundamental for the microbial ecology (Chapter 3) and biogeochemical processes found within marine systems.

2.2 Photosynthesis

Algae (micro- and macroalgae) and cyanobacteria such as *Synechocccus* and *Prochlorococcus* are **photoautotrophs**, as they use light as their energy source and carbon dioxide (or one of its various forms in the water, see 2.2.1) to produce new organic matter.

● Photoautotrophs use light as an energy source and carbon dioxide to produce new organic matter.

The photosynthetic reaction can be divided into a **light** reaction and a **dark** reaction. The light reaction converts light into metabolic energy and reducing power. The dark reaction utilizes these to convert (**fix**) carbon dioxide and form organic material. The overall reactions are:

$$\text{Light reaction: } 2H_2O + \textit{light} \Rightarrow 4[H^+] + \textit{metabolic energy} + O_2$$

$$\text{Dark reaction: } 4[H^+] + \textit{metabolic energy} + CO_2 \Rightarrow [CH_2O] + H_2O$$

where $[CH_2O]$ is used as a general symbol for organic material.

Fig. 2.4 A concentrated water sample from the Arabian Sea, Indian Ocean, stained by the fluorescent DNA stain DAPI, and viewed by epifluorescence microscopy. Above: Under excitation by UV light, individual bacteria and flagellates are visible by their DAPI-induced blue fluorescence. Below: The same preparation under blue light excitation. Yellow (*Synechococcus* sp.) and even smaller red fluorescing picoplankton cells (*Prochlorococcus* sp.) are visible. Unlike in the upper picture, only the auto-fluorescence of the natural photosynthetic pigments can be seen. The large, very bright cell is a dinoflagellate (*Gymnodinium* sp., about 20μm in diameter). (Photographs: Marcus Reckermann).

Photosynthetic organisms contain specialized light sensitive pigments such as chlorophylls to absorb light energy, which is subsequently transferred to reduce CO_2 to organic compounds. CO_2 is fixed within the **Calvin cycle**, and the first step of the reduction of CO_2 in this cycle is catalysed by the enzyme **ribulose bisphosphate carboxylase/oxygenase (RUBISCO)**. It is important that there is a high supply of CO_2 at the active site of RUBISCO. Many species of cyanobacteria and algae have

Box 2.2 The diatom frustule

(a)

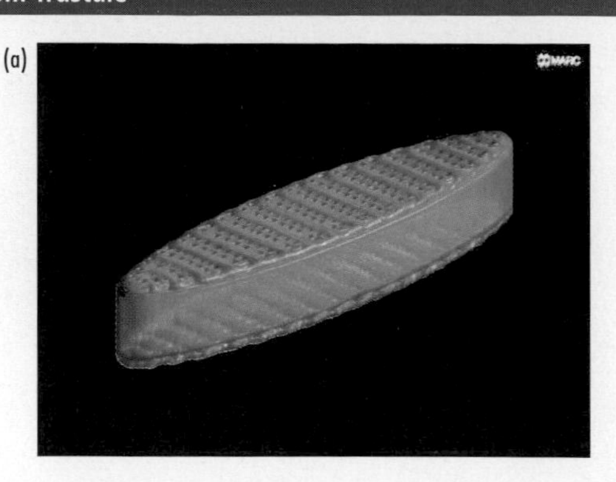

Sophisticated models of diatom frustules, such as this computer-generated pennate diatom, are helping researchers understand how the structures of the silica cell walls are related to their ecology and evolution (Image: Christian Hamm, Alfred Wegener Institute).

Diatoms, a type of alga, are major contributors to the phytoplankton of marine and fresh water. There are also many diatom species that live within and on top of sediments, as well as species that grow as epiphytes on the surfaces of animals, plants, and macroalgae. The characteristic of all diatoms is that they produce cell walls made of silicate, which are not only very beautiful to look at, but also apparently very strong. Diatoms can form dense blooms in coastal waters, and are an important food source for protozoan and zooplankton grazers. However, once formed, the diatom cases, or frustules, dissolve only slowly, and in some regions of the world's oceans, the sediments are characterized by diatomataceous or siliceous oozes: massive accumulations of diatom frustules that have sunk from the surface waters over eons of time.

Many diatom species have highly ornate frustules, with spines, spikes, hooks, and other protrusions. Many of these adaptations are thought to resist sinking and aid colony/chain formation, but also deter grazers from attempting to eat them. Spined diatoms have even been known to clog fish gills and pierce delicate membranes in gill tissues.

The strength of the diatom frustule seems to be a most remarkable feature of these algae. Although diatoms are grazed they are clearly not an easy food source to break into. The unique architectures and design of the frustules give them immense mechanical strength, and the diatom silica has material properties comparable to cortical bone with greater elasticity than glass. In fact, the grazers must exert tremendous force, and therefore expend extra energy, to break the frustules open and get to the cell contents. Recent studies using microscopic **crash tests** as well as computer-based simulations suggest that the diatom frustules have arisen from an **evolutionary arms race** in which the capability of grazing organisms to break open its prey has been pitched against the evolution of very strong elastic diatom frustules (Hamm et al. 2003). Both copepods and euphausiids, major consumers of diatoms, have silica-edged mandibles and gizzards lined with sharp crushing structures that function like teeth. It is likely that these structures have co-evolved with the development of the diatom frustule, just as the anti-grazing silica spicules in grasses have co-evolved with the evolution of teeth in animals that graze on land.

- The Strength of diatom frustules is possibly linked to an anti-grazing strategy.

continues

BOX 2.2 continued

(b)

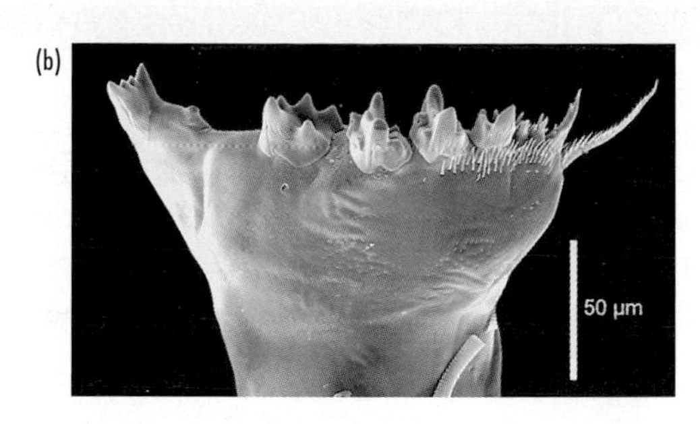

Scanning electron micrographs of the mandibular gnathobases of the copepod *Calanoides acutus*. The gnathobases of this species have strong tooth-like structures that consist of a different material from the rest of the gnathobases. These structures are very suitable for cracking hard diatom frustules. (Photograph: Jan Michels, Alfred Wegener Institute).

active **carbon concentrating mechanisms** (CCMs) that maintain the required high levels of CO_2 within the cells (Box 2.3).

● Carbon concentrating mechanisms maintain high carbon dioxide levels at the RUBISCO enzyme.

2.2.1 Carbon dioxide and photosynthesis in the sea

Dissolved inorganic carbon occurs as several forms in seawater: carbon dioxide (CO_2) gas, carbonic acid (H_2CO_3), bicarbonate ions (HCO_3^-), and carbonate ions (CO_3^{2-}). The proportions of these forms in seawater are in an equilibrium that is primarily governed by the acidity (pH), salinity (Chapter 4), and temperature of the water.

● pH, temperature and salinity govern the form of inorganic carbon in marine systems.

$$HCO_3^- + H^+ \Longleftrightarrow H_2CO_3 \Longleftrightarrow H_2O + CO_2$$

In seawater of a salinity of 35 and a 'typical' pH of 8.1 to 8.3 (Box 2.4; Fig. 2.5) approximately 90% of the inorganic carbon occurs as HCO_3^-, with 2 mM of HCO_3^- and only about 10 µM in the form of CO_2. RUBISCO requires CO_2 as a substrate. CO_2 is taken up directly by marine algae but this is dependent on diffusion from low external concentrations compared with the high amount of carbon present in the HCO_3^- pool. Surprisingly, there is still debate as to which inorganic carbon form is the predominant form used by marine algae (Burkhardt et al. 2001). HCO_3^- is taken up by some algal species and this is converted within the cell to CO_2 by the enzyme carbonic anhydrase (Elzenga et al. 2000). In many algae carbonic anhydrase activity has also been measured on the outer cell surfaces converting HCO_3^- to CO_2, which is then taken up into the cells.

● Note that in chemistry we talk about a substrate, whereas later in this book we discuss seabed habitats: the substratum.

● The enzyme carbonic anhydrase converts HCO_3^- to CO_2 within and on the outside of some cells.

Generally photosynthesis (both light and dark reactions) is represented by the following equation:

$$6CO_2 + 6H_2O + 48\text{photons of light} \Rightarrow 6O_2 + C_6H_{12}O_6$$

This highly simplified equation masks the great complexity of the photosynthetic process. However, it does show that for these organisms

● RUBISCO is an enzyme – ribulose bisphosphate carboxylase/ oxygenase.

Box 2.3 RUBISCO

RUBISCO is an enzyme of very high molecular weight and in planktonic algae amounts to 50% of the protein of the cell. It is the most abundant protein on the planet (estimates of 40 million tonnes). Given this estimate and taking global productivity at 100×10^9 tonnes per year, 1 g of RUBISCO fixes about 2500 g C per year, *i.e.* about 10 g per day (this is equivalent to 70 CO_2 molecules per RUBISCO molecule per second (enzymes characteristically are much more reactive, catalysing 10 000 to 100 000 molecular reactions per second).

The carboxylase reaction catalysed by RUBISCO is as follows:

$$\text{Ribulose-1, 5-Bisphosphate} + CO_2 \Rightarrow 2(\text{3-Phosphoglycerate})$$

Then follows a complex sequence of reactions, from which the sugars glucose ($C_6H_{12}O_6$) and sucrose ($C_{12}H_{22}O_{11}$) are common initial products. RUBISCO is unusual as an enzyme in that it has a second and quite different function as an **oxygenase.** In this reaction rather than adding carbon dioxide onto ribulose-1,5-bisphosphate, oxygen is added:

$$\text{Ribulose-1, 5-Bisphosphate} + O_2 \Rightarrow \text{3-phosphoglycerate} + \text{phosphoglycolate}$$

This reaction, known as **photorespiration**, results eventually in the loss of carbon and the formation of CO_2. It is an important loss reaction in plants, especially tropical plants. The balance between the two alternative reactions is controlled by the ratio of O_2 and CO_2 concentrations at the enzyme:

Carboxylase (CO_2 fixing) reaction is **high at high** CO_2/O_2 ratios

Oxygenase (CO_2 releasing) reaction is **high at low** CO_2/O_2 ratios

Therefore:

At high CO_2/O_2 ratios $\Rightarrow$ carboxylase function $\Rightarrow$ CO_2 fixation & O_2 production

At low CO_2/O_2 ratios $\Rightarrow$ oxygenase function $\Rightarrow$ CO_2 production & O_2 utilization

By having carbon-concentrating mechanisms (CCMs – see above) the algae maintain high concentrations of CO_2 inside the cells where RUBISCO is situated, thereby ensuring that the carboxylase function dominates (Raven 1997).

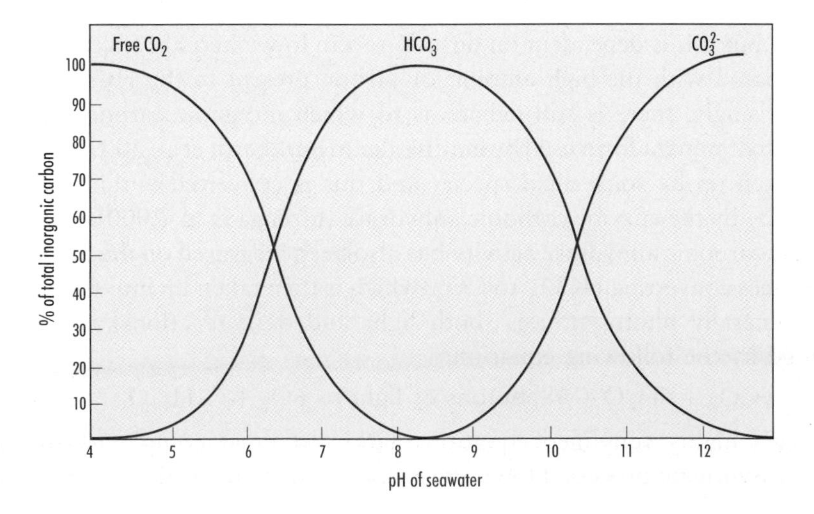

Fig 2.5 Relationship between pH and the relative proportions of dissolved inorganic carbon species in seawater.

Box 2.4 Photosynthesis and stable carbon isotopes

Carbon is present in nature in various isotopic forms: the radioactive isotope ^{14}C and the stable carbon isotopes, ^{12}C and ^{13}C. Photosynthetic carbon assimilation results in discrimination against ^{13}C, and as a result the produced biomass becomes enriched in ^{12}C whereas the remaining total dissolved inorganic carbon pool becomes relatively enriched in ^{13}C (Burkhardt et al. 1999a, 1999b; Riebesell et al. 2000).

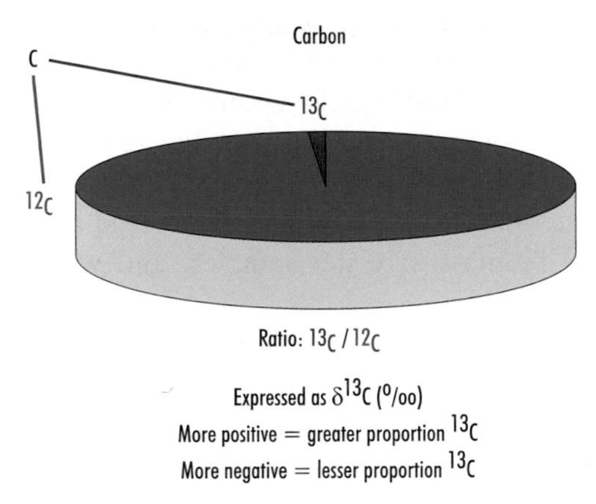

Schematic illustrating the relative proportions (not to scale) of the stable carbon isotopes ^{13}C and ^{12}C.

This is reflected in their stable isotope ratio, which is commonly reported on a ‰ (parts per thousand) basis in the δ notation relative to the international standard Vienna Pee Dee Bellemnite (VPDB) as, $\delta^{13}C = 1000((R/R_{VPDB}) - 1)$, with $R = {}^{13}C/{}^{12}C$. Negative values indicate ^{12}C enrichment and more positive values indicating ^{13}C enrichment.

The isotope effect attributable to RUBISCO is about −27‰ in photosynthetic algae. In other words, the RUBISCO effect alone will result in a $\delta^{13}C$ of the photosynthetically assimilated organic carbon ($\delta^{13}C_{POC}$) that is more enriched in ^{12}C (i.e. isotopically lighter) than the CO_2 available for assimilation by an equivalent amount. This can be observed during growth in CO_2-replete conditions. However, the overall (net) biological isotope fractionation during photosynthesis and, hence, the final $\delta^{13}C$ values are a complex function of a number of factors, such as the dissolved CO_2 concentration, passive and/or active dissolved inorganic carbon transport into the cell, growth rate, cell size, and cell geometry. The departure from the chemical equilibrium value has been used by geochemists and palaeobiologists to identify the earliest evidence for life on the planet.

The $\delta^{13}C$ signatures have become a valuable tool for researchers interested in food webs (see later chapters). The signature tends to be conservative and maintained. Therefore an organism eating the primary producer will tend to assimilate carbon with the same isotopic signature (in line with the proverb 'You are what you eat'). By measuring the isotope signatures valuable information can therefore be gleaned for determining food web dynamics.

water is used as an electron donor to produce the reducing power in the overall metabolism, with oxygen produced as an end product. Because oxygen is produced these organisms are referred to as being **oxygenic photoautotrophs**.

2.2.2 Photosynthetic pigments

● Some types of photosynthetic bacteria only photosynthesize in anoxic conditions.

There are groups of photosynthetic bacteria that use other reductants, such as hydrogen, hydrogen sulphide, and ferrous iron, and do not produce oxygen. Oxygen can inhibit photosynthesis in these organisms, which include purple sulphur bacteria, purple non-sulphur bacteria, and green sulphur bacteria among others.

● A wide range of pigments, including chlorophylls are used by photosynthetic organisms to trap light.

There is a great variety of chlorophylls and other light absorbing pigments in photoautotrophs. Chlorophyll-*a* (Chl*a*), is the major chlorophyll of the algae, and is green because it absorbs blue (maximally at 430 nm) and red wavelengths (maximally at 680 nm) of light and reflects green wavelengths (Fig. 2.6). Because Chl*a* is such a ubiquitous pigment in a wide diversity of photosynthetic organisms it is used as a measure of algal biomass in water samples, and when photosynthetic

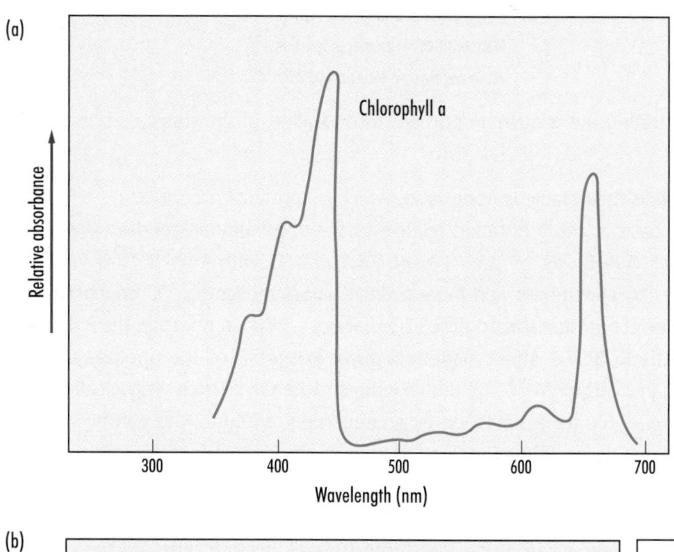

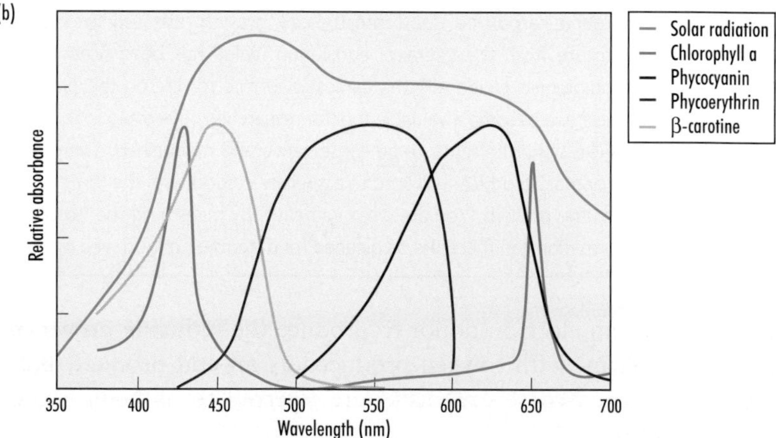

Fig. 2.6 (a) The absorption spectra of chlorophyll-*a*. (b) Absorption spectra of chlorophyll-*a* (green) and some accessory pigments: Phycocyanin (blue), Phycoerythrin (red) and *β*-carotene (yellow). The spectrum of photosynthetically active radiation (PAR) is overlain.

rates are calculated (see 2.13) they are often referred to the amount of chlorophyll present in the sample.

Carotenoids such as β-carotine and fucoxanthin, as well as chlorophyll-*b*, absorb in the green part of the light spectrum (400 to 520 nm), whereas phycoerythrin absorbs in a different range of the green region (490 to 570 nm) (Fig. 2.6b). Phycocyanins and allophycocyanins absorb light in the green-yellow (550 to 630 nm) and orange red (650 to 670 nm) parts of the spectrum respectively. These pigments are examples of **accessory pigments.** The wavelengths of light that range from 400 to 700 nm are called the **photosynthetically active radiation (PAR)**, and photosynthetic organisms adjust their light harvesting pigments to absorb various components of this spectrum of light which varies with water depth.

● Photosynthetically active radiation is the wavelengths 400 to 700 nm.

Among the prokaryotes the cyanobacteria have chlorophyll-*a*, as in eukaryotic photoautotrophs. However, other phototrophs such as the purple and green bacteria contain any of a large number of bacteriochlorophylls. These have different absorption characteristics than the algal chlorophylls. e.g bacteriochlorophyll-*a* absorbs maximally between 830 to 925 nm, whereas algal chlorophyll-*a* absorbs maximally at 430 and 680 nm.

2.3 Respiration

The other major metabolic pathway that needs to be introduced here is **aerobic respiration** (also see Chapter 3). When oxygen is present, compounds are oxidized using O_2 as an electron acceptor to produce CO_2 and adenosine triphosphate (ATP). *All* organisms carry out respiration, and unlike photosynthesis that can only take place in the light, respiration takes place continuously. Effectively the process of respiration can be exemplified by the reverse of the equation given for photosynthesis above:

$$6O_2 + C_6H_{12}O_6 \rightarrow 6CO_2 + 6H_2O + \text{energy}^*$$

(*When glucose is oxidized during respiration 2870 $kJ\,mol^{-1}$ of energy is produced.)

The amount of energy produced is dependent on the nature of the starting material: one mole of fat (e.g. palmitic acid) results in the production of 9959 kJ of energy compared to the production of 2870 kJ when the same amount of glucose (a carbohydrate) is respired. For a given weight of material, fats yield at least twice as much energy as carbohydrates (hence their value as storage material and their occurrence in the eggs of most animals).

● The oxidation of fats yields far more energy than the oxidation of the same amount of simple sugars.

Where oxygen does not occur in sufficient concentrations for normal aerobic respiration, some organisms can use **anaerobic respiration**, and use nitrate, sulphate, carbonate, and organic compounds as electron acceptors.

Nitrate respiration is key to the **denitrification** process (Chapter 3), in which **denitrifying bacteria** reduce nitrate and nitrite to nitrous oxide or dinitrogen gas:

$$NO_3^- \rightarrow NO_2^- \rightarrow NO \rightarrow N_2O \rightarrow N_2$$

- Anaerobic respiration can result in the production of potentially toxic products such as hydrogen sulphide and methane.

Sulphur-reducing bacteria use sulphate, thiosulphate, and elemental sulphur as electron acceptors in respiration. The metabolism produces sulphide or hydrogen sulphide, the rotten egg smell, characteristic of many anoxic waters and sediments. **Methanogenic bacteria** use CO_2 as an electron acceptor producing methane (CH_4) as an end product.

Many of the products of anaerobic respiration are toxic (e.g. H_2S, which is even more toxic than hydrogen cyanide). This toxicity has very great ecological significance in water management, since plankton, zooplankton, and higher organisms (e.g. fish) are particularly susceptible to this toxicity and are at risk in environments that are prone to produce these compounds. Oxygen is always the preferred electron acceptor, and providing there is oxygen in the water, these toxic compounds will not be produced.

2.4 Heterotrophic Metabolism

- Heterotrophic organisms assimilate carbon derived from the oxidation of organic matter.

Organisms that use chemical compounds as an energy source, rather than light, are called **chemotrophs**. The **chemo-organotrophs** include those bacteria and fungi that live via the oxidation of organic compounds and in oxygenated habitats, they catabolize organic matter via aerobic respiration. Because these organisms do not get their carbon from CO_2, but rather from the oxidation of organic matter, they are called **heterotrophs** (c.f. Box 2.5). Those species that live in aerobic conditions use oxygen (**oxygenic heterotrophs**) as the external electron acceptor, whereas other species in anaerobic conditions (**anoxygenic heterotrophs**) can use other oxidized substrates such as nitrate and sulphate instead of oxygen as the terminal electron acceptor. The anaerobic heterotrophs are particularly important in anoxic sediments and the biogeochemical transformations that take place within these. These bacteria and fungi assimilate low molecular weight organic compounds such as sugars, amino acids, pyruvate, ethanol, and acetate, which are transported directly into the cells. These organisms release hydrolytic enzymes to break down larger organic compounds into low molecular weight substrates that can then be transported into the cell and then respired and built up into new biomass.

- Anoxic means without oxygen.

- Heterotrophic organisms obtain their carbon from the oxidation of organic matter, whereas autotrophic organisms obtain carbon from CO_2.

There are other groups of prokaryotes that use inorganic chemicals as their energy source (**chemolithotrophs**). Most of these organisms obtain their carbon from CO_2, and so are autotrophs. There are many sources of inorganic electron donor used by these prokaryotes which include both bacteria and archaea. These include hydrogen sulphide, sulphur, ammonium, nitrite, and ferrous iron. Again some of these organisms can only survive in anaerobic conditions whereas others are tolerant of oxygen.

Examples of chemolithrophic bacteria include the **sulphur-oxidizing bacteria**, which grow in the tissues of hydrothermal vent organisms

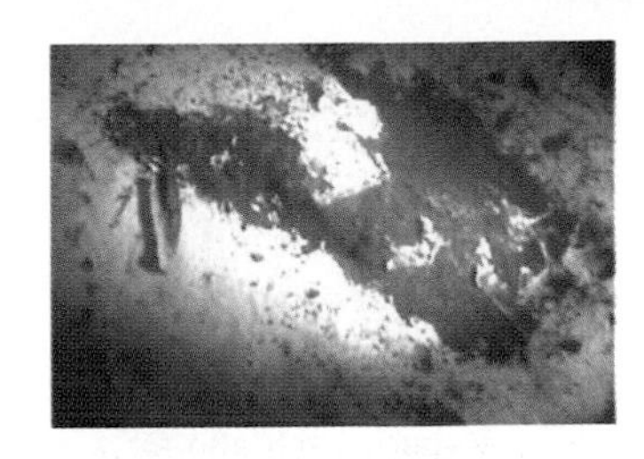

Fig. 2.7 *Beggiatoa* form filaments that twine together to form the white mats shown here. *Beggiatoa* is found in habitats that have high levels of hydrogen sulphide, including deep hydrothermal vents, sulphur springs, sewage-contaminated water, and mud layers. (Photo: Paul Dando.)

● Mixotrophs use both phototrophic and heterotrophic means for assimilating energy.

Box 2.5 Switching metabolism

There are groups of organisms that can also switch metabolism. Good examples are the **anoxygenic phototrophic bacteria**. These bacteria are capable of utilizing organic carbon when it is available, but capable of photosynthetic light utilization and CO_2 metabolism when organic carbon sources are low. These organisms are abundant in the upper oceans and are estimated to make up to 11% of the microbial community.

There are also organisms, **mixotrophs**, which combine the use of phototrophic and heterotrophic nutrition (e.g. Stoecker 1999). Many phytoplankton species, **phagotrophs**, have been shown to be able to ingest particulate organic material to meet part of their nutritional requirements. These range from small nanoflagellates that ingest bacteria and cyanobacterial sized particles through to photosynthetic dinoflagellates that can consume phytoplankton and small ciliates more than 10 μm in diameter. Many phagotrophic algae increase their rates of particle ingestion in response to nutrient limited conditions in order to obtain growth limiting compounds and elements.

Some marine organisms, including some molluscs, foraminiferans, helizoa, ciliates, and dinoflagellates retain chloroplasts that they have ingested when grazing on photosynthetic organisms. The chloroplasts, although not fully integrated into the metabolism, can be a useful source of energy. In general the chloroplasts do not function for long periods of time in the new 'host', and their function gradually declines and they are lost. However, in some species, such as the ciliate *Mesodinium rubrum*, which is a major species in some 'red tides', are truly photosynthetic organisms (Dolan & Pérez 2000).

(a)

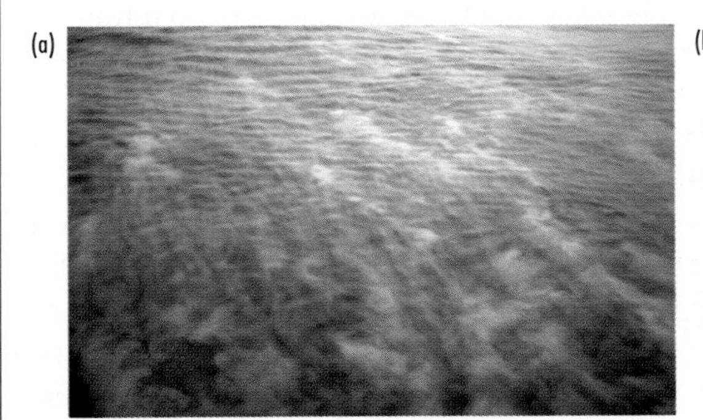

(b)

(a) Bloom of the autotrophic ciliate *Mesodinium rubrum* in surface waters of the North Sea. This an obligate phototrophic ciliate that contains endosymbiotic cryptophyte chloroplasts. (b) The relationship between corals and zooxanthellae are one of the best-known symbiotic relationships in marine systems. (Photographs: David Thomas).

continues

BOX 2.5 continued

One of the best-known symbiotic relationships in the marine world is that between photosynthetic dinoflagellates (zooxanthellae) and benthic corals. However, symbiotic relationships are also widespread in marine pelagic communities. A variety of algal species from several classes have been observed to form symbiotic associations with protozoans, medusae, turbellarians, and siphonophores. In most cases these relationships are highly species specific, most hosts only having one species as a symbiont. Certain species of heterotrophic dinoflagellates have cyanobacteria attached to their surfaces that sometimes reside within specialized pockets of the host's cell wall. Planktonic radiolarians and foraminifers can contain up to tens of thousands of symbiotic algae per individual.

It is clearly not an easy task to unravel the complexities of the metabolic pathways that are vital for the productivity in the oceans. With ever-increasing analytical tools the current trend for the discovery of new microbial components of the marine microbial world is likely to continue.

- Symbiotic relationships are vital for coral species, but also for planktonic radiolarians, foraminifers, turbellarians, molluscs, and siphonophores.

where sulphides are introduced into well-oxygenated seawater. Mats of *Beggiatoa* also grow on the reduced sulphur from the vents (Fig. 2.7). **Purple sulphur bacteria** are another example of organisms that oxidize H_2S and elemental sulphur:

$$H_2S + 2O_2 \Rightarrow SO_4^{2-} + 2H^+$$
$$S + H_2O + 3/2O_2 \Rightarrow SO_4^{2-} + 2H^+$$

- Nitrifying bacteria are important for nitrate regeneration in marine systems.

Nitrifying bacteria are vital for nitrogen cycling and the regeneration of nitrogen forms that can be utilized for growth in other organisms (2.10, Chapter 3). Two groups of chemolithotrophic **nitrifying bacteria** exist, one group (including *Nitrosomonas*) oxidizing ammonium to nitrite and another group (including *Nitrobacter*) oxidizing nitrite to nitrate. The reactions in this vitally important process of **nitrification** are:

$$NH_4^+ + O_2 \Rightarrow NO_2^- + 4H^+ + 2e^-$$
$$NO_2^- + H_2O \Rightarrow NO_3^- + 2H^+ + 2e^-$$

2.5 Light in water

- Without light, photosynthesis cannot take place.

Although many factors interact to determine the net primary production of photoautotrophs in the oceans, naturally it is light that is the dominant factor that determines the rate and extent of photosynthetic activity (Kirk 1994). It is both the quality of the light, and the quantity of the light that reaches the chloroplasts within the cells that control these reactions. As anybody who has dived or snorkelled in open waters can testify, the penetration of light can vary greatly (Fig. 2.8a).

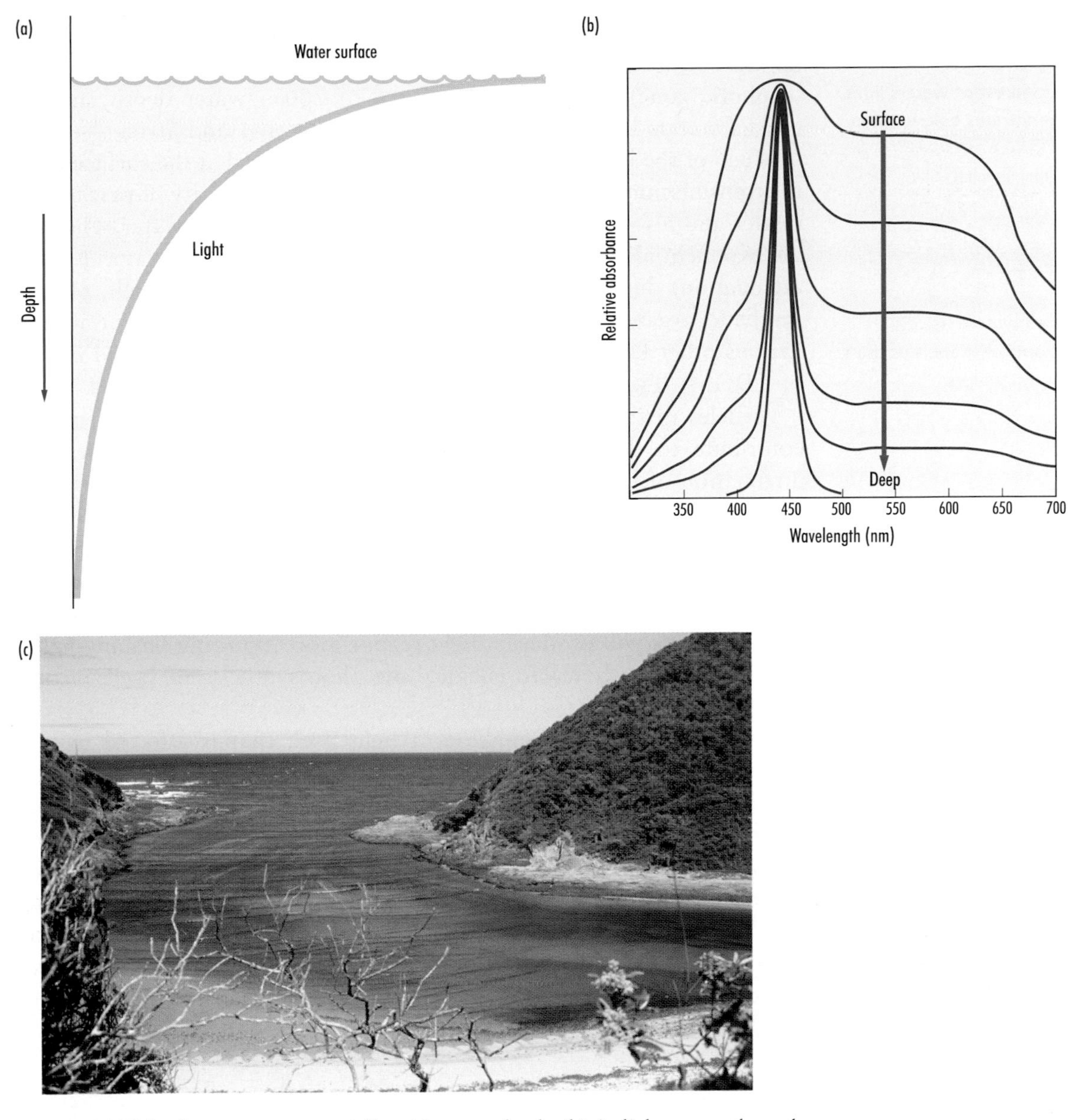

Fig 2.8 (a) Light decreases exponentially with water depth. (b) As light passes through a body of water, it is not just the amount of light, but also the spectral quality of light that changes. (c) River waters often contain high amounts of dissolved organic matter (DOM). In humic-rich rivers the DOM turns the waters brown. Where they mix with seawater the coloured waters can greatly influence the optical properties of coastal waters. (Photograph: David Thomas.)

In clear waters light can penetrate many hundreds of metres and in waters heavily laden with sediment or particulate matter it can be difficult to see a few metres. No matter how clear the water, below 1000 m water depth essentially no light penetrates and hence most of the world's oceans with an average depth of 4000 m are permanently in total darkness (**the aphotic zone**) (Chapter 8). In general the **photic zone** (**euphotic zone**) seldom extends down to 200 m water depth, and in coastal waters light penetrates to depths between 1 and 50 m.

- In the clearest of waters light seldom penetrates below 200 m.

Much of the light incident on the water is reflected at the surface, and the transmission of light is then dependent on the quantity of particulate matter and also dissolved organic matter (DOM) in the water. There is an exponential loss of light as it passes through the water (called **attenuation**) due to absorbance of the light by the water itself, photosynthetic organisms, particles in the water, and humic material and various other DOM compounds that colour the water (**coloured DOM** or CDOM) (Fig. 2.8c).

- Coloured water has very different optical properties than non-coloured waters.

Particles in the water, such as bacteria, plankton, and sediment, all contribute to the scattering of light in the water, which causes the attenuation of light. Therefore the distribution of such particles is intimately linked to the transmission of light through the water. Even bubbles, ripples of water on the surface, and waves will have dramatic effects on the underwater light regime. These features may result in short-term fluctuations in irradiance, with focusing and defocusing of the light. This will produce a light regime more akin to a flashing light, which has been shown to enhance the photosynthesis of some phytoplankton species.

- Bubbles, particles, and surface ripples all greatly alter the light field underwater.

It is not just the transmission of light itself that is affected by the scattering and absorption of light, but also the spectral quality of the light (Fig. 2.8b). Water absorbs strongly in the red and infrared part of the spectrum, and so at deeper water depths the light is reduced in this part of the spectrum and effectively enriched in the blue and blue-green wavelengths. Water looks blue because of this differential absorption of the blue and red parts of the spectrum. Coastal waters have a large input of humic DOM or **yellow substances** (sometimes called ***Gelbstoffe***). These reflect in the yellow-red part of the spectrum, and hence the characteristic yellow/brown colour of some river and coastal waters (Fig. 2.8c).

- Humic-rich dissolved organic matter often colours coastal waters.

2.6 Light and Photosynthesis

- *E* is the recognized symbol for irradiance.

The relationship between photosynthesis and irradiance is described by the characteristic ***P/E* curve** (Fig. 2.9) At low irradiance the photosynthetic rate is linearly proportional to increases in irradiance. At a particular irradiance the photosynthetic rate is equal to the respiration rate, the **compensation irradiance, E_c**. This irradiance is species specific, and within a single species can vary with season and even on shorter timescales. As irradiance increases the trend becomes gradually nonlinear and a point is reached where further increases in irradiance do not

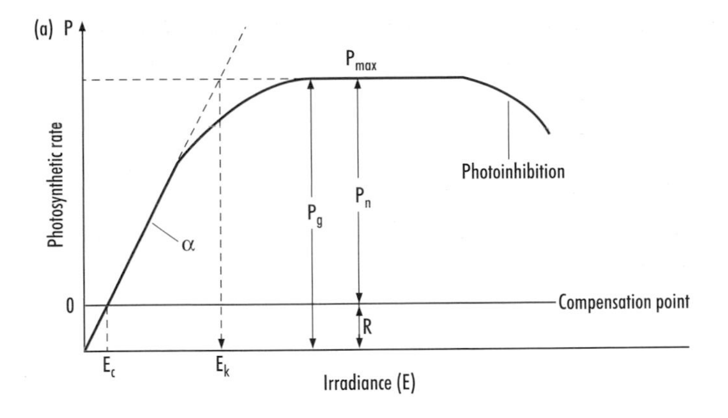

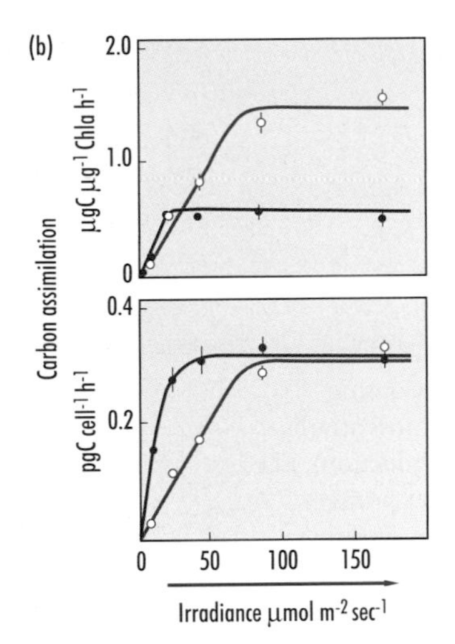

Fig. 2.9 (a) The response to photosynthesis (P) in response to changes in irradiance (E) – a P/E curve. With increasing light, photosynthesis increases linearly and the slope of the increase is α. At the compensation irradiance E_c the photosynthetic rate is equal to the respiration rate (R). With increasing irradiance the linear trend ceases, and at the saturation irradiance (E_k) the rate of photosynthesis is saturated (P_{max}). In some organisms, there can be a decrease in photosynthetic rates at high irradiances (photoinhibition). Respiration typically does not change with increasing irradiance, and gross photosynthesis is indicated by P_g and net photosynthesis by P_n (after Lalli & Parsons, 2004).

(b) P/E curves for phytoplankton cells grown in high (-o-) and low light (-•-). In the top set the rate of photosynthetic carbon assimilation is expressed as a function of chlorophyll concentration of the phytoplankton. In the bottom set, the same data is expressed on a per cell basis. The low-light algae have acclimated to the low light by increasing cellular concentrations of chlorophyll, and on a per cell basis reach the same P_{max} as the high-light acclimated algae, although their value of α is greater (i.e. more efficiently utilizing the lower irradiances).

result in increases in the photosynthetic rate. In other words, the rate of photosynthesis is light saturated (**P_{max}**). The slope of the linear part of the P/E curve is denoted by the symbol α. The **saturation irradiance, E_k,** is calculated from the intercept between α and P_{max}. In some organisms, there can be a decrease in photosynthetic rates at high irradiances. This decrease is a result of **photoinhibition.** This results from damage to components of the photosystems such as cellular membranes or electron-transport proteins.

● At high light levels maximum rates of photosynthesis can be inhibited, this is called photoinhibition.

Just as in the terminology used for primary production, **gross photosynthesis** is equivalent to the total photosynthesis, and **net photosynthesis** is equal to gross photosynthesis minus respiration (Fig. 2.10).

The characteristics P_{max}, E_c, E_k, and α are all species dependent, and also vary within a particular species depending on environmental conditions of light, nutrient status, and temperature. Generally P_{max}

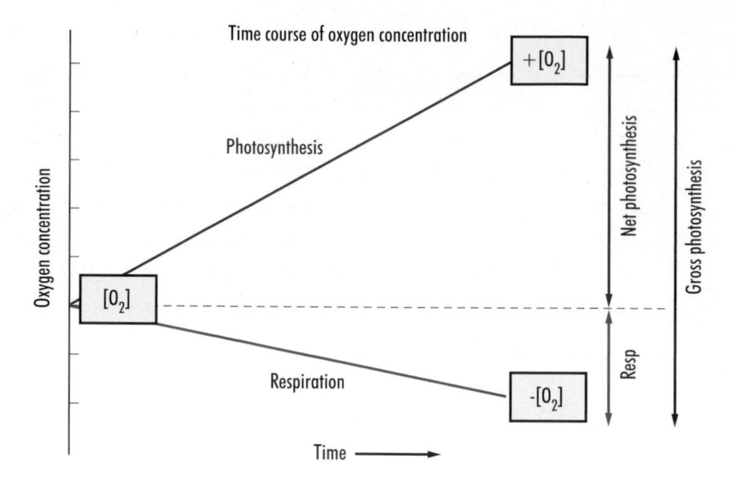

Fig. 2.10 The relationship between gross photosynthesis (gross primary production), net photosynthesis (net primary production), and respiration. (Image: Peter J. le B. Williams).

increases with increasing temperature (up to physiological limits), and is higher in organisms that grow at high rates in nutrient replete conditions compared with growth-limited cells in poor nutrient conditions.

2.6.1 Light acclimation

Within certain limits, most photosynthetic organisms are able to acclimate to a given light regime (MacIntyre et al. 2002; Raven & Geider 2003), mostly by altering the concentration of chlorophyll and/or accessory pigments per cell or per unit cell area. This process is called **photoacclimation**, and in low light conditions the Chlorophyll : Carbon (Chl*a* : C) ratio can be considerably greater than that for the same organism acclimated at higher light level (i.e. in the low light cells there is more chlorophyll per cell than in the cells at high light). Under nitrogen, phosphorus and trace metal (e.g. iron) limitation the Chl*a* : C ratio of a particular photosynthetic organism tends to decrease. Likewise in lower temperature acclimated cells, the Chl*a* : C ratio is lower than in cells acclimated at higher temperatures. The changing pigment concentrations of cells obviously have significant effects on the *P/E* characteristics of a particular algal species (Fig. 2.9b).

● By changing pigment content algae can photoacclimate to changing light regimes.

A Chl*a* : C ratio of 0.02 : 1 is often cited in the literature and used in models describing phytoplankton dynamics. However, because chlorophyll concentrations within phytoplankton cells can be altered due to external stimuli (temperature, irradiance, and the growth status of the algae) within time periods of hours, the utility of this ratio is rather dubious. Furthermore, ratios in the order of 0.01 : 1 to 0.005 : 1 are not uncommon in field populations (Lefevre et al. 2003).

Another feature, exhibited by some algal cells when exposed to changing light conditions, is an obvious movement of the chlorophyll-containing chloroplasts within the cells. The chloroplasts move along cytoplasmic strands in a process known as **karyostrophy**. Generally chloroplasts are distributed within the cell so that efficient light absorption can take place. However, in high light, clumping of chloroplasts, often around the nucleus, is frequently observed, and is thought to be associated with mechanisms to protect cell organelles from damaging light effects. Although not universal within aquatic photosynthetic organisms such mechanisms are known from terrestrial higher plants. It is likely that chloroplast movement processes are important for coping with rapidly changing light environments in turbulent surface waters or diurnal changes in light.

● Karyostrophy is a process by which chloroplasts move in reaction to changes light conditions.

Box 2.6 Compensation and critical depths

The euphotic zone is the upper part of the water column that supports photosynthesis. The bottom of this zone is generally defined as the depth at which 1% of the surface irradiance is measured. However, a better representation of the bottom of the euphotic zone is the **compensation depth**. This is the depth at which the gross photosynthetic carbon assimilation by phytoplankton equals the respiratory carbon losses, or when the net photosynthesis is 0.

● Compensation depth is the depth in a water column at which net photosynthesis is 0.

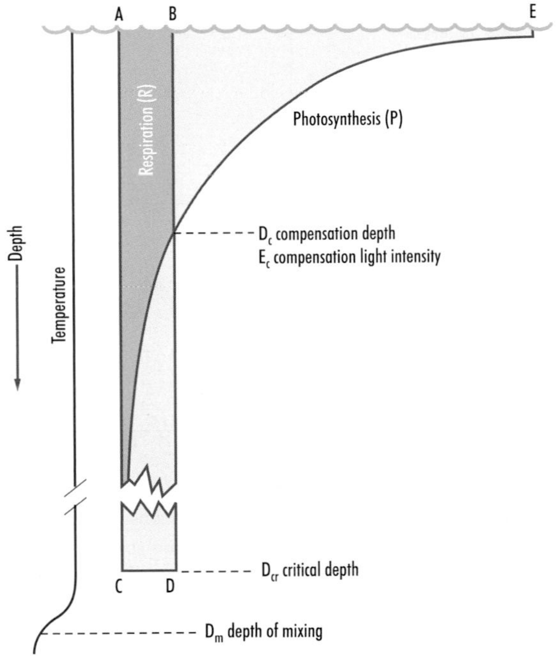

At the compensation depth (D_c) the phytoplankton photosynthesis is equal to the respiration, i.e. the compensation light intensity E_c. Phytoplankton is mixed in the water column, above and below the compensation depth, down to the depth of mixing (D_m). The critical depth is the water depth where the integrated water column photosynthesis is equal to the integrated water column respiration. In this diagram the area bounded by the points A, B, C, & D represents respiration, and the area A, C, & E represents the photosynthesis. At the critical depth these two areas are equal. When the depth of mixing is deeper than the critical depth, no net growth takes place. When, however, the

continues

BOX 2.6 continued

depth of mixing is shallower than the critical depth net phytoplankton growth occurs. (After Lalli and Parsons 2004).

It is worth noting that there is often a reduction in the photosynthesis rates measured at the surface of the water. This represents an often observed lowering of the photosynthetic rate due to photoinhibition in the very topmost meters of the water column.

Of course phytoplankton cells are not at a static depth as they and/or the water may move. In fact they are mixed either throughout the whole water column or, where water stratification takes place, within surface mixed water layers (Chapter 6). Because of this phytoplankton cells will be mixed above and below the compensation depth, to depths as deep as the **mixed layer depth**. When considering net phytoplankton growth it is therefore more pertinent to relate the daily integrated photosynthetic gains to the integrated respiration losses over the water column (day and night) to the mixed layer depth.

The **critical depth** is the water depth where the integrated daily photosynthetic carbon assimilation is balanced by the integrated daily respiratory carbon losses. As long as sufficient nutrients are present, net phytoplankton growth occurs when the mixed layer depth is shallower than the critical depth. When the mixed layer extends below the critical depth algal growth is limited by light, and there is no net phytoplankton growth (Sverdrup 1953; Smetacek & Passow 1990).

● The critical depth is the water depth where the integrated daily carbon assimilation is balanced by daily respiratory carbon loss.

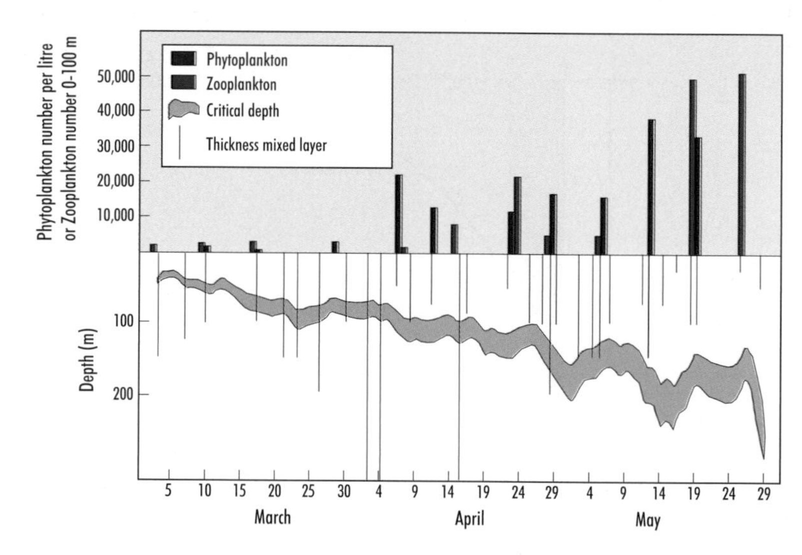

Data from the Norwegian Sea in 1949 showing the relationship between mixed layer depth, critical depth, and phytoplankton and zooplankton abundance. Growth occurs only when the depth of mixing is consistently above the critical depth. (Illustration adapted from Sverdrup 1953).

The critical depth theory was first proposed by Sverdrup in 1953. In this theory the respiration losses are not just the algal respiration losses, but also losses due to grazing organisms and the respiration of bacteria and other heterotrophic organisms. Interestingly it led to a major advance in the design of freshwater reservoirs, where preventing algal blooms is an important part of their management.

The seasonal changes of mixed layer depth and incident light play a key role in the seasonal dynamics of phytoplankton (see below). These are discussed below in conjunction with the inorganic nutrient demands of growing phytoplankton populations.

2.7 Supply of Inorganic Nutrients

In addition to carbon, oxygen, and hydrogen, the plant must incorporate other elements into organic material. This arises as a consequence of the elemental composition of the various macromolecules, notably proteins and nucleic acids. The principal additional requirements are nitrogen and phosphorus and in aquatic ecology, these elements are commonly referred to as **nutrients** or **inorganic nutrients.**

2.7.1 Nutrient status of water

Waters that have low concentrations of essential nutrients for algal growth are called **oligotrophic** and are regions of low primary productivity. In contrast **eutrophic** waters have high concentrations of nutrients, and generally support high levels of primary production. Waters between the two states are referred to as **mesotrophic** waters, and these sustain intermediate levels of primary production.

Most marine systems are classified on the basis of the annual primary production, which is another way of expressing the supply or production of organic matter in the water body:

Fig. 2.11 Increased inorganic nutrient supply can result in excessive algal growth as shown by the mass of *Enteromorpha sp.* on clam cultivation plots in the River Exe, England. (Photograph: Brian Spencer.)

Organic Carbon Supply	
Oligotrophic	$<100\ g\,C\,m^{-2}\,year^{-1}$
Mesotrophic	$100–300\ g\,C\,m^{-2}\,year^{-1}$
Eutrophic	$300–500\ g\,C\,m^{-2}\,year^{-1}$
Hypertrophic	$>500\ g\,C\,m^{-2}\,year^{-1}$

The process of **eutrophication** can be defined as **an increase in the rate of supply of organic matter to an ecosystem.** This occurs when there is a change in the concentration of a factor (can be more than one) that limits algal growth. This is often an increase in inorganic nutrients such as nitrogen or phosphorus, often associated with the run-off of artificial fertilizers from agricultural land (Chapter 14). The resulting increase in algal growth (both phytoplankton and/or macroalgae) if excessive (Fig. 2.11) can have deleterious effects for the whole ecosystem (Skei et al. 2000).

● Eutrophication can be defined as an increase in the rate of supply of organic matter to an ecosystem.

It is important to stress that eutrophication is a process or change and is not a trophic state. For example, an estuary may have been mesotrophic and is now classified as eutrophic, but it is not necessary undergoing further eutrophication. Although eutrophication is generally perceived as a detrimental process, it is also important to stress that it can be a reversible process. There are also instances when low levels of eutrophication can even be considered as being a positive state for increasing the productivity of a specific water body.

● Eutrophication is a reversible process and not always detrimental.

Strictly speaking eutrophication is a process by which the productivity of an aquatic system is increased, and can therefore be caused by factors other than nutrient input. These factors include reducing the suspended material in a water body and therefore increasing the light levels available for photosynthesis, or changing the residence time of water within a particular system. Therefore eutrophication is also a natural phenomenon, and is not always associated with anthropogenic activities. Coastal regions can receive high dosages of nutrients both directly via marine outfalls and by discharges from estuaries. This, coupled with their relatively long residence time, makes coastal ecosystems especially vulnerable to eutrophication (Chapters 7 and 14).

2.7.2 Supply of nutrients

Photoautotrophs require a diverse range of elements for balanced growth. These include nitrogen, phosphorus, silicon, sulphur, potassium, and sodium (all known as macronutrients). Many trace elements (micronutrients) are also required, including iron, zinc, copper, and manganese, as well as vitamins such as B_{12} (cyanocobalamin), biotin, and thiamine.

● A wide range of macro- and micronutrients are needed for algal growth.

Although each nutrient has the potential to limit the growth of photoautotrophs, in most marine environments it is either nitrogen or phosphorus that is generally the limiting element (c.f. Box 2.7 shown in section 2.9). It is actually the supply of the nutrient to the organism that is critical. A nutrient can be present in low concentrations, but if the uptake rate by the organism is low, only a low supply rate is required. Naturally growth can be limited by the supply of more than one nutrient at any one time: nutrients can be either biomass limiting or rate limiting. In the case of the former the nutrients are exhausted so that no more biomass can be produced. In contrast rate-limiting nutrients simply limit the rate of new biomass production by their rate of supply.

● Resupply of nutrients primarily takes place by molecular diffusion.

When a nutrient is taken up by an organism, there is immediately a reduction in that nutrient in the micro-environment surrounding the cell or organism. The resupply of nutrients takes place primarily by molecular diffusion from the bulk medium of water. Surrounding each cell or surface in water is a **diffusive boundary layer** (DBL) in which water movement and molecular diffusion is restricted. The thickness of the DBL surrounding the organism is therefore critical to determining the rate at which nutrients are transported to cell surfaces. The smaller the organism, the smaller the DBL due to the surface area : volume relationship (Chapter 3). This gives smaller organisms a physiological advantage at low nutrient concentrations and is presumed to be why small species of phytoplankton prevail in oligotrophic waters.

● Small phytoplankton species have a greater surface area : volume ratio than larger species.

It has been estimated that for cells less than 1 μm in diameter molecular diffusion is adequate for the resupply of nutrients, but for

Box 2.7 Iron and high-nutrient, low-chlorophyll regions

In some areas of the world's oceans the supply and assimilation of nitrogen, phosphorus, and carbon appear not to be linked. In these waters the phytoplankton standing stocks are never large enough to assimilate the N and P in the surface waters fast enough to deplete them at any time throughout the year. These are the 'high-nutrient, low-chlorophyll' (HNLC) waters of the subarctic Pacific, the Southern Ocean, and the equatorial Pacific.

● HNLC = 'high-nutrient, low-chlorophyll'.

Several hypotheses have been proposed to explain HNLC regions, including suggestions that in these regions light (either low or damaging high light intensity) limits production to a degree that inorganic nutrients are not utilized or that grazing pressure limits the standing stocks of phytoplankton. Whereas these factors clearly do play a role to varying degrees, the most compelling explanation is that the rate of supply of iron, an essential trace element for phytoplankton growth, is limited.

● Lack of iron may limit the growth of phytoplankton in 'high-nutrient, low-chlorophyll' regions.

Dissolved iron concentrations in offshore areas are extremely low, since the primary source of iron to the surface waters of the oceans is from the land, either via atmospheric dust deposition in offshore areas, or direct depositions from land masses. Atmospheric dust deposition in the two major HNLC areas – the Antarctic and equatorial Pacific Oceans – are the lowest in the world. Conversely, in the equatorial North Atlantic, which receives large amounts of dust from the Sahara, iron concentrations are sufficient for the complete assimilation of available nitrates and phosphates.

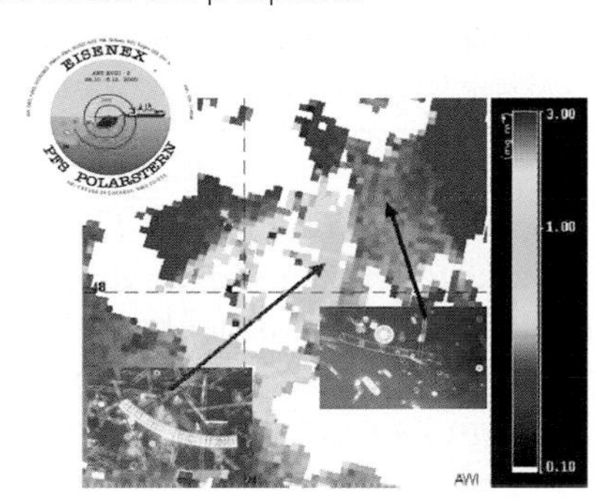

A satellite image of an iron fertilized patch during the Eisenex expedition to the Southern Ocean in 2000. The sparse phytoplankton outside of the patch is striking compared with the abundant growth of phytoplankton within the patch following fertilization. The satellite image shows the increase in chlorophyll (orange/red) compared to the waters surrounding the patch (blue). (Image: Philipp Assmy & Joachim Henjes, Alfred Wegener Institute).

In the late 1980s John Martin developed the idea that a lack of iron is the cause and laboratory experiments confirmed how vital iron is for phytoplankton growth. A series of experiments in the Equatorial Pacific Ocean, where large areas of the ocean (hundreds of square kilometres) were seeded with iron, led to substantial increases in phytoplankton growth. In particular diatoms grew and it appears that not all phytoplankton species are equally iron limited (Martin et al. 2002). Several oceanographic expeditions showed that during spring in the Southern Ocean phytoplankton bloom in iron-rich waters, but do not in waters with limited iron reserves (de Baar et a1.1995). It was pertinent to therefore extend the Pacific iron-fertilization experiments to the Southern Ocean. Several studies have now 'fertilized' Antarctic water bodies and in all of these diatoms did bloom in

continues

BOX 2.7 continued

response to the added iron. This growth was in turn responsible for the absorption of significant quantities of carbon dioxide from the water during the experiment (Boyd et al. 2000; Buesseler et al. 2004; Coale et al. 2004).

It is this link between the phytoplankton growth and drawdown of atmospheric carbon dioxide that fuels a vigorous debate about these experiments. There is a concern that these results may be viewed as providing a simple answer for mopping up excess carbon dioxide, thereby curbing the effects of increasing greenhouse gases. It is thought by some that by spreading iron over huge swathes of the ocean, enhanced phytoplankton growth would effectively trap carbon dioxide. Such ideas about large-scale ecological engineering have little to do with the work of the scientists conducting the experiments. Iron fertilization is in fact a poor way to tackle greenhouse gas problems. Calculations show that iron fertilization of the Southern Ocean would not in fact be an effective mechanism for carbon dioxide removal. Levels of carbon dioxide are increasing at such a rate that even by maximizing biological uptake in these oceans by adding iron there would still be a net increase in atmospheric carbon dioxide (Chisholm et al. 2001; Buesseler et al. 2004).

The real interest of this work comes from the implications for the understanding of the atmospheric carbon dioxide levels in past climate history. This new evidence supports the theory that low amounts of atmospheric carbon dioxide (measured in ice cores taken in the Arctic and Antarctic) during past ice ages may be linked to high amounts of iron in Antarctic waters that supported large standing crops of phytoplankton.

● Iron-fertilization experiments have resulted in increased phytoplankton growth within the fertilized patches.

● Fertilization of the oceans with iron will not be an easy fix for combating rising atmospheric carbon dioxide concentrations.

larger organisms it is a major limiting factor. Therefore mechanisms for reducing the DBL around the organism are key to the nutrient metabolism of aquatic organisms. Movement through the water, either by sinking or swimming, means that the nutrient depleted boundary layer is dragged with the organism causing fluid from the layer to be sheared away. This can then be replaced by nutrient-replete water. Clearly the velocity and magnitude of distance travelled will affect the degree of replacement. However, it is estimated that swimming significantly reduces diffusion limitation only in organisms greater than 100 μm diameter. In many of the organisms between 1 and 100 μm diameter, movement is a means of relocating into regions of higher nutrient concentration rather than addressing diffusion limitation by altering DBL properties. Any property of an organism that alters the organism's size and/or shape will alter the properties of the DBL. Therefore organisms that form colonies or chains, such as the diatoms, change their shape and therefore sinking rate, and also the characteristics of the DBL surrounding the colony or chain (Fig. 2.12).

● The diffusive boundary layer is key to determining nutrient supply to a cell, or surface of a macroalga.

In the case of macroalgae and seagrasses, relief or structures on the thallus or frond surface will cause eddy formation and turbulent motion of water passing over the surface (see also Chapter 7). This will have the effect of reducing DBLs and therefore enhancing nutrient exchange. As macroalgae and seagrasses cannot move in the water column by

● Ridges and structures on the surface of a macroalga can increase the rate of exchange of nutrients and dissolved gasses.

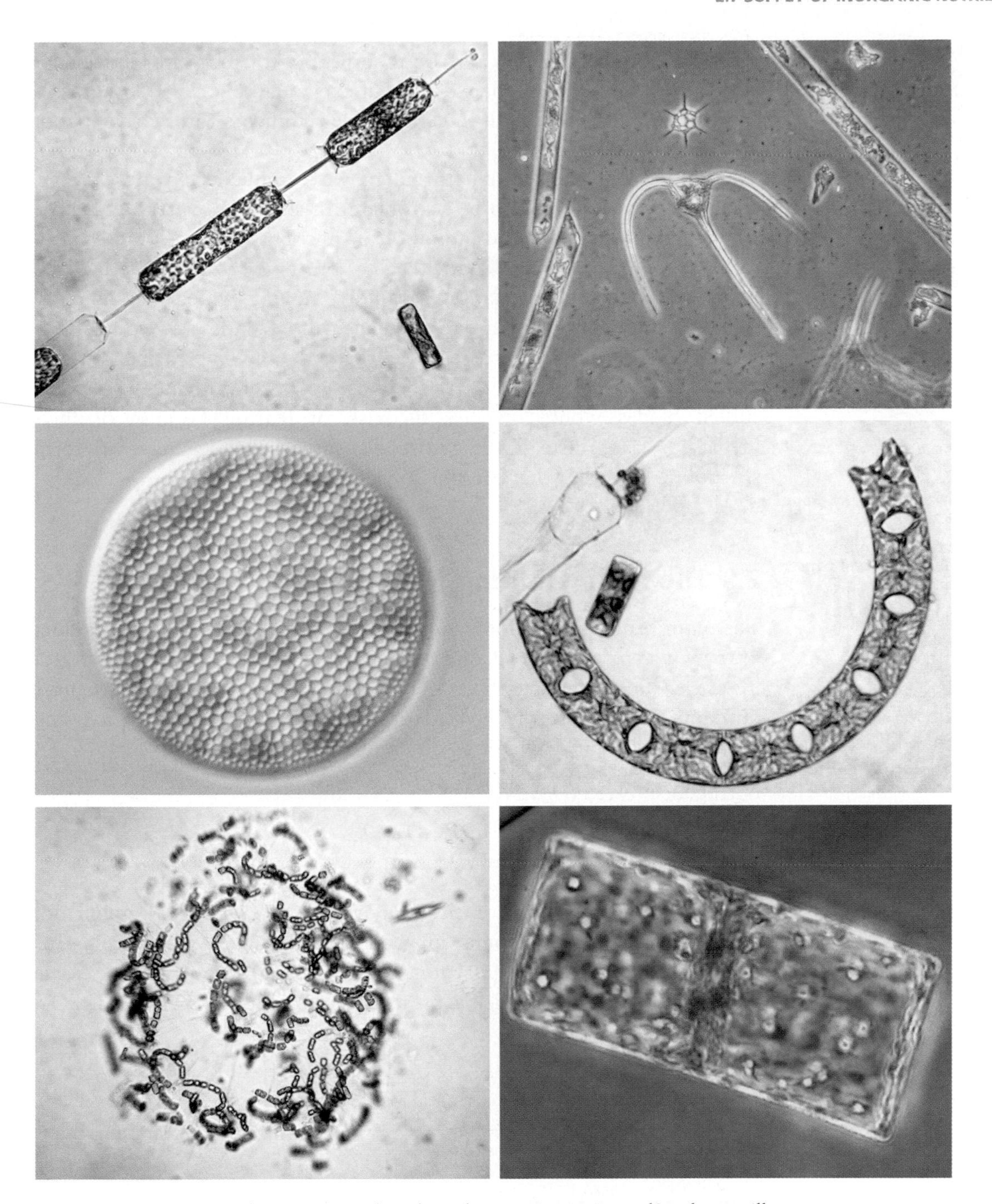

Fig. 2.12 Any property of an organism that alters the organism's size and/or shape will alter the properties of the diffuse boundary layer (DBL). Therefore the variety of shapes and forms of phytoplankton shown here will have very different sinking rates as well as different characteristics of the DBL surrounding the cells, colonies, or chains. Starting from top left and moving clockwise the images are: 1) *Ditylum brightwellii*, 2) *Ceratium tripos & Rhizosolenia* sp., 3) *Eucampia zodiacus*, 4) *Guinardia flaccida*, 5) *Chaetoceros socialis*, 6) *Coscinodiscus* sp. (Photographs: Ian Lucas).

swimming or sinking, they rely on modification of the water movement across their surfaces to enhance nutrient exchange (Fig. 2.13). Seagrasses are also able to take up nutrients from the sediments through their root and rhizome systems.

Naturally on exposed and turbulent wave influenced waters, water exchange is never going to be a problem. In sheltered waters with little water movement, diffusion limitation of nutrients becomes more of an issue. For example, the giant Pacific brown seaweed (*Nereocystis luetkeana*) has smooth blades in rapidly moving water, but in more sheltered waters its fronds are ruffled. The ruffled blades serve to increase the turbulence as water passes over them thereby increasing nutrient supply and gas exchange to the fronds. In faster moving waters the ruffled blades would increase the risk of tearing because of their increased drag; instead the seaweed's smooth blades tend to form streamlined bundles that are not so easily damaged.

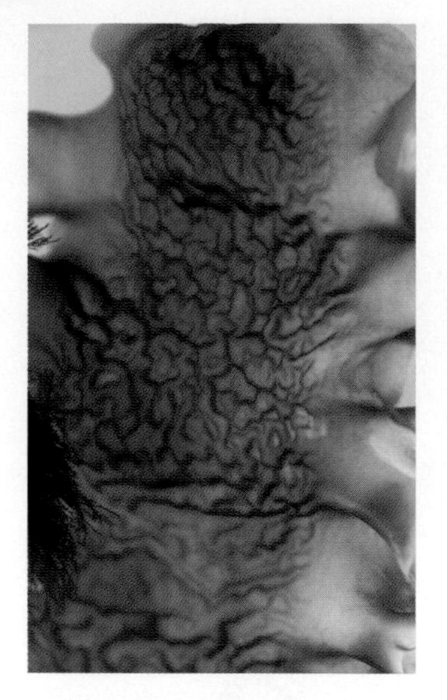

Fig. 2.13 Structures on the surfaces of macroalgae can cause turbulent water movements over the surfaces. This will increase the exchange of gases and inorganic nutrients compared with when undisturbed lamina flow passes over the surface. (Photograph: David Roberts).

2.8 The Main Limiting Nutrients for Growth

The main products of photosynthesis are sugars, reductant (the product of reducing enzymes), ATP, and oxygen, which are themselves substrates in further biosynthetic pathways (Table 2.2). The other major inorganic nutrients needed for the myriad of molecules that make up a living organism occur in seawater as different chemical species:

Nitrogen: NO_3^-, NO_2^-, NH_4^+, NH_3, N_2, and urea

Phosphorus: HPO_4^{2-}, PO_4^{3-}, and $H_2PO_4^-$

Sulphur: SO_4^{2-}, H_2S

● Nitrogen and phosphorus are the main growth-limiting nutrients in marine systems.

Particular nutrients are critical for certain organisms, although not limiting for photoautotrophs in general. An example is silicate, for diatoms and silica-scaled *prymnesiophytes* (and some cysts of some dinoflagellate species), which is present in several forms as well: H_4SiO_4, $H_3SiO_4^-$ and $H_2SiO_4^{2-}$.

Table 2.2 Typical percentage biochemical and elemental composition of algal cells.

Biochemical	% of an algal cell	% C	% H	% O	% N	% P	% S
Carbohydrate	40	44	6	49			
Protein	40	53	7	23	16		1
Lipids	15	69	10	18	1	2	
Nucleic acid & nucleotides	5	36	4	33	17	10	

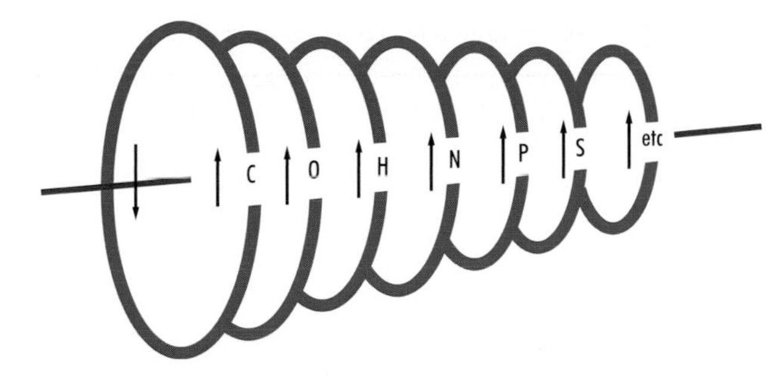

Fig. 2.14 There is a link between nutrient cycles and uptake of nutrients by primary producers. Ultimately these cycles and relative magnitudes are based on the biochemical composition of the organisms themselves. (Image: Peter J. le B. Williams.)

Table 2.3 Examples of nitrate, nitrite, ammonium, and phosphate concentrations in surface and deep waters of the Atlantic, Pacific, and Indian Oceans.

Ocean	Depth (m)	NO_3^- (μmol l^{-1})	NO_2^- (μmol l^{-1})	NH_4^+ (μmol l^{-1})	$PO4^{3-}$ (μmol l^{-1})
Atlantic	5	0–20	0–0.5	0–3	0.1–1
	4000	20	<0.5	<2	2
Indian	5	0–20	0–0.5	0–3	0.1–1
	4000	30	<0.5	<2	3
Pacific	5	0–20	0–0.5	0–3	0.1–1
	4000	35	<0.5	<2	3

The balance of these forms of any one element is highly dependent on many complex biogeochemical processes. Many of the trace nutrients are actually only mostly found in complexed forms with organic compounds in seawater (Table 2.3). The nutrient demands of individual species are naturally a reflection of its biochemical demands and composition. Clearly the proportion of the relative macromolecules, e.g. lipid (cf. nucleic acid), will determine the elemental composition of the cell and the relative demands for C, N, and P for growth. Although there is great variation in the composition of the cell, the functioning of the cell (i.e. the requirements for metabolism (enzymes – proteins) and reproduction (nucleic acid)) sets constraints on the relative proportions of the various macromolecules, and consequently the relative requirements for carbon, nitrogen, and phosphorus during photosynthesis.

2.8.1 Elemental composition of algae

There are obvious nutrient demands that are virtually universal, and these will be discussed here. However, it must be stressed that in many instances it is not the major inorganic nutrients (Table 2.4) that may

● A typical algal cell is 40% protein, 40% carbohydrate, 15% lipid, and 5% nucleic acids.

Table 2.4 The major functions of some selected inorganic nutrients.

Nutrient	Examples of functions
Nitrogen	Major metabolic importance, structural amino acid, protein metabolism
Phosphorus	Structural in particular membranes and energy metabolism
Potassium	Osmotic regulation, protein stability
Calcium	Ion transport, enzyme activation, structural function
Magnesium	Ion transport, enzyme activation, pigments such as chlorophyll
Sulphur	Structural function, active in enzyme activity
Iron	Active in enzyme activity
Sodium	Ion transport, osmoregulation, enzyme activation
Manganese	Electron transport and membrane structure

limit growth, but rather the rate of supply of trace elements that restrict the growth rates (Box 2.7).

Typically marine phytoplankton are comprised of more than 40% protein, 5% nucleic acids and nucleotides, 40% carbohydrates, and 15% lipids. These proportions can vary greatly depending on the inorganic nutrient supply, age of the organism, temperature, and irradiance conditions. It is straightforward to categorize the average composition of the organic materials within the phytoplankton.

If we rewrite the simplified photosynthesis equation to take into account the need for nitrogen (mostly supplied as nitrate) and phosphorus (supplied as phosphate) we get:

$$106CO_2 + 16NO_3^- + HPO_4^{2-} + 122H_2O + 18H^+ \Rightarrow C_{106}H_{263}O_{110}N_{16}P + 138O_2$$

Typically from the equation above, the ratio of **carbon : nitrogen : phosphorus** in healthy, actively growing algal cells is **106 : 16 : 1**. This ratio is referred to as the **Redfield ratio**, after the oceanographer A.C. Redfield. Therefore the typical C : N ratio is 6.6 : 1. This is a commonly used parameter to measure the physiological status of algae, since when nitrogen is limited, or the algal cells are senescent or dying the ratio increases considerably (Burkhardt & Riebesell 1997; Lenton & Watson 2000; Geider & la Roche 2002).

● The Redfield ratio, carbon : nitrogen : phosphorus is 106 : 16 : 1.

The quotient of CO_2 to O_2 from the above equation is 1.3. This is known as the **photosynthetic quotient (PQ)** (Williams 1998). This quotient is highly dependent on the state of oxidation/reduction of the nitrogen source. When nitrate is taken up by an algal cell it has to be reduced within the cell to ammonium before it can be utilized in cellular metabolism (Fig. 2.17). Algae can also take up ammonium (NH_4^+) directly as a nitrogen source, avoiding the energy in the reduction stage when nitrate is assimilated (see below). Therefore growth on ammonium

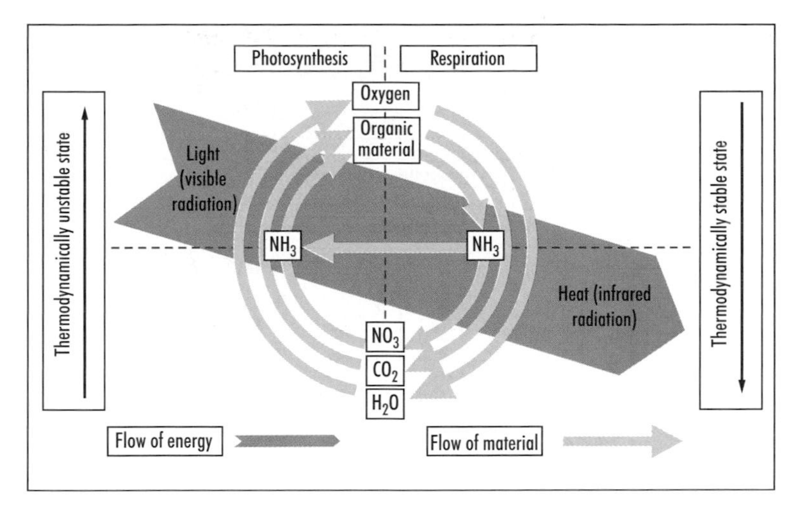

Fig. 2.15 The grand biological cycle is driving the biogeochemical pathways within marine systems. To simplify the diagram nitrate (NO_3^-) and ammonia are not shown in their charged forms. See further discussion in Chapter 3.

is about 19% more efficient on comparing the PQ values. The resulting PQ is lower at around 1.09.

● PQ = the moles of O_2 evolved per moles of CO_2 assimilated.

It must be stressed that this discussion of elemental ratios and typical cell composition is oversimplified. The cell composition (and therefore elemental ratios) will vary greatly at different life-history stages, and with changes in the prevailing temperature, light, and nutrient status. For example in older phytoplankton cells there is a marked switch from carbohydrate production to the accumulation of lipid reserves. When there is a lack of nitrogen in the surrounding water, protein synthesis is suppressed and the relative proportion of lipid and carbohydrates increase (Fig. 2.15).

2.8.2 Nitrogen

Nitrogen is present in seawater as dissolved molecular N_2, ammonium (NH_4^+), nitrite (NO_2^-), nitrate (NO_3^-), and as organic forms such as urea, amino acids, and a diverse range of complex dissolved organic nitrogen (DON) compounds (Fig. 2.16). In seawater ammonia (NH_3) exists as a mixture of the ammonium ion (NH_4^+) and NH_3. At seawater pHs (approx. 8) over 95% is in the form of NH_4^+. With increasing pH the relative contribution of NH_3 increases (e.g. at pH 9 NH_4^+ is about 75%).

● Only some species of cyanobacteria can fix nitrogen gas directly.

Nitrogen is the element that most frequently limits primary production in the oceans. Only some cyanobacteria, such as *Trichodesmium* species can reduce (fix) nitrogen gas, and these species thrive in waters where other forms of nitrogen are limited and thus restrict the growth of other phytoplankton (Berman-Frank et al. 2001). However, they still need sources of other nutrients such as phosphate.

Blooms of nitrogen fixing cyanobacteria are a feature of oligotrophic waters (particularly oceanic tropical coastlines), where nitrogen is

● Nitrate is the primary form of nitrogen assimilated by marine primary producers.

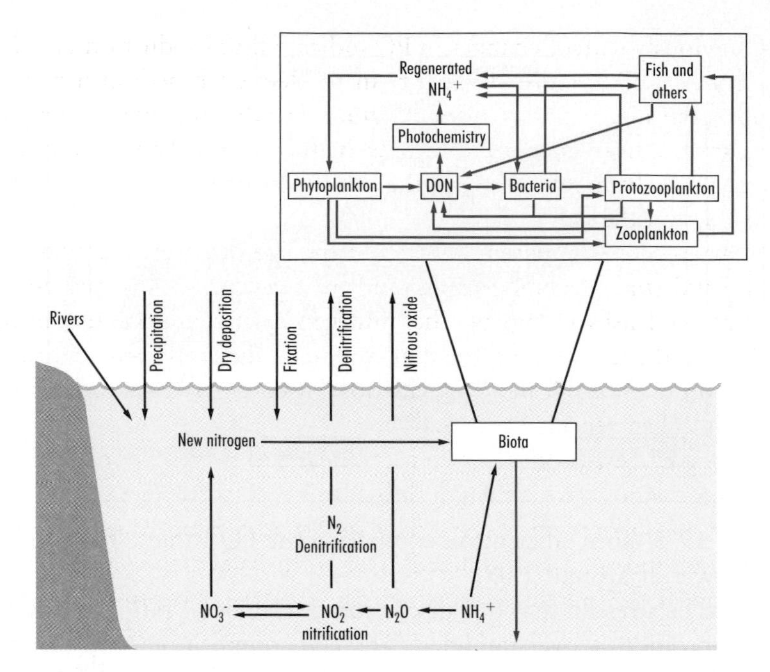

Fig. 2.16 Schematic of nitrogen in the ocean including input, transformation, and loss terms.

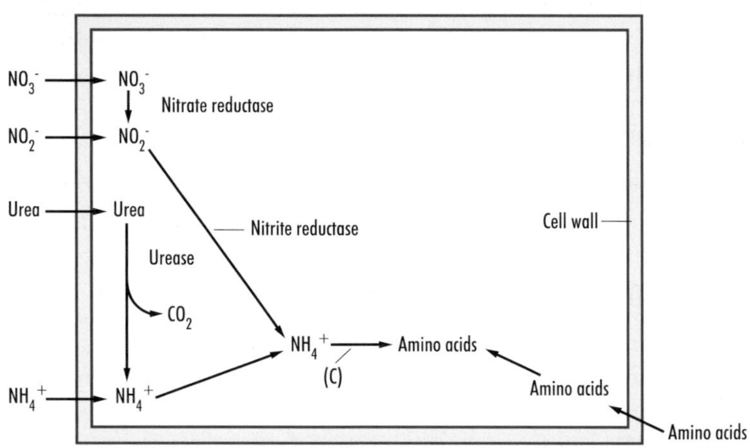

Fig. 2.17 Highly simplified schematic of major routes of nitrogen uptake and cellular transformation of inorganic nitrogen into amino acids by algae.

present in the water in very low concentrations, thereby restricting the algal growth. Other nitrogen fixers are also found growing on and in sediments in coastal regions, saltmarshes, and estuaries as well as in the roots of seagrasses and other saltmarsh grasses (e.g. *Spartina* spp.) where the nitrogen fixed by the cyanobacteria also supports the growth of the plants. In general, NO_3^- is the primary source of nitrogen utilized by algae, although NO_2^- and NH_4^+ can also be taken up (Fig. 2.17). Whatever the original source of nitrogen, NH_4^+ is the form utilized in cell metabolism and NO_3^- and NO_2^- have to be reduced by the enzymes **nitrate reductase** and **nitrite reductase** within the cell:

$$NO_3^- \rightarrow \text{nitrate reductase} \rightarrow NO_2^- \rightarrow \text{nitrite reductase} \rightarrow NH_4^+$$

As previously stated, changes in PQ indicate that production based on ammonium is 19% more efficient than production based on nitrate. In fact in some algal species nitrate uptake is inhibited when there are significant ammonium concentrations in the water. This inhibition is thought to be brought about by the inhibition of the nitrate reductase activity within the cells.

Primary production based on nitrate is referred to as **new production**. The breakdown, decomposition, and respiration of organic matter (Chapter 3) releases a range of other nitrogen species, ammonium, nitrite, and urea, and even amino acids that can be used as nitrogen sources for primary production. Primary production based on non-nitrate nitrogen sources is called **recycled production**.

The ratio of new production to total (new and recycled) production is called the **f-ratio**. The average global value is between 0.3 and 0.5. However, in oligotrophic deep ocean sites where there is little input of fresh nitrate into the upper mixed layer from below, the values can be well below 0.1 (Chapter 8). In contrast, in coastal regions or sites of coastal upwelling where input of nitrate can be high, the f-ratio can be up to 0.8 (Chapters 6 and 7). The inverse of the f-ratio gives the number of times the element recycles per year i.e., an f-ratio of 0.3 implies that an average nitrogen atom cycles round 3 times per year.

- The f-ratio is the ratio of new production to total (new and recycled) production.

2.8.3 Phosphorus

Following nitrogen, phosphorus is the second most common limiting nutrient in marine systems. Phosphate occurs in several forms in seawater, HPO_4^{2-}, PO_4^{3-}, and $H_2PO_4^-$, although at pH 8 and a temperature of 20 °C HPO_4^{2-} accounts for 97% of the free ions. Phosphorus is also present in a diverse range of organic compounds (organic phosphates). These can be broken down by enzymes (phosphatases) that are located in the membranes of many algal species. Phosphate limitation may also result in algae releasing alkaline phosphatases into the surrounding water to break down organically bound phosphorus. When external concentrations of inorganic phosphate are high, the production of alkaline phosphatases is repressed.

- Phosphorus can also limit primary production in some marine systems.

2.8.4 Sulphur

Sulphur is rarely a limiting nutrient as seawater is rich in sulphate, but it is a vital nutrient primarily for amino acid and protein synthesis. Curiously, in some macroalgae, such as the brown *Desmarestia*, the cells contain such high concentrations of sulphuric acid within their vacuoles that the low pH of the tissues deters grazers. Many species of macroalgae also contain large quantities of sulphated polysaccharides in their cell walls.

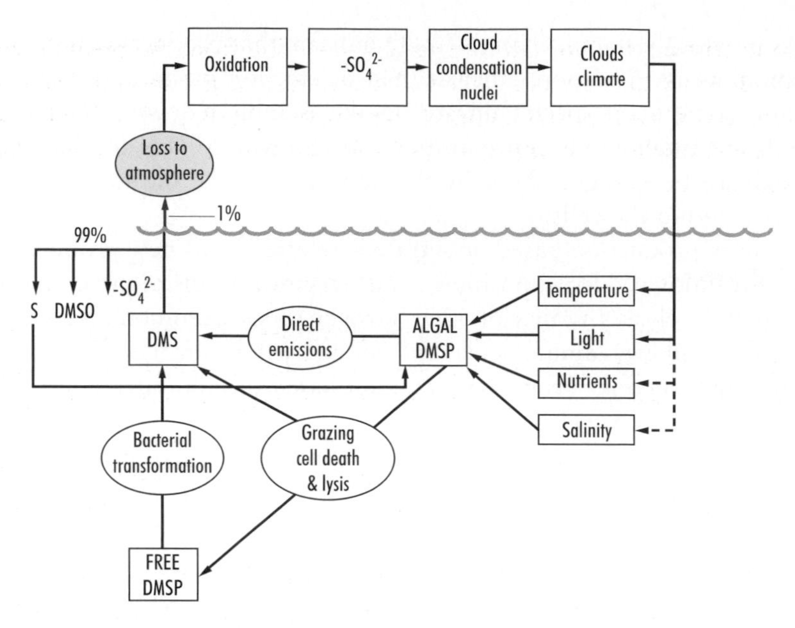

Fig. 2.18 Dimethyl Sulphide (DMS) released from phytoplankton and macroalgae (which contain DMSP) accounts for most of the non-marine salt sulphate in the atmosphere, and the oxidation of DMS to sulphur dioxide and the subsequent formation of aerosol particles and cloud condensation nuclei is part of a complex system of localized and global climate control. (NB Most of the DMS in the water is converted to DMSO (dimethyl sulphoxide), SO_4^{2-} and S, only 1% being exported to the atmosphere.) Image from Gunter Kirst.

● DMSP is cleaved into DMS and acrylic acid.

● DMS is important in the production of cloud condensation nuclei.

Several phytoplankton species, as well as red and green macroalgae, produce a specialized compound, dimethylsulphonioproprionate (DMSP), which is used as an osmolyte in osmoregulation, a storage product, and possibly an antifreeze compound. In elevated external salinities intracellular concentrations of DMSP are raised in order to restore osmotic balance, and when external salinities are lowered DMSP is broken down proportionally into dimethyl sulphide (DMS) and acrylic acid (Fig. 2.18).

DMSP is also broken down through the action of the enzyme DMSP-lyase and from grazing by proto- and metazoans, as well as viral infection. DMS released from phytoplankton and macroalgae accounts for most of the non-marine salt sulphate in the atmosphere, and the oxidation of DMS to sulphur dioxide and subsequent formation of aerosol particles and cloud condensation nuclei is part of a complex system of localized and global climate control (Malin & Kirst 1997). High concentrations of DMSP are also known to deter grazing of phytoplankton by protozoans (Wolfe 2000).

One of the most prolific producers of DMSP in coastal waters is the planktonic colonial alga, *Phaeocystis*. When these algae bloom, there is a characteristic stench of DMS in the air, resulting from the breakdown of DMSP to DMS. Likewise intertidal green macroalgal species *Enteromorpha* and *Ulva* have high concentrations of DMSP, and following a rain shower during low tide, the DMS released from the *Enteromorpha* can be quite pungent.

2.8.5 Carbon

The supply of inorganic carbon for photosynthesis and algal growth is seldom (if ever) limiting in marine systems. However the role of oceans in the global carbon cycle has been the focus of intense study in the past decades, especially in relation to increasing carbon dioxide in the atmosphere as a result of anthropogenic activity (Siegenthaler & Sarmiento 1993; Takahashi et al. 1993; Feely et al. 2004; Sabine et al. 2004; Takahashi 2004).

The biggest pool of carbon in the oceans is that locked up in the sediments where globally about 10 million Gigatonnes (Giga is $\times 10^9$). In comparison there are about 39 000 Gigatonnes of dissolved inorganic carbon in the various forms discussed above. The next largest pool is that contained within the dissolved organic carbon (DOC) pool at about 700 Gigatonnes. It is striking to compare these numbers with the 30 Gigatonnes of carbon contained within the particulate organic carbon (POC) pool, which contains all of the organisms from bacteria to blue whales within the world's oceans (Fig. 2.19).

● 30 Gigatonnes of carbon is contained in all of the particulate phase biology compared to 700 Gigatonnes in the dissolved organic carbon pool and 39 000 Gigatonnes in the inorganic carbon pool.

There are many algae that deposit calcium carbonate ($CaCO_3$) in their cell walls, sometimes together with smaller amounts of magnesium and strontium carbonates. In the phytoplankton, the most conspicuous

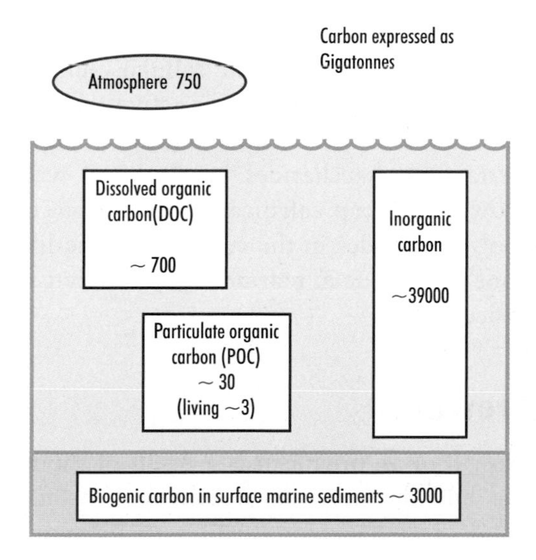

Fig. 2.19 Schematic to show the major pool of carbon within the oceans. Particulate organic carbon (POC) includes all organisms from bacteria-sized particles to whales. Dissolved organic carbon (DOC) is generally considered to be all carbon that can pass through a 0.2 μm filter. (Image: Rubén Lara.)

group of algae to exhibit calcification are the coccolithophorids that produce external 'shells' composed of calcium carbonate plates called coccoliths. These small phytoplankton species are common in all seas, although not as abundant in polar oceans. They can form extensive blooms where the ocean surface turns a milky white. One of the better-known species of coccolithophorid is *Emiliana huxleyi*, which can form blooms in the North Atlantic covering an area of the ocean equivalent to the size of Great Britain (Fig. 2.20). The coccoliths sink and are incorporated into sediments, where they can accumulate in huge amounts locking up $CaCO_3$ (Young & Ziveri 2000).

● Some microalgal and macroalgal species have calcified cell walls.

Normally when phytoplankton bloom there is a drawdown of CO_2 due to the assimilation through photosynthesis. However, the production of calcium carbonate structures by coccolithophorids can actually result in CO_2 being released to the atmosphere due to the formation of the calcium carbonate:

$$Ca^{2+} + 2HCO_3^- \Rightarrow CaCO_3 + CO_2 + H_2O$$

There are also many examples of calcareous brown, red, and green macroalgae, that is, species that have deposited calcium carbonate in the form of calcite or aragonite crystals within their tissue. This can be so extensive that these species can be important in the formation of tropical atolls and help to cement coral reefs together (Chapter 10). Calcite and aragonite never occur together in the same alga, and there is still some debate as to the metabolic processes that actually lead to the deposition. In some calcifying species the crystals are laid down outside the cells, such as in the green *Halimeda*, which causes aragonite crystals to precipitate in the spaces between the cells. In the fan-shaped brown *Padina*, aragonite is precipitated in concentric bands on the outer surface of the thallus. The corallines such as *Lithothamnion* and *Lithophyllum*, on the other hand, deposit calcite within their cell walls to the extent that the cells become encased, except for the cellular connections.

● In some coral reef systems over 70% of the reef can be made from calcified macroalgae.

Calcification is related to photosynthetic activity, and in particular the effects of pH on the dynamics of calcium and carbonate and bicarbonate in seawater. Certain polysaccharides in algal cell walls can actually block crystal growth and stop calcification taking place. The ability to produce these polysaccharides in the cell walls is the likely reason why calcification is not ubiquitous in marine algae and that relatively so few species are calcified.

2.9 Algal Growth

The ultimate growth of an organism is a result of the balance between the energy input and the necessary energetic costs for cell processes to take place and of course reproduction to maintain the following

Fig. 2.20 (a) Planktonic coccolithophorids ('round-stone-bearers'), such as this *Coccolithus pelagicus*, synthesize exquisitely sculptured calcium carbonate cell walls known as coccoliths. (Photograph: Jeremy Young, Natural History Museum, London). (b) Several species of macroalgae have calcified cell walls, such as those comprising this maërl bed. Several species of coralline algae form maërl beds (Chapter 7), which are often found at depths of 0–35 m, although in some parts of the world free-living calcified macroalgae have been found at depths down to 200 m. (Photograph: Bill Sanderson).

Box 2.8 Ultraviolet radiation damage

Ultraviolet radiation (UV; UVA wavelength 320–400 nm and UVB 280–320 nm) can damage RNA transcription and DNA replication, and in particular UVB radiation can damage the photosystem II of photosynthetic organisms severely limiting photosynthetic carbon assimilation and therefore growth (Buma et al. 2001; Hebling et al. 2001). Some accessory pigments have a photoprotective role, and these are used to protect cells from damaging high light levels and also harmful UV radiation. Concentrations of these are usually low in light-limited algae, but form rapidly when the algae are transported into high light environments or high levels of UV radiation.

The effect of UV stress on many phytoplankton and macroalgal species is to produce UV screening agents such as β-carotene, mycosporine-like amino acids (MAAs), and photoprotective carotenoids (Hannach & Siglo 1998). These effectively act as sunscreens preventing damage to cell structures and DNA (Aguilera et al. 2002; Hoyer et al. 2002). The general response to increased UV radiation is a combination of a complex suite of cellular mechanisms including protection, repair, cell size, growth rates, and photo-acclimation. Interspecific differences in response to this complex of factors will dictate any changes in phytoplankton composition. Therefore, the most likely scenario due to this environmental change is a shift in species composition or succesional patterns (Vincent & Roy, 1993).

- A range of UV-absorbing compounds are produced by algae in high UV conditions to prevent cell damage.

One of the few generalizations that seem to be possible to make is that small cells are more vulnerable to UV radiation damage (Karentz et al. 1991; Buma et al. 2001). This is due to their surface area to volume ratios and the low effectiveness of screening pigments in small cells, although Helbing et al. (1992) found that UV radiation inhibited microplankton more than the nanoplankton.

It is not simply the amount of UV radiation that is damaging. This is because extremely low-light adapted algae are more susceptible to UV damage than algae grown in high light environments. Therefore algae growing in shade or low-light habitats suddenly exposed to high levels of UV radiations are highly susceptible to UV radiation damage. This may occur when subtidal stands of macroalgae are exposed on very low spring tides or phytoplankton cells are transported from deep waters to surface waters.

generation (Box 2.8; Fig. 2.21). The energetic gains, investments, and losses of individual organisms include the following:

Material and energy gains: Photosynthesis or chemosynthesis

Material investments: Skeleton formation, production of energy storage compounds and formation of reproductive material

Energy and material losses: Movement, buoyancy, excretion, osmoregulation, nutrient uptake and respiration.

Growth is the expression of the integration of all of these losses and gains, within an organism. Thus, in general form:

$$\text{Growth} = \text{Material and energy gains} - \text{Material and energy losses}$$

This growth can be measured in **material** (mass) terms, e.g. dry weight, wet weight or in **energy** terms, kJ.

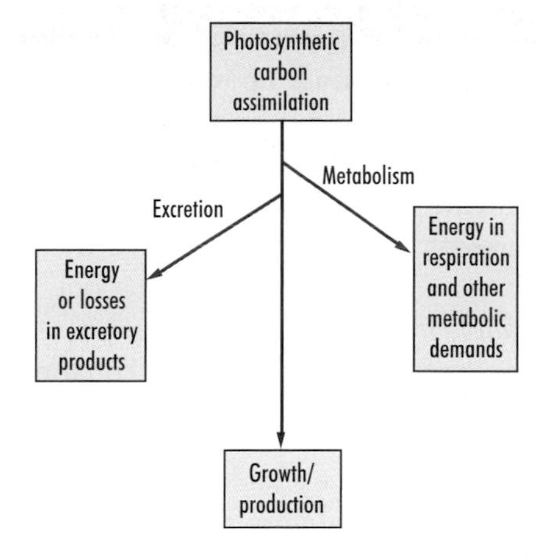

Fig. 2.21 The net carbon used for growth or production of a primary producer is a product of the photosynthetic assimilation minus several loss terms.

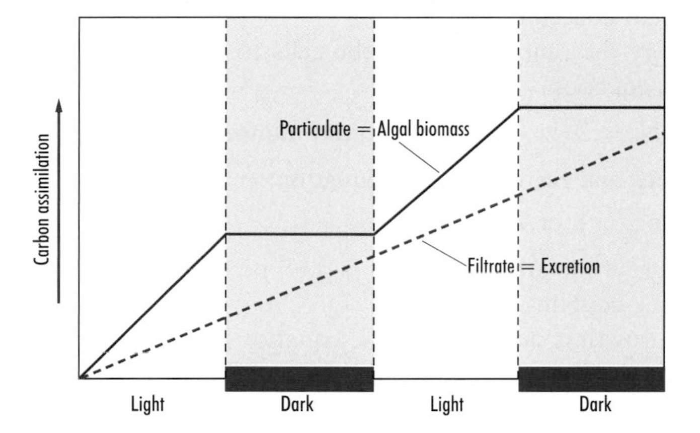

Fig. 2.22 Carbon assimilation by phytoplankton during light and dark periods. In the light, carbon is assimilated by the cells (particulate fraction), but not during the dark. In contrast carbon continues to be excreted by the cells (collected in the filtrate) in both the light and dark.

One of the striking features of algal growth is that significant amounts of the carbon assimilated during photosynthesis are released as **excreted** organic matter (see dissolved organic matter, Chapter 3). This varies greatly in composition from organic acids such as glycollate to vitamins, polysaccharides, fatty acids, amino acids, proteins, and simple carbohydrates. There are reports that up to 70% of carbon assimilated during photosynthesis can be excreted, but in general most values lie between 0 and 20%.

- Up to 20% of the carbon assimilated during photosynthesis can be excreted as dissolved organic carbon (DOC).

Some algal species increase the excretion of organic matter towards the end of a bloom, or when light and/or nutrients become limited. In contrast laboratory experiments have also shown that excretion of organic matter continues at a constant rate of about 10%. This excretion also continues during darkness and so is uncoupled from photosynthetic activity (Fig. 2.22).

2.9.1 Nutrients and growth

Algae can grow at high rates, given an adequate supply of light and no limitation in nutrient supply. Phytoplankton growth is usually expressed as the rate of cell division or increase in biomass per unit of time. In ideal conditions, small picoplankton can divide to produce up to 3 generations per day (cf. bacteria growing in ideal conditions can divide every 20 minutes, or 72 generations per day). For most phytoplankton maximum growth rates of 0.3 to 1 generations per day are more usual. However, in natural conditions the growth rate is better reflected by the net rate of change in numbers or biomass including the gains due to reproduction and losses due to mortality or export from the system.

● Typically phytoplankton divide at 0.3 to 1 generations per day depending on the temperature and nutrients available.

The growth of phytoplankton can be expressed by the following equation:

$$N_t = N_o e^{\mu t}$$

where N_o is the starting number of cells, N_t is the number of cells after time t (the original N_o + the cells produced) and μ is the growth rate (which is species specific and depends on temperature, irradiance levels, and nutrient concentrations).

Therefore the time needed for the cells to double is given using the following equation:

$$N_t / N_o = 2 \text{ (i.e. double the initial number)} = e^{\mu t}$$

Taking logs and rearranging the equation:

$$t = \log_e 2 / \mu \text{ or, } t = 0.69 / \mu$$

Therefore if the growth rate (μ) is 0.69 per day the doubling time (t) would be 1 doubling per day.

When growth is described by the equation $N_t = N_o e^{\mu t}$, the growth of the population is exponential, and if natural logarithms of both sides of the equation are taken, the equation describes a straight line (Fig. 2.23a), the slope of which is μ:

$$\ln N_t = \mu t + \ln N_o$$

However, no population can continue growing exponentially for an indefinite period. Typically exponential growth ends following the utilization of one or another essential nutrient. When this happens a stationary phase is reached where there is no net increase or decrease in cell numbers (Fig. 2.23b). It is important to note that although there is no growth of the algal population, the cells continue to metabolize and produce and/or turnover cellular products. There may even be some cell division (growth), which is balanced by the numbers of cells dying during this phase. Ultimately as the stationary phase extends, the percentage of the cells dying increases and thereafter the population enters a death phase.

● Ideal algal growth includes a lag phase, exponential growth phase, stationary phase, and death phase.

If light and temperature conditions remain unchanged and losses due to grazing are not important, growth of algae is mostly limited by the availability of nutrients. The effects of nutrients on algal growth is described by the following equation:

$$\mu = \mu_{max} ([S/K_N] + S)$$

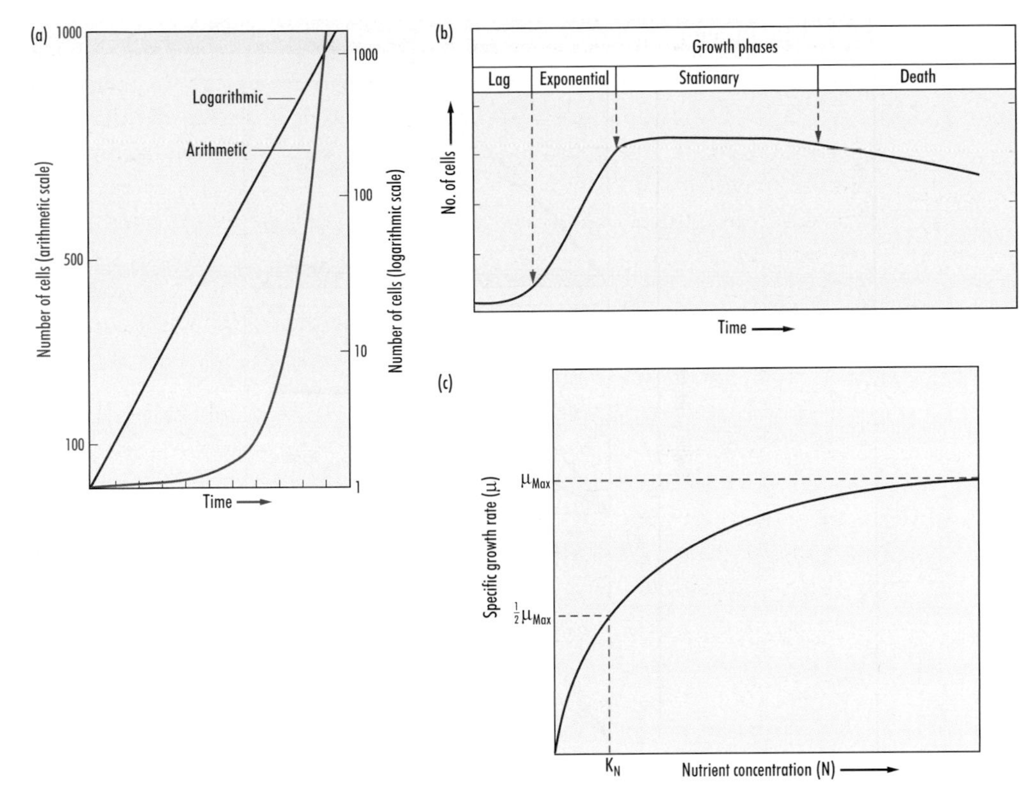

Fig. 2.23 (a) The growth rate of a phytoplankton culture expressed on both arithmetic and logarithmic scales. (b) Idealized growth curve for a phytoplankton population. (c) Relationship between nutrient concentration (N) and the growth rate of a primary producer. μ_{max} is the maximum growth rate, K_N the nutrient concentration at which growth rate is half the maximum (0.5 μ_{max}) – the half-saturation constant.

where S is the concentration of the nutrient and K_N is the half-saturation constant, or the concentration of the nutrient at which $\mu = \mu_{max}/2$.

This is a **Michaelis-Menten** relationship (as was the relationship between photosynthesis and irradiance, see above), and so when at low nutrient concentrations the rate of increase in growth rate increases rapidly with rising external nutrient concentration (Fig. 2.23c). At higher concentrations, increases in nutrient concentration add progressively less to the growth rate until the maximum growth rate is reached and further increases in nutrients do not affect the growth rate.

Each species of phytoplankton present in a body water (may be many tens to hundreds of species at any one time) has species-specific growth rate (μ and μ_{max}), and also a species-specific half-saturation constant (K_N) for all of the nutrients it needs to assimilate. This is one of the reasons why many different species of phytoplankton can coexist in a water body (Box 2.9).

Previously in the chapter it was described that small algae such as the picoplankton most efficiently take up nutrients because of diffusive

Box 2.9 **Nutrient dynamics and phytoplankton growth**

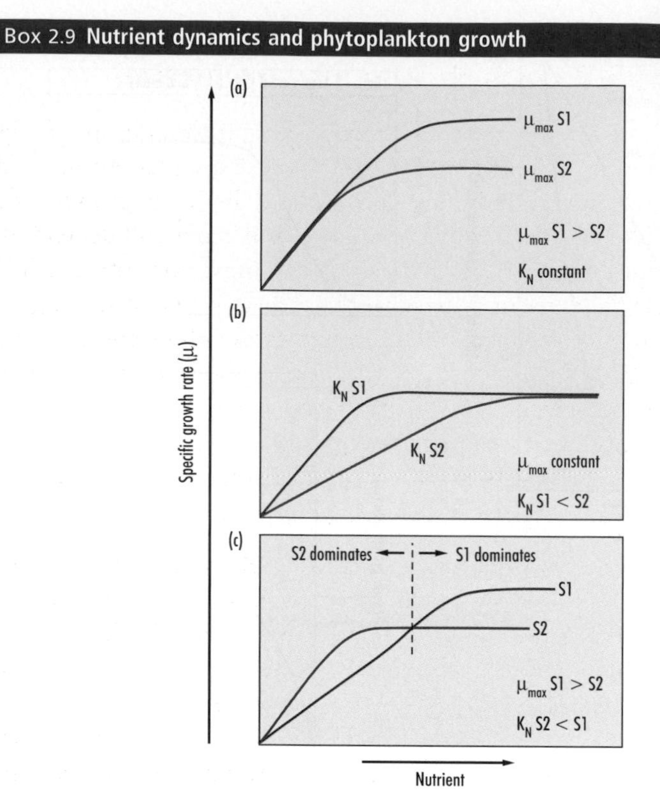

Examples of possible variations in nutrient growth curves of competing pairs of phytoplankton with different specific growth rates (μ), maximum growth rates (μ_{max}) and half saturation constants (K_N) for nutrient uptake (from Lalli & Parsons 2004).

In the first example (a) species 1 has a higher μ_{max} than species 2, but both have the same K_N. In this case both species grow at the same rate until a certain level of nutrients, after which species 1 continues to grow further until it reaches its maximum growth rate. In this case species 1 will dominate at nutrient concentrations greater than those at K_N.

In (b) both species 1 & 2 have the same value of μ_{max}, however, species 1 has a lower value of K_N than species 2. In this case species 1 reaches its maximum growth rate at lower nutrient concentrations than species 2. Therefore in low nutrient concentrations species 1 will dominate, although at higher nutrient conditions both species will grow equally.

In the third example (c) species 1 has a higher μ_{max} than species 2, but the latter has a lower K_N than species 1. At lower nutrient concentrations species 2 grows faster and dominates because it reaches its maximum growth rate at lower nutrient concentrations. However, at higher nutrient concentrations species 1 dominates because of its greater maximum growth rate.

boundary effects. Small photoautotrophs such as these also tend to grow at faster rates. Therefore why isn't there nothing else but the picoplankton? The answer is that in the discussions here, we have ignored a major controlling factor limiting algal standing stocks, namely, grazing by protozoans and zooplankton. In aquatic systems, grazing pressure exerts a major control on the dynamics and distribution of photosynthetic organisms.

2.10 Seasonal Trends in Primary Production

In nature it is impossible simply to consider the limitation of algal growth in terms of a single factor, as in the examples given above (Jickells 1998). Although only one nutrient may limit the growth of an alga at any one time, whether or not an alga will grow will depend also on other growth-limiting factors of which the most important is light. No matter what the nutrient concentration, if there is not enough light to reach maximum photosynthetic rates, growth rates will be compromised. By contrast, if there is plenty of light to saturate photosynthetic systems, but there are no nutrients, growth will not take place. This is most vividly shown in the seasonal phytoplankton growth dynamics in temperate (mid to high latitude) waters in which seasonal thermal stratification of the water column takes place (Fig. 2.24; Chapters 3 and 6).

● Thermal stratification of water masses is central to the seasonal phytoplankton dynamics in temperate waters.

Winter: The sun is low in the sky and much of the thermal energy is reflected. Winds tend to mix surface waters deep into the water column. With no **thermocline**, the water column is fully mixed (**isothermal**) and the inorganic nutrients are evenly distributed throughout the water column. Although phytoplankton are present throughout the water they do not grow because short days and the low angle of the sun mean that light levels are too low to support high rates of photosynthesis and growth.

● The thermocline is the boundary between dense cooler bottom water and warmer less dense surface water.

Spring: In spring the sun is higher in the sky and day length increases. This results in more solar energy absorption by the water and the development of a thermocline, which effectively traps phytoplankton in the surface waters. If the mixed layer depth is shallower than the critical

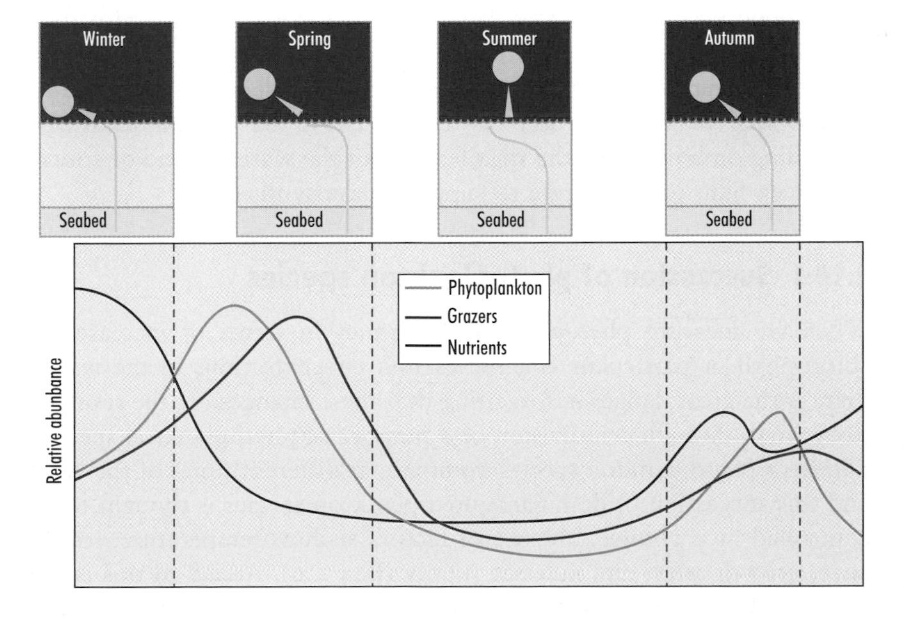

Fig. 2.24 Seasonal changes in phytoplankton, grazers, and inorganic nutrients (upper mixed layer) in temperate waters in relation to the seasonal thermal stratification of the water column.

depth (see above) phytoplankton will bloom, since there are abundant nutrients and adequate light to support maximal growth rates. This is termed **new production.**

Summer: The sun is high in the sky and day length is long. This results in the surface waters warming even more and the development of a strong thermocline that prevents any mixing between the surface and waters below the thermocline. Once the nutrients in the surface waters have been used up (typically nitrate and/or phosphate) phytoplankton growth ceases, despite the fact that light conditions are good enough for maximum photosynthetic rates to take place. Some growth will continue based on regenerated nutrients. This is termed **recycled production.**

Autumn: The sun is lower in the sky and day length shortens. The surface waters begin to cool and autumn winds tend to mix surface waters deeper leading to a breakdown in the thermocline. This results in nutrients from deeper waters mixing above the thermocline and this supports an autumn bloom of phytoplankton while light levels are still high enough to support photosynthesis. The autumn bloom is not as great as the spring bloom, because of lower light levels and also because concentrations of nutrients are not as great as at the beginning of spring. This is termed **new production.**

The depletion of nutrients in surface mixed water layers results in the establishment of a **nutricline** between the surface waters and deeper water: a gradient of low to high nutrients with increasing depth. There is therefore a potential flux of nutrients from below to above, and the maximum flux will be in the region of the nutricline. If the nutricline occurs above the critical depth in stable stratified waters, sub-surface chlorophyll layers form. These can occur in seasonally stratified waters (e.g. in summer), but are more characteristic of more permanently stratified waters found in tropical and subtropical oceans (Fig. 2.25). Chlorophyll maxima as deep as 20 to 100 m have been measured depending on how stable the stratification of the water is, and of course how deep light can penetrate to support photosynthesis.

2.10.1 Succession of phytoplankton species

When we measure phytoplankton dynamics in terms of increases in chlorophyll or particulate organic carbon concentrations in the water, there is the great danger in forgetting that these changes are the result of the growth through cell division of a number of phytoplankton species. Different phytoplankton species dominate at different times of the year, and this succession of dominant phytoplankton species is thought to be controlled by a complex mosaic of factors such as temperature, irradiance, growth rates, and nutrient supply (Box 2.6). Added to this is the

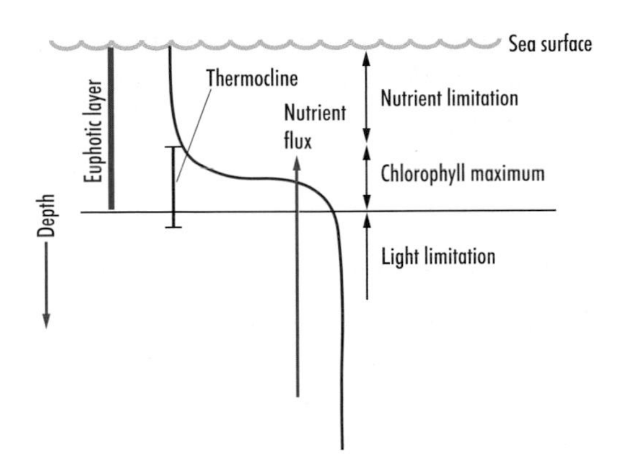

Fig. 2.25 In stratified waters nutrients may become limiting in the surface waters. However, nutrients may diffuse from deeper waters upwards across a thermocline. If this layer is in the euphotic zone (above the critical depth) phytoplankton growth can occur leading to the formation of sub-surface chlorophyll layers.

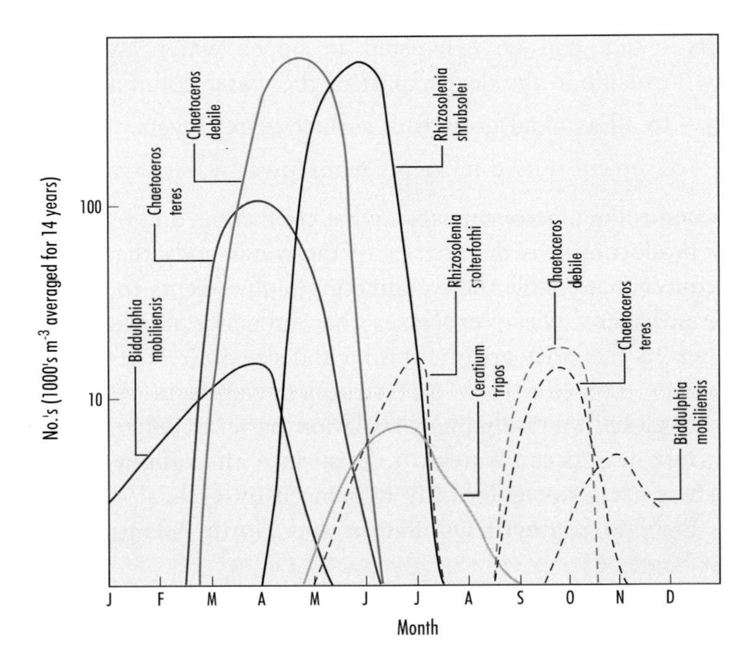

Fig. 2.26 Seasonal succession of dominant phytoplankton species in the Irish Sea averaged over 14 years (from Barnes & Hughes 1999).

influence of grazing by protozoan and zooplankton species, which can have a major role. It is important to note that this succession is *not* the same as that described for terrestrial systems where species succession results in a climax community (Sommer 1989; Roelke & Buyukates 2002; Worm et al. 2002).

● Many species make up the phytoplankton, and there is a succession of species throughout a bloom or from one season to the next.

Even within a 'bloom' event, there is a succession of species, even though blooms normally start by the explosive growth of an individual species. During the course of a diatom bloom silicate is taken up by the diatoms to build silicate frustules. As the concentrations of silicate fall in the water due to this uptake there can be a progression from large diatom species to small diatom species that have less demanding silicate requirements for growth.

2.11 Global Trend in Primary Production

When considering primary production on a global scale, the advent of satellite colour images of chlorophyll distribution around the globe have been very successful in showing us that primary production is far from uniform at a large scale (Chapter 1, Fig. 1.6). Indeed, large-scale patterns in primary production are partly used to define distinct regions or biomes of the world's seas (see Chapter 7, Table 7.2). Ultimately there are four major factors that govern primary production in marine systems (Falkowski et al. 1998):

Light – only available in the upper part of the water column (max 200 m).

Nutrients – that can be exhausted in upper water layers, but are generally available in the deeper part of the water column.

Stability – to allow algal growth in surface water layers.

Mixing – to replenish used nutrients from lower water layers to surface.

● Ultimately it is the physics of a water body that controls the primary production.

These controlling factors are somewhat conflicting, and in areas of **high** primary production it is the physics of the water body that gives rise to circumstances that enable these conflicting requirements to be met.

There are many ocean processes that influence nutrient supply to surface waters that vary greatly in time and size scale. These range from storm events that effectively mix stratified waters in shallow waters through to global thermohaline circulation patterns (Chapters 6, 7 and 11) that mix deep ocean waters to the surface after about 500 years or more. These are also significantly influenced by cyclical oceanic events such as El Niño Southern Oscillation and North Atlantic Oscillation events (Chapter 6).

● Tropical waters have generally low primary production due to low surface water nutrient concentrations.

In **tropical and sub-tropical waters**, there is normally permanent thermal stratification due to the high degree of solar heating of the surface waters. Irradiances in these waters are high, and the clear waters result in light passing deep into the water column (>200 m). These conditions are of course conducive to high primary production, but because of the lack of mixing of inorganic nutrients from the deeper waters, tropical waters generally only support low primary production, although at a rather constant level throughout the year. When storm events do mix surface and deeper waters, phytoplankton blooms can result from the input of nutrients. Likewise tropical waters are sensitive to eutrophication processes, since when nutrients are introduced the high light levels support a rapid build up of primary producers (Fig. 2.28).

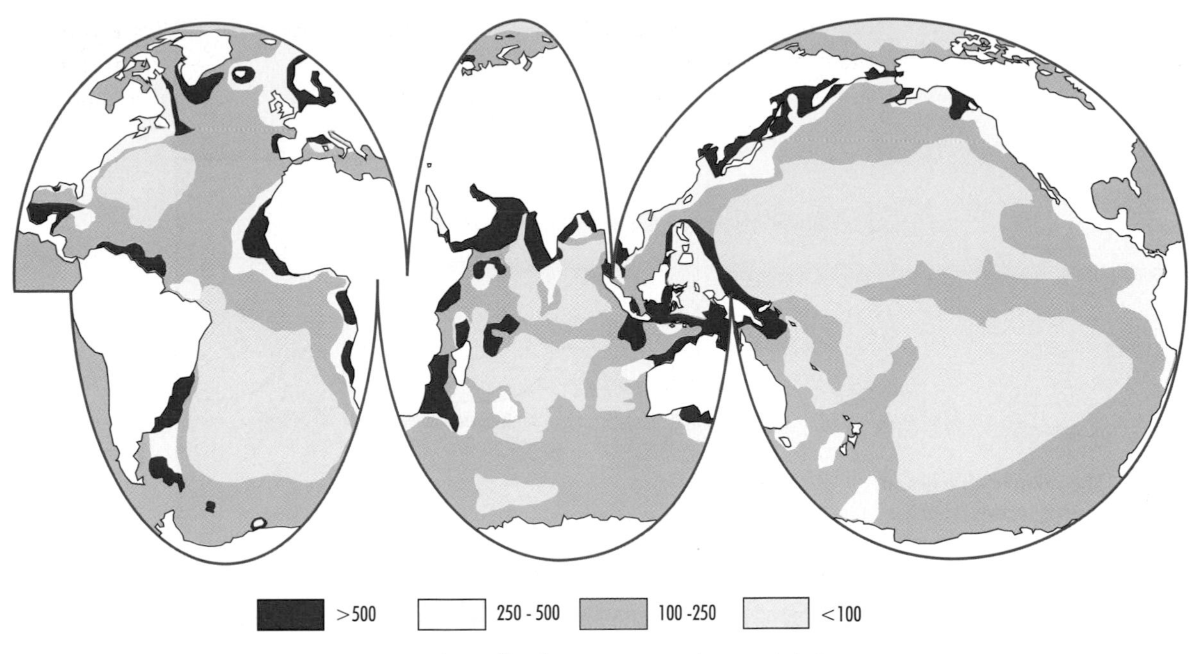

Fig. 2.27 The annual primary production as $gCm^{-2} y^{-1}$ of the oceans from a global perspective. Adapted from Barnes and Hughes 1999. (See also Chapter 1)

In contrast, in the **polar oceans** there is significant mixing of nutrient rich lower waters with the surface waters. However, these regions are characterized by long times of the year when day length is short and sun angles shallow. Therefore, primary production in these waters is limited by irradiance, and generally there is a single peak of primary production when light is high enough to support net gain in algal growth. There are differences in the nutrient status of the Arctic and Southern Oceans. In the latter the major nutrients (nitrogen and phosphate) are in excess, but primary production is restricted due to limitation of iron. In the Arctic, nutrient limitation does occur following the annual late-spring/summer plankton bloom (Fig. 2.28).

● Primary production in polar waters is restricted to short seasonal windows when light is available.

As described above, **temperate waters** are characterized by a suite of complex of seasonal dynamics of light and thermal stratification. These seasonal variations impart a distinctive seasonality in primary productivity: spring and autumn blooms of phytoplankton, with low standing stocks of phytoplankton in summer and winter (Fig. 2.25).

Large **oceanic gyres** are a conspicuous feature of the Atlantic, Pacific, Indian, and Southern Oceans. In the northern hemisphere these gyres move in a clockwise direction and anticlockwise in the southern hemisphere. These **anticyclonic gyres** tend to deepen the thermocline, moving water towards the centre of the gyre. Therefore nutrient replenishment into surface waters does not take place, and these anticyclonic gyres tend to be regions of low primary productivity (Chapter 6).

● Anticyclonic gyres are regions of low primary production whereas cyclonic gyres sustain higher rates of primary production.

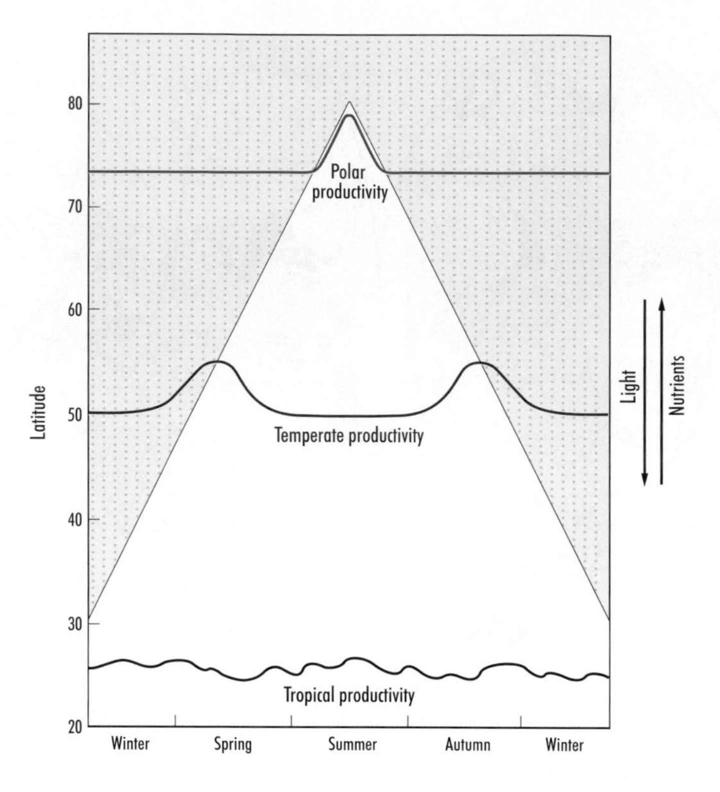

Fig. 2.28 The relative changes in seasonal primary productivity in polar, temperate, and tropical latitudes as a result of light- and nutrient-limited growth. The relative differences in light (unshaded) and nutrients (shaded) at the different latitudes are represented. (After Lalli & Parsons 2004).

However, **cyclonic gyres** (anticlockwise in northern hemisphere and clockwise in southern hemisphere) actually result in water being mixed from below the thermocline into surface waters, due to water being transported outwards from the centre of the gyre. Therefore in such gyres higher rates of primary production are supported, due to the mixing of nutrient rich water into surface waters.

Upwelling of nutrients in nutrient rich waters is characteristic of several coastal regions. Offshore winds give rise to offshore transport of nutrient-depleted surface water that is replaced by upwelling nutrient-rich water, supporting high rates of primary production. Upwelling also occurs at major ocean frontal systems in the open oceans, such as the equatorial regions where north-east and south-east trade winds generate two westerly flowing surface currents, the North and South Equatorial Currents. The **Coriolis effect** causes the currents to be deflected northwards in the northern hemisphere and southwards in the southern hemisphere. The divergent flow of these surface waters from the equator promotes nutrient-rich water upwelling, supporting higher rates of primary production (Chapter 6).

● Regions of upwelling are important sites of high primary production.

Coastal waters and waters overlying continental shelves (<200 m) support the greatest primary productivity. This is because many coastal waters are shallower than the critical depth (see Box 2.6), and also

(a)

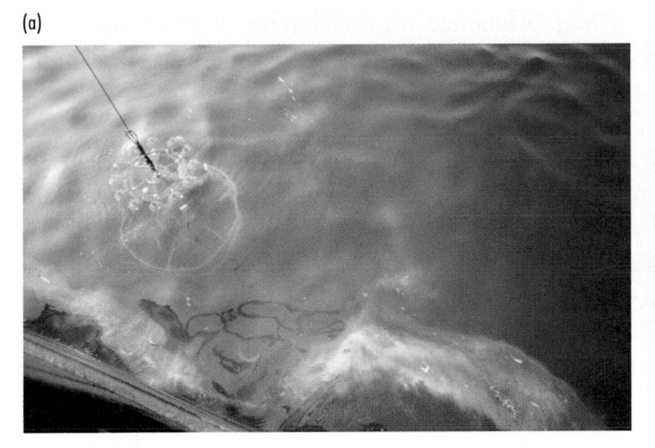

(b)

Fig. 2.29 (a) Fronts can be zones where concentrations of biology are found, either as a result of physical concentration or by increased growth of organisms. Here there is a huge concentration of photosynthetic ciliate *Mesodinium rubrum* in the surface waters of a front in the North Sea, as well as bird feathers, and green macroalgae concentrated on the water's surface. (Photograph: David Thomas) (b) Frontal systems are enhanced zones of biological activity, as seen here by the whales and birds feeding on plankivorous fish at a front. (Photograph: Michel Kaiser).

because coastal waters receive large amounts of growth limiting nutrients from river inputs into the coastal zone (Jickells 1998). There are many different types of frontal systems in shelf seas, and these tend to be associated with increased primary production, e.g. fronts associated with river plumes and shelf-sea fronts (tidal fronts) where the water column changes from being tidally mixed to being stratified. Typically the dense accumulations of plankton that occur at fronts attract feeding aggregations of fish, birds, cetaceans, and fishers (Fig. 2.29; Chapters 6 and 7).

● Frontal systems tend to be regions of enhanced biological activity, either from increased primary production or a concentration of organisms through physical processes.

The high standing stocks of plankton at tidal fronts may simply result from physical concentration by the water dynamics in the frontal zone. However, the conjunction of two separate water bodies can result in an exchange of nutrients (or other components) from one body of water to the other.

2.12 Primary Production in Seaweeds and Seagrasses

Much of this chapter has concentrated on primary production in planktonic photoautotrophs, since these contribute the greatest proportion to the primary production in the oceans. However, in coastal waters, the intertidal seagrasses and macroalgae are a very dominant feature. These have a huge influence on the dynamics of coastal inorganic nutrients and cycling of organic matter (Chapter 9). More general discussions about the ecology of macroalgae and seagrasses will be found in later chapters (Chapters 5, 7, and 9).

Nevertheless seaweeds on a shore can grow at impressive rates, amassing large biomass (Fig. 2.30a). The productivity of seaweeds and seagrasses is equal to, or in many cases greater than, that of terrestrial plant systems. For instance *Laminaria* dominated communities have annual productivity rates of approximately 2 kg carbon $m^{-2}y^{-1}$. Seagrass meadows are extremely productive with annual production rates up to 6 kg carbon $m^{-2}y^{-1}$ measured at some tropical sites, far exceeding estimates for even tropical rain forests and monocultures of crops (Chapter 9).

● The annual primary production of seaweed and seagrass meadows can be in excess of rainforest or crop species.

In water, seaweeds obtain the carbon needed for photosynthesis from CO_2 or from HCO_3^-. When they are exposed to air, there is no bicarbonate and photosynthesis can only take place by the uptake of carbon dioxide from the air. As long as the seaweeds do not dry out, many species photosynthesize in air at rates similar to those measured when they are fully submerged. However, when they begin to dry out there are considerable differences in capacities for photosynthesis. Photosynthesis

(a)

(b)

Fig. 2.30 (a) Infrared image enabling a comparison to be made of terrestrial and intertidal primary producers. The trees in the background show up red (due to the fluorescence of chlorophyll) in the image, as do the large standing stock of macroalgae on the shore in foreground. (Photograph: David Roberts.) (b) Air bladders at the end of each *Macrocystis pyrifera* blade help to support the giant algae in the water, and maximize exposure to the incident light. (Photograph: David Thomas.)

in the low shore kelps such as *Laminaria*, even when mildly desiccated, is greatly reduced, whereas the green alga *Ulva*, often found in intertidal zones, continues photosynthesis down to 35% water loss. *Fucus vesiculosus*, *F. serratus*, and *F. spiralis* can all photosynthesize even when desiccated down to a point where the remaining tissue water content is less than 30% (Dring & Brown 1982).

● Macroalgae continue to photosynthesize when exposed to air, and some species even when harshly desiccated.

When submerged, considerable energy is expended in maximizing the amount of light getting to the chloroplasts. This is especially important in turbid coastal waters full of plankton and suspended particles that reduce light penetration to a few metres or less. There are several morphological features common to many species of seaweed that help to address this problem. The first is the **stipe** that supports the bulk of the photosynthetic tissue at the surface of the water where incident light is at a maximum. *Nereocystis luetkena* is a superb example of this, with a stipe of over 30 m long supporting up to 100 blades each several metres long. In this case the stipe acts very much like the trunks of canopy species of trees in a rain forest, getting the maximum surface area of photosynthetic cells as close to the incident light as possible (Fig. 2.30b).

The long stipe of *Nereocystis* is rather flexible, and the blades only sit on the surface because the apex of the stipe terminates in an approximately 15 cm diameter float, or **pneumatocyst**. This large gas-filled bladder ensures that the blades are floating as high in the water as possible. In contrast, the giant kelp (*Macrocystis pyrifera*) has a different arrangement with a small gas bladder at the base of each of its thousands of blades, which connects them to the highly branched stipe.

● Gas-filled bladders are important in many macroalgal species for maintaining photosynthetic tissues as close to the incident light as possible.

Although not as impressively large, many other species use gas-filled bladders in a similar fashion, e.g. *Ascophyllum*, *Sargassum*, *Fucus*, and the bead-like *Hormosira*. The gases within the bladders contain oxygen and nitrogen in roughly the same proportion as in air, and varying amounts of carbon dioxide. Curiously the huge pneumatocysts of *Nereocystis* contain up to 10% carbon monoxide, although why this should be remains unclear.

As in trees that form terrestrial forests, the stipes of the seaweeds also act as surfaces for epiphytes to grow on. These can vary from other macroalgal fronds through to biofilms made from bacterial and/or microalgal assemblages. Colonial animals such as Bryozoa can also form dense cover on the outside of large macroalgal species both on the stipe and blades. Epiphytic growth can reach such densities that they restrict the exchange of gases and inorganic nutrients and cut down on light that reaches the photosynthetic cells of the host. They can also increase drag to such an extent that they make their hosts more prone to being washed away in strong water currents (Fig. 2.31). To avoid such problems, 'skin shedding' is widespread among large seaweed species

● Epiphytes growing on the surfaces of seaweeds restrict gas exchange and the supply of inorganic nutrients, as well as increasing drag.

Fig. 2.31 Macroalgae, like many other structures, can quickly become host to other algae and sessile animals that grow on their surface. Here a frond of *Fucus serratus* is nearly overgrown by the epiphytic brown *Ectocarpus sp*. (Photograph: David Roberts.)

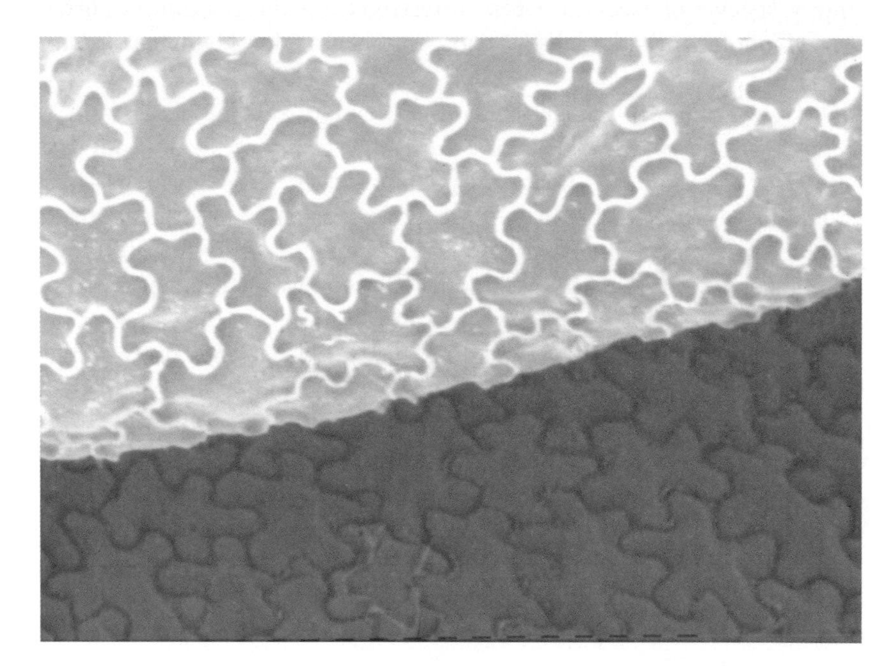

Fig. 2.32 One of the mechanisms by which several seaweed species overcome problems derived from epiphyte growth on their surfaces, is the periodic shedding of outer layers of cell walls. In this scanning electron microscope image, the outer layer of cell walls of the brown alga *Himanthalia elongata* can be seen lifting off, with the pattern of the underlying cells imprinted. (Photograph: George Russell.)

(Russell & Veltkamp, 1984). Layers of cell walls are shed from the outer surfaces of the seaweed carrying any attached epiphytes along with them (Fig. 2.32).

Seagrass blades are also renowned for supporting dense epiphytic growth. The biomass of epiphytes can range from 25 to 80% and production from 50 to 100% that of the seagrasses. Skin shedding is not carried out by seagrass species, but rather they produce large numbers of shoots, and blades heavily laden with epiphytes break off and are lost.

2.13 Measurement of Primary Production

> *The act of observation alters the object being observed* (rough interpretation of Schrödinger's Cat thought experiment).

One of the key issues facing biological oceanographers trying to get good measurements of primary production rates is how to translate the measurements they can make to the conditions from where their field samples are taken. For the past 50 years the key techniques to measure primary production have been to measure directly the carbon assimilation during photosynthesis and/or the oxygen production and oxygen uptake during photosynthesis and respiration respectively.

- Primary production is mainly estimated by measures of the fluxes of carbon dioxide and/or oxygen.

The amount of carbon or oxygen flux per unit of time is then related to the biomass of the photosynthetic organisms creating the fluxes. Commonly used biomass terms are amount of chlorophyll, or the weight/concentration of the organic carbon contained within the photosynthetic organisms. Therefore the production is expressed as: **change in concentration of O_2 or CO_2 per unit of time per unit of mass or concentration of algae.**

The combinations of units are diverse, depending on the types of measurements being made and/or whether it is phytoplankton or macroalgae/seagrasses that are under investigation. For phytoplankton, rates of gas flux are often expressed per unit of chlorophyll or particulate organic carbon (POC). Measurements with macroalgae and seagrasses are normally expressed on the basis of fresh or dry weight, since these are easy to determine.

- Rates of primary production are usually based per unit of chlorophyll, particulate organic carbon or in seaweeds on a dry or wet weight basis.

To be most realistic, the incubations during which the fluxes of gas are measured should be made under the exact light regime that would be experienced in nature (Fig. 2.34). In a few instances it is possible to take a sample of phytoplankton, or macroalgae, from a particular water depth, put it in a container and then perform the incubations back at the same depth and then subsequently measure the gases produced or consumed over the incubation. However, the simple act of using a closed vessel introduces changes to the system that some researchers claim result in unrealistic measurements.

- How realistic 'bottle incubations' are is open to debate.

It is logistically difficult to deploy *in situ* incubations, especially if incubations are made throughout the euphotic zone, which may extend down to 200 metres in the open ocean. It is also preferable to make incubations last for whole diurnal periods so that the carbon and/or

● Although ideal, *in situ* incubations are often difficult to employ.

oxygen changes can be integrated over daily periods, thereby including periods of darkness where photosynthesis is absent, but respiration continues. These measurements are onerous; hence compromise is necessary, as discussed below.

2.13.1 Radiocarbon labelling – the '^{14}C method'

Primary production is basically the measure of the amount of inorganic carbon that is assimilated through photosynthesis to create new organic matter over a period of time and over a specified area. Intuitively therefore direct measurements of the decrease in the external inorganic carbon pool or the increase in the organic carbon pool would seem the most obvious way to measure primary production.

In the1950s the Danish scientist Einer Steemann Nielsen introduced the use of radiolabelled isotope ^{14}C for measuring carbon assimilation in phytoplankton. Since then, the '**^{14}C method**' as it is often referred to has become probably the most standard technique for measuring primary production in both phytoplankton and macroalgal studies.

● Uptake of radioactive ^{14}C is a widely used way to estimate primary production, although it is not clear whether net or gross production is being measured.

^{14}C in the form of $NaH^{14}CO_3$ is added to the incubation water surrounding a macroalgal sample or into a sample of phytoplankton. The photoautotrophs assimilate the ^{14}C during photosynthesis and at the end of an incubation period the quantity of radioisotope within the photoautotrophs is measured. This is directly converted into the amount of carbon that has been photosynthetically assimilated during the incubation.

Technically the method is rather straightforward, however, there is still much controversy about what it actually measures in terms of net or gross production. The other major disadvantage of the method is that it is not possible to measure dark respiration rates and, therefore, estimate gross primary production.

2.13.2 Oxygen determinations

It is possible to measure both photosynthetic oxygen production and respiratory oxygen consumption during incubations. The most widely used accurate oxygen determination method is based on Winkler titrations for determining the oxygen concentration of water. When performed with care, this chemical titration determination method is able to measure precisely very small changes in oxygen concentration, vital where respiration and photosynthetic rates are low.

● By measuring oxygen fluxes both photosynthesis and respiration can be measured, enabling both gross and net photosynthesis to be estimated.

The method involves measuring the increase in oxygen concentration in glass bottles (ranging in size from 10 to 200 ml) in which photosynthesis has taken place and the decrease in oxygen concentration in darkened bottles in which only respiration has taken place. One major difference between this and the '^{14}C method' is that the oxygen fluxes due to bacterial, protozoan, and zooplankton respiration present in the sample will be included, because in reality it is impossible to screen these out of field samples without completely disrupting the photo-autoptrophs. Therefore the oxygen methods tend to measure community metabolism, whereas in general the ^{14}C method is measuring activity of only the photoautotrophs.

The advantage of the oxygen system for the estimation of primary production is that both rates of respiration and photosynthesis can be directly measured, and so there is no ambiguity about whether or not net or gross production is being measured. The major difficulty comes in the conversion of the oxygen measurements into carbon terms. The **photosynthetic quotient** (PQ) is the term that describes the moles of oxygen evolved per moles of CO_2 assimilated, and is often quoted as 1.25. However, as described above, the PQ varies with different nitrogen and/or phosphorus supply. If lipids are being predominantly produced the PQ is significantly different than if carbohydrates are being produced. **Respiratory quotients** (RQ) are the inverse of the PQ and describe the respiratory relationships between oxygen consumption and CO_2 production, and are typically taken to be 1.0. Although less variable than PQ values, there will be variations around this value, making standardized conversions less certain.

● Photosynthetic and respiratory quotients are important to convert oxygen fluxes into units of carbon in primary production studies.

2.13.3 Electrodes

There is an increasing array of electrodes that can measure pH, O_2, CO_2, as well as sulphur and nitrogen species (de Beer 2000). These are not typically used to measure the fluxes of O_2 or CO_2 in incubations as described above, but rather are used in field deployments to measure temporal trends and estimate fluxes. Since the 1980s there have been great advances in technology that have enabled reliable electrodes with tip diameters of less than 50 μm to be constructed. These have been revolutionary in measuring small-scale fluxes in photosynthesis and respiration at surfaces such as biofilms and sediments. Estimates of primary production using these techniques have been greatly enhanced by the engineering of miniature light sensors with tip diameters of approximately 100 μm. Therefore the prevailing light regime can be measured at the same time as the chemical fluxes (Fig. 2.33).

● Increasingly electrodes are used to measure pH and gas fluxes, especially for fine-scale work on the surfaces of sediments.

The latest generation of equipment capable of measuring small-scale fluxes are the micro-optodes. In these, light of a specific wavelength is conducted via fine glass fibres to the measuring tip, which can be less than 20 μm. The tip contains a fluorescent dye that fluoresces at a different wavelength to the exciting light. The intensity of fluorescence depends on the concentration of the substance being measured.

Fig. 2.33 (a) Microelectrodes are now available for several parameters including oxygen and carbon dioxide. These have incredibly small diameter tips (insert) enabling scientists to profile mini-scale (μM) differences in primary production and respiration, in particular in benthic systems. The images shows a 'lander' equipped with several electrodes, which it automatically pushes into the sediments. It logs the chemical data measured by the electrodes for downloading when the lander is retrieved. (Photographs: Dirk de Beer, Microsensor Group, Max Planck Institute for Marine Microbiology, Bremen and Christian Lott, Hydra institute.) (b) *In situ* fluorometer (window in centre) that can be deployed for long periods of time (months) to measure concentrations of chlorophyll in the water. Note that instruments left in the sea become fouled by growth of sessile organisms, requiring the antifouling washing arm shown to clean the fluorometer window periodically. (Photograph: David Thomas.) (c) Measuring photosynthetic activity on algae, corals, and sponges (symbiotic algae) in the Red Sea with the *in situ* chlorophyll fluorometer DIVING-PAM. (Photograph: Camillo Weis, Heinz Walz GmbH, Germany, **www.walz.com**.)

2.13.4 Fluorescence measurements

The measurement of changes in biomass is one route to estimate primary production. One of the most commonly employed methods for measuring chlorophyll is by measuring its fluorescence either within the cells directly or after extraction from the organism by a solvent. The chlorophyll is excited by carefully defined wavelengths of light and the resulting fluorescence measured directly (Krause & Weis 1991). Since the 1970s reliable instruments have been available for measuring the chlorophyll fluorescence in open waters. These send out pulses of light and the fluorescence generated is measured. Many systems that profile for salinity (**conductivity**), temperature, and pressure (depth) (**CTD sensors**) can also be easily adapted to have *in vivo* chlorophyll fluorescence sensors attached and this information relayed back to the ship together with the temperature and salinity data. Such instruments have become invaluable for determining the distribution profiles of phytoplankton through the water column. For instance, such profiles often help scientists determine from which depths they should take water bottle samples for their oxygen or ^{14}C incubations for primary production determinations.

● Fluorescence sensors are used for the determination of *in situ* concentrations of chlorophyll, and therefore phytoplankton biomass.

Such sensors can also be deployed for the long-term measurement of fluorescence on moorings, **automated underwater vehicles** (AUVs), or on seabed platforms (see below). These can be left for periods of over a year and the data logged to allow the seasonal dynamics of phytoplankton biomass to be elucidated in fine detail for that particular site (Chapter 6).

It is possible to derive more information from the fluorescent characteristics of a photoautotroph than from the simple measures of *in vivo* chlorophyll concentrations. **Pulse Amplitude Modulated** (**PAM**) fluorometry is becoming a widespread tool for measuring *in situ* photosynthesis (Fig. 2.33). PAM fluorometers use a range of flashing (pulsed) lights to measure the photosynthesis from the fluorescense induced by the flashes of light (e.g. Glud et al. 2002). Generally the fluorescence is excited at high repetition rates by microsecond pulses of different wavelengths of light from light emitting diodes (LED).

● Fluorometric methods are used for direct measurements of primary production.

The accurate measurement of the variable fluorescence characteristics of phytoplankton is the basis of profiling devices that can measure 'real time' primary production as they are lowered through the water. These are known as **Fast Repetition Rate Fluorometers** (FRRF). The fluorescence excitation system generates excitation flashes at rates exceeding 200 kHz. The stimulated fluorescence and excitation flashes as well as the photosynthetically active radiation are all measured simultaneously. These measurements are then used to calculate various biophysical parameters that allow rates of photosynthesis and therefore primary production to be calculated (Suggett et al. 2003).

2.13.5 Remote sensing

Probably one of the most significant changes in biological oceanography over the past twenty years has been the development and deployment of satellite- and aircraft-borne colour sensors that can record the colour of water masses and, using sophisticated algorithms, can allocate the colour to concentrations of dissolved constituents such as coloured dissolved organic matter (CDOM), suspended solids, and (most importantly for primary production) chlorophyll and other algal pigment concentrations.

● Increasingly sensors on satellites provide information on primary production on a global scale.

Such satellite ocean colour sensors provide the only means we have for looking at the large-scale distributions of phytoplankton so that monthly and annual distribution patterns can be created (Fig. 2.3).

These large-scale distribution studies are fundamental to our understanding of the large scale processes in the oceans and how these affect primary production, e.g. the interannual variability due to massive cyclical phenomena such as the North Atlantic Oscillation (NAO) or the El Niño Southern Ocean Oscillation (ENSO). Naturally as more satellites are deployed carrying such sensors, and more years of information are collected, these methodologies will greatly enhance our understanding of large-scale ocean processes.

● Surveys of water bodies using aircraft give rapid information about phytoplankton distribution in surface waters.

Ocean colour sensors can also be deployed on aircraft and have been used successfully to record the dynamics of phytoplankton blooms, especially in coastal waters within the operational ranges of the aircraft. **Light Detection and Ranging (LIDAR)** flights use a variety of pulsed lasers to stimulate phytoplankton chlorophyll and other pigments to fluoresce, and the fluorescent signals are then detected by sensors on the aircraft.

One advantage of such data is that it can be collated rapidly by scientists on board the aircraft and relayed to colleagues on the ground or on research vessels. In this way the research vessels can be directed to particular areas of research interest such as developing phytoplankton blooms, or phytoplankton dynamics associated with water frontal systems. Increasingly research expeditions will rely on information such as this, as well as 'on-line' satellite information to plan and modify cruise tracks to conserve expensive ship time.

● Many remote sensing methods only give information in surface waters, they cannot give information about sub-surface phytoplankton distribution.

The major disadvantage of ocean colour sensors, either aircraft- or satellite-borne, is that it is only the colour of the very top few metres of the oceans that is measured. As has been discussed above (section 2.10), this is unlikely to be a reflection of much of the phytoplankton biomass in any one body of water. In particular features such as sub-surface chlorophyll maxima and accumulations of chlorophyll associated with sub-surface processes are not accounted for using these sensors.

2.13.6 Automatic measuring devices

● The typical daily cost of an ocean going research vessel is between $US20 000 and $US50 000 depending on size and sophistication.

Naturally such ground truth exercises are limited to the rather restricted activities of research vessels. Although many millions of dollars are spent each year on maintaining research ships at sea, the coverage of global oceans at any one moment of time is minimal. Many research vessels now have the technology to measure the chlorophyll content of surface waters continually during the whole of their cruise track by pumping surface waters past a sensor that measures the fluorescence, which can then be converted to phytoplankton biomass. This technology is robust enough to deploy on other ships, such as commercial freighters, container ships, and ferries. These vessels tend to keep to well-defined routes, and so information can be collected that will document seasonal changes with high precision.

As well as chlorophyll concentrations, automated measuring systems are being developed for measuring temperature, salinity, and inorganic nutrients such as nitrate, phosphate, and silicate, as well as dissolved gases. Increasingly such devices will be deployed on a variety of ships (including ferries and commercial shipping), together with fluorescent sensors for phytoplankton determination (e.g. Cooper et al. 1998). This will enable the determination of seasonal biological and chemical dynamics along specific routes to be well defined. Naturally these measurements still suffer from the limitation that they are only measuring surface waters, since the water is normally pumped from seawater inlets in the keel of the ships, only a few metres below the water surface.

● Coupled biological and chemical sensors are deployed on ships that have regular routes, enabling large-scale collection of data over long periods of time.

The opportunity to profile the concentrations of phytoplankton and the parameters that determine primary productivity over greater water depths is clearly the next step. Very important information may be gained using towed platforms that undulate during towing from the surface to depths of 500 m (Fig. 2.34). These **undulators** can be towed at speeds up to 12 knots ($c.6\,\mathrm{m\,s^{-1}}$) and their 'flight-paths' can be set by the operators. These platforms can support a multitude of sensors such as fluorescent

● Towed instrument platforms that undulate between the surface and depths of 500 m can provide valuable information about the biological and chemical characteristics of a water body.

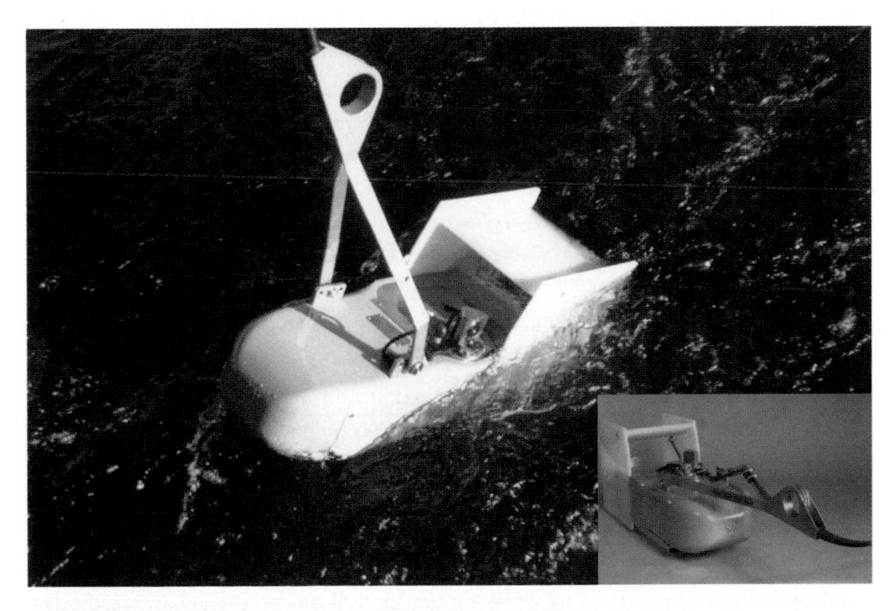

Fig. 2.34 Various profilers are now available that can be towed from a ship. They are designed to undulate over a water column profiling a range of parameters using a range of sensors (insert) carried on the vehicle. Data can either be logged on the profiler or, more normally, be sent back through cables so that scientists get real-time data on board the ship. (Photographs: Chelsea Technologies Group, **www.chelsea.co.uk**.)

chlorophyll sensors, light sensors, salinity, temperature, pressure, nitrate and nitrite, and particle concentration sensors. The data from the sensors can be sent to the operators by the connecting cable, giving the scientists on board 'real time' data about a whole suite of information.

These technologies are still restricted to the places that ships travel, and therefore large part of the worlds' oceans will be visited only infrequently, if at all. What is needed is roaming platforms that are able to measure physical, biological, and chemical parameters across depths and for long transects. A whole range of Autonomous Underwater Vehicles (AUV), gliders, and floaters have been under development since the late 1990s. These battery-powered devices are designed to 'roam' pre-programmed tracks over large regions of the ocean, collecting data. Periodically these devices surface to send the collated data to satellites, which then transfer the data to base stations from where scientists can download the information. To date, many of these devices carry sensors for salinity, temperature, and pressure among others. However, as technology progresses fluorescent and nutrient sensors will be routinely deployed on such platforms.

● In the future Autonomous Underwater Vehicles will be used to roam the oceans sending information about primary production and chemistry to satellites and back to researchers at their desks.

CHAPTER SUMMARY

- About half of the global primary production takes place in marine systems. Most of the primary production in the world's oceans is due to microscopic phytoplankton, since macroalgae are restricted to a rather narrow band on coastlines.
- Photosynthetic algae can vary in size from just a few μm to giant macroalgae 50 m long. Growth of primary producers is the difference between the gains from photosynthesis and losses to respiration, excretion and the construction of skeletal material and storage products.
- Rates of primary production are mainly controlled by light and inorganic nutrient supply. The amount of light available for photosynthesis depends upon water depth, and the amount of light-scattering particles that occur in the water.
- The main limiting inorganic nutrients are nitrogen and phosphorus, while in certain marine systems trace elements such as iron are limiting.
- Seasonal dynamics of algal growth are controlled by a complicated suite of interactions between irradiance and nutrient supply, ultimately driven by the physical dynamics of the system.
- Eutrophication of a water body is a reversible process and can be caused by factors other than solely increased inorganic nutrient loading of a system.
- Frontal systems, gyres, river plumes and coastal upwelling all influence the rate of primary production, as they influence the transport of nutrient-rich waters to the sea surface.
- Primary production in Polar oceans is restricted to a short summer season, in contrast to temperate waters where two peaks in production are often observed. Primary production of tropical waters is generally consistently low.

- The measurement of small-scale primary production can be made using oxygen and carbon dioxide tracers, or electrodes and various fluorometric techniques. On a global scale primary production is measured using satellite-borne colour sensors that are used to estimate the concentrations of plankton in the water.

FURTHER READING

A classic introduction to phytoplankton dynamics and constraints on growth is presented by Fogg (1991) as well as short essays by Smetacek (1999, 2000, 2001). The evolution of modern phytoplankton is discussed by Falkowski et al. (2004).

Both Falkowski and Raven (1997) and Williams et al. (2002) give comprehensive overviews of primary production in aquatic systems, whereas del Giorgio and Williams (2004) deal specifically with respiration. However, these primarily deal with phytoplankton, while Lobban and Harrison (1997), Lee (1999), and Graham and Wilcox (2000) give good overviews of factors influencing primary production of macroalgae. Microbial processes and the underlying biochemistry of photosynthesis, respiration, and associated metabolism is given by Madigan et al. (2002), who also give a good overview of microbial physiology and ecology. Bacterial metabolism and ecology is comprehensively covered by Dyer (2003); as is general marine microbial dynamics by Kirchmann (2000).

The influence of physical processes on primary production is dealt with by Mann and Lazier (1996). Bigg (2003) summarizes many of the large-scale ocean processes that influence primary production.

It is important to set the topics covered by this chapter into a wider context of biological oceanography, and Lalli and Parsons (2004), Libes (1993), Mann and Lazier (1996), and Millar (2004) comprehensively link aspects of physics, chemistry, and biology. There are three excellent books published by The Open University (1989, 1995, 2000), that together make a superb companion text to discussions about issues related to marine primary production.

For a more global consideration of primary production and its role in global biogechemical cycles, comprehensive overviews are given by Andrews et al. (1996), Black and Shimmield (2003), Longhurst (1998), and Schulz and Zabel (2000).

Arrigo (2005) gives a comprehensive overview on the role of microorganisms, nutrient cycles and biogeochemical cycling in marine systems.

- Andrews, J. E., Brimblecombe, P., Jickells, T. D. & Liss, P.S. 1996. *An Introduction to Environmental Chemistry*. Blackwell Publishing, Oxford.
- Arrigo, K. R. 2005. Marine microorganisms and global nutrient cycles. *Nature*, 437: 349–355.
- Bigg, G. 2003. *The Oceans and Climate*. Cambridge University Press, Cambridge.
- Black, K. D. & Shimmield, G. B. 2003. *Biogeochemistry of Marine Systems*. Blackwells Publishing, Oxford.
- Del Giorgio, P. & Williams, P. J. le B. 2004. *Respiration in Aquatic Systems*. Oxford University Press, Oxford.
- Dyer, B. D. 2003. *A Field Guide to Bacteria*. Comstock/Cornell paperbacks. Cornell University Press.

- Falkowski, P. G. & Raven, J. A. 1997. *Aquatic Photosynthesis*. Blackwell Science, Oxford.
- Falkowski, P. G., Katz, M. E., Knoll, A. H., Quigg, A. & Raven, J. A. 2004. The evolution of modern eukaryotic phytoplankton. *Science* 305: 354–60.
- Fogg, G. E. 1991. Tansley Review No. 30. The phytoplankton ways of life. *New Phytologist* 118: 191–232.
- Graham, L. E. & Wilcox, L. W. 2000. *Algae*. Prentice Hall.
- Kirchman, D. L. 2000. *Microbial Ecology of the Oceans*. Wiley-Liss, New York.
- Lalli, C. M. & Parsons, T. R. 2004. *Biological Oceanography. An Introduction* (2nd Edition). Butterworth-Heinemann.
- Lee, R. E. 1999. *Phycology*. Cambridge University Press, Cambridge.
- Libes, S. M. 1993. *An Introduction to Marine Biogeochemistry*. John Wiley & Sons. Inc.
- Lobban, C. S. & Harrison, P. J. 1997. *Seaweed Ecology and Physiology*. Cambridge University Press, Cambridge.
- Longhurst, A. 1998. *Ecological Geography of the Sea*. Academic Press. London, New York.
- Madigan, M. T., Martinko, J. M. & Parker, J. 2002. *Brock Biology of Microorganisms* (10th Edition) Prentice Hall International.
- Mann, K. H. & Lazier, J. R. N. 1996. *Dynamics of Marine Ecosystems*. Blackwells Publishing, Oxford.
- Miller, C. B. 2004. *Biological Oceanography*. Blackwell Publishing, Oxford.
- Open University. 1989. *Ocean Chemistry and Deep-Sea Sediments*. Butterworth-Heinemann.
- Open University. 1995. *Seawater, Its Composition, Properties and Behaviour*. Butterworth-Heinemann.
- Open University. 2000. *Waves, Tides and Shallow-Water Processes*. Butterworth-Heinemann.
- Schulz, H. D. & Zabel, M. 2000. *Marine Geochemistry*. Springer Verlag.
- Smetacek, V. 2002. The ocean's veil. *Nature* 419: 565.
- Smetacek, V. 2001. A watery arms race. *Nature* 411: 745.
- Smetacek, V. 1999. Revolution in the ocean. *Nature* 401: 647.
- Williams, P. J. le B., Thomas, D. N. & Reynolds, C. S. 2002. *Phytoplankton Productivity: Carbon Assimilation in Marine and Freshwater Ecosystems*. Blackwell Science, Oxford.

Chapter 3

Microbial Production and the Decomposition of Organic Material

CHAPTER SUMMARY

Micro-organisms play the major role in the recycling, decomposition, and remineralization of organic material in the plankton. This chapter will reveal why they play such a dominant role. The ecological context into which these micro-organisms are embedded is developed and arrives at a view that the food web may be seen as having two bases, photosynthesis being one, the other starting from detritus. The production and composition of dissolved and particulate organic detritus is discussed, as is the basic process of decomposition. The organisms that comprise the microbial community are briefly introduced and their relative abundances and interactions explored. The chapter then considers the dynamics of bacterial growth and its measurement. The process of respiration, its very basic biochemistry, and the concept of net community production that results from the balance between respiration and photosynthesis is then discussed together with their variations in space and time.

3.1 Introduction

The Microbe is so very small
You cannot make him out at all,
But many sanguine people hope
To see him through a microscope.
His jointed tongue that lies beneath
A hundred curious rows of teeth;
His seven tufted tails with lots
Of lovely pink and purple spots,
On each of which a pattern stands,
Composed of forty separate bands;
His eyebrows of a tender green;
All these have never yet been seen –
But Scientists, who ought to know,
Assure us that they must be so. . . .
Oh! let us never, never doubt
What nobody is sure about!

Hilaire Belloc, 'The Microbe' from *More Beasts for Worse Children* (1900)

Sustainable systems are cyclic. In the previous chapter we have considered the **production** of organic material by the marine algae. The production process utilizes energy in the form of light, water, and inorganic nutrients: salts such nitrates and phosphates, and carbon dioxide. For the cycle of nature to be completed, this energy must be released as heat and the elements assimilated into organic material during photosynthesis, then recycled back to their inorganic state. This process, the complement to photosynthesis, is **respiration**. In its narrowest meaning, respiration is simply the biological oxidation of organic material typically by oxygen and the formation of water and carbon dioxide; the reformation of the inorganic nutrients is an intimately associated process, known by a number of synonyms: **mineralization** and **remineralization** (Chapter 2). As the process also results in the breakdown of organic material it also carries the name **decomposition**. All living organisms respire; however, in the oceans this process is dominated by the micro-organisms, notably the bacteria.

3.2 The Microbial Powerhouse

Whereas Belloc's ideas of microbial anatomy are more than bizarre, the sentiments in the first and last four lines of his delightful poem are as relevant today as they were a century ago. Still the non-specialist has to accept the details about micro-organisms as an act of faith. We are aware of the presence of whales, fish, and large invertebrates in the seas as we can see them and so make some qualitative assessment of their abundance. The sanguine person mentioned in Belloc's verse would be disappointed if they viewed a drop of seawater under a high-powered microscope, the chances are that he or she would see not a single micro-organism. The oceans are mainly (99.99999%) salty water; the biota occupy only approximately 0.00001% of the total volume.

● Organisms occupy only about 0.00001% of the volume of the oceans.

It is difficult to convince even expert marine scientists that the biomass of micro-organisms is comparable to that of whales or fish, and that their activity, as we shall see, exceeds that of the combined metabolism of vertebrates many times. But why is this so, and why are the micro-organisms so important in the turnover of organic material in the seas?

3.2.1 Why are small organisms important?

There are broadly two answers. First, the primary production of organic material in the seas is associated with microscopic organisms (see Chapter 2) and the large number of steps between the primary producers and the marine vertebrates results in the loss of almost all the organic material produced by photosynthesis such that there is very little for large organisms to live on (3.3.2). Second, and related to this, is the consequence of size. The marine pelagic community spans a massive range of size from marine bacteria ($c.0.1\ \text{pg} = 0.1 \times 10^{-12}\ \text{g}$), arguably the smallest free-living organisms, to the whales ($c.100\ \text{tonnes} = 10^{8}\ \text{g}$), the largest of animals. This means that the total span of biomass is 21 orders of magnitude.

● The total span of biomass in the seas is 21 orders of magnitude.

Body size has a major consequence for metabolism (Fenchel 2005). All free-living organisms take in requirements for their growth and dispose of their waste products through the body wall. This transfer is the principal limitation to maximum growth, thus surface area, rather than biomass, is the main determinant of metabolic rate. If an organism, such as a whale, were to achieve a metabolic rate comparable to a bacterium, the metabolic heat it produced would be so difficult to dissipate that its body temperature would rise to 1500 °C – it would be white hot.

● Body size has a major consequence for metabolism; surface area, rather than biomass is the main determinant of metabolic rate.

We can explore the consequences of this simply by considering the organism as a simple sphere.

$$\text{Then } \frac{S}{V} = \frac{4\pi r^2}{4/3\pi r^3} = \frac{3}{r}$$

where r is the radius, S is the surface area (which will reflect metabolism), and V is volume (which will reflect biomass). Thus, in the case of our simplistic organism, the specific metabolic rate (metabolism/biomass = S/V) is proportional to $1/r$, i.e. inversely proportional to the organism's linear dimensions. If we take our end members of the marine (the marine bacterium and the whale) to have linear dimensions of 0.5 μm (0.5×10^{-6} m) and 10 m, then if this simple rule applied the metabolic rate of a given mass of bacteria would be 2×10^{7}, i.e. about 10 million times greater than the same biomass of a whale. The results of this simple calculation are shown in Figure 3.1. Now clearly the calculation is simplistic, as organisms are, particularly larger ones, not spheres; further they overcome the S/V limitation in part by increasing the surface areas available for exchange, by convoluted structures such as gills and lungs. Nonetheless there is a limit to this and so we end up with a broad compliance to the S/V rule, but it is not as simple or exact as the formula above.

3.2.2 How is metabolic activity distributed? – What are the important organisms?

Metabolic activity is constrained by **metabolic capability** and abundance. A surprising observation made by Sheldon et al. (1972) is that if estimates of biomass are assembled in logarithmic categories there is

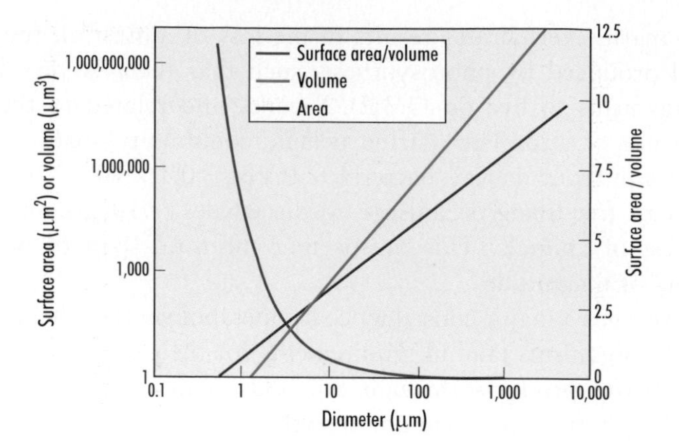

Fig. 3.1 The relationship between diameter, volume, and surface area for a sphere.

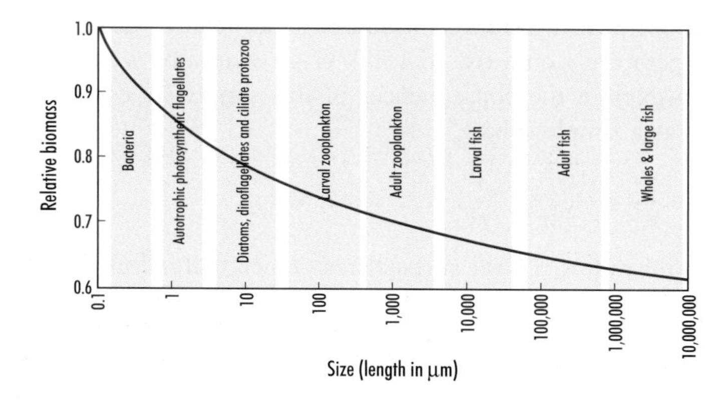

Fig. 3.2 The red line shows the theoretical distribution of biomass with size. The approximate size bands occupied by various planktonic groups are shown.

comparatively little change with body size. Food web theory suggests that a small decrease in biomass occurs with an increase in the size fraction, which is broadly upheld by field observations. Models that give the distribution of biomass through logarithmic increments of size (as length) suggest a small decrease with body size, which complies with the following relationship

$$\text{Biomass} = \text{k} \times (\text{Size})^{-0.22}$$

A plot of the above relationship is given in Figure 3.2, along with the planktonic groupings that are mainly associated with the various size categories.

● Specific metabolic rate (metabolism/biomass) is inversely proportional to the organism's linear dimensions.

In Figure 3.3 the distributions of activity and biomass given in Figures 3.1 and 3.2 are combined. Because the change of metabolism with size (Fig. 3.2) is much less than biomass (Fig. 3.1), the resultant line is very similar the surface to volume relationship given in Figure 3.1.

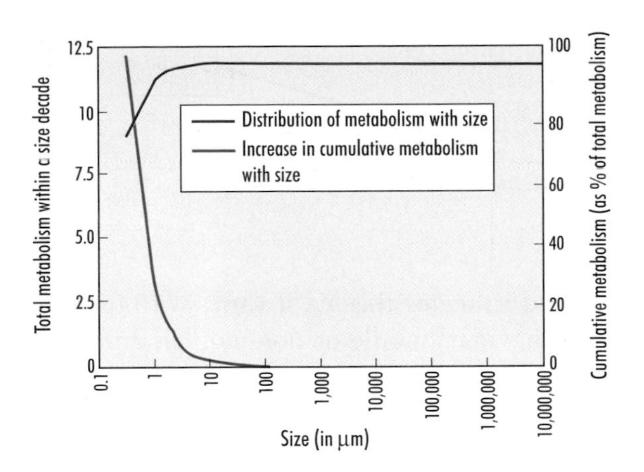

Fig. 3.3 The size distribution of metabolism, and its progressive integral, with size. The rates are determined as the product of the size distribution of specific metabolic activity (Fig. 3.1) and biomass (Fig. 3.2) with size. The figure shows that almost all of the metabolism is associated with organisms less than 100 μm.

Thus we have a basis for understanding the major part played by micro-organisms in the planktonic community.

3.3 The Ecological Context in which the Micro-organisms are Embedded

Before we can explore the role and functioning of the microbial community, we need to understand the ecological context into which it is embedded. As micro-organisms play a central role in the metabolism of organic material in the oceans, as in most other ecosystems, it is logical to outline the rationalizations surrounding the flow of organic material through an ecosystem.

3.3.1 The concept and consequence of growth yield

An organism cannot grow (convert food into body tissue) with 100% efficiency. The laws of thermodynamics tell us that work must be done to create new tissue and that this requires energy. This energy derives from respiration, which occurs at the expense of some of the energy derived from assimilated food (see Fig. 3.4). This introduces the notion of the efficiency of growth, usually expressed as a **growth yield** or, since it is often expressed as a percentage, **growth efficiency**.

● An organism cannot convert the energy derived from food into new tissues with 100% efficiency.

$$\text{Growth yield} = Y_g = \frac{\text{growth}}{\text{food intake}} \text{ or } \frac{G}{G+R}$$

where G = growth and R = respiration.

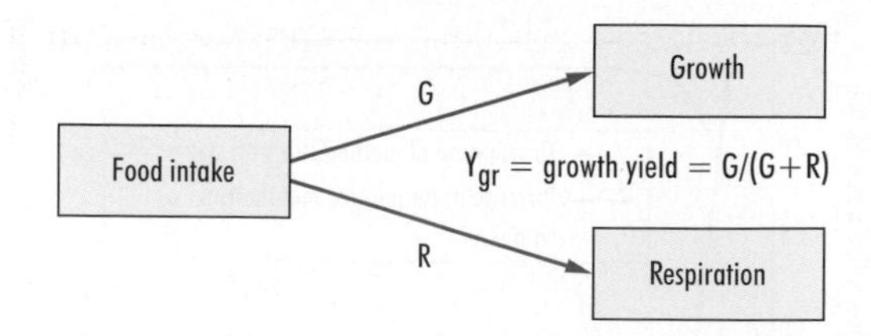

Fig. 3.4 Schematic of the apportionment of overall metabolism of a heterotroph into respiration (R) and growth (G).

There is no fixed value for this Y_g, it varies with organism type, level of complexity, behaviour (motile or non-motile), and stage of development (larva, juvenile, adult). Characteristically values lie in the range of 10 to 30% (Chapters 12 and 13).

3.3.2 Trophic yield and food chain efficiency

● Trophic levels indicate the position of an organism in a food chain or food web, hence primary producers make up the lowest trophic level, with predators towards the top.

One organism's growth is another organism's meal. If we consider a simple linear food chain, comprised of physiologically similar organisms within each **trophic** grouping, we can extend the growth yield concept to a trophic yield. In this case the production (growth) at one level is taken as the food intake for the subsequent one (Chapters 6 and 7).

$$\text{Trophic yield} = Y_t = \frac{\text{production at trophic level}_{i+1}}{\text{production at trophic level}_i}$$

In this way we may link together the type of flow diagram shown in Figure 3.4 into a string of trophic yields (see Fig. 3.5).

Given this as a concept we may calculate the overall yield of our simple food chain as

$$\text{Overall efficiency} = Y_{t=2} \times Y_{t=3} \times Y_{t=4} \times \text{etc.},$$

where $Y_{t=2}$, $Y_{t=3}$ etc are the yields at the various trophic levels.

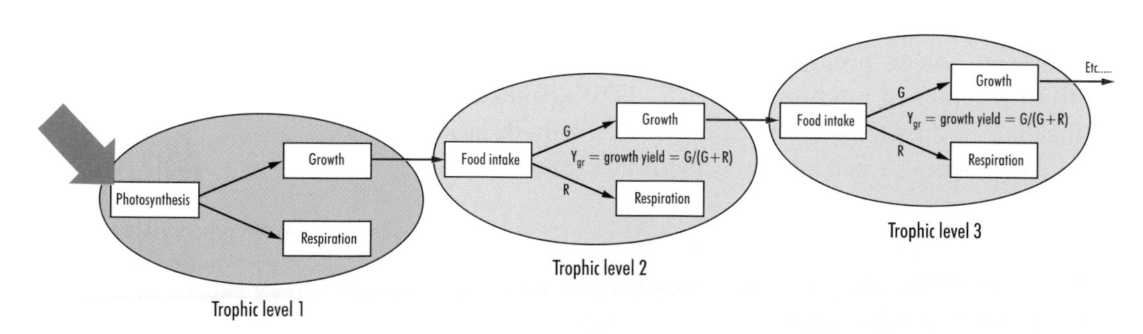

Fig. 3.5 The simple scheme shown in Fig. 3.4 embedded in a linear heterotrophic food chain Photosynthesis (the green ellipse) is taken to be the starting point.

If we make the further simplifying assumption that in a given food web the values of Y can be treated as the same, then the above equation simplifies to

$$\text{Overall efficiency} = (Y_t)^n$$

where Y_t is a simple representative trophic yield and n is the number of steps (note, one less than the number of trophic levels) in the food chain. This simple concept was the basis of the seminal paper in 1969 by the biological oceanographer John Ryther, who used the concept to predict the fishery potential of the oceans (Ryther 1969). This concept allows us to explore the significance of growth (trophic) yield and the number of steps in the food chain.

This sort of analysis can be extended to more realistic branched food webs, with different yields at various trophic levels, but the broad outcome is much the same, and at this level the additional complexity is not justified. The point made by these calculations (Table 3.1) is that both properties have a profound effect upon the overall efficiency of the food web and most natural food chains (and in particular food webs) are, and need to be, **highly inefficient**. The inefficiency is a consequence of the very large fraction respired and recycled back (remineralized) to form the inorganic carbon, and more importantly inorganic nitrogen and phosphorus; originally used during phytoplankton production. The phrase 'earth to earth; ashes to ashes, dust to dust' (*Book of Common Prayer*) just about sums it all up. Without this, the food web in the oceans, as in any other ecosystem, would rapidly stagnate.

● Most natural food chains are, and need to be, highly inefficient; if they were not, the ecosystem would rapidly stagnate.

We may now build this recycling into our growing conceptual model of the food chain.

Table 3.1 Calculation of the effect of trophic yield and the number of trophic levels on overall losses and yields in a simplified food chain.

Trophic yield (as %)	Number of trophic levels in the food chain	
	2	5
	Yield of food chain (as %)	
10	1	0.001
30	9	0.24
	Loss through the trophic levels (as %)	
10	99	99.999
30	91	99.8
	Yield from 1 tonne of phytoplankton production	
10	10 kg	10 g
30	90 kg	2.4 kg

3.4 The Decomposition Process

So far we have explored a simple linear food chain, supported by photosynthetic production. The losses due to respiration have been built into this sequence in Figure 3.6. We now need to consider the organic losses – the formation and decomposition of organic detritus. This leads to a realization that there is a second base to the food web; a food chain leading from detrital organic material. Although decomposition is seen to be the prerogative of the micro-organisms (the group referred to by ecologists as the **decomposers**), it is a process contributed to by all heterotrophic organisms, from bacteria to whales, although for the reasons discussed in 3.2.1 bacteria and the protists play the dominant role. Decomposition, the breakdown of organic material by heterotrophic metabolism as a consequence of respiration, results in the production of inorganic carbon (as CO_2), nitrogen (as ammonium), and phosphorus (as phosphate), the process is collectively referred to as **remineralization** (or sometimes simply as **mineralization**).

● Decomposition is a process to which all heterotrophic organisms contribute, although the microbiota play a dominant role.

3.4.1 Marine detritus

Although, as we will see later, there is direct release of soluble organic material from marine organisms during feeding and growth, it is convenient to consider the decomposition process as starting from particulate material. The initial stage is the conversion of the insoluble, non-diffusible material (e.g. cellulose, proteins and fats) to low molecular weight material by extracellular hydrolytic enzymes (digestive enzymes). In principle, the first stage must occur outside the cell, in the gut, in vesicles or in the external environment. As the process of hydrolysis is biochemically simple it does not require complex organization. The products

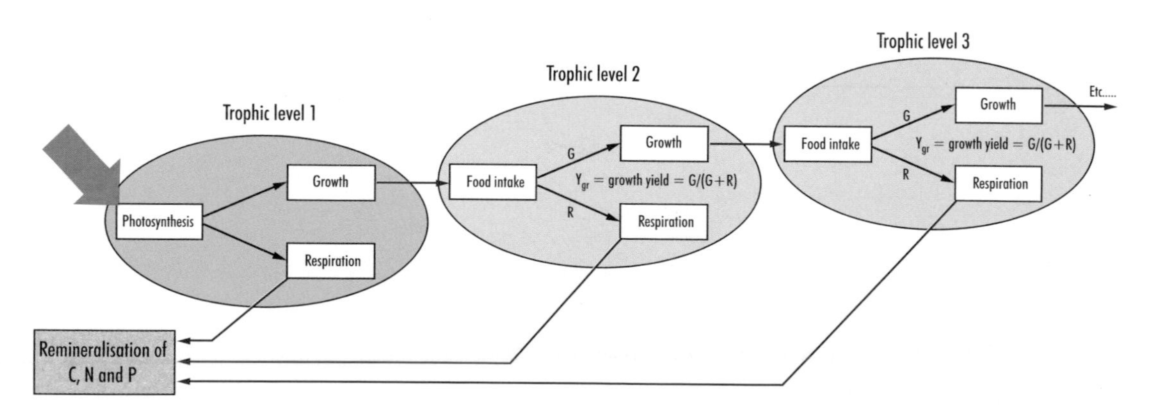

Fig. 3.6 The flow of respired nutrients has been added to Fig. 3.5.

Fig. 3.7 An extended version of Fig. 3.4, showing the two stages in decomposition.

(amino acids, sugars, fatty acids) may now diffuse into the cell, through the gut wall in the case of most metazoans. The subsequent stages of metabolism, and in particular the very final stages (respiration), are complex processes that require a high level of organization that can only occur within the cell. As these events are more or less universal in all heterotrophic organisms, we can now separate the first stage in Figure 3.4 into these two steps (see Fig. 3.7).

- Hydrolysis is biochemically simple and does not require complex organization.
- Respiration is a complex process that requires a high level of organization that can only occur within the cell.

No food chain is entirely closed, there is always some wastage of food along the way. This may come as faeces, from incomplete digestion, release of soluble organic material from either the algae or heterotrophic organisms (**exudates**), or damage of, or incomplete feeding upon, the prey (so called '**munchates**') – the odd leg torn off or piece of cell not consumed. This (both soluble and particulate material) is collectively known as **marine detritus** (see Fig. 3.8), the nature and origins of which are dealt with in the following section.

- No food chain is entirely closed, there is always some wastage of food.

3.4.2 The nature and production of marine detritus

Marine organic detritus is partly composed of the scraps left over from the various meals of the marine heterotrophs. As such it has no particular composition and thus will vary with the type of food organism and the feeding mechanism. Although it may be of poor nutritional value to the organism giving rise to its production, the micro-organisms have the metabolic and nutritional versatility to make a meal out of

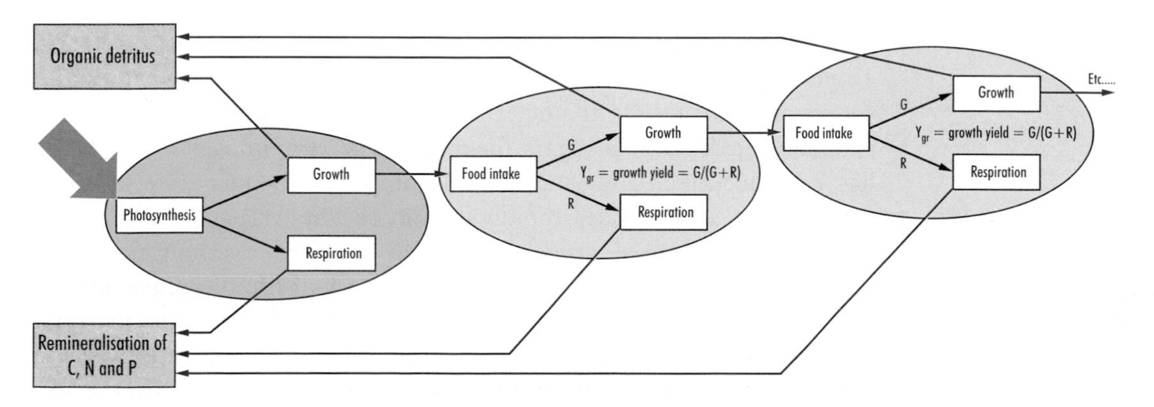

Fig. 3.8 Incorporation of remineralization and detritus formation into Fig. 3.6.

it: indeed 'one person's trash is another's treasure'. In addition to the uneaten remains, there are two additional sources of marine detritus. Various marine organisms secrete mucilage or slough off polysaccharide material during their growth and development; for example, coral are a rich source of mucilaginous material. Appendicularians (a class of chordates also known as larvaceans) surround themselves with a mucilaginous 'house', which contains filtering apparatus with which food is garnered. When the filters become blocked they discard the 'house' and construct a new one, this process takes a few minutes and can be repeated up to a dozen times per day. A quite different source of organic detritus comes from the algae themselves. The mechanism and causes of release of this dissolved organic material are far from clear. It occurs in active as well as moribund cells, so it may be regarded to be a 'normal' part of the metabolism of the algae. It would appear to be high when the cells are under high irradiances and in such situations a metabolic pathway (photo-respiration), closely associated with the primary CO_2 fixation, is known to produce compounds such as glycolic acid, and the amino acids serine and glycine as by-products. However, this is probably only one of a series of processes giving rise to the release of these compounds. The extent of release in actively growing, unstressed cells, is probably in the range of 5 to 15% of photosynthetic production (Chapter 2), but may rise to 50 to 80% if the algae are stressed by high light or low nutrient conditions.

● Marine detritus is in part the scraps from the various meals of the marine heterotrophs.

Because of its multiplicity of sources and formation mechanisms, marine detritus is a heterogeneous mix of compounds and not readily amenable to detailed chemical analysis. The normal procedure is to divide it into two categories based primarily on size, using fine filters, into so-called **dissolved** and **particulate organic material** (DOM and POM). The point of separation is not exact but lies in the region 0.2 to 1 μm, determined by the nature, properties, and loading of the filter used as well as the morphology of the particles themselves. The dissolved organic carbon (DOC) fraction constitutes by far the largest organic pool in the sea, and one of the largest in the planet (see, Fig. 2.19, Table 3.2). DOC has been subject to detailed chemical examination and yet we are only able to characterize 5 to 10%, as simple molecules such as amino acids, simple sugars (hexoses). The remaining 90% of DOC has been identified only into broad categories (Benner 2002). While some of the DOC cycles rapidly within the upper part of the water column, a substantial proportion of the DOC is resistant to decay. **^{14}C dating** of the resistant material gives it a half-life of approximately 6000 years. It was previously thought that this recalcitrant material consisted of complex high molecular weight material and was given the name '**marine humus**' but recent work has shown that the material has a molecular weight less than 1000 daltons and thus does not conform

Table 3.2 Major organic carbon pools sizes in the biosphere.

Pool	Pool size (as mol C)
Oceanic pools	
Dissolved organic material	1×10^{17}
Particulate organic material	3×10^{15}
Marine plankton	1×10^{14}
Fish	5×10^{12}
Terrestrial pools	
Terrestrial plants	5×10^{16}
Soil	2×10^{18}
Global pools	
Fossil fuels	5×10^{17}

to these earlier notions. A major challenge facing the marine chemist and microbiologists is the chemical nature of this material, it is resistant to decay, and the mechanism of its eventual decay remains unknown. The current view is that its decomposition involves photochemical and microbiological reactions working in consort, but without doubt there is a lot to be learned about this aspect of carbon flow in the oceans.

- Marine detritus is a heterogeneous mix of compounds and not readily amenable to detailed chemical analysis.
- DOC constitutes by far the largest organic pool in the sea, and one of the largest in the planet.

The particulate organic fraction comprises a melange of small living organisms, bits and pieces left over from the various meals and particulate material created by physicochemical processes from soluble detritus. The material (Fig. 3.9) has the general overall appearance of snowflakes (although it lacks their characteristic symmetry) and is commonly referred to as '**marine snow**'. Early bathysphere explorers of the ocean depths reported marine snow to occur at the scale of a terrestrial snow blizzard on occasions. The first step in their creation is the agglutination by shear forces of hydrated mucilaginous fibrils released from the plankton to form **transparent exopolymer particles** or TEPs (Wotton 2004). The marine snowflakes arise by random collision of TEPs. In the northern Adriatic these flakes can reach massive size (Fig. 3.10). As discussed later in this chapter (3.5.4) they are a habitat for marine micro-organisms and possibly oases in an otherwise nutritional desert. It is thought that these flakes are the main vehicle for the transport of organic matter from the upper ocean to the ocean floor and, as such, a vital step in the carbon cycle of the oceans. The fate of marine snow and the rate at which it is decomposed by its bacterial hitchhikers is critical to the behaviour of the oceans as a potential sink for carbon.

- 'Marine snow' has been reported to occur on blizzard scales.

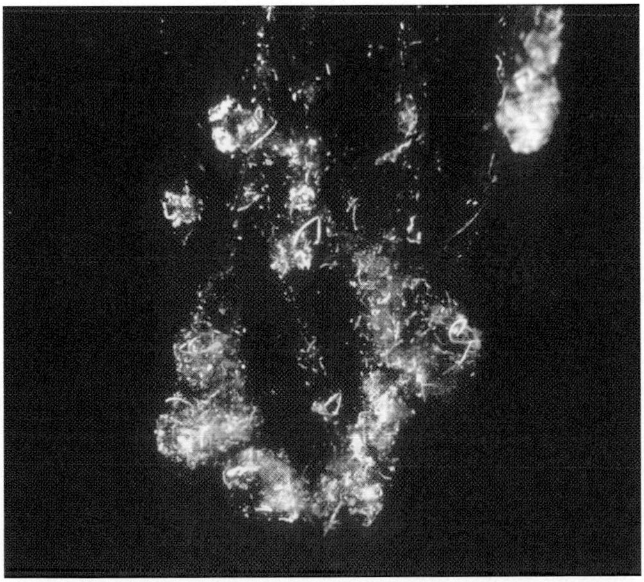

Fig. 3.9 Marine snow particle. (Photograph: Alice Alldredge.)

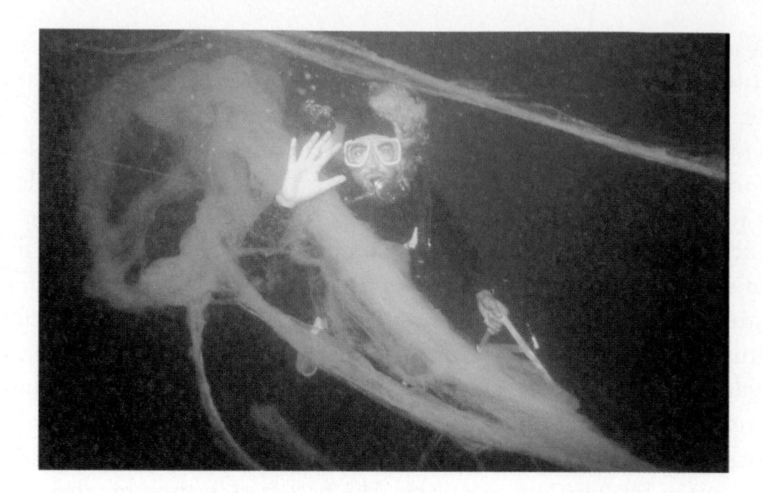

Fig. 3.10 A massive marine 'snowflake' in the northern Adriatic. (Photograph: Michael Stachowitsch.)

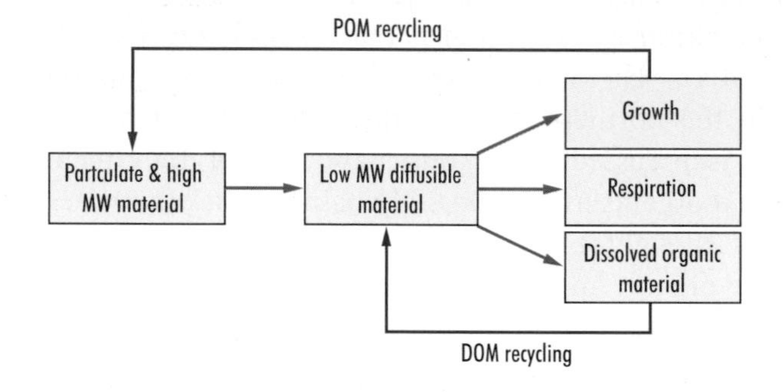

Fig. 3.11 Extension of Fig. 3.7, showing the recycling of particulate and dissolved organic material back into the food chain.

In a simplified form Figure 3.11 builds on Figure 3.7 to show how this 'waste' material will feed back into the food chain. As shown in Figure 3.12 (3.4.3) the flows to the detrital section are multiple. They would appear to equal to 50% of the primary photosynthetic formation of organic material.

3.4.3 The utilization and recycling of marine detritus

● Detrital material is potentially a valuable food resource.

Detrital material is potentially a valuable food resource that is utilized by a complex of organisms (the decomposers). To fit this into our developing model of the marine plankton community and explore the consequences of the recycling of organic material, we need to depart from our simple linear food chain, with a single starting point (the photosynthetic autotrophs), to two initially parallel food chains with separate starting points: one based on photosynthesis, and the other starting from detritus, which involves the microbial community. In Figure 3.12 these separate food chains are shown to merge at one point; the reality is more like that depicted in Figure 3.14 – that the whole system is more of a network.

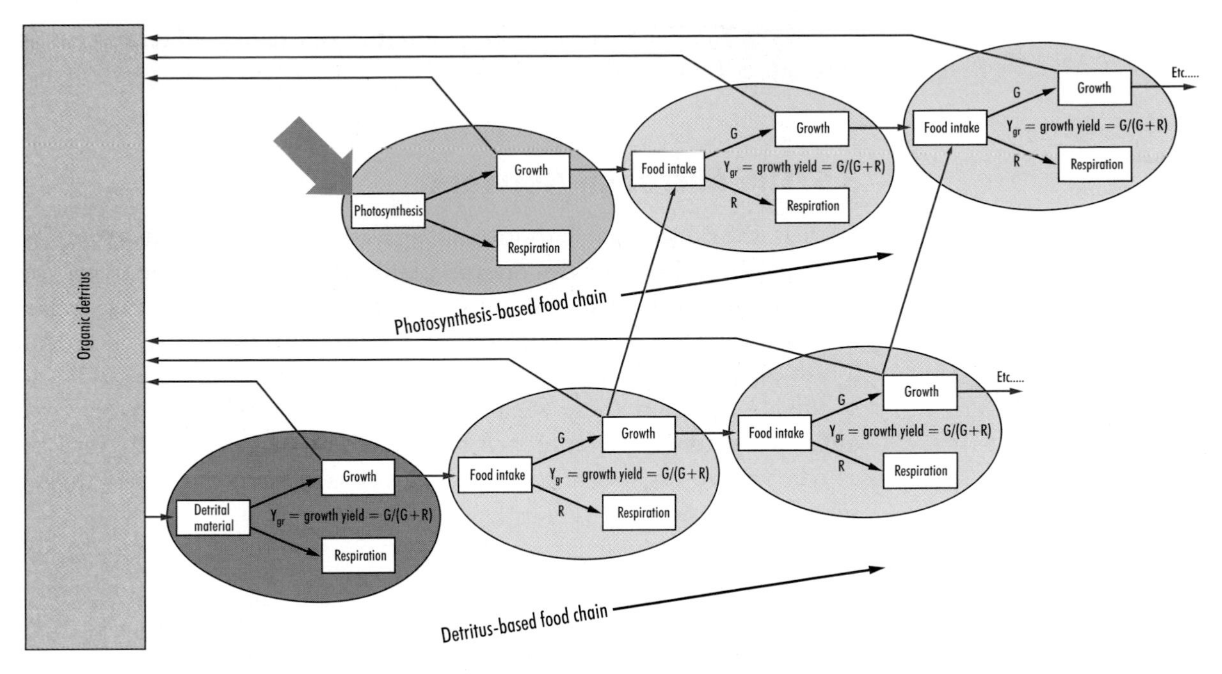

Fig. 3.12 The dual food chain, showing the two starting bases: photosynthesis (the green ellipse) and organic detritus (the red ellipse).

Thus, whereas organic material can only pass through respiration once, a single carbon atom (also a nitrogen or phosphate atom) in organic material can be taken up by marine organisms several times before it is respired. This introduces the concept of **organic recycling** within the plankton community. The consequence of recycling is that the organic material used by the planktonic, or any community, will be greater than the primary introduction of organic material. If we know the fraction recycled (let us call this Y_{det}) it is a simple matter to calculate the total amount of material passing through the food web. If we take a commonly estimated figure of about 50% recycled through each cycle (Nagata 2000), then of the original amount, 50% will be recycled on the first pass, 50% of that (25% of the original) on the second pass, etc, and the total amount passing through will be in this case:

$$1 + 0.5 + 0.25 + 0.125 \ldots \text{or } 0.5^0 + 0.5^1 + 0.5^2 + 0.5^3 + \cdots,$$

or expressed more generally:

$$Y_{det}^0 + Y_{det}^1 + Y_{det}^2 + Y_{det}^3 + \cdots$$

This is a mathematical series with a surprisingly simple solution:

$$\frac{1}{1 - Y_{det}}$$

Thus if $Y_{det} = 0.5$, then the total amount of organic material passing through is $1/(1 - 0.5) = 2$.

3.5 The Microbial Network

Such is the conceptual background against which we need to set the organismal biology. In the marine plankton microbial community we have traditionally identified four major categories – bacteria, algae, protozoa, and the larval zooplankton. Comparatively recently we added a fifth category: the marine viruses. There are a number of approaches to subdividing within these separate groups and typically a taxonomic approach is used. The approach used here is based principally on size, as when micro-organisms acquire food, size is in many cases the primary selection criterion.

3.5.1 Trophic components

Viruses

It is only in the late 1980s that we have been aware of the ecological importance of viruses. They are not organisms in the normal sense, but their impact on the marine community as a whole appears to be substantial. They occur in numbers of $10^7 cm^{-3}$ and they are known to infect organisms from bacteria to seals (Fuhrman 1999). There is evidence that they can play a major role in terminating blooms of coccolithophorids (Chapter 2) and we may expect the same is true for other groups of phytoplankton as well.

- Marine viruses infect organisms from bacteria to seals.

Bacteria

As with viruses the full understanding of the role of bacteria developed late also, in the 1970s and early 1980s (Sherr and Sherr 2000). There is an enormous variety of metabolic types within the bacteria, far greater than any other group of organisms. Here attention will be restricted to two major physiological types: **organotrophs** and the **chemoautotrophic** nitrifying bacteria.

- There is an enormous variety of metabolic types within the bacteria.

The former group are the classical heterotrophs, which can use a vast range of organic compounds for their growth. The initial attack on the organic detritus in the natural environment (and unnatural environments such as sewage works) is led by these organisms. It is axiomatic that all material produced naturally, and many novel synthetic organic chemicals, are broken down by this group of organisms. They are termed heterotrophs as they need organic material for their growth; they gain the energy for growth and other activities such as motility from the respiration of the assimilated organic material.

- It is axiomatic that all materials produced naturally are broken down by bacteria.

The other group we need to consider are a highly specialized group, the so called '**nitrifying**' bacteria (see also Chapter 2). This group of organisms gain their existence from oxidizing ammonia to nitrite, and then nitrite to nitrate. We can broadly recognize two groups within the nitrifiers, with one carrying out the former reaction and a second carrying out the latter. As they are organisms that utilize inorganic compounds exclusively to provide the carbon and energy for growth, they are accordingly termed **autotrophs**. Autotrophs lack the capability to use organic material for growth and have to produce the organic material themselves by '**fixing**' CO_2 as do the algae, and both groups of organisms use the same enzymatic pathways.

Algae

The categorization of algae in this context is in relation to their availability as a source of food. The algae span a zone where the combination of fluid dynamics and size has a considerable bearing on the mechanisms used to collect food. In the case of the smaller algae we enter a realm characterized by a value known as **Reynolds number** (Box 3.1), where the behaviour of particles gives the impression that the water is very

Box 3.1 Life at low Reynolds number – where our intuitions fail

Reynolds number is a coefficient and gives us a scale that allows us to anticipate how a moving organism will experience the fluid physics of its environment. It is calculated from the size of the organism, its rate of movement and a property known as the kinematic viscosity, which is the normal viscosity times the density of the fluid. We occupy high Reynolds numbers, e.g. 10^6, whereas micro-organisms occupy low numbers, e.g. 10^{-5}. Our everyday experience gives us a poor understanding of the circumstances at low Reynolds number and our intuitions can be very misleading. If we were to experienced life at the scale of micro-organisms we would be in for big surprises. Consider the basic component of Reynolds number, the kinematic viscosity. The kinematic viscosity of air is *greater* than water, thus it means that a particle the size of a bacterium falling through the atmosphere would settle faster once it entered the water, not slower as we would experience on our scale. At low Reynolds numbers, the fluid gives the appearance that it is highly viscous, and if we look at organisms under a high powered microscope (when we have entered the world of low Reynolds numbers) we can actually see this phenomenon. Scaled up to the human scale, the environment has a viscosity somewhere between molasses and tarmac. But, in truth the viscosity is the same, but the organisms have virtually no momentum. This lack of momentum means that when micro-organisms cease swimming they stop almost immediately, a bacterium coasts for 10 microseconds, only the length of a hydrogen bond (0.1 nm). On our scale such a deceleration would be lethal, many orders of magnitude worse than driving a Formula One car at full speed into a concrete wall. These properties at small scales have a controlling influence of the mechanisms for motility and feeding of organisms living at these scales. A delightfully entertaining account of life at low Reynolds numbers is given by Purcell (1977).

viscous; although it is important to stress that this is in appearance only, as it is a feature of their **lack of momentum** rather than any change in the viscosity of water. This means that feeding mechanisms that rely upon straining or collecting the particles by sweeping them into funnel-shaped gullets simply do not work for algal flagellates and bacteria; whereas with larger algae it is very effective. There is no sharp boundary but a practical dividing line that lies in the vicinity of 5 μm body size is commonly used. The small phytoflagellates lie below this boundary, the dinoflagellates and many of the diatoms above, while the coccolithophorids span the boundary.

Protozoa

Currently the protozoa are grouped with the algae into a single category: the protista (Sherr and Sherr 2000), although here we deal with them separately. In functional ecology, protozoa are not so much classified by their size as by the nature of their food, although in a broad way this will also reflect their size. Bacterivores essentially define themselves functionally, but they comprise a diverse range of different protistan groups. The common feature is that they are flagellated, collectively know as the **heterotrophic nanoflagellates**, and span a size range of 2–20 μm. In this group, the flagellum serves a dual function by providing motility and as a device for collecting bacteria. It is broad grouping in which the separation between photosynthetic and heterotrophic forms is least clear and a number of species within this collection of phylogenetic groups, the **myxotrophs**, take advantage of both forms of nutrition (see Chapter 2). The other functional grouping are the **microzooplankton protists**, characteristically they fall in the size range 20–200 μm. They include major groupings such as the ciliates and the larger dinoflagellates. These organisms are characteristically herbivores that prey mainly on the smaller phytoplankton. They will also prey upon bacteria, as is so often with these small organisms, sharp functional divisions simply do not exist.

● Modern thinking places the protozoa with the algae into a single category – the protista.

Larval zooplankton

Although taxonomically very distinct from the microzooplankton protists, the larvae of many metazoan plankton are functionally similar to them and the two groups are commonly lumped into a single category, the **microzooplankton** (see also Chapter 6).

3.5.2 How much do we find?

In Figure 3.2, we produced profiles of biomass and activity developed extensively from theory. However theory can only take us so far. Field observations of biomass and calculated surface area are broadly consistent with the theoretical distribution given in Figure 3.2 (Fig. 3.13 and Table 3.3). The biomass profile shows that there is not a marked difference in biomass between the small and large forms, and the high biomass of phytoplankton is probably in part an artefact as much of the sampling from which these compilations are derived, as have been made during the bloom period. The striking feature is the span of surface areas, dominated by the bacteria (Table 3.3).

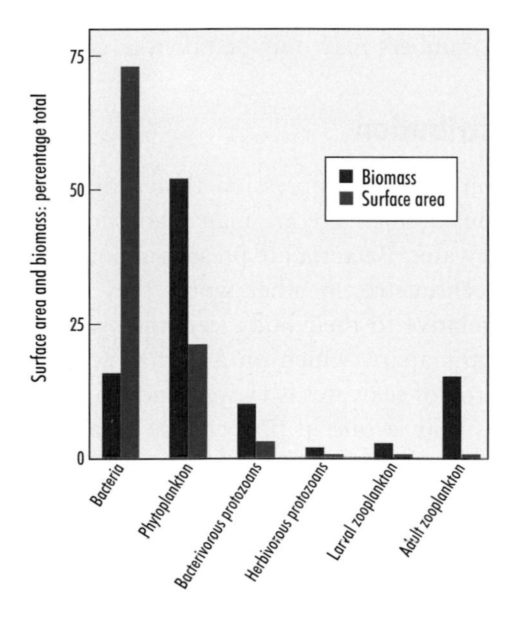

Fig. 3.13 The typical distribution of surface area and biomass of the major planktonic groupings.

Table 3.3 Typical values for abundance and biomass and surface area of the major planktonic types.

	Number (per m^3)	Biomass (mmol C/m^3)	Biomass (%)	Surface area (m^2/m^3)	Surface area (%)
Viruses	c.10^{13}	–	–	–	–
Bacteria	c.10^{12}–10^{13}	1.5	16	1.00	73
Phytoplankton	c.10^{8}–10^{9}	5.0	52	0.30	22
Bacterivorous protozoans	c.10^{10}	1.0	10	0.05	4
Herbivorous protozoans	c.10^{6}	0.2	2	0.01	1
Larval zooplankton	c.10^{5}	0.3	3	0.005	0.4
Adult zooplankton	c.10^{3}	1.5	16	0.01	1
		9.5		1.37	

3.5.3 Trophic arrangement

We now need to put the pieces together. Figure 3.14 summarizes our current understanding of the major flows within the microbial network and provides a pictorial summary of the statements in 3.5.1. This diagram can be reorganized to give the twin source food chains developed in 3.4.3. As the intensity of the various flows will vary with the location and time of year (considered in 3.7.3), no attempt is made to generalize on them here; this will be done in 3.8. The relative distribution of the biomasses is summarized in Table 3.3, but again these figures vary spatially and temporally and the numbers may only be taken as guides.

3.5.4 Spatial and phase distribution

Popular accounts of the plankton give the impression that a drop of water is teeming with life. Without doubt there are many thousands of individual organisms in a drop of water. Bacteria are present in numbers of a million or more per cubic centimetre, in other words they occur about 100 μm apart. Expressed relative to their body size, they will be on average about 200 body lengths apart, which on a human scale is about 0.5 km apart. Hence, if a drop of seawater is viewed under a high-powered microscope there is less than a one in fifty chance of a bacterium occurring in the field of view.

● If a drop of seawater is viewed under a high-powered microscope there is less than a one in fifty chance of a bacterium being in the field of view.

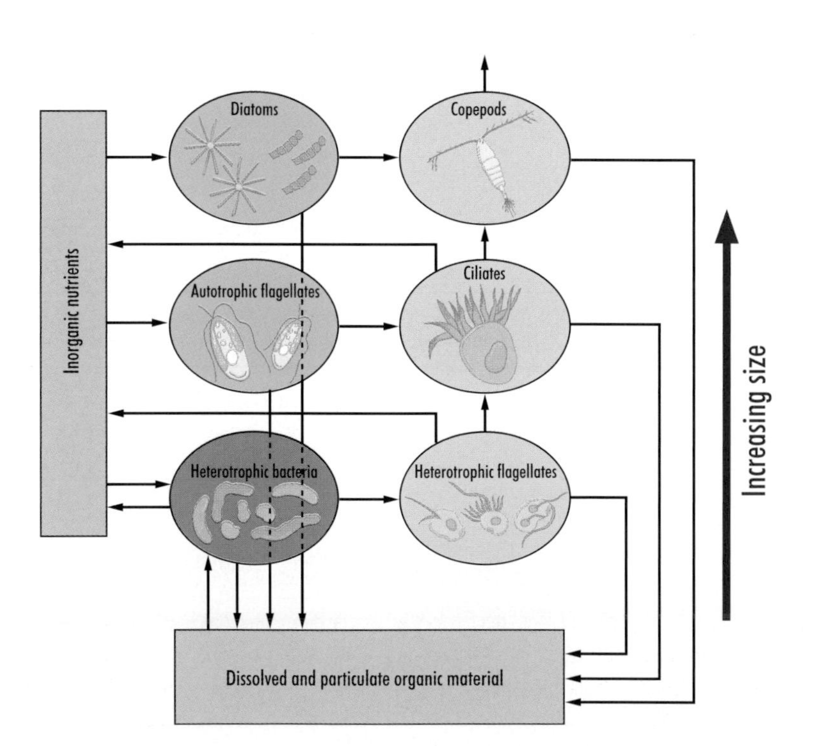

Fig. 3.14 The trophic connections within the microbial network.

Micro-organisms are almost certainly not homogenously dispersed though the water. The simple dichotomy of organic material into dissolved and particulate used in chemical analysis is a poor representation of the reality in seawater, where there is a continuum of sizes of organic material. Micro-organisms can be found attached to living algae, where they are thought to play a key role by pruning mucus from the surface of the algal cell thereby reducing its stickiness, inhibiting aggregation and the consequent enhancement of settling of the algae out of the illuminated part of the ocean. Bacteria and other micro-organisms colonize all forms of detritus including fragments of body parts and faecal pellets, thereby facilitating their decomposition. Marine snow (3.4.2) would appear to provide a rich habitat for micro-organisms of all types, including a range of bacteria and autotrophic and heterotrophic protists. These types of associations have been discussed by Azam (1998) and are show in cartoon fashion in Figure 3.15. The benefits and negative aspects of colonizing particles are complex. The particle may provide a source of organic material for microbial growth if the basic matrix is organic, but less so if it is primarily an inorganic matrix, for example a diatom test. The physical chemistry of adsorption leads us to expect that the availability of organic material will be lower in the vicinity of a surface than in the surrounding water (Box 3.2). The potential

● Bacteria and other micro-organisms colonize all forms of detritus including fragments of body parts and faecal pellets, thereby facilitating their decomposition.

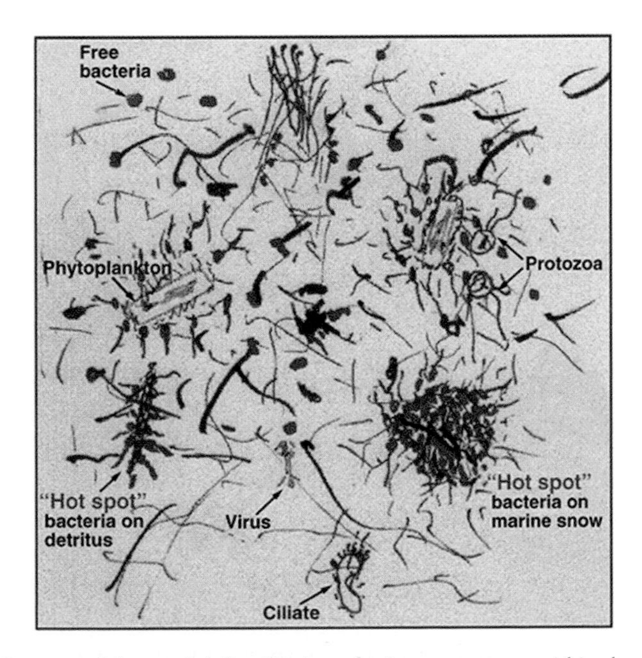

Fig. 3.15 Cartoon of the spatial distribution of micro-organisms within the various phases present in the oceans. The abundance of the organisms is exaggerated for presentational purposes. Reprinted with permission from Azam, F, 'Microbial Control of Oceanic Carbon Flux: the Plot Thickens', *Science*, 280: 694–6. Copyright 1998 AAAS.

Box 3.2 Adsorption and the concept of activity

A common misunderstanding is that as many surfaces adsorb organic material, they are areas that are favourable for the growth of micro-organisms. However, an understanding of the basis of adsorption indicates the opposite. The physical chemist defines a property '**activity**', which is the fraction of the molecules present that are reactive. It is only the active molecules that can engage in chemical and biochemical reactions and set up diffusion gradients. Adsorption occurs because the adsorbing surface reduces the activity of the absorbed molecule below that in the bulk water. This sets up a diffusion gradient, which decreases towards the surface (in contrast to the concentration which increases). The consequence of this is that material diffuses towards the surface hence the elevated concentration at the surface (concentrations at surfaces however is a questionable concept). Thus, the availability of the compound for a micro-organism to assimilate is less at the surface, because its activity is lower than in the bulk medium. This notion was resisted for some considerable time but now it has been shown that bacteria grow more slowly on surfaces, which is consistent with these arguments.

disadvantages of being attached to particles are very real: the presence of the protist community will increase the threat due to grazing by these forms; further the large size and the rich microbial community on marine snowflakes make them choice food-items; these attached organisms are now available to metazoans, which would not have the capability to feed on individual microbial cells. Finally, hitching a ride on a rapidly sinking particle on its way to the cold organically sparse ocean depths would not seem a strategy that benefits the attached organisms. We should be wary of seeking for the benefits of attachment and be aware that the forces that give rise to attachment, **van der Waal's forces**, are extremely powerful on the microbial scale. Thus the micro-organisms simply may not be able to detach themselves – rather like the drunk on the barroom floor, who is there for no better reason than that he lacks the ability to remove himself.

● The benefits and negative aspects of colonizing particles are complex.

3.6 The Dynamics of Bacterial Growth and its Measurement

The 1/diameter rule (see 3.2.1) developed from the surface to volume relationship to account for the relationship between metabolism and size applies equally to microbial growth and the consequence is that bacteria can have very rapid growth rates. The record is a division every 10 min, the norm for cultures is 20 min to an hour, for natural populations it is nearer a day. Binary division gives rise to exponential growth: 1, 2, 4, 8, 16, etc., thus growth is literally explosive. Simply expressed, the change in cell numbers after 'n' divisions is 2^n. The consequence of this is impressive. An organism that divides at a rate of once every 10 min, would result in $24 \times 60/10 = 144$ divisions after 24 h. From an initial single cell, this would give rise to $2^{144} = 2.2 \times 10^{43}$ cells. They would occupy a space of 1.5×10^{15} km^3, 1000 times the volume of the earth! One has to be honest that the outcome of the calculation is highly dependent upon the growth rates assumed and the time scale considered; nonetheless exponential growth is clearly explosive. Even in the case of more slow-growing bacteria that occur in the sea, we would arrive at the same position after 4 months. This explosive nature of exponential growth is the problem that concerned the Victorian social scientist Malthus. He was concerned with the growth of the human population, but broadly the same rules apply. As with human populations, a variety of factors contrive to prevent the mathematical apocalypse (3.6.2).

● The explosive nature of exponential growth was the problem that concerned the Victorian social scientist Malthus.

3.6.1 The measurement of bacterial growth

In a natural population, where organisms are constantly being grazed or eaten by predators, one cannot, as in a culture, simply measure growth from an increase in numbers; much more subtle methods need to be employed. Two somewhat similar approaches are used to measure bacterial growth.

The first follows the incorporation into DNA of **tritium-labelled thymidine** added to seawater. Subsequent to the incubation, which can be as short as 30 min, the sample is filtered through a fine filter and the DNA extracted, purified and its radioactivity measured. This provides a rate of incorporation of thymidine into DNA. As thymidine represents close to ¼ of the bases in DNA, it is a very simple matter to calculate the quantity of DNA produced. Given the assumption over the total DNA content of a single bacterial cell and the mass of the cell, one can calculate the rate of formation of bacterial biomass. Nevertheless, it is important to note that the validity of the technique is intimately linked to the validity of the assumptions. The second approach follows the incorporation of **tritium-labelled leucine** into protein. The basic approach is much the same: the calculation in this case requires knowledge of the ratio between cell biomass and its leucine content (or the percentage leucine in proteins and the percentage protein in the cell). Again the uncertainties in these assumptions limit the accuracy of the techniques. The two techniques measure fundamentally different properties: the [^{3}H]-thymidine technique is associated with the rate of cell production, the [^{3}H]-leucine technique with the rate of biomass production. Any attempt to compare the two methods requires an accurate knowledge of the biomass of a single bacterial cell, which in a natural population is an extremely difficult property to determine (Ducklow 2000).

Table 3.4 Bacterial and Phytoplankton Production and Growth Rates in the Euphotic Zone of Various Oceanic Areas. (recalculated from Ducklow, 2000).

	Equatorial Zone	Subtropical Gyre	Temperate (N. Atlantic)	Polar (Ross Sea)
Euphotic zone depth (m)	120	140	50	45
Average biomass (as mg C m^{-3})				
Phytoplankton	11	2	20	5
Bacteria	9	4	90	254
Average prodution rate (mg C m^{-3} d^{-1})				
Phytoplankton	2	1	6	0
Bacteria	11	3	22	28
Average growth rate (d^{-1})				
Phytoplankton	0.1	0.1	0.3	0.3
Bacteria	0.7	0.8	0.3	0.1

Table 3.4 contains a summary of data for bacterial biomasses, production and growth rates in the euphotic zone of major zones of the oceans, along with comparable data for the phytoplankton. At low latitudes, bacterial and algal biomasses are comparable, but in cold waters, the bacterial mass may dominate. This may represent a community shift to compensate for reduced rates of bacterial metabolism at low temperatures. Consistent with this, in the warm water of low latitudes, bacteria grow much faster than the phytoplankton. The production rates draw attention to a major contemporary problem in biological oceanography: balancing the books. As a consequence of recycling (3.4.3) organic material can pass through the bacteria more than once, so in principle bacterial production can exceed primary production, this can only occur to a limited extent and the heterotrophic processes in the water column cannot exceed production by large amounts. However, when attempts are made to produce a budget for the oceans, most frequently we end up with a presently unexplained deficit of organic material (see also 3.8). This is currently a matter of active debate and without doubt there are issues of the errors due to temporal and spatial averaging as well as questions over the accuracy of the methods for measuring both photosynthetic and heterotrophic rates of production.

● A major contemporary problem in biological oceanography is balancing the books on autotrophic and heterotrophic processes.

3.6.2 The dynamics of micro-organisms

Given the potential for explosive growth, changes in bacterial numbers through the seasons are surprisingly muted (Fig. 3.16). In the case of

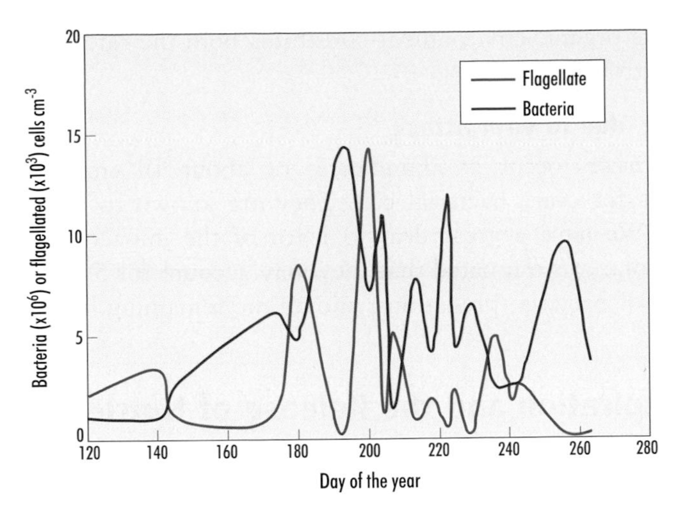

Fig. 3.16 Seasonal changes in the abundance of bacteria and heterotrophic nanoflagellates in a shallow marine fjord. Redrawn from Strom (2000), original material from Anderson and Sorensen (1986).

bacteria, probably three things constrain bacterial growth in the oceanic water column: (1) control by grazing, (2) resource limitation, and (3) mortality.

● Bacterial numbers through the season are controlled by grazing, resource limitation and mortality.

Grazing

Bacteria are grazed by small protozoa, in particular nanoflagellates. The dynamics between this predator–prey relationship and that of the phytoplankton–mesozooplankton coupling are different and a critical factor in the determination of the dynamics of the various plankton communities. In the latter case, the growth rates are very different; the zooplankton reproductive rate is about tenfold slower than that of phytoplankton. This mismatch allows the phytoplankton to bloom before their predators can increase in numbers or size sufficient to graze down the bloom. In the case of the bacteria–protozoan coupling, their growth rates are more comparable (approximately a day) so the bacteria have no opportunity to bloom, at least for any extended period, as their increase in abundance is quickly followed by an increase in the numbers of the protozoans. The latter graze down the bacteria again, giving rise to Lotka–Volterra **limit cycle oscillations** (Fig. 3.16). There is a secondary feature of the control by grazing that is related to prey-size selection. The protozoan predators tend to select the larger bacterial cells preferentially, so the size distribution of the bacterial population, when it is grazed, is shifted down from a modal diameter of 0.75 μm to 0.5 μm, resulting in a threefold reduction in cell volume from 0.22 μm^3 to 0.065 μm^3.

● Protozoan-bacterial predator–prey dynamics lead to Lotka–Volterra limit cycle oscillations.

Resource limitation

The seas are an organically dilute environment. Characteristically the concentration of individual molecules is in the region of 10^{-7} molar (10 grains of sugar per ton of seawater would give a similar concentration).

● The seas are an organically dilute environment.

This dilute organic environment constrains both the rate and extent of bacterial growth.

Mortality due to viral attack

Marine viruses occur at abundances of about $10^7 cm^{-3}$, i.e. about 10 viruses for every bacterial cell. They are known to infect and kill bacteria. We have a great deal to learn of the impact of viruses on marine biota: it is estimated that they may account for 5 to 40% of the mortality of bacteria (predation removes the remaining fraction).

● Virus infection may account for 5–40% of bacterial mortality.

3.7 Respiration and the Release of Nutrients

The ecological context and nutritional circumstances of respiration were dealt with in 3.3, but there was no consideration of rates. The measurement of respiration has proven a major challenge for the biological oceanographer, and there is no easy solution using radioisotopes as with photosynthesis. We are very much dependent upon very careful chemical measurements of *in vitro* oxygen consumption in the dark. The changes are commonly in the order of 0.25% per day, so to measure them with any precision is a daunting analytical task and one has to work at the limits of the analytical methods and often under difficult circumstances at sea. An alternative approach (Packard, 1985) is to extract the respiratory enzymes from bacteria caught on fine filters and assay them. This has a far greater sensitivity but as so often with the non-chemical methods, the price paid for the gain in sensitivity is uncertainty over interpretation. The uncertainty however is no worse than the methods for measuring bacterial growth.

3.7.1 The biochemical basis of respiration

Respiration is the last step in the decomposition process (3.4). At its most fundamental level this involves the transfer of a **proton** (a hydrogen ion, H^+) and an accompanying **electron** (ϵ^-) from a **proton donor** to a **proton acceptor**. The proton gradient created gives rise to the formation of **adenosine triphosphate** (ATP), which is a primary energy currency in biology (Fig. 3.17). In most present-day circumstances, organic material is the principal proton donor and oxygen is the principal proton acceptor, water and carbon dioxide are the end products (Fig. 3.17).

● Respiration fundamentally involves the transfer of a proton and an accompanying electron from a proton donor (e.g. organic material) to a proton acceptor (e.g. oxygen).

Organic material and oxygen are not the only participants in this reaction. When oxygen is depleted a number of other proton acceptors may replace oxygen, the list of which is quite long with 14 compounds identified to date. Some of the proton acceptors are well-known, such as nitrate and sulphate, but others are more obscure and in many

(a) The basic redox couple inherent in respiration

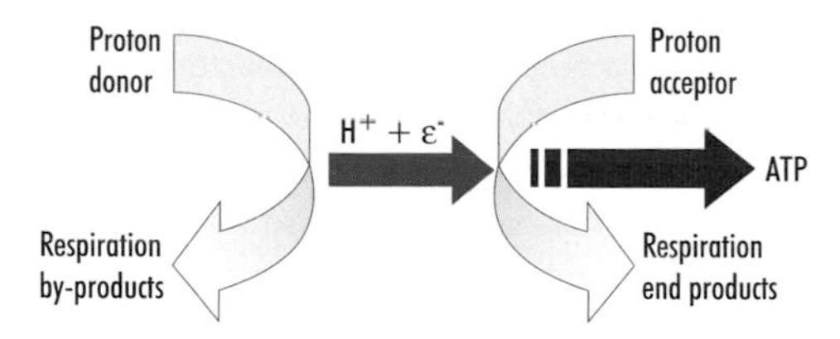

(b) The organic material-oxygen respiratory couple

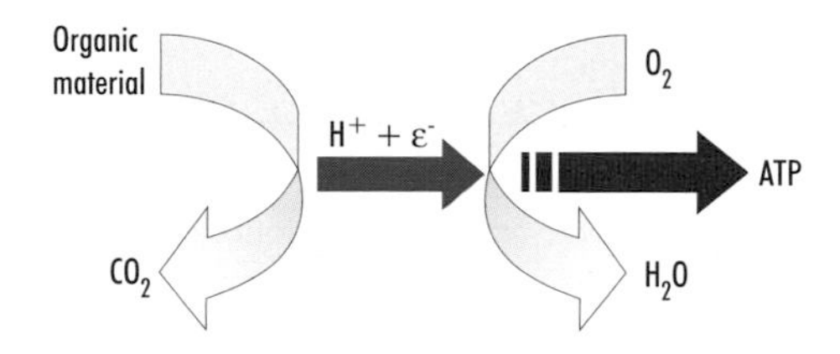

(c) The carbon dioxide-water photosynthetic couple

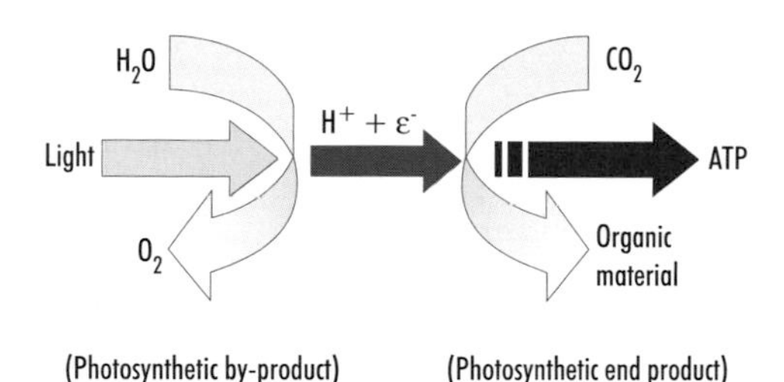

Fig. 3.17 A simple schematic of the flow of protons (H^+) and electrons (ϵ^-) during respiration and photosynthesis. (A) the fundamental flow from a proton donor and a proton acceptor during respiration, (B) the specific example of the respiration of organic material by oxygen, (C) the flow during oxigenic photosynthesis.

respects surprising; for example, uranium (see Box 3.3), perchlorate, and arsenic can also be used by micro-organisms (see King 2005). There is a broad generalization that if a reaction is thermodynamically favourable some micro-organism will exploit it; it is a generalization more commonly respected than broken. As an environment depletes its oxygen then these alternative reactions become important. The proton acceptor and its reduced form (e.g. oxygen and water, nitrate and nitrite, sulphate and sulphide) are known as **redox couples**. There is a particular voltage (the **redox potential**) at which the two halves of the couple are in balance: +0.82 V in the case of the important O_2/H_2O couple, +0.43 V in the case of the nitrate/nitrite couple, −0.22 V in the case of the sulphate/sulphide couple. This determines the sequence in which a particular proton donor comes into play after the oxygen initially present is used

Box 3.3 Bacterial-run nuclear reactor

The utilization of uranium salts by bacteria as electron acceptors gave rise to a remarkable phenomenon; a bacterial run uranium reactor! This sounds the stuff of science fiction but here is an instance where fact is every bit as strange as fiction. About 1800 million years ago in the Precambrian, in the Oklo region of what is now the Gabon in West Africa, the fissile form of uranium U^{235} was concentrated to the extent that the deposits went critical. The likely explanation is that the metabolism of uranium by bacteria concentrated the element to the extent that the nuclear reaction went ahead, although at least one web site attributes this unhesitatingly to an early advanced civilization. Whatever the truth may be, Enrico Fermi and his Chicago team were forestalled by 1.8 billion years in producing a uranium reactor. One would prefer to believe that it was bacteria that beat them to it.

Table 3.5 The basic reactions of the some proton donors and acceptor reactions involved in respiration. $[CH_2O]$ as symbolism for organic material, for simplicity of presentation all compounds are shown in their non-ionized form.

Proton donor		Proton acceptor	Respiration end product(s)		Respiration by-product	Specific bacteria group responsible
Aerobic metabolism						
$[CH_2O]$	+	O_2	$\rightarrow CO_2 + H_2O$			No specific group
Anoxic metabolism – alternative proton acceptors to oxygen						
$[CH_2O]$	+	$2HNO_3$	$\rightarrow CO_2 + H_2O$	+	$2HNO_2$	Denitrifying bacteria
$[CH_2O]$	+	HNO_2	$\rightarrow CO_2$	+	NH_3	Denitrifying bacteria
$[CH_2O]$	+	$\frac{1}{2}H_2SO_4$	$\rightarrow CO_2 + H_2O$	+	$\frac{1}{2}H_2S$	Sulphate reducing bacteria
Chemoautotrophic metabolism – additional proton donors to organic material						
HNO_2	+	$\frac{1}{2}O_2$	$\rightarrow$		HNO_3	Nitrite oxidizing nitrifying bacteria
NH_3	+	O_2	$\rightarrow H_2O$	+	HNO_2	Ammonia oxidizing nitrifying bacteria
H_2S	+	$2O_2$	$\rightarrow$		H_2SO_4	Sulphur bacteria

up and as the environment becomes more reduced, in other words when the redox potential of the environment becomes more negative. Thus, in order to attain maximum benefit, the couples present in the environment with the most positive redox potential are used first: thus we see a sequence in which oxygen is first used up, then nitrate, then nitrite, then later sulphide. The sequence, with the products and reactants is shown in the upper two parts of Table 3.5. The by-products of two of these reactions are toxic, ammonia and sulphide, and thus one of the fundamental aims in water management is to avoid the onset of anoxia, which promotes the production of these products. The proton and electron flow in oxygenic photosynthesis is depicted in Figure 3.17c, and it may be seen that in essence it is the pattern in Figure 3.17b, simply rotated round a diagonal axis.

● There are alternative proton acceptors to oxygen.

As there are alternative proton acceptors to oxygen, so there are alternative proton donors to organic material (see Chapter 2 and Table 3.5). The by-products of the reaction of the alternative proton acceptors are: nitrite in the case of nitrate, ammonia in the case of nitrite, sulphide in the case of sulphate, the list also includes the reduced forms of iron and manganese, methane, and also elemental sulphur; all are potential sources of energy in the presence of oxygen (bottom section of Table 3.5). There is an interesting common property of the micro-organisms carrying out these reactions: they are all **autotrophs**, in that they are unable to use external organic material for their energy and carbon for growth. As they gain their energy from the oxidation of inorganic compounds, they are commonly known as **chemoautotrophs** (although it is better to describe them as **lithoautotrophs**, as the heterotrophs use organic chemicals for their growth; the prefix 'litho' makes clear that inorganic substrates are utilized to provide the energy). This group of organisms is the basis of the food chain in the **deep-sea vents**. These chemoautotrophic bacteria were almost certainly among the earliest free-living forms that developed on our planet and their metabolism we believe reflects the then prevailing environment. Their ancient origins and phylogenetic similarities have resulted in them being classified in a group of their own, the **Archaea**, distinct from the second major group the **Bacteria**; the third group is the **Eukaryotes**, which contains all remaining living organisms: fungi, protozoa, plants, and animals. Among the chemoautotrophs is a group of organisms known as the **nitrifying bacteria** – the reactions they carry out are shown in the lower part of Table 3.5. It should be noted in passing that although the nitrifiers are autotrophs they are placed in the Bacteria, rather than the Archaea, thus an exception to the above generalization. They have been referred to already and the major role they play in the cycling of nitrogen in the terrestrial and marine environment. Indeed every molecule of nitrate used during photosynthesis on the planet will eventually pass through these organisms.

● There are alternative proton donors to organic material.

3.7.2 The ecology of respiration and photosynthesis

When we consider the ecology of respiration (the utilization and decomposition of organic material) it is logical to do this in conjunction with the companion process of photosynthesis, which is responsible for its production. The two processes represent the two halves of the massive cycle of nature, wholly dependent upon one another (Chapter 2). Photosynthesis provides the organic material for respiration, while for its part respiration provides the inorganic nutrients for photosynthesis. To a large extent they represent the yang and yin in the cycle of life, the Tao, the basis of much of Chinese philosophy. Interestingly, essential to

● Photosynthesis provides the organic material for respiration, while for its part respiration provides the inorganic nutrients for photosynthesis.

Box 3.4 Definition of Productivity Terms

There is a set of definitions used in ecology that derive basically from three physiological processes, namely:

(1) Photosynthesis = **P**

(2) Algal respiration = $\mathbf{R_a}$

(3) Whole community respiration = $\mathbf{R_c}$

The primary photosynthetic event is termed **Gross Production = P**
The difference between production and respiration in the algal community is termed **Net Primary Production** (also **Net Photosynthesis)** = $\mathbf{P-R_a}$
The difference between production and respiration within the whole community is termed **Net Community Production** (also **Net Ecosystem Production)** = $\mathbf{P-R_c}$

Net Primary Production tells us how much energy and organic material is available to the heterotrophic community, whereas **Net Community Production** tells us about the balance within the community as a whole, i.e. whether the community is growing or contracting. It is important not to confuse these terms and the use of net production, which is commonly found in the literature and could refer to either is to be avoided.

this understanding is that both the yin and yang have the seeds of the other in itself (these are the dots in the yin–yang symbol), and we see the same in the biogeochemical cycle: photosynthesis produces organic material – the seed for respiration and respiration the inorganic nutrients – the seed for photosynthesis. For a number of reasons the two processes of photosynthesis (P) and respiration (R) are out of phase, so there will be places and times when and where P > R and R > P. The difference between these two processes (P–R) is termed **net community production** (NCP) (see Box 3.4). Unlike photosynthesis and respiration, net community production is not a process in its own right: there is not a dedicated set of enzymes or a single metabolic pathway that gives rise to NCP; rather it is an arithmetic difference. Nevertheless it is arguably the most valuable planktonic rate measurement we can make, and is the best descriptor of the waxing and waning of the population. Net *community* production should be distinguished from the similar term **net *primary* production** (NPP – also referred to as net photosynthesis: see 2.8), which is photosynthesis minus autotrophic respiration. Arithmetically NPP ≥ NCP.

● There are at least five distinct forms of respiration in algae.

Whereas the single structure for respiration shown in Figure 3.17 – with proton acceptors and donors and the transfer of protons and electrons between them – gives a largely satisfactory description of that occurring in most heterotrophic organisms, photosynthetic forms have a number of additional mechanisms of respiration; in all there are at least five distinct forms of respiration in algae (Raven and Beardall 2005). Algal respiration generally does not give rise to the release of inorganic nitrogen or phosphate from the cell (certainly not in large quantities) presumably they are recycled and used for photosynthesis. Two forms of algal respiration occur (the **Mehler reaction** and the **photorespiration** pathway) only or mainly in the light. The Mehler reaction appears to serve to remove excess reductants and oxygen produced at high photosynthetic rates; the photorespiration pathway would appear to serve a similar function and it has been noted earlier (3.4.2) that by-products of this reaction, glycine, serine, and glyoxylic acid, may add to the dissolved organic pool.

3.8 The Seasonal Cycle of Production and Consumption

We now have all the pieces in place to look at the dynamics of the microbial network. In this and the following section we explore the changes that occur with time and in space. The seasonal cycle of temperate waters gives the best example to consider as it encompasses most of the characteristic situations seen elsewhere in the surface oceans.

The biology of the cycle of seasons, embedded in the overall biogeochemical cycle, introduced in Chapter 2 is shown again in Figure 3.18. The cycle involves a **throughput of energy**, entering as visible radiation and departing as heat in the form of long-wave radiation, and a

● The biogeochemical cycle involves a throughput of energy and a cycle of nutrients.

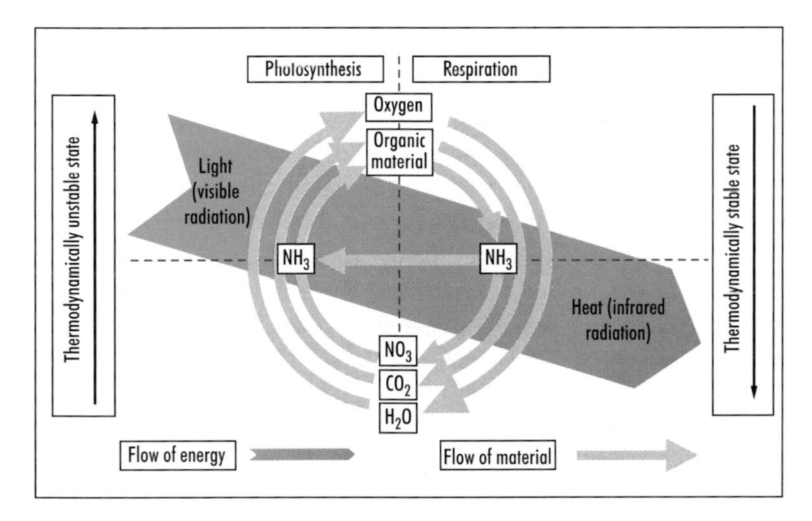

Fig. 3.18 A simple schematic of the principal reactants of the photosynthesis–respiration cycle. The broad, multicoloured arrow depicts the flow of energy through the system, entering as high grade, visible radiation, leaving as low grade energy – heat, i.e. long-wave infrared radiation. The cycle of nitrogen shows the intermediate production of ammonia and the short-circuiting of the nitrogen part of the cycle. The left-hand sector of the diagram represents the photosynthetic process, the right respiration. The food web interactions are not shown. (To simplify the diagram nitrate (NO_3) and ammonia are not shown in their charged form.)

cycle of nutrients. Thus, whereas an atom of say C, N, or P can cycle round indefinitely, perhaps a number of times in one season (the reciprocal of the *f*-ratio, see 2.8.2, gives the number of times a nitrogen atom is cycled per year,) energy passes through the system once only, to be lost eventually as heat to outer space. As it is a cycle, one can begin anywhere, but it is logical to begin at the point of photosynthetic formation of organic material. This event, and the circumstances initiating it, is discussed in Chapter 2. The photosynthetic process involves the absorption of light and the splitting of water to provide the reductants (a proton and an electron) needed for the reduction of carbon dioxide. The reductants are also used to reduce nitrate to ammonium, prior to their incorporation into organic molecules as amino acids. About 20 to 25% of the energy used in photosynthesis can be spent on reducing nitrate to ammonium; small amounts are also used for the reduction of sulphate and other minor elements. Oxygen is appropriately viewed as a toxic waste product of the photosynthetic reaction. Phosphate is also incorporated into organic material at his stage, but unlike C, N, and S there is no change in the oxidation state and for this reason it is not included in Figure 3.18.

● Oxygen is appropriately viewed as a toxic waste product of the photosynthetic reaction.

Some of the formed organic material will be used to fuel algal respiration, but by far the major part will pass into and cycle round the food web (Fig. 3.14). As it flows and cycles through the food web it is respired. The principal respiration products are carbon dioxide, water, and ammonium. To complete the cycle, the ammonium needs to be oxidized back to nitrate by the nitrifying bacteria (the lower right-hand quadrant in Fig. 3.18). However, this does not automatically occur for there is a short circuit, as the algae may use ammonium, indeed for energetic reasons they use it in preference to nitrate. This gives a cycle (the two upper quadrants in Fig. 3.18) within the overall cycle. As we shall see, over the seasons the flows through the four quadrants vary in intensity through the season.

The broad temporal sequence of events in the seasonal cycle of the plankton is shown in Figure 3.19 and the primary driving forces are discussed in Chapter 2. It is convenient to consider the seasonal events as four phases:

Phase I – the initial development of the spring phytoplankton bloom,

Phase II – the demise of the phytoplankton bloom,

Phase III – a mid-summer recycling phase and finally

Phase IV – the regenerative period, lasting until the bloom the following spring.

This sequence ignores the autumn bloom. This is done for the sake of simplicity, the autumn bloom can be considered as a brief re-run of Phases I and II. The relative flows over the seasons of the various biological and chemical pools is shown as a mandala in Figure 3.20.

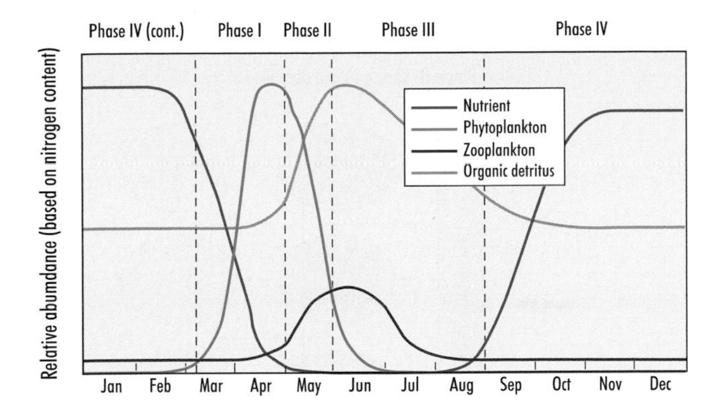

Fig. 3.19 The relative fluctuations of inorganic nitrogen, phytoplankton, zooplankton, and organic detritus (dissolved and particulate organic material) through the season in temperate waters. The phases (I–IV) are described in the text.

3.8.1 Phase I: the initiation and growth of the phytoplankton bloom

The warming of the upper water column, the development of the thermocline, and the consequent shallowing of the mixed layer, along with the concomitant deepening of the compensation and therefore the critical depth, trigger the initiation of the phytoplankton bloom. This results in an initially uncontrolled bust of algal growth (Figs. 3.20a and 3.19). The bloom is characteristically dominated by diatoms, as the high levels of nutrients present favour their growth. Their predators, traditionally thought to be the calanoid copepods, do not have the metabolic capacity to respond to the banquet offered to them; hence the bloom initially is uncontrolled. There will be, as a consequence of the algal growth, some release of dissolved organic detritus and some particulate detritus as a consequence of zooplankton activities. This will result in an increase of the pool of non-living organic matter and consequent bacterial growth. In this phase, the autotrophic phase, the left-hand quadrants of Figure 3.18 dominate (see Fig. 3.22a). As a result of the assimilation of the organic matter released by the algae, the flow of organic material through the microbial sector of the food web increases. Through this phase, the major part of the respiration is associated with the algae; that associated with the heterotrophic population is minor. This is a period of positive net community production, as depicted in Figure 3.21. The algal growth results in a drawdown of the pool of inorganic nutrients and this, coupled with the consequential growth of the zooplankton, eventually results in a slowing down of the rate of phytoplankton growth leading onto to the next phase.

● As a result of the assimilation of the organic matter released by the algae the flow of organic material through the microbial sector of the food web increases.

3.8.2 Phase II: the demise of the spring bloom

The combination of the events above results in the demise of the spring bloom and a progressive growth of the heterotrophic parts of the

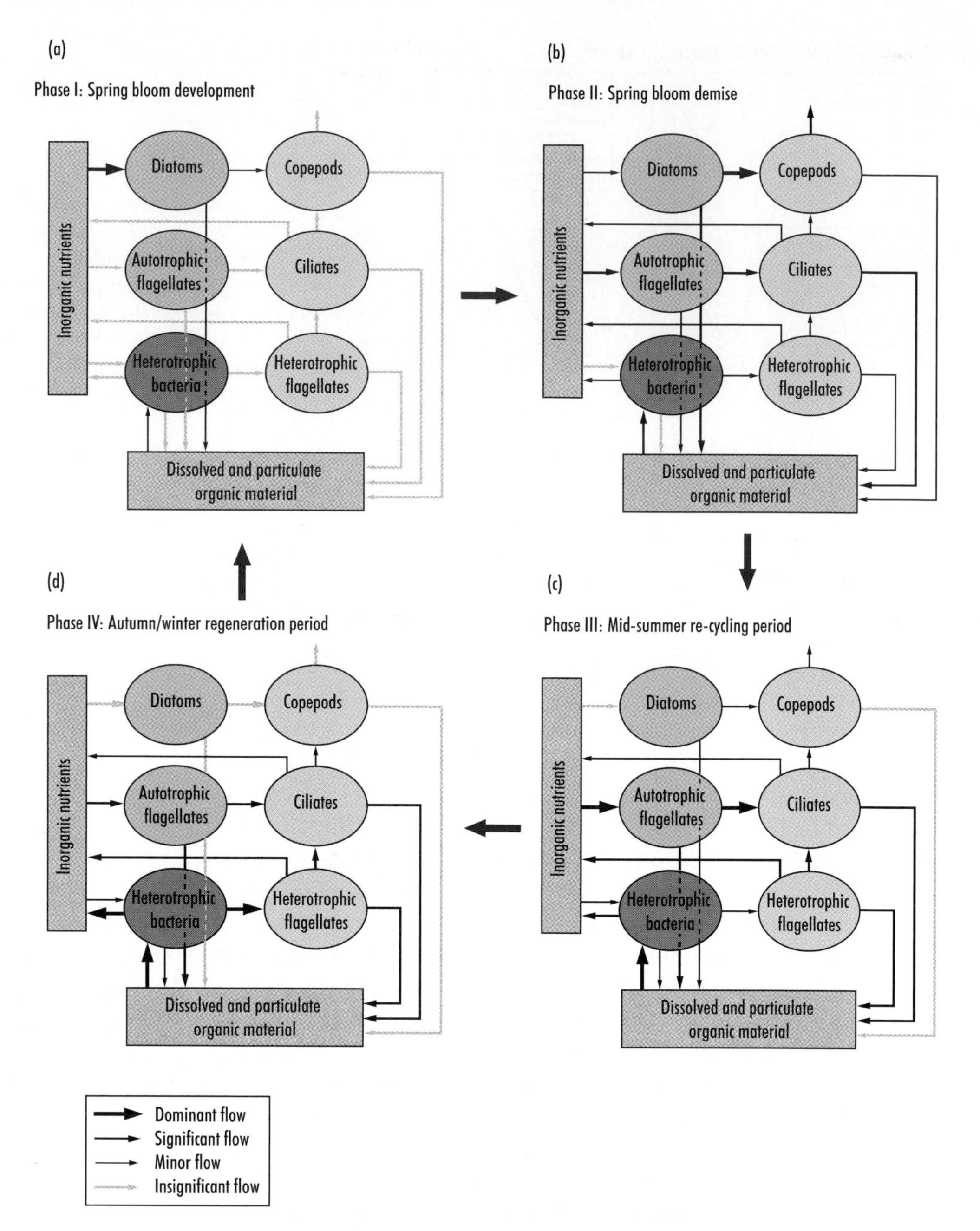

Fig. 3.20 The variations in the flows of material through the microbial network shown in Fig. 3.14 through the four phases of the year shown in Fig. 3.19. The width of the arrows indicates the intensity of the flow.

community, especially the microbial sector, the bacteria, the heterotrophic nanoflagellates, and the microzooplankton (see Fig. 20b). There is, during this period, a change in the balance between production and respiration and a consequent fall in net community production

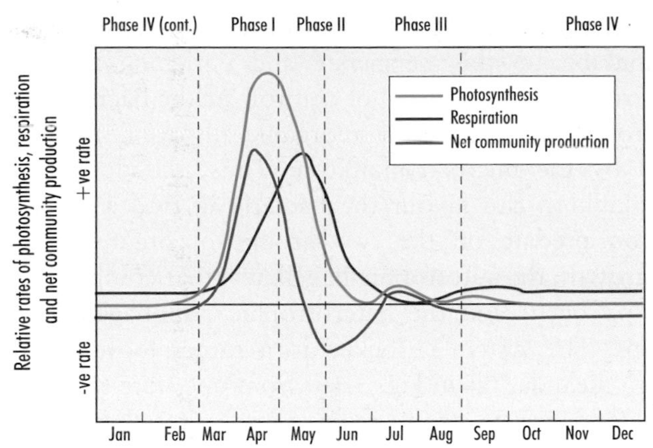

Fig. 3.21 The distribution of respiration and photosynthesis rates through the season. Also shown is net community production (the difference between photosynthesis and respiration).

becoming negative on occasions (Fig. 3.21). Importantly there is a downward shift in size of the primary producers from the diatoms to the autotrophic flagellates. This is driven in part by selective grazing of the larger forms by the zooplankton, but more importantly by the fall in nutrient concentration. The smaller sized algae have a narrower diffusion boundary layer (see 2.7.2) and so are able to out-compete the diatoms for nutrients when the concentrations are low. This drives the whole centre of metabolism downwards in size (Fig. 3.20b) and so the growth in emphasis of the microbial network. In many respects this is a **transitional phase** leading to the establishment of the next phase.

3.8.3 Phase III: the mid-summer recycling period

As the nutrients fall in concentration and the recycled nutrients become more important we enter a phase of very closely coupled metabolism and a food web tightly controlled by grazing. The reason for this is that the organisms that now constitute the main nub of metabolism all have comparable growth rates and so as one group increases in abundance so its predator responds by increased grazing and its abundance increases. Eventually the prey becomes diminished and so the grazing activity of its predator is curtailed. There are two further layers of interaction. First, the growth of the predator gives rise to an increase of its own predator and so its own demise. To give a concrete example: the increase in abundance of the bacteria will give rise to an increase in the abundance of the heterotrophic nanoflagellates, which pin down the bacteria. The subsequent increase in the nanoflagellates gives rise to an increase in the microzooplankton, which will graze down the nanoflagellates, releasing the control on the bacteria, which then can flourish a second time. This gives rise to the sort of perturbations seen in Figure 3.16.

- As the nutrients fall in concentration and the recycled nutrients become more important we enter a phase of very closely coupled metabolism and a food web tightly controlled by grazing.

The second form of interaction derives from the fact that much of the organic detritus the bacteria use (e.g. mucilage) is nitrogen-deficient and therefore nutritionally poor. Bacteria, however, have the facility to

- Much of the organic detritus the bacteria use (e.g. mucilage) is nitrogen-deficient and therefore nutritionally poor.

assimilate inorganic nitrogen and incorporate it into organic material thereby making up this deficiency and enriching the food quality. This gives rise to a further level of control. As the bacteria compete with the autotrophic flagellates for inorganic nitrogen, these in turn are controlled by the microzooplankton. Thus, the development of the microzooplankton can favour the bacteria in two ways as the microzooplankton predate on the two groups of organisms that control bacterial growth; the autotrophic flagellates (that compete with them for inorganic nitrogen) and the heterotrophic nanoflagellates (that graze upon them). The system is locked in a rather loose equilibrium and will tend to meander (as in Fig. 3.16) from one state to the other. There will be a tendency to swing from a net autotrophic state to a net heterotrophic one, thus NCP will shift from positive to negative.

● The system is self-sustaining and in principle could oscillate in this manner indefinitely.

The system, shown as the upper two quadrants in Figure 3.22b, is **self-sustaining** and in principle could cycle and oscillate in this manner indefinitely. It is in fact the basic structure of many **low latitude oligotrophic planktonic communities.** This is a stable situation and will persist until some external force disturbs it sufficiently to break down the controls (see also **alternative stable states and phase shifts** in Chapter 7). Part of this stability derives from the recycling of ammonium.

● There is in the plankton a three-way competition for ammonia.

There is a **three-way competition for ammonium** in the plankton as the planktonic autotrophs, the heterotrophic bacteria and the nitrifying bacteria all compete with one another for ammonium. The competition between the bacteria and the autotrophs is part of the self-sustaining mechanism, however the nitrifiers will produce nitrate and its production would move the system out of this state. The growth of the nitrifying bacteria is inhibited by light, whereas that of the photo-autotrophs is of course light-dependent, and thus light may provide the basis of the switch. In the mid-summer period, when the sun is high, the nitrifiers compete poorly against the micro-algae for ammonium (the short circuit in Fig. 3.22b). As the autumn period is entered and the sun angle decreases, the balance between the nitrifiers and the algae shifts in favour of the nitrifiers and the ammonium produced by the micro-heterotrophs is not recycled but converted to nitrate. Nitrate has to be reduced first before it can be incorporated into organic material, as this is an energy-consuming reaction it is a less satisfactory nitrogen source for the bacteria and to a large extent also the algae. This is probably one of a number of mechanisms that jerk the community out of its tightly coupled structure. Another will be the breakdown of the thermocline and the formation of the autumn bloom, essentially a re-run of Phase I. When the autumn bloom wanes, the circumstances may no longer favour the re-establishment of the structure seen in mid-summer. Thus, for a number of reasons the planktonic community enters the last phase of the seasonal cycle: the heterotrophic phase of regeneration.

(a) Autotrophic phase

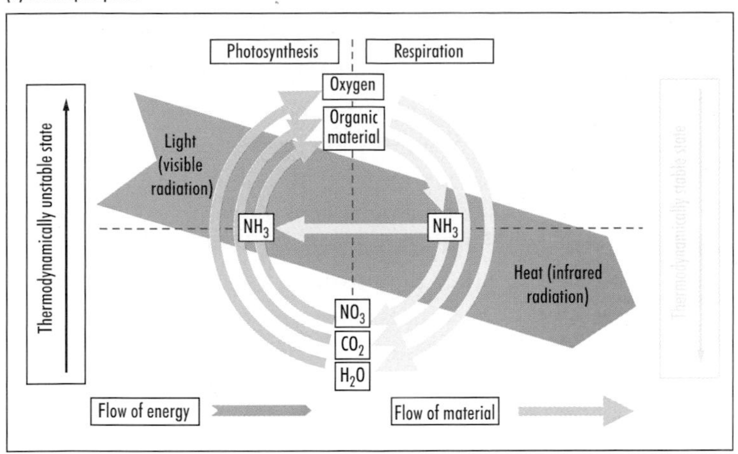

(b) Recycling phase

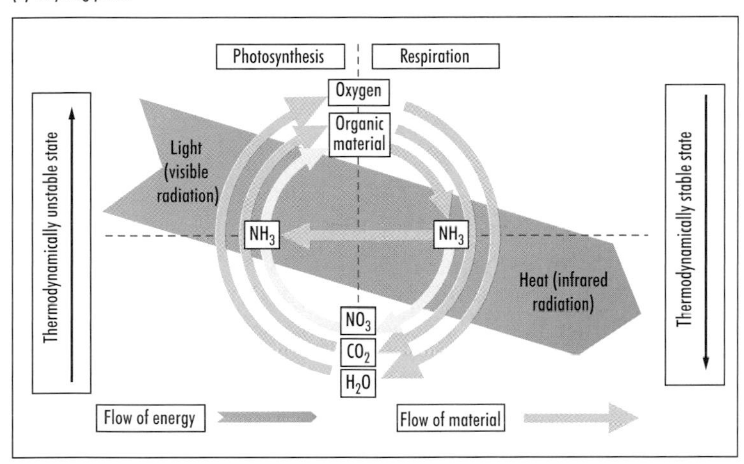

(c) Heterotrophic phase

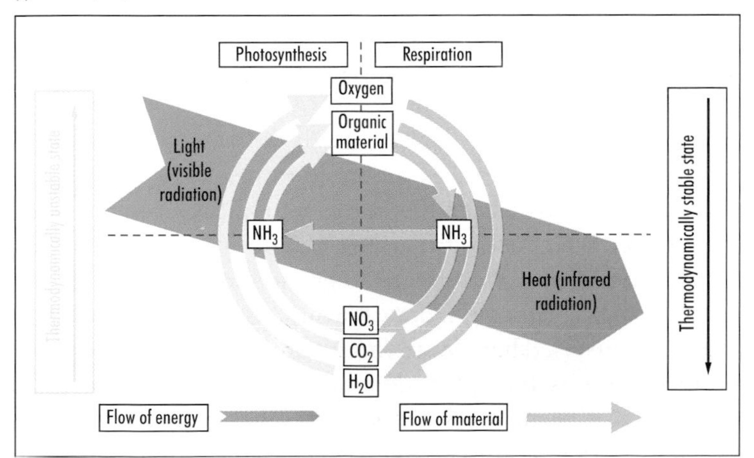

Fig. 3.22 Figure 3.18 redrawn to show the changes in the dominant flows during three phases of the year: the autotrophic phase (Phase I), the recycling phase (Phase III), and the regeneration phase (Phase IV). Phase II is not depicted as it is transitional between Phases I and III.

3.8.4 Phase IV: the final regeneration of nutrients

Whereas, as one enters the autumn period, phytoplankton activity decreases, the particulate and especially the dissolved organic detritus, left as a result of their growth, persists and is a source of food for the microbial heterotrophic community. Phosphate regeneration is essentially a one-step process as it is produced in its final form by respiration.

As we have seen, the case of nitrogen is more complex, as ammonium, the immediate product of respiration, in the presence of oxygen is a potential source of energy that the nitrifiers can utilize. The conversion back to nitrate is a two-step process (see the lower third of Table 3.5), carried out by two different types of nitrifying bacteria and the intermediate, nitrite, shows a temporary rise is concentration in the autumn period prior to its conversion to nitrate. Slowly, through the late autumn and the winter period, the heterotrophs use up the detrital material and **regenerate the nutrients**, eventually reinstating the circumstances that existed at the onset of the spring bloom. This, the upper and lower right-hand quadrants in Figure 3.22c, is a period of predominant heterotrophy and a sustained period of negative net community production (Fig. 3.21).

● Through the late autumn and the winter period, the heterotrophs use up the detrital material and regenerate the nutrients.

Thus, by the time winter gives place to spring, the great cycle of the plankton is complete, ready to start again. The Chinese philosophers would see this last phase of the cycle as the yin taking over from the yang and sowing the seeds for the yang to rise once again, and completing the Tao – an ancient understanding very much in tune with contemporary ecological thinking – or perhaps the logic is more appropriate the other way round, we now understand the 3000-year-old teaching of the Taoists!

So far we have dealt with the processes that occur within only a small proportion of the oceans. Hence finally we need to consider what happens in the remaining 95% of the oceans outside the euphotic zone.

3.9 The Distribution of Respiration in the Global Ocean

Famously it was Edward Forbes, in many respects the founder of biological oceanography as a scientific study, who, in 1843, put forward the **azoic hypothesis**, which stated that below 300 fathoms (550 m) no life existed. The hypothesis was refuted beyond doubt when in 1860 a submarine cable, brought up from 1830 m, was found to be encrusted with living organisms. There is still however a tendency to regard great areas of the ocean as biological deserts (Chapter 8). This particularly applies to the *c*.4000 m of water that lies between the base of the euphotic zone and the deep-sea sediments, which is often seen as merely a transit zone. As there is no photosynthesis beyond about 200 m, metabolism at these depths must be supported primarily by **transport of organic material out of the euphotic zone**, and marine snow (3.4.2) plays an important role in this transfer. The organic content of these settling particles is depleted during their descent so we may expect a general decrease in activity with time and therefore depth. This is indeed what occurs (Fig. 3.23).

● Beyond about 200 m, metabolism must be supported primarily by transport of organic material from the euphotic zone.

We conventionally divide the oceans into five depth zones: the **epipelagic zone**, from the surface to 150 m, for all intents and purposes synonymous with the euphotic zone; the **mesopelagic zone** (150 to 1000 m), the **bathypelagic zone** (1000 to 4000 m); and the **abyssopelagic** and the **hadal** zones (4000 to 6000 m and 6000 to 10 000 m respectively) (see Chapter 8). We have no information of the rates of respiration in these two latter zones but we are beginning to build up a picture of the upper three zones (Robinson and Williams 2005; Arístegui et al. 2005). A summary of rates for these three oceanic zones, the oceanic sediments, as well as the coastal zone is given in Table 3.6. Not surprisingly, close to 90% of metabolism occurs in the epipelagic zone of the open ocean, of the 10% or so that passes out of the epipelagic zone, all but a small fraction is respired away during its transit to the sediments. The amount accumulating in the sediments is minute by comparison: 1 and 12 Tmol C^{-1}, in the oceanic and coastal sediments respectively; only 0.1% of the total oceanic carbon flux. As one moves from the surface into the deep water, there appears to be a shift in the balance of metabolism between the micro-organisms and the larger organisms. The present estimates are that in the euphotic zone, the respiration

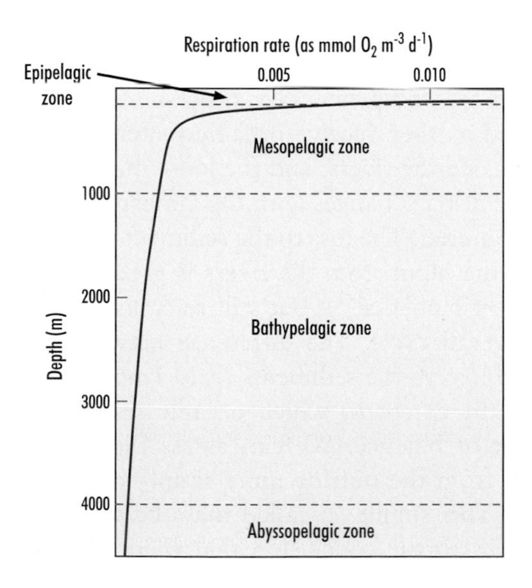

Fig. 3.23 The vertical distribution of respiration in the dark ocean. Redrawn from the equation given in Arístegui et al. (2005).

Table 3.6 Estimates of respiration for major zones of the oceans. Data from del Giorgio and Williams (2005).

Depth zone	Depth range (m)	Global annual respiration ($T^{\dagger}mol\,C\,a^{-1}$)		Contribution to open ocean respiration* (%)
		Coastal zone	Open ocean	
Epipelagic	0–150	1200	9000–12 000	88
Mesopelagic	150–1000	–	600–1400	8.5
Bathypelagic	1000–4000	–	160–230	2
Sediments		12	130–180	1.5
Total		1200	10 000–14 000	

† T = tera = 10^{12}.
* numbers rounded off.

associated with the metazoans as a whole is about 10% of the heterotrophic micro-organisms, whereas in the mesopelagic and bathypelagic zones the fraction is nearer 30%. This perhaps at first is surprising as these zones would seem to the zone of the detritivores, until one realizes that, unlike the micro-organisms, the metazoans (the fish, copepods, and the euphausids) are highly mobile and are not confined to this zone and undertake extensive vertical migrations, in many cases on a diel frequency (Chapter 8), so they are able to visit the rich feeding areas of the upper ocean during their nightly excursions. As they will defecate and lose organic material by other routes during their time in deep water, the **zooplankton are a vector for transport of organic material** from the surface to depths, thus supplementing the passive rain of organic particles.

● Zooplankton are a vector for transport of organic material.

Finally, we should take stock of our two assessments of the **overall oceanic carbon cycle**: that derived from estimates of photosynthetic production and the other from the complementary process of respiration. The oceans are not closed as they receive organic material from land and freshwater ecosystems, via the rivers, and the loose material to the marine sediments. There are also exchanges with the atmosphere but they are very small and can be ignored. The loss to the sediments is small compared to overall turnover, the input from the rivers is greater (current estimates give a figure of 34 Tmol C a^{-1}), but still remains a small component (*c.*0.3%) of the overall cycle. The difference between the input along with the rivers and loss to the sediments (*c.*20 Tmol C a^{-1}), i.e. the net supply, represents the extent to which oceanic respiration and photosynthesis can be out of balance. We can assess the overall gains and losses to the oceans from the outside quite simply and with more than adequate accuracy. This simple so-called **mass balance** calculation (see Table 3.7) leads us to the conclusion that there is a net input of about 20 Tmol C a^{-1} to the oceans. So if the oceanic system is in some form of equilibrium, as we believe it to be, then respiration will exceed photosynthesis by this amount, as this net import of material, along with marine photosynthesis, has to be consumed within the oceanic system. Thus, **the oceans are most probably net heterotrophic** with respiration exceeding photosynthesis by about 0.2%.

What we find, when we compare the consensus estimates of oceanic photosynthetic production, given in Chapter 2, section 2.1.2, of 50 to 60 $PgCa^{-1}$ (4000 to 5000 Tmol C a^{-1}) with the sum of the respiration estimated given in Table 3.6 (10 000 to 15 000 Tmol C a^{-1}), a very different story: a massive difference between estimates of photosynthesis and respiration. To some small extent we are not comparing like with like, as the estimate of respiration (the total removal of organic material) should be compared with the total production of organic material (gross production) whereas the estimate of production from ^{14}C technique is of net primary production (i.e. gross photosynthesis minus autotrophic respiration). Thus we need to add on the material lost by autotrophic respiration to the net primary production estimate in order to derive an estimate of gross production. We can make a fairly safe assumption that this form of respiratory loss is no more than 40% of net primary production so we need to add this amount on to the figures from 4000 to 5000 Tmol C a^{-1}, to give 5000 to 7000 Tmol C a^{-1} (see Table 3.7). This still, however, leaves a deficit of almost a factor of two. There is almost certainly no overlooked source of carbon to the oceans of a scale that would bridge the gap, so we have to acknowledge massive errors in one, or more likely both, of our global estimates of the processes of oceanic production and respiration. The basis of this discrepancy is currently a matter of active debate among biological oceanographers and biogeochemists.

Table 3.7 Budget sheet for the organic fluxes into, out of, and within the oceans (taken from del Giorgio and Williams, 2005, Chapter 14). All fluxes are in T mol Ca^{-1} ($T = 10^{12}$), †Corrected for algal respiration—see text.

Sources and sinks	Inputs	Outputs
	External	
Rivers	34	
Sediments		14
	Internal	
Photosynthesis	3000–5000	
Photosynthesis (Corrected†)	5000–7000	
Respiration		10 000–15 000

CHAPTER SUMMARY

- The small size and large surface to volume ratio of micro-organisms endows them with extraordinary high metabolic rates.
- Due to the high organic losses at each trophic step, marine food chains transfer a minute amount of primary production to the top predators.
- Ecosystems consequently are highly efficient at recycling organic matter and nutrients.
- Most of the metabolism in the ocean plankton is associated with single-celled organisms.
- The utilization of 'waste' organic detritus by the micro-organisms provides the basis of a second food chain that originates from marine bacteria.
- The seasonal cycle of the trophic metabolism may be considered in four phases:

  Phase I – the initial development of the spring phytoplankton bloom.
  Phase II – the demise of the phytoplankton bloom.
  Phase III – a mid-summer recycling phase.
  Phase IV – the regenerative period, lasting until the bloom the following spring.

FURTHER READING

The 1974 paper by Pomeroy is seminal. It described what at that time was a radical view of the marine food web, turning established, and entrenched, ideas on their heads. It is very readable and still very relevant and it is much recommended reading.

Chapter 2 by the Sherrs in the book 'Microbial Ecology of the Oceans' (Sherr and Sherr 2000) gives a very informed and valuable account of the various planktonic micro-organisms, and is an advanced account of out knowledge of microbial processes in the sea. Many of the other chapters will provide valuable specialist information and accounts.

'Biogeochemistry of Marine Dissolved Organic Matter' by D. A. Hansell and C. A. Carlson gives an up-to-date account of our knowledge of dissolved organic material in the oceans. The chapter by Benner describes our current understanding of the chemical composition of dissolved organic material, and the concluding chapter by Hansell is a valuable account of the biogeochemistry of dissolved organic material.

The book on 'Respiration in Aquatic Ecosystems', edited by del Giorgio and Williams, contains chapters on the process of respiration in coastal and oceanic waters (Robinson and Williams), the marine sediments (Middelburg et al.) and the dark ocean (Arístegui et al.). In Chapter 14, del Giorgio and Williams summarize our present understanding of the scale and role of respiration in aquatic ecosystems as a whole.

A valuable account of the structure of marine microbial food webs is given in the paper by Legendre and Rassoulzadegan (1995). They offer a slightly different perspective to the one given is this chapter. Most valuable is that they review the supporting evidence.

A comprehensive review of the role of viruses in the oceans is given by Suttle (2005).

Increasing genomic and molecular techniques are being used at the forfront of biological oceanography and these up to date commentries are given by Delong & Karl (2005) and Giovannoni & Sting (2005).

- del Giorgio, P. A. and P. J. le B. Williams 2005. The Global Significance of Respiration in Aquatic Ecosystems: From Single Cells to the Biosphere. In P. A. del Giorgio and P. J. le B. Williams (eds) *Respiration in Aquatic Ecosystems*, Oxford University Press, Oxford, pp. 267–303.
- DeLong, E. F. and D. M. Karl 2005. Genomic perspectives in microbial oceanography. *Nature*, 437: 336–342.
- Giovannoni, S. J. and U. Sting 2005. Molecular diversity and ecology of microbial plankton. *Nature*, 437: 343–348.
- Hansell, D. A. and C. A. Carlson 2002. *Biogeochemistry of Marine Dissolved Organic Matter*. Academic Press, Amsterdam.
- Legendre, L. and F. Rassoulzadegan 1995. Plankton and nutrient dynamics in marine waters. *Ophelia* 41: 153–72.
- Pomeroy, L. R. 1974. The ocean's food web, a changing paradigm. *Bioscience* 24: 499–504.
- Sherr, E. and B. Sherr 2000. Marine Microbes: an overview. In D. L. Kirchman (ed.) *Microbial Ecology of the Oceans*, John Wiley and Sons, Inc. New York, pp. 13–46.
- Suttle, C. A. 2005. Viruses in the sea. *Nature*, 437: 356–361.

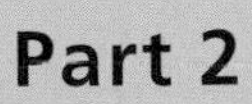

Part 2

Systems

Chapter 4

Estuarine Ecology

CHAPTER SUMMARY

Estuaries represent the great transition between freshwater and marine biomes, and as such are influenced by both aquatic realms, yet they have a distinct, exceptionally variable, and fascinating environment and ecology of their own. The perception of estuaries as low diversity, muddy and inhospitable places is misguided and stems from a narrow appreciation of the estuarine system. Due to the incredibly varied conditions present in estuaries they are inhabited by a wide range of invertebrates, often in very high numbers, and are vital nursery systems for fish and feeding grounds for birds. They have also been referred to as the most degraded habitat type on earth, so are perhaps the finest example for studying the conflicts between ecology, conservation and human use. Such study, however, requires a broad knowledge of the ecosystem of an estuary and how it functions as a whole.

4.1 Introduction

4.1.1 What is an estuary?

An estuary is commonly thought of as the place where a river meets the sea, but this interaction of the two major aquatic realms results in a wide variety of systems of which an estuary is just one type. However, while most people would not have trouble picturing what an estuary looks like, providing a definition of what exactly an estuary is has proved exceptionally difficult. Over 40 definitions of an estuary have so far been attempted, using a range of criteria such as dilution of seawater, tidal regime, and geomorphology. Salinity levels traditionally have been used to define estuaries, but have perhaps led to a focus on the role of salinity at the expense of other environmental and biological variables, as will be explored later. Additionally, such definitions can produce anomalies, an extreme example being the Amazon estuary. In the Amazon estuary, tidal influence extends 735 km from the estuary mouth, but due to the vast river flow there is no significant seawater penetration into the

● An estuary is an inlet of the sea reaching into a river valley as far as the upper limit of tidal rise, usually divisible into three sectors: (a) lower, free connection with the open sea; (b) middle, subject to strong salt and freshwater mixing; (c) upper, characterized by fresh water but subject to daily tidal action.

estuary, while mixing of river and seawater is still evident 1000 km seaward of the river mouth. Tidal influence, however, is extremely important in defining the difference between, for example, deltas and estuaries, and consequently the ecology of these systems. The definition in the margin seems to be the most suitable for the global study of estuarine ecology and highlights how potentially limited much of the research on estuaries has been. The vast majority of work has been concentrated in the middle regions of temperate systems, resulting in a rather narrow perception of estuarine ecology that pervades their study.

Defining the upper and lower limits of estuaries can be difficult, particularly if specific levels of salinity are used as a guide. In unmanaged estuaries, from our definition the upper limit of an estuary can be considered to be the limit of daily tidal influence, which may be truncated in urban estuaries by the construction of barriers such as weirs. In this case, the upper limit of the estuary will always be fixed spatially regardless of river flow or tidal incursion. The outer limits of estuaries are much more difficult to define. Large rivers produce a plume of reduced salinity water far out into the open sea. Hence, fixing the outer limit of the estuary as the location where the dilution of seawater is no longer measurable can be unhelpful. In practice, the boundaries of most estuaries tend to be subjectively defined by coastal morphology or management criteria (if an agency only has jurisdiction to 3 nautical miles from the coastline this is often used as a cut-off point). What is clear is that an estuary is not an isolated entity and any consideration of estuarine ecology needs to place the system in a continuum from the river to the sea. Both of these major aquatic biomes have strong and controlling influences over the ecology of estuaries, as shall be seen in this chapter.

● Estuaries should be treated as part of a continuum from river to sea and not as an isolated system.

4.1.2 Evolution of estuaries

Estuaries are very different from many other marine systems in that they are ephemeral over geological time, particularly in temperate regions where they have been successively influenced by the process of glaciation. They are also areas of high sediment deposition, so through time can demonstrate a degree of infilling and succession. At such timescales, they are therefore more similar to lakes than other marine systems. At the height of the last glaciation (17 000 BP) sea level was over 130 m lower than it is now (Fairbridge 1980), the retreat of the ice sheet resulted in a comparatively rapid elevation of sea level (Box 4.1). This initiated flooding of the exposed continental shelf (for example the North Sea which was either dry or a freshwater lagoon) and eventually

● BP = Before Present.

Box 4.1 Past Sea Level Rise

There is much current concern about the consequences of global warming induced sea level rise (see Chapter 14), but there have been dramatic changes in global sea level during the last 1.8 million years that have moulded many of the marine systems we have today, such as coral reefs and in particular estuaries. Many temperate estuaries were completely covered with ice only 17 000 years ago, while most others did not exist in their current form due to sea levels being close to the edge of the continental shelf. Consequently, estuaries have successively appeared and disappeared as part of coastal systems. Following the retreat of the last (Weichselian) glaciation during the Holocene, sea levels rose rapidly – around 20 m in 2000 years as the ice melted and changed the face of earth's coast. Around 6000 years ago sea level was more similar to current levels, but there have been minor sea level variations over this time, which can be detected, for example, in estuarine saltmarsh sediments.

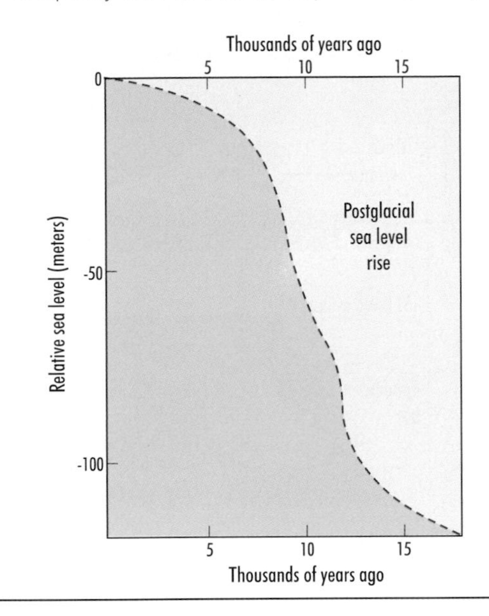

the inundation of river valleys by 6000 BP when modern estuaries were created. Consequently, the majority of the world's estuaries are exceptionally new systems in geological and evolutionary terms, which has significant consequences for their colonization, diversity and ecology.

● Estuaries are very new systems, most having only existed for the past 6000 years.

4.1.3 Classification of estuaries

A problem contributing to the difficulty in defining estuaries is that globally there are a wide range of estuary types, demonstrating large differences in geomorphology, tidal range and salinity characteristics. Consequently, a range of classification schemes have been devised in order to provide a comparative framework for the study of estuarine systems.

A popular way of classifying estuaries is to use their topography (Fig. 4.1), which has arisen in general from their method of formation following post-glacial sea level rise. Estuary Types 1–3 all result from the

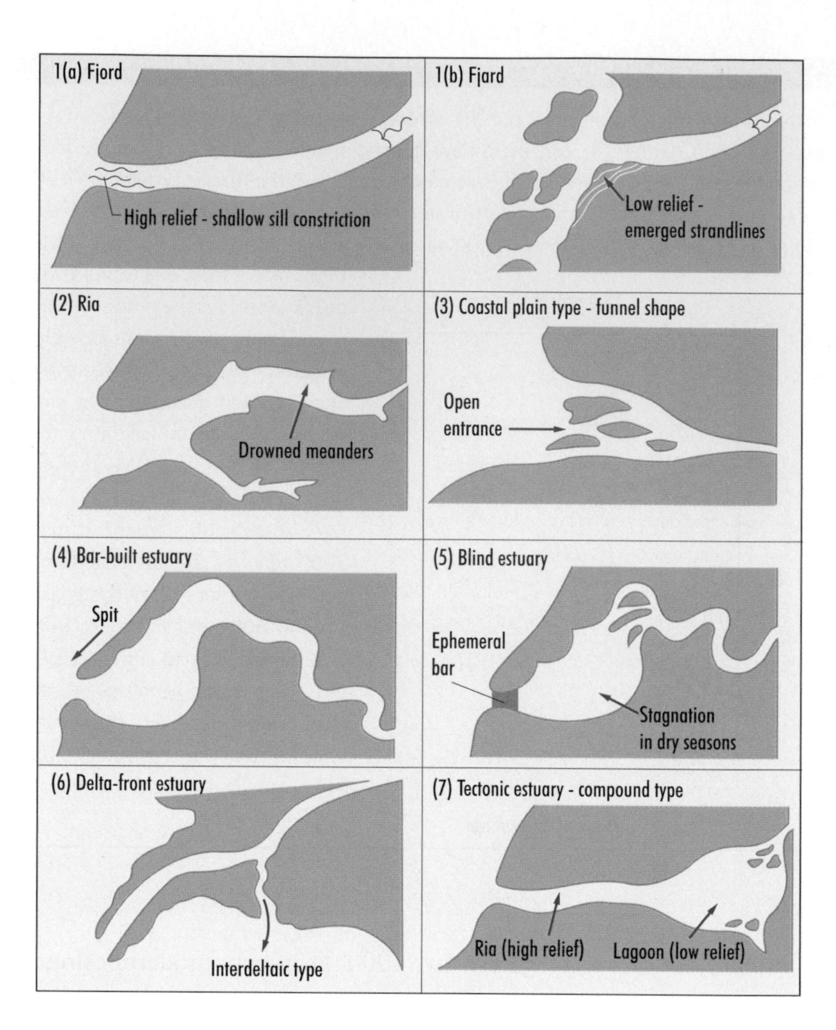

Fig. 4.1 Diagrams of the basic estuarine types as classified using their topography.

flooding of river valleys following **eustatic** sea level rise, the differences among them arise from the topographic relief surrounding the estuary. This ranges from high mountains and deep sided estuaries formed from glacial U-shaped valleys (fjords, sea lochs, Fig. 4.1(1)) through moderate relief estuaries with a V-shaped valley (rias, Fig. 4.1(2)) to funnel-shaped estuaries spreading through low-lying land (coastal plain estuaries, Fig. 4.1(3)). Fjords tend to be very deep, but have a shallow sill at their mouth formed from glacial deposits. This limits the exchange of seawater and thus circulation within the fjord. Consequently, despite their spectacular appearance, fjords can have quite anoxic basins and few organisms living within their depths. Estuaries can also be created by offshore deposits, forming barriers across bays and inlets into which rivers flow. These barriers, which can be joined to the coast (longshore) or out to sea (offshore) restrict the flow of seawater, allowing the build up of diluted water in the resulting bay and form an estuary. These are known as bar-built estuaries (Fig. 4.1(4)) and, as in the case of the south-east of the United States, can be extensive features of the coast. Further deposition can result in the estuary being cut off in dry seasons (blind estuary, Fig. 4.1(5)). Some river systems deposit material at their mouth, again restricting tidal flow (delta-front estuary, Fig. 4.1(6)). Finally, estuaries can be created by a completely different mechanism: isostatic variation, in areas of tectonic activity (Fig. 4.1(7)). The classic example of land sinking, flooding the valley is San Francisco Bay. The San Andreas fault-line runs adjacent to San Francisco, with a history of tectonic activity. A past large earth movement lowered the land seawards of the fault, enabling the Pacific Ocean to spill over the entrance (where the Golden Gate bridge is located) and into the basin we call San Francisco Bay.

● Sea level change – change in sea level can be either *eustatic*, due to variations in volume of the ocean or *isostatic* due to variations in the level of the land (Chapter 7).

While such a classification on topography is perhaps the most obvious method of defining estuaries, it has little real ecological relevance in terms of the conditions provided for associated organisms except on a broad biogeographical scale. For example fjords tend to be boreal whereas blind estuaries are more prevalent in tropical regions. Estuaries have therefore also been classified in terms of tidal range (Box 4.2), a factor that has huge consequences for the dynamics and ecology of estuaries and which, for example, makes them separate from brackish water seas. Globally, estuaries demonstrate a wider range of tidal amplitude than other marine systems, ranging from less than a metre in many tropical systems (e.g. Vellar estuary, India) to over 16 m in the Bay of Fundy, Nova Scotia, Canada (Fig. 4.2). The scale of water movement into estuarine systems can be huge. The tidal prism of the Bay of Fundy is estimated to be 100 billion tonnes! Four categories of estuary therefore have been suggested depending on tidal range: microtidal (<2 m range), mesotidal (2–4 m), macrotidal (4–6 m), and hypertidal (>6 m). The tidal range in estuaries can be artificially enhanced by urbanization through a combination of encroachment and flood defence. A classic case is the Thames estuary in central London, where the channel has been narrowed through centuries of construction. The tidal range at London Bridge is now over 10 m.

● The difference in the volume of water in an estuary between high and low tide is known as the tidal prism.

Tidal movement is therefore a key indicative descriptor of estuaries. However, tides have a greater importance by varying a range of physico-chemical parameters that in turn affect organism function, survival, and distribution. While organisms can adapt well to constant extremes, such as temperature, acute environmental variability is exceptionally stressful. Due to this high level of variability, estuaries are perhaps the most naturally stressful of all aquatic systems. Several key environmental parameters vary spatially and temporally in estuaries and thus influence the organisms that inhabit them.

● The influence of tides in estuaries causes high variability in physico-chemical parameters, resulting in extremely stressful conditions.

Box 4.2 Explaining the tides (a very simplified starting point . . . imagine a world with no land mass)

The daily movement of the tide is caused by the gravitational pull of the moon. At any one point in time, water is pulled towards the moon forming a bulge that relates to local high tide. Areas of the earth perpendicular to the line of the moon will experience low tide. As the earth rotates, the bulge of water appears to move over the earth surface, where in practice the earth is moving below the bulge. High tide is therefore really one very large wave moving over the earth's surface.

The size of high and low tide varies over a range of timescales. The most familiar is the monthly cycle of spring and neap tides. As the moon orbits the earth, its position relative to the sun changes (resulting in the phases of the moon). When the moon and sun are in alignment, their gravitational pull is complementary, resulting in higher raising forces and therefore greater high tides and subsequent lower low tides (see figure). When moon and sun are perpendicular to the earth, the pull of the sun counteracts, to a small degree, the gravitational force of the moon resulting in low amplitude (neap) tides. In addition to the spring tide cycle, there are much longer tidal patterns. The moon's orbit is not consistent, so at times it moves closer to the earth, resulting in greater gravitational pull. The relative distances and positions of moon and sun result in long-term tidal cycles, such as the 18.6 year lunar nodal tide (reflected in tree rings, for example) up to an 1800 year cycle that has a strong influence on global climate (Keeling & Whorf 2000).

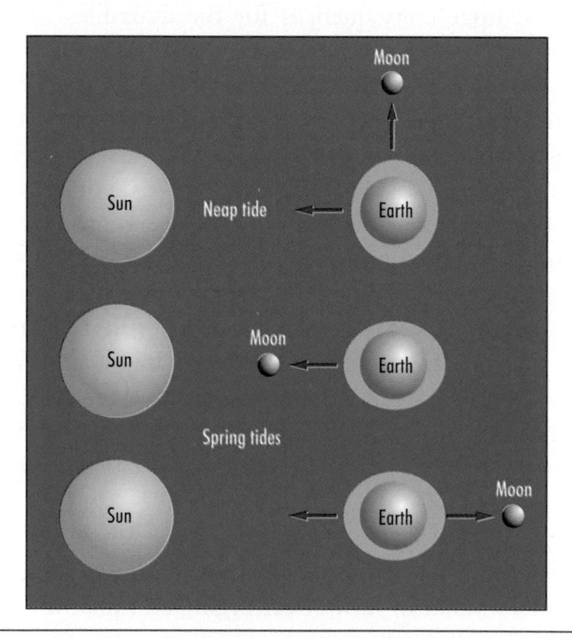

4.1.4 Important environmental variables

Salinity

Classically, salinity has been considered to be the dominant variable influencing the survival and distribution of estuarine organisms due,

Fig. 4.2 Extreme tidal range in the Bay of Fundy, Nova Scotia. The two pictures were taken 4 hours apart, a student is circled at the water's edge in the second photo. (Photographs: Martin Attrill.)

perhaps, to it being the most obvious feature that defines an estuary. While the role of salinity may have been overplayed in the past, it is still one of the most important factors that affects organisms and demonstrates a wide variation within estuaries. This variability has two components. First, there is spatial, longitudinal variability, salinity ranging from 0 at the river end to around 34 at the mouth. In tropical estuaries, evaporation can cause salinity levels to exceed those of seawater (see 4.5). However, the most important aspect of salinity variation that defines estuaries is that salinity at any one point varies considerably over time. With every tidal cycle, an organism could potentially be exposed to near fresh water at low tide and near marine conditions at high tide. In the mid-Ythan estuary, Scotland, salinity ranges from 3 to 29 over a tidal cycle.

● Salinity is now presented as a dimensionless measure known as Practical Salinity Units (PSU). It is unitless when presented, i.e. just the number.

However, such dramatic variation between low and high tide salinities is only indicative of the mid-estuary area, salinity range decreasing considerably towards both ends of the estuary (Attrill 2002, Fig. 4.3). The salinity of water trapped in the sediment of intertidal areas (**interstitial water**) is much less variable over a tidal cycle (<2 in the Ythan), the sediment filling with heavier seawater, the lighter fresh water flowing over the surface of the sediment. In addition, many intertidal areas will not experience low salinity water as they are exposed by the receding tide. The salinity of interstitial water in subtidal sediments appears to be more similar to that of the overlying water.

● Salinity variation can be very high over a tidal cycle. However, this classic feature of estuaries is only true of mid-estuary water. Salinity range is much lower in the sediment and in other parts of the estuary.

Salinity levels can also vary seasonally due to changes in freshwater flow. This aspect of temporal salinity variation in estuaries has been comparatively overlooked, but can have dramatic implications for the ecology of estuaries (see 4.2.3). However, freshwater flow rates affect the current speed of river water entering the estuary and the way fresh water reacts with saltwater entering with the rising tide, and the topography of

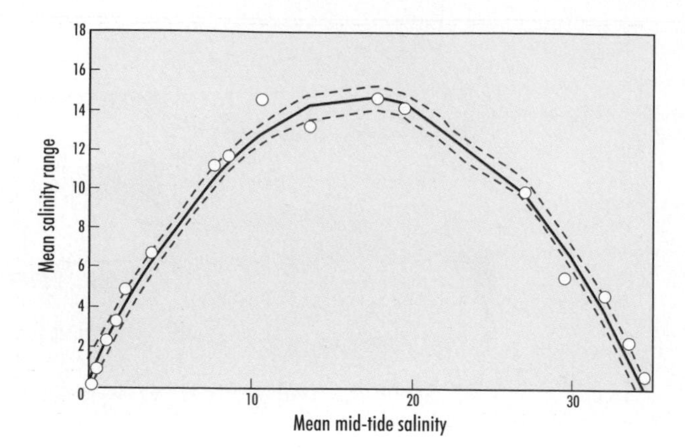

Fig. 4.3 Salinity range along the length of an estuary, in this case the Thames.

● River and tidal currents are strongest in mid-channel – the friction of estuary banks slows down the water speed.

the estuary, can produce estuaries of different vertical salinity structure. This has allowed an alternative classification of estuaries based on their **salinity profile**. A variety of types have been defined, but basically depend on how well mixed the two water sources become and these can vary over tidal cycles and with seasonal input of fresh water (Fig. 4.4).

While salinity variation is an inherent feature of estuaries, it is not necessarily the only important stressor influencing organisms. Three other factors can potentially have equally significant roles on the ecology of estuaries.

Sediment

Perhaps the most evocative feature of estuaries are the expansive intertidal mudflats composed of soft, fine grey silt, extremely difficult to walk over and giving off a distinctive sulphurous odour when the sediment is disturbed. This fine, mobile sediment is a highly difficult habitat to colonize. There are no points of attachment for macroalgae or sessile organisms, while organisms intolerant of mechanical disturbance would not be able to deal with the mobility of the substratum. Additionally, the fine particles can easily clog delicate feeding and respiratory structures and the sediment makes locomotion difficult. It has therefore been suggested that organisms that thrive in estuarine mudflats need to be more adapted to the sediment than to the salinity variation (Barnes 1989). However, despite being symbolic of estuaries, under normal conditions large fine-grained mudflats only tend to form in the middle reaches of the estuary and are comprised mainly of marine-derived material, even though they may be many kilometres from the sea. The deposition of sediment in an estuary is mainly due to comparative tidal and river current speeds (Box 4.3), although the highest suspended sediment concentration tends to be further up the estuary (the **turbidity maximum**) due to a dynamic interaction between fresh and saline water

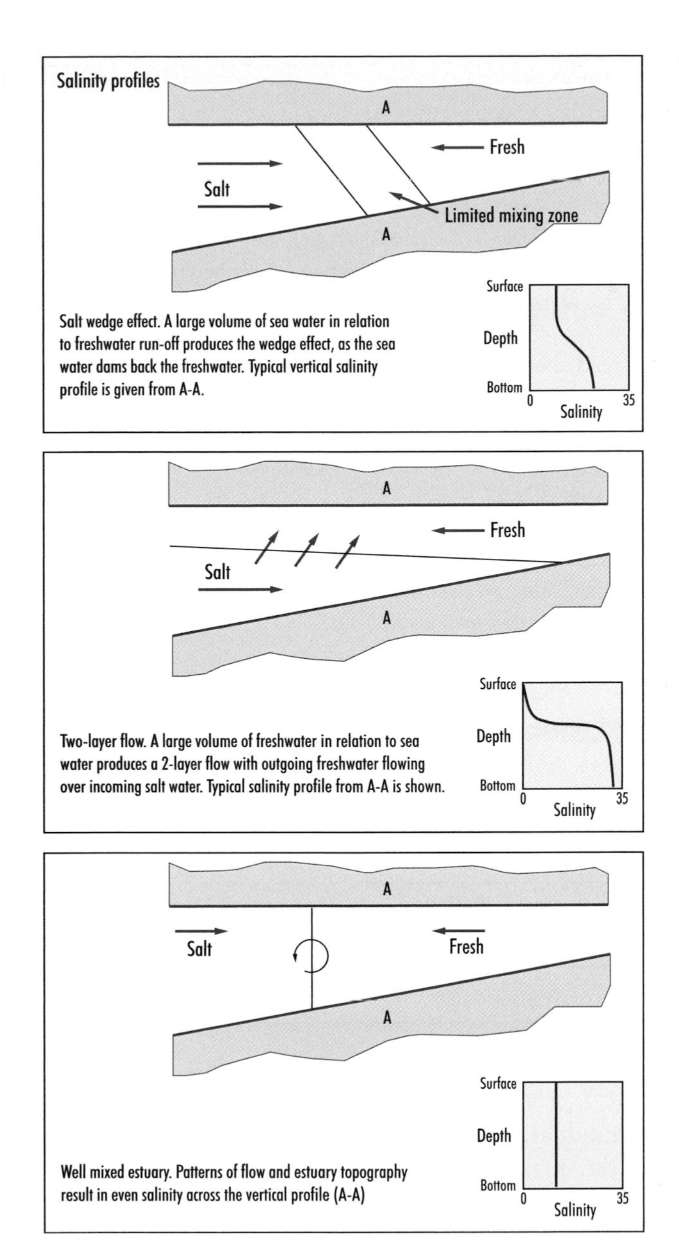

Fig. 4.4 Stylized salinity profiles of estuaries.

(see Dyer 1997). Consequently, sediments near the mouth and heads of estuaries tend to be much more coarse as tidal and river flows are great here, thus preventing settlement of finer particles (see also Chapter 7).

● Adaptation to the fine sediment of mudflats is as important for survival in estuaries as adaptation to salinity variation.

The extensive mudflats found in estuaries are exceptionally rich in fine organic material that has settled out with silt particles in the water column. Consequently, estuaries have one of the highest levels of secondary production in aquatic systems (see 4.3). If a deposit-feeding organism can tolerate the extremely stressful conditions prevalent in a

Box 4.3 Sediment deposition in estuaries

Estuaries are typically associated with extensive banks of very fine mud, but these only really occur in the middle of the estuary and are mainly composed of marine sediment. How do they form here, often tens of kilometres from the sea? Sediment deposition is related to current speed – when tide or river currents are fast they can hold on to much of their sediment. However, in the middle estuary, river and tidal currents meet and slow down, the reduced velocity allowing the settlement of the finest silt particles carried in the water. As either river or tide slows as it moves along the estuary, successively smaller particles settle out, creating a distinct sediment profile (see figure). Generally, tidal movement is much greater in terms of volume than river input, so the mid-estuary mud is mainly of marine origin. This is clearly a very simplified version of sedimentation in estuaries, which is further influenced by many factors such as salinity structure, local topography and tidal movement, etc., but provides the general, large-scale principles. For full details see Dyer (1997).

Due to the large amount of suspended material, estuaries are naturally turbid. In the 1960s the Kinks sang about the Thames being a 'dirty old river' – this may have been the case at the time, but now the comparatively unpolluted estuary still looks dirty due to the high natural suspended solid load.

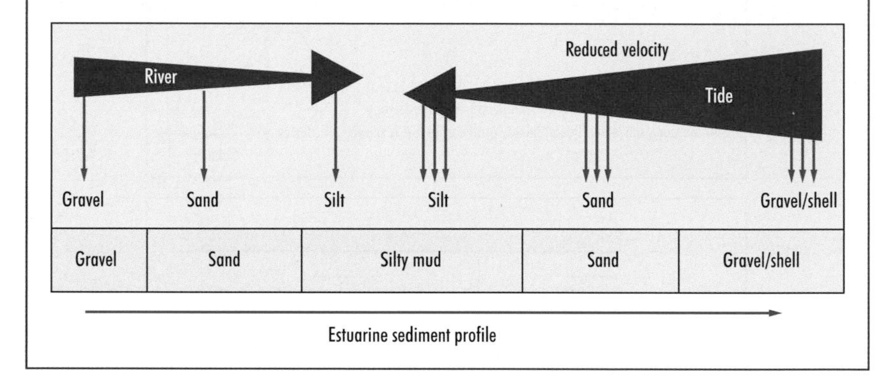

mid-estuary mudflat, then growth rates and production can be extremely high due to the quantity and quality of available food.

Dissolved Oxygen

Very few aquatic systems have naturally low levels of dissolved oxygen (DO) and appropriate oxygen levels in water are vital to sustain life. Both estuarine waters and sediment have extremely high levels of organic material (see 4.3.1), so consequently have very high levels of bacterial activity. Aerobic bacteria dominate the breakdown of matter in estuarine waters, and consequently use up dissolved oxygen in the process, resulting in reduced oxygen levels in mid-estuary water. Oxygen levels tend to be normal at the head and mouth of an estuary, so a longitudinal profile of DO would show a notable dip in the mid-estuary.

This is known as a DO sag. Oxygen removal is more severe from interstitial water in mudflats. The fine nature of the sediment prevents re-aeration by diffusion. Hence, below the surface of the mud, bacterial action removes oxygen, with subsequent breakdown of matter undertaken by anaerobic bacteria. This process results in potentially toxic by-products, such as methane and hydrogen sulphide, the latter having the effect of turning the mud black and producing the 'rotten eggs' smell. This is a natural process in estuarine mud (it would occur in unpolluted estuaries) and adds a further physiological stress for organisms attempting to live in mudflats.

● Bacterial breakdown of organic material in mudflats results in removal of oxygen beneath the mud surface and the production of anaerobic by-products. The resultant black, smelly mud is entirely natural and nothing to do with pollution.

Generally, the level of oxygen removal from estuarine water will not have a dramatic effect on the associated biota. However, many estuaries worldwide have been favoured places of human settlement, with large cities having been constructed on their banks. Human activity (e.g. sewage disposal) can supplement the amount of organic matter in the estuary and thus stimulate excessive removal of oxygen by bacterial action. In extreme cases (e.g. the Thames (Attrill (1998)), estuarine waters can become exceptionally depleted of oxygen, with dramatic and severe consequences for the ecology of the system (see also Chapter 14). The naturally low levels of oxygen in mid-estuarine water makes it exceptionally vulnerable to human impact.

Temperature

Temperature does not vary greatly over a tidal cycle and changes in temperature longitudinally are fairly minimal. However, in higher latitudes, seasonal patterns in temperature change can be the major trigger for migration and reproduction of estuarine organisms, particularly mobile species such as fish and large crustaceans (see 4.2.2). This influence of temperature can be more important than salinity variation, so needs to be considered as a major environmental factor in estuarine ecology. Temperature can also influence the speed of microbial breakdown in estuaries. Bacteria process organic matter much faster in the tropics or during the summer in temperate systems. This has consequences for DO levels and temperate urbanized estuaries are more vulnerable to severe DO sags in summer than in winter. Human activity can also affect the temperature of estuaries, with knock-on effects for oxygen levels, through, for example, the discharge of warm water effluent from power stations. It has been estimated that at the height of power station activity, the water temperature in the Thames Estuary was raised by over 3 °C.

● Temperature can affect the speed of breakdown of organic material and thus influence the levels of dissolved oxygen in estuarine water.

The conditions within, and surrounding, estuaries are directly linked to human activity. Few, if any, estuaries in the world could be regarded as totally free from human influence, whether this is an impact on the water quality of the system, interference from fisheries and aquaculture, or more commonly modification of the surrounding bankside environment. Therefore, estuaries arguably represent the most anthropogenically degraded habitat type on earth, a factor that must be kept in mind when investigating their ecology (Edgar et al. 2000).

4.1.4 Division of estuaries into zones

To enable further sub-classification of estuaries, they have been traditionally divided up into zones reflecting the overall environmental conditions apparent in that particular region. Several schemes have been devised, generally based around salinity levels (e.g. the Venice system), but all are subjective to a high degree and zones are hard to spatially define in a dynamic system that is varying over tidal and seasonal cycles.

● The majority of estuarine studies have been concentrated in the upper and middle reaches.

A typical classification is as follows:

Zone 1: **Head.** Estuary dominated by river flow, salinity generally <5, strong river currents resulting in coarse sediment of stones and gravel.

Zone 2: **Upper Reaches.** Main area of mixing of fresh and saline water, salinity highly variable (e.g. 5–18), currents can be negligible, resulting in fine, muddy sediment in intertidal areas.

Zone 3: **Middle Reaches.** Flows dominated more by tidal currents, salinity 18–25, extensive intertidal flats mainly muddy but with more sand present.

Zone 4: **Lower Reaches.** Faster tidal currents, salinity 25–34. Sediment now mainly sand.

Zone 5. **Mouth.** Where estuary meets the sea, strong tidal currents and generally fully saline conditions. Sediment clean sand with shell fragments, or even rock in some cases.

4.2 Estuarine Organisms

4.2.1 Origin of estuarine organisms

Estuaries are exceptionally new systems in geological terms, particularly in temperate and boreal regions. Therefore, they have been open to colonization by organisms only since the retreat of the last ice age (4.1.2). Tropical systems potentially have been more stable through time, although they will have been affected to some degree by sea level changes. The effective extinction of estuaries at higher latitudes during the last glaciation event would have had severe consequences for the fauna and there is much debate as to whether brackish water conditions existed during this period to act as refugia for estuarine organisms (Barnes 1994). Three scenarios for the recolonization of estuaries following glacial retreat are therefore possible. If local refugia were present, then recolonization by similar suites of organisms present prior to glaciation might have occurred. Organisms may have remained in brackish water systems at lower latitudes and recolonized using successively higher latitude systems as stepping stones. Alternatively, some species may have become extinct if no refugia existed, so a third scenario is that organisms have evolved into estuaries from suitable marine species during each post-glacial period. Evidence is equivocal, although some estuarine species recorded as sub-fossils are known to have disappeared from the Northern European fauna after the last ice age. A good example of the latter is the bivalve *Mya arenaria* (Box 4.4). Barnes (1994) suggested that the diversity of very similar *Gammarus* species existing in brackish water represents successive species evolving into estuaries after each ice age.

● High latitude estuaries have been recolonized after each ice age. What happened to estuarine species during ice cover is open to debate, but potentially successive recolonization may have stimulated speciation.

Whatever the evolutionary pattern of estuarine colonization, estuarine organisms comprise species that originate from two possible sources: species of freshwater origin extending into estuaries from rivers and marine species colonizing estuaries from the sea. The dynamic, stressful nature of estuarine conditions produce strong evolutionary

Box 4.4 Extinction and reintroduction of *Mya arenaria*

What happened to estuarine species during glaciation? Mollusc shells provide a clue. The common estuarine bivalve *Mya* appeared to become extinct in Europe during the last ice as its habitat disappeared. It did not naturally recolonize from North America, but most likely was reintroduced by human movement across the Atlantic from the fourteenth century. Subsequently, the bivalve has spread through brackish water in Europe, only reaching the Black Sea in the 1960s and would be regarded by most as a classic species of these habitats. Did other species also suffer this extinction? We have little evidence to go on, but the fate of estuarine species during glaciation and the subsequent method of recolonization may be important in helping to understand diversity patterns and speciation in estuaries.

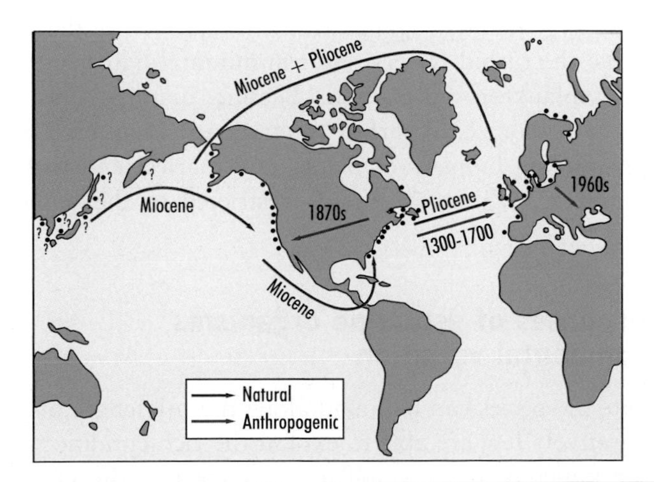

selection forces, so speciation in such habitats is potentially rapid. Indeed, genetic research on estuarine organisms, such as amphipods, has revealed significant genetic differences between populations within the same estuary but that are found in different habitats (Stanhope et al. 1993).

The head of an estuary is dominated by organisms of freshwater origin, with most freshwater taxa represented. Insects generally do not penetrate far into estuaries, in parallel with the full marine environment, with the possible exception of dipteran larvae such as the Chironomidae. The most abundant freshwater taxa in estuaries tend to be fish (e.g. catfish in tropical estuaries), pulmonate molluscs (such as the Lymnaeidae) and, in particular, oligochaetes. These worms are extremely successful colonizers of estuaries, with several species found only in the upper and mid-estuarine areas, often in vast numbers. The majority of oligochaetes belong to two families, the Enchytraeidae and the Tubificidae. In particular, tubificids have been recorded at densities of over 1 million m^{-2} in mid-estuary mudflats.

● Estuarine organisms are either of marine or freshwater origin. Oligochaetes are the most successful freshwater group to colonize estuaries.

The majority of other common organisms that penetrate far into estuaries are of marine origin, in particular polychaetes, nematodes, crustaceans, fish (see 4.2.4), and bivalve molluscs. Mudflats in the middle of an estuary will tend to be inhabited by large numbers of polychaetes, oligochaetes and small crustaceans, in particular amphipods. In outer estuary areas, large beds of bivalve molluscs, such as mussels and oysters, are not uncommon. In parallel with the head of an estuary, examples of the majority of marine taxa can be found in the outer estuary area. Other species utilize estuaries during migration, either from fresh water to the sea (e.g. the Chinese mitten crab (*Eriocheir sinensis*), the eel (*Anguilla anguilla*)) or vice versa (e.g. salmon species, such as the chinook (*Oncorhynchus tshawytscha*)).

● Organisms that migrate from the sea to fresh water to spawn are known as anadromous, whereas those that migrate from rivers to the sea to spawn are catadromous.

Plants are poorly represented below the high tide mark in estuaries (coastal plants such as mangroves are dealt with in Chapter 9). Few hard substrata are available for algal attachment (apart from those of human origin), while the turbid waters result in minimal light penetration and restrict phytoplankton production. Benthic diatoms and ephemeral green algae can colonize the surface of mud flats (Chapters 2 and 3). The only large standing biomass of plants in estuaries tend to be seagrass beds (Chapter 9), which are usually restricted to the outer estuarine regions.

4.2.2 Responses of estuarine organisms to environmental variation

A large suite of species can be found along the full length of an estuary, but comparatively few are able to exploit the rich conditions present in the mid or upper-estuarine mudflats, perhaps due the high degree of environmental variability detailed in 4.1.3. Traditionally, the physiological ability to respond to reduced salinity has been seen as the key to the survival of marine species in estuaries and determines the realized distribution of any organism. A reduced salinity in the external medium exerts an osmotic stress, resulting in uptake of water by the organism, which has an internal fluid concentration greater than that of the surrounding water (**hyperosmotic**). To prevent lethal dilution of body fluids, water uptake has to be mitigated or prevented by, for example, active pumping of water or internal movement of ions. One might then expect that good powers of osmoregulation would be a distinctive feature of all successful estuarine organisms. In fact, they exhibit a very wide range of abilities to osmoregulate. Many organisms, such as molluscs, have little, if any, osmoregulatory ability and are thus termed **osmoconformers**. They survive and thrive in estuaries by exhibiting exceptionally high tolerance to diluted blood concentrations, which is sufficient over most of their range in estuaries. At the other extreme, crustaceans are amongst the most efficient osmoregulators. Species such as the mysid *Praunus flexuosus* are able to maintain a relatively constant internal medium across a very wide range of external salinities. Many other animals fall between these two extremes in their osmoregulatory ability. Some algae can also survive extreme salinity variation; the green alga *Enteromorpha* is perhaps the most tolerant, changing ionic and metabolic concentrations in response to variations in salinity. Estuarine *Enteromorpha* also possess thinner and stretchier cell walls than fully marine species, allowing changes in tissue water content and cell volume.

● An organism with blood concentration the same as the external concentration is isosmotic, blood more concentrated than external is hyperosmotic, blood less concentrated than the external medium is hyposmotic.

Thus, while osmoregulation is clearly important to aid survival of many species that encounter reduced salinity conditions, considering this to be the major determinant of species distribution in estuaries is an oversimplification. Organisms found in the most variable conditions within an estuary are not necessarily the most sophisticated osmoregulators, but often simply the most tolerant. The majority of experimental work on osmoregulation has considered organism response to fixed salinity values, whereas organisms in estuaries (particularly in subtidal areas) are subject to highly fluctuating conditions on a tidal and seasonal basis. Therefore, an experiment determining the minimum salinity at which a species can survive may have limited application in the estuarine environment. Estuarine organisms also demonstrate a marked **physiological plasticity**, enabling them to adapt to different salinity regimes even if they have minimal osmoregulatory abilities, such as the mussel *Mytilus edulis* (Tedengren et al. 1990). Effectively, the majority of estuarine species are able to tolerate a wide range of salinities, and while their absolute upstream limit will be affected by their ability to deal with fluctuating, reduced salinities (amongst other variables), their physiological tolerance would generally be greater than that experienced in the environment over

● The physiological tolerance of most estuarine species is generally greater than the range of conditions they actually experience in an estuary.

their realized range. Other environmental factors therefore potentially control the survival and distribution of estuarine organisms on a day-to-day basis, and recent work supports this:

- The distribution of many infaunal organisms is determined by the availability of mud habitat rather than salinity levels (Barnes 1989). The polychaete *Nereis diversicolor*, for example, is found practically throughout an estuarine system wherever stable mudbanks occur. As such a habitat is not found above the upper estuary, *Nereis* is not present here; settling neochaetes have been recorded, however, from samples in near freshwater conditions.
- The mysid *Neomysis integer* is an excellent osmoregulator and is found across large parts of Northwestern European estuaries. However, avoidance of high flow areas appears to be the key factor in determining its local distribution within estuaries, rather than salinity (Lawrie et al. 1999).
- The shore crab (*Carcinus maenas*) is a classic estuarine organism and an efficient osmoregulator. However, modelling work suggests that salinity appears to be unimportant in determining its movements and distribution, with temperature the key factor (Attrill et al. 1999). Below approximately 8 °C the crabs become comparatively inactive. Dissolved oxygen concentration is the only physicochemical parameter significantly related to distribution patterns of the brown shrimp (*Crangon crangon*). Findings for both of these species are supported by other laboratory results.
- Temperature appears to be the dominant variable influencing the movement and distribution of many estuarine fish species (see 4.2.4).

● *Carcinus* populations found in estuaries appear to be quite different from those on a rocky shore, with different behaviour, physiological responses and migration patterns. The orange-coloured mature crabs are rare in estuaries.

In summary, while an important environmental factor, salinity should not necessarily be considered the major influence over distribution patterns in estuaries. Overall environmental variability is probably the limiting factor for most organisms that penetrate estuaries, with other environmental factors influencing their local distribution and survival.

4.2.3 Impact of freshwater inflow on estuarine organisms

Much focus on environmental variation in estuaries has been on the short-term impact of tidal cycles. However, conditions within an estuary vary over a longer, seasonal time period, particularly in higher latitudes and in tropical areas with seasonal monsoons, as the amount of fresh water entering the estuary varies considerably. Most organisms respond to tidal variations in conditions by simply tolerating the variability for this short time period or by avoiding contact with lower salinity water by burrowing, bivalves closing their shells, or fish moving in and out of estuaries with the tide. However, large changes in the amount of fresh water entering the estuary due to increased, seasonal precipitation can alter the conditions in any one part of the estuary quite dramatically (e.g. salinity, Fig. 4.5), changes that are effective over much longer periods of time. However, the influence of such variations in flow has been neglected, despite the potentially large impact they may have on estuarine ecology and the spatial distribution of estuarine organisms.

● Changes in freshwater flow entering an estuary can result in contrasting salinity profiles in different seasons.

Estuarine organisms can demonstrate three responses to increases in flow, and subsequent large decreases in salinity, particularly in the upper estuary where seasonal inputs can lower the salinity to practically fresh water. Either the organisms tolerate the changes, they migrate or drift to avoid low salinity water, or they do not survive at all. In the first scenario, temporal samples of the community in an estuary would demonstrate few changes, whereas the latter two responses would result in the removal of the organism from the assemblage and possible shifts in community up and down the estuary. Clearly, long-term studies are needed to elucidate responses of organisms to seasonal flow changes, but unfortunately few such data sets exist.

● How do benthic organisms disperse in estuaries? In rivers, the drift of organisms in the flow is well documented, but such movements by estuarine organisms are less well known. Drift could allow both upstream and downstream dispersal in estuaries in response to changing conditions.

In the Fraser River, Western Canada, flow is highly seasonal, but due to ice formation and melting in the Rocky Mountains. Lowest flow rates into the estuary occur during winter when water is locked up as snow in the mountains, dramatically increasing over spring and summer as melt-water flows to the coast. The responses of benthic invertebrates were recorded over this seasonal change (Chapman & Brinkhurst 1981). There were notable patterns in organism occurrence at the more fresh-water estuarine sites, with freshwater species, such as the oligochaete *Tubifex tubifex*, present in large numbers for much of the year, but absent during winter when flow levels dropped. Similarly, the freshwater species *Paranais frici* was replaced in winter by a species more adapted

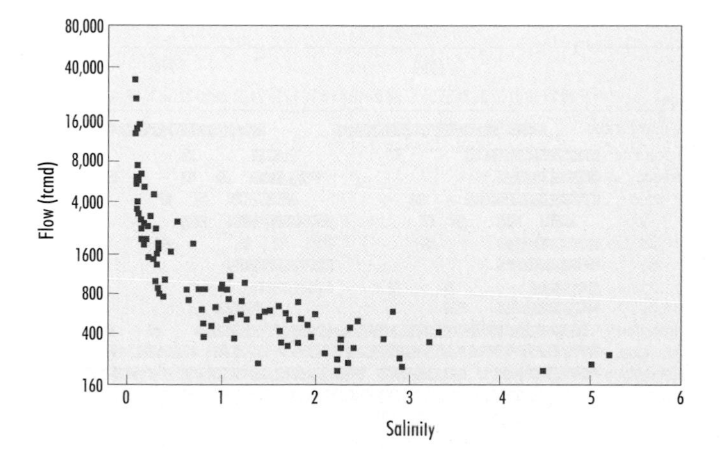

Fig. 4.5 The relationship between freshwater inflow and salinity levels in an upper estuary.

to estuarine conditions, *P. litoralis*. Populations of these highly sedentary worms demonstrated movements both up and down the estuary on a seasonal basis. Similar changes to ranges within the estuary were noted in the Thames, UK (Attrill & Rundle 2002), where severe reductions in flow (due to drought conditions and abstraction of water) resulted in the upstream movement of many sedentary species. Changes in the community at the very head of the estuary were even more marked (Attrill et al. 1996). Despite very small changes in salinity (0.25) many species, particularly insects, disappeared rapidly from the community (Fig. 4.6) leaving a low-diversity assemblage of tolerant species. Traditionally, the freshwater part of an estuary is considered to be the area <5, but this study demonstrated that very subtle changes in the salinity regime can have dramatic consequences for upper estuarine communities. Either species that inhabit this area of the system are far more susceptible to salinity variation than traditionally thought, or other factors associated with variation in flow have large impacts on these organisms.

● Abstraction – the planned removal of water from rivers or lakes to supplement drinking supply. Generally water is pumped to reservoirs to maintain their level. This action can dramatically affect the amount of water entering estuaries.

Drought-induced changes in freshwater inflow can also have large consequences for the water quality of the mid-estuary, and subsequently impact the mobile organisms present there (Attrill & Power 2000a, b). Despite being over 60 km from the river, low-flow conditions over a 4-year period were shown to significantly affect a range of parameters in the mid-Thames estuary (including dissolved oxygen, temperature, salinity) that are the major influences on mobile estuarine organisms. The long-term nature of this study (17 years) allowed the populations of mobile invertebrates to be modelled and demonstrated significant shifts in their abundances and distribution patterns following the onset of drought (Fig. 4.7).

● Modelling of estuarine populations allows their abundance to be predicted from measured environmental parameters.

There is also evidence that different inflow rates can have significant effects on the productivity of estuaries. Montagne & Kalke (1992) studied two estuaries in Texas, USA, which were very similar apart from the size of the river entering the system, providing a spatial comparison of the effect of freshwater inflow. The Guadalupe estuary had nearly

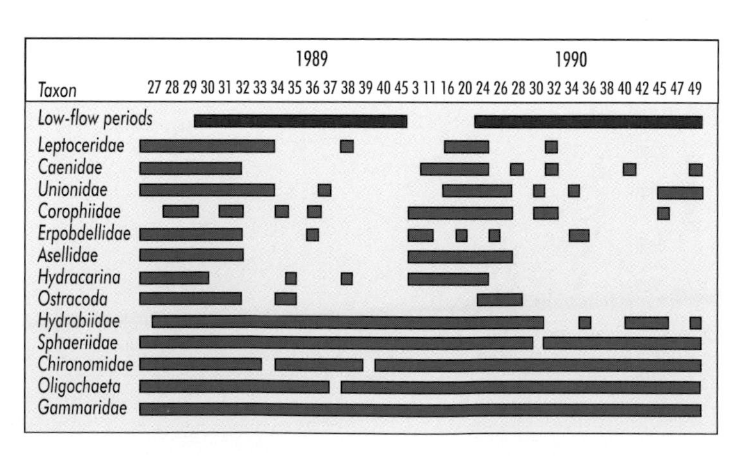

Fig. 4.6 Disappearance of upper estuarine taxa following reductions in freshwater flow into an estuary.

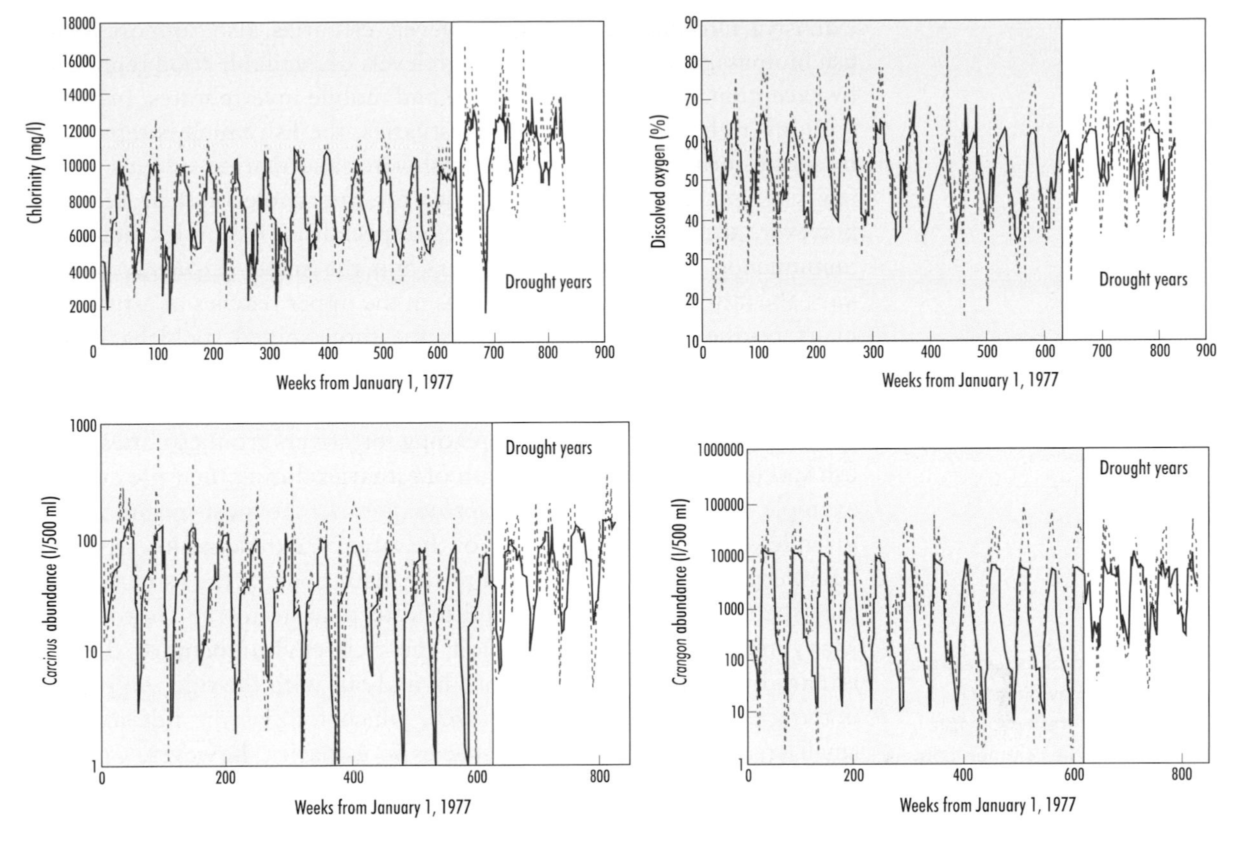

Fig. 4.7 Examples of models constructed for water quality and mobile invertebrates in the Thames estuary demonstrated marked changes in seasonal patterns following the onset of drought.

80 times the freshwater inflow than the Nueces estuary, and had a much higher macrofaunal density and shellfish production, but a lower density of **meiofauna**. The difference in production was explained by higher nutrient input to the Guadalupe estuary via river flow that boosted primary production in the oligohaline zone, the higher resulting increased macrofaunal densities creating greater predation pressure on meiofauna and so reducing their numbers. The impact of inflow on estuarine production was also apparent in a study of phytoplankton in the Swan River Estuary, Australia (Chan & Hamilton 2001), and led to the conclusion that alterations to freshwater discharge can have a significant impact on phytoplankton biomass and species composition, with regulation of the dominant group (e.g. green algae and diatoms) by inflow.

● Meiofauna – benthic organisms that generally pass through a 500 μm sieve, such as nematodes, copepods, and mites. Animals retained on this sieve are macrofauna.

● Higher freshwater inflow can boost estuarine productivity through higher dissolved nutrient input.

4.2.4 Estuarine fish communities

Due perhaps to their relative ease of sampling, much work on the ecology of estuaries has focused on the invertebrates inhabiting the extensive intertidal mudflats. However, estuaries also support a large fish biomass, which exploits the high levels of available food represented by exceptional numbers of infaunal and mobile invertebrates. In parallel with invertebrate taxa inhabiting estuaries, the fish fauna is represented by species originating from both freshwater and marine systems. Only a few fish species depend on estuaries for the whole of their life cycle, however, and can be termed estuarine specialists. Examples include the mummichog (*Fundulus heteroclitus*) and the smelt (*Osmerus eperlaunus*), the latter species laying its eggs in the upper reaches of estuaries. In many northern temperate estuaries the three-spined stickleback (*Gasterosteus aculeatus*) is common, a species more generally associated with fresh water. However, the stickleback has only colonized freshwater systems since the last ice age, spreading into rivers from estuaries. Other fish species utilize the whole length of estuaries during their life cycle. An example is the flounder (*Platichthys flesus*), the post-metamorphosis juveniles inhabiting the very top of the estuary, extending into rivers. As flounders grow they move down to the main body of the estuary, where they spend most of their life, and tend to migrate to the sea to spawn. Many marine fish species exploit the rich environment of the mid-estuary by undertaking migrations in and out with the tide. An example is the grey mullet (*Chelon labrosus*), which follows the tide into even small creeks. Unlike most fish species in estuaries, however, *Chelon* is principally a grazer, feeding off diatom films on the mud surface and leaving distinctive feeding traces in the mud. However, probably the most important role of estuaries for fish species is as a nursery ground for marine fish species, one fourth of all fish species in Virginia/Carolina region of the USA being estuarine-dependent (Ray 1997). The majority of fish found in temperate estuaries are juveniles within the first two years of life. Vast numbers of such fish can move into temperate estuaries at certain times of the year (particularly winter), utilizing the high food supply and the generally less harsh, warmer conditions than present in the open sea. The high turbidity of estuaries may also provide a certain level of refuge from visual predators. Therefore, estuaries are exceptionally important in providing recruits to many coastal fisheries. Seasonal use of estuaries by fish also occurs in tropical regions, here dominated by wet and dry seasons. While these systems are much less studied than temperate areas, higher densities and species richness have been recorded during the rainy season. While, like invertebrates, the overall salinity regime can influence the longitudinal distribution of fish species, their comparatively high tolerance to a range of salinities coupled with their motility, which enables them to avoid unfavourable conditions, means that salinity appears unimportant in determining the usage patterns of estuaries by fish. The majority of large-scale studies have concluded that water temperature is the most important parameter

- At the head of estuaries, 'marine' fish may directly compete with freshwater fish. For example, the common goby (*Pomatoschistus microps*) with the minnow (*Phoxinus phoxinus*).

- As well as being important nursery grounds for young marine fisheries species, many estuaries support their own fisheries, particularly for bivalves such as oysters, mussels, and cockles.

related to the distribution of fish over the year, initiating migration movements for example. However, other parameters are also highly influential for certain species and at a more local level. Adequate dissolved oxygen levels are vital for fish survival in estuaries, and historically pollution-induced low DO levels have reduced, or completely eliminated, the nursery roles of estuaries such as the Tyne, north-eastern England. Some fish species may also follow their prey. For example whiting (*Merlangius merlangus*) are closely associated with the brown shrimp (*Crangon crangon*), upon which they exert strong predation pressure (Berghahn 1996). On a more local level, habitat complexity may be a highly important influence on the size and diversity of the fish assemblage, particularly in tropical systems where seagrass beds and mangroves are prevalent.

● Dissolved oxygen sags in estuaries can prevent the migration of fish species to their freshwater breeding grounds, in particular salmonids, which require high oxygen concentrations.

Estuarine fish assemblages may also be influenced by much larger-scale phenomena. Attrill & Power (2002) reported that diversity, abundance, and growth of fish species in the Thames estuary over a 16-year period were primarily influenced by variations in climate known as the North Atlantic Oscillation (Fig. 4.8, Box 4.5). For a large number of fish species, this climatic influence appeared to be greater than factors such as water temperature, salinity or freshwater flow. How could a variable climate pattern have such a controlling influence? A significant relationship was evident between the NAO and the difference in water temperatures between the North Sea and estuary. At low phases of the NAO (cold, dry winters) the estuary was significantly warmer than the open sea, providing highly favourable conditions for many juvenile fish. Consequently, more individual fish appeared to be using the estuary during this phase of the climatic cycle. Growth and diversity were higher during wet, warmer phases, and increased species number was related to the appearance of southern species in the estuary.

● The appearance of warm water southern species in northern temperate regions has been highlighted as a consequence of global warming, but some species may just be responding to climate cycles.

4.3 Productivity and Food Webs

4.3.1 Energy sources – where does the carbon come from?

Overall plant biomass and primary production are low in estuaries (4.2.1), with the exception of extensive seagrass beds that can form in the mouths of estuaries (Chapter 9). Production in estuaries is therefore dominated by the utilization of detritus; amorphous organic matter present in very high levels in estuarine sediment and water. Typical levels of dry organic matter (measurement of the biomass of all organic material) in an estuarine water sample can be 110 $mg\,l^{-1}$, compared with 1–3 $mg\,l^{-1}$ for open seawater. Similarly, the organic carbon in an

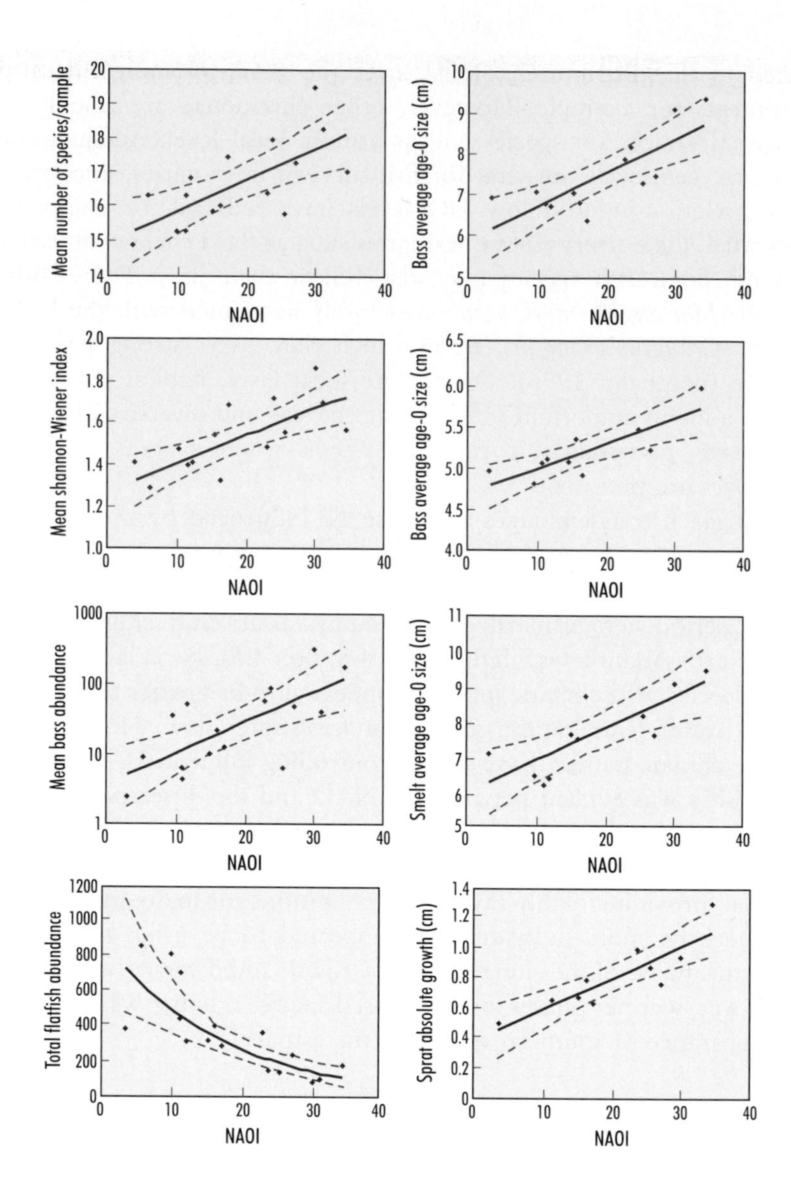

Fig. 4.8 Relationships between the North Atlantic Oscillation Index (NAOI) and fish diversity, abundance and growth in the Thames Estuary.

estuarine mudflat can result in a biomass of detritivores ten times that of offshore sediments. These invertebrates in turn support high numbers of predators, including birds and fish. The carbon from this detrital source is therefore the main foundation for the estuarine food web, so where does it mainly come from? Detritus finds its way into mudflats from the full range of surrounding habitats (**allochthonous**), including fully terrestrial (e.g. leaf litter brought down by rivers), marine (material from algae, seagrass, and animal material brought in with the tide), semi-terrestrial bordering systems (e.g. saltmarshes, mangroves), and from production generated within the estuary itself (**autochthonous**). So how

Box 4.5 The North Atlantic Oscillation (NAO)

The climate in the North Atlantic undergoes a relatively unpredictable cycle in terms of its weather pattern, this cycle being known as the NAO and its fluctuations are measured by a simple atmospheric pressure differential between Iceland and the Azores (the NAO Index). At one extreme, high pressure over Iceland will result in cold, dry winters in Northern Europe (a low NAOI), whereas comparatively high pressure over the Azores results in dominant winds from the SW and consequently warm, wet and windy winters. The oscillation between these two states is shown in the figure. This climatic cycle is now recognized as having an extremely important influence on the weather, the position of the Gulf Stream, sea temperatures, and consequently the ecology of terrestrial, freshwater, and marine systems in NW Europe. Correlated long-term cycles have been observed in a disparate range of marine organisms from plankton, through to benthic communities and fish catches.

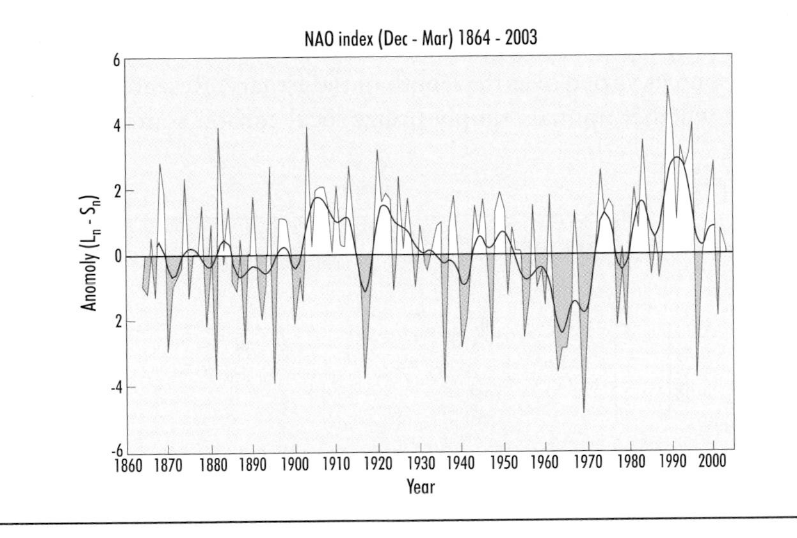

important are these respective sources in estuarine food webs? Are particular carbon sources favoured by estuarine organisms? Until comparatively recently, such questions were hard, if not impossible, to answer, but the development of stable isotope techniques have enabled estuarine ecologists to gain an insight into these questions.

● Organic carbon is the bioavailable carbon present in a sediment, so does not include mineral forms like $CaCO_3$ found, for example, in mollusc shells. It is generally assessed by the loss of material after combustion in a furnace.

4.3.2 The use of stable isotope analysis in estuarine food webs

Due to different modes of photosynthesis plants accumulate heavy isotopes of carbon at different rates (see Chapter 2). Therefore, it is possible to distinguish between, for example, a leaf from a tree and green algae by determining the stable isotope ratios within each plant's tissues. Therefore, a signal can also be obtained from detritus, which

● Carbon has two stable isotopes, ^{12}C and ^{13}C. Analysis measures the proportion of the heavy isotope and compares this with the standard. The difference is expressed as $\delta^{13}C$ (‰). This value varies depending on the source carbon.

indicates its origin. The resolution is improved considerably by also determining the stable isotope ratios of nitrogen, or preferably sulphur, within the material, giving a second axis allowing separation of carbon sources that overlap when carbon alone is used (Fig. 4.9). The isotopic ratios of C and S are conserved when plant material is ingested and incorporated into the bodies of animals, so analysing the tissues of estuarine animals will enable information to be obtained on the carbon source utilized by these organisms. This has proved a very important tool in aiding our understanding of how estuaries function.

One of the most comprehensive studies was undertaken in the Plum Island Sound Estuary, north-eastern USA (Deegan & Garritt 1997). Samples of plant material and a large range of organisms were obtained from the full length of the estuary and analysed for stable isotope ratios in order to determine which carbon sources were most important for the estuarine organisms and, additionally, whether the relative importance of these sources varied over the length of the estuary. Researchers wanted to know whether animals simply utilize local carbon sources or if they

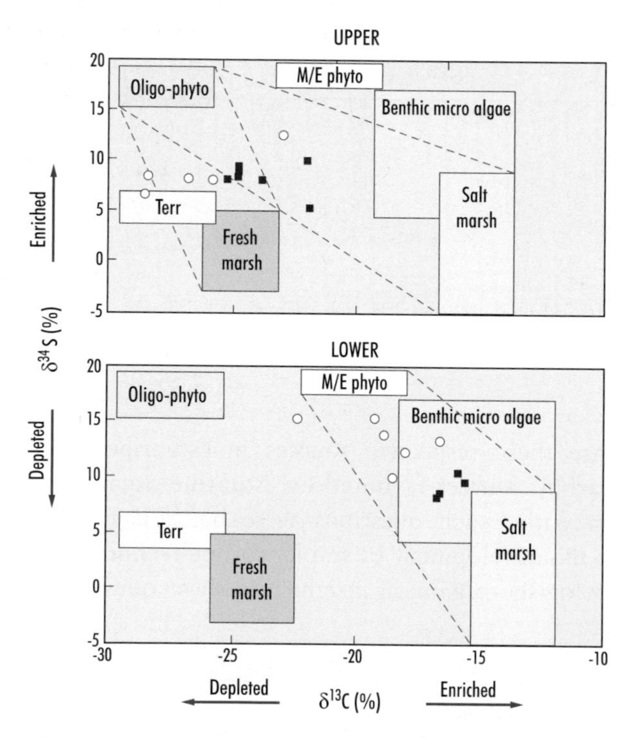

Fig. 4.9 Example ranges of stable isotope ratios for carbon and sulphur in a range of plants contributing to the detritus of upper and lower regions of an estuary. Symbols on the figure refer to the comparative isotopic ratios within the tissues of estuarine animals (open circles – pelagic, closed squares – benthic), enabling the carbon source on which they feed to be assessed. M/E = mid-estuary, oligo = oligohaline, Terr = terrestrial material, Fresh = freshwater, Phyto = phytoplankton.

were more selective. Organisms were found to rely heavily on local sources of carbon (Fig. 4.9). Upper estuarine food webs were based on fresh water, saltmarsh, and phytoplankton carbon, in contrast to the lower estuary where marine phytoplankton, benthic algae, and perhaps saltmarsh material were the most important. The results were supported by notable shifts in the carbon sources utilized by species present along a large length of the estuary, such as the mummichog (*Fundulus heteroclitus*). Interestingly, despite the large seasonal input of terrestrial leaf matter in New England, this carbon source seemed unimportant in the food web of the Plum Island Sound Estuary.

● Current evidence suggests that carbon from terrestrial sources is not important in estuarine food webs.

A similar, though perhaps even more unexpected, result was found in the tropical Embley River estuary in Australia (Loneragan et al. 1997). Stable isotope analysis was utilized to investigate whether seagrass or mangrove derived carbon were important carbon sources for prawn species in the outer estuary. Organic matter in the sediment was found to correspond to the main local source, but this pattern was not reflected in the tissues of the prawns. While prawns inhabiting seagrass beds did seem to rely completely on seagrass-derived material, mangrove-inhabiting prawns did not depend on carbon from mangroves, utilizing either algae or seagrass material. The conclusions from work so far undertaken is that animals in estuaries do not appear to be able to utilize terrestrially derived material to any great degree.

Analysis of nitrogen isotope ratios can provide further evidence helping to determine the position of organisms within food webs. The heavy isotope of nitrogen (^{15}N) concentrates up the food chain at approximately 3‰ per step. Therefore, such analysis of animal tissue (together with other components of the food web) can enable the **trophic level** of an organism to be determined. This technique was put to interesting use in the Exe estuary, southwest England, where there is a large bed of blue mussels (*Mytilus edulis*). A copepod, *Mytilicola intestinalis*, occupying the guts of *Mytilus* had previously been blamed for population crashes in commercial mussel stocks on Europe, but evidence that it was in fact a harmful parasitic was equivocal. Stable isotope analysis determined that the copepod had a nitrogen isotope ratio 2.8‰ above that of the host gut (Gresty & Quarmby 1991), suggesting it therefore feeds on host tissue (although not directly, perhaps sloughed off cells) rather than sharing the gut contents.

● Stable isotopes of nitrogen have also been used to investigate the impact of introduced organisms on the trophic status of native species.

4.3.3 The importance of birds in estuarine food webs

The majority of books and chapters on estuaries include diagrams of food webs, indicating the major links from detritus up to large predators, often with attempts to quantify energy flow along these links. However, such constructions tend to be oversimplified in the context of estuaries due to the variability in species distribution over tidal and seasonal cycles. Such food webs can only ever be a generalization of trophic interactions within estuaries, which will vary considerably over both spatial and temporal scales, and provide minimal information beyond the intuitive. Many large and diverse groups tend to be lumped in one box within food webs, indicating their trophic level but masking any detail about their specific interactions. Fish in estuaries, for example, feed on all possible trophic levels from primary production to other fish, and their diet potentially varies considerably with time of year, position in the estuary, and the body size of the fish. This is even more true for birds, which have been comparatively neglected in estuarine food web analysis (which concentrate on 'marine' species), or assigned to a single compartment at the top of the web. Both approaches oversimplify the role of birds in estuaries to such a degree that an erroneous impression can be given of their importance, and thus potentially how estuaries function.

Estuaries are exceptionally important habitats for birds, particularly those in temperate latitudes, which provide key feeding areas for birds undertaking migration. Principally estuaries are vital for huge populations of wading birds (Charadriformes) and wildfowl (Anseriformes); 4.5 million shorebirds migrate to just the British Isles every year, while >1 million western sandpiper (*Calidris mauri*) stop over on the Fraser River Estuary, west Canada. Among the birds that use estuaries we can find examples of species feeding at all trophic levels from geese grazing on seagrass (e.g. Brent geese, *Branta bernicla*) to piscivorous species such as cormorants (e.g. *Phalacrocorax carbo*) and herons (e.g. *Tigrisoma mexicanum*). However, the vast majority of birds (particularly waders) feed on the rich supply of invertebrates inhabiting the mudflats of estuaries, with a distinct niche separation in terms of prey item evidenced by different bill lengths, feeding methodology, and feeding position on the shore. Overall, the dynamic interaction between the two groups of organisms contributes to the size of populations of both invertebrates and birds present in an estuary. The number of birds feeding on an estuary is directly related to the size of available mudflat, a simple area effect (Fig. 4.10a) with important consequences for estuarine habitat loss. However, at a smaller scale the abundance of feeding birds is dependent on the amount of available food in the mud, a higher biomass of invertebrates sustaining a larger population of birds (Fig. 4.10b). The distribution of invertebrates across both estuaries and individual mudflats can vary considerably, so the supply of food for foraging waders is not uniform and influences the distribution of birds on a mudflat. Clearly, the optimum patch for a bird to feed in is one with the highest density of prey, but this also depends on the number of other competing birds foraging in the patch. A good patch with lots of

● Grazing Brent geese can have a dramatic impact on intertidal seagrass beds as they eat the roots as well as the leaves. Huge flocks can therefore be major bioturbators in outer estuaries, reworking the top cm of sediment 8 times in 3 months.

● Many estuaries have suffered extensive habitat loss, reducing the available area for overwintering birds.

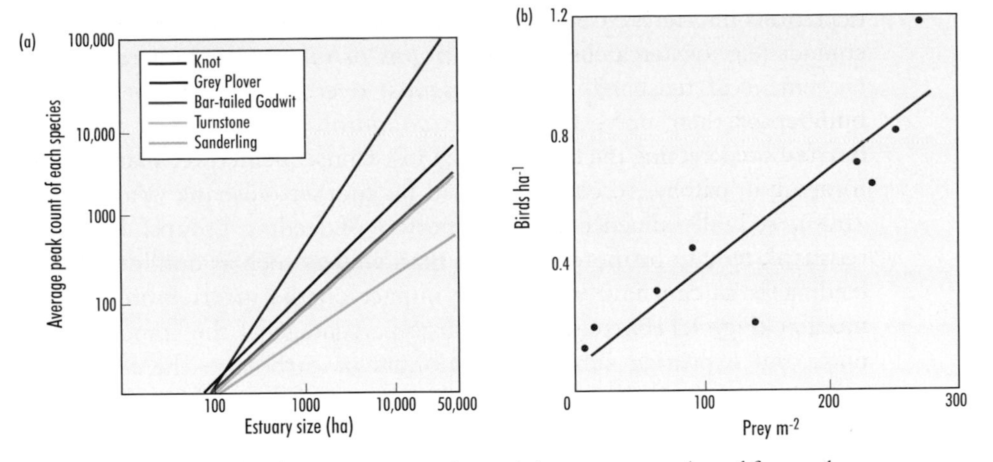

Fig. 4.10 (a) Relationship between estuary size and the average number of five wader species present. (b) Relationship between prey density (*Nereis, Scrobicularia*) and wader density (curlew, *Numenius arquata*).

competitors will eventually be less suitable than a less-rich patch with fewer competitors due to increasing interference with competitors, so birds will move to exploit other areas in which the rate of reward is greater. Eventually, birds should be distributed across the mudflat so that the proportion on a patch will approximate the proportion of resources in that patch, and all individuals receive equal rewards. This is known as the 'input matching rule' of the **ideal free distribution** (see Sutherland 1996) and is an example of **game theory**, which demonstrates that the pay-off of a strategy depends on which strategy is adopted by others. If birds conform to the IFD, the strategy is considered to be stable as no individual can gain a higher food intake by moving.

● Ideal Free Distribution: Ideal refers to birds choosing to forage where rewards are highest, free refers to the fact that there are no impediments to their movement.

However, other factors tend to moderate the ideal free distribution so that it rarely applies on mudflats. One factor is **territoriality**; if one species is dominant and actively defends its foraging patch then other bird species are no longer free to move between patches, so suitability will vary between patches. This is known as the **ideal despotic distribution**. On estuarine mudflats, for example, aggressive species of gull or crow can highly influence wader distribution. A further factor highly relevant to estuarine birds is the tendency for many waders (such as dunlin, *Calidris alpina*) to forage in flocks, mainly to reduce predation risk. Newly arriving birds tend to join existing flocks rather than risk foraging alone; despite the potential higher rewards in areas without flocks, predation risk reduces the suitability of these habitats. This is known as the **Allee effect** and can dramatically shift the theoretical ideal free distribution. Bird distribution can also be affected by **interference** and

depletion. Interference between birds can be through direct aggressive contact (e.g. oystercatchers, *Haematopus ostralegus*) or indirectly. The movement of redshank (*Tringa totanus*) over the mud influences the number of their prey, *Corophium volutator*, which retreat into the burrow on detecting the bird (Fig. 4.11a). Consequently, redshank avoid foraging in patches recently exploited by another redshank (Yates et al. 2000), so will influence the distribution of feeding groups. Foraging redshank tend to be more dispersed than waders such as dunlin. Finally, feeding birds can have a dramatic impact on the invertebrates in the mudflat (Fig. 4.11b), removing a large percentage of the standing biomass and exporting substantial amounts of carbon to the terrestrial environment through their faeces (Post et al. 1998). This removal of prey resource can have a major impact on bird distribution as previously rich patches become depleted, altering their suitability. Simple depletion models have been shown to accurately predict shorebird distribution over a range of different scales (Gill et al. 2001).

● Gulls and crows often attempt to take food from other birds. This resource stealing is known as kleptoparasitism.

4.4 Diversity Patterns in Estuaries

4.4.1 Diversity trends along estuaries

Estuaries are generally considered low diversity systems, with few species able to survive in the harsh environment they present. This is certainly true for mid-estuary mudflats, which tend to be used to represent the whole estuarine system; this habitat contains fewer species than in bordering marine or freshwater systems. This image of estuaries, however, is perhaps a little simplistic as over the full length of the system many species from a wide taxonomic range can be found. Few systems on earth will present such a wide phyletic diversity due to the combination of marine

● Estuaries are considered low diversity systems, but often only the mid-estuarine mud flats are considered. Across the whole estuary there can be a high diversity – over 750 species of invertebrate have been recorded from the Thames for example.

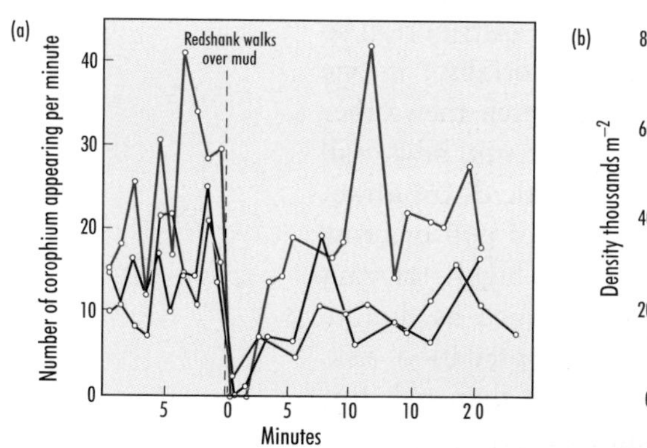

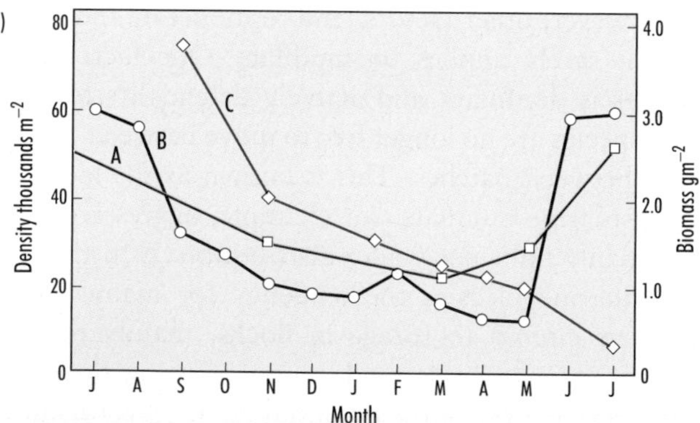

Fig. 4.11 (a) The response of *Corophium* to redshank foraging. (b) Depletion of prey by bird foraging over the year. A = *Nereis* and *Nephtys* in the Wash, eastern England; B = *Corophium* and C = *Mytilus* in the Ythan Estuary, Scotland.

and freshwater taxa. Additionally, sites in outer estuaries have been noted to have higher species richness than adjacent marine systems, generally due to the provision of complex structural habitats such as seagrass meadows and mussel/oyster beds. However, trends in benthic macroinvertebrates have provided the paradigm for estuarine diversity trends, exemplified by the **Remane** diagram, which is generally used to describe how diversity varies with salinity along an estuary (Fig. 4.12a). This trend shows a dramatic decrease in number of species from river into the estuary, with a species minimum around 7 (known as the **artenminimum**), then a gradual increase in diversity towards fully marine conditions.

- The Remane diagram is named after Adolf Remane, a German marine scientist who worked primarily in the Baltic in the early twentieth century.

The diagram is subdivided into freshwater species, marine species and true estuarine species (vertical hatched sector in Fig. 4.12a). The existence of these estuarine, or brackish water, species has been questioned by several authors and three completely independent studies have redrawn the Remane diagram, removing this group of organisms (Fig. 4.12b). Barnes (1989) argued from a physiological angle, suggesting these true estuarine organisms are simply marine opportunists able to exploit the particular conditions in the mid-estuary, but in terms of salinity tolerance could be found practically anywhere in the estuary. Sediment provision is perhaps a more important factor determining where these organisms are found. Attrill & Rundle (2002) examined faunal patterns in estuaries at the landscape scale concluded that species present in the mid-estuary were simply marine, or freshwater taxa at the extreme edges of their range, the pattern of species change into an estuary from each end representing an **ecocline**.

- Ecocline and ecotone are two forms of landscape boundary between systems. Ecoclines tend to be gradual, with a transition of communities, whereas ecotones are more abrupt (see Kent et al. 1997 for more details).

The relevance of the Remane diagram as a model for diversity trends in estuaries has also been questioned as it was developed in the 1930s as a hypothetical description of trends in the Baltic (see 4.5.1), a practically tideless sea. As described earlier (4.1), the environmental variability in estuarine systems is the major constraint to organism colonization, so an axis with fixed salinity may not be very meaningful. What does a salinity of '7' actually mean in an estuary? If salinity range is used as a measure of environmental variability in estuaries (Attrill 2002), a linear model can be constructed to describe diversity trends in estuaries (Fig. 4.12c). It is often assumed that the classic model of diversity trends for macroinvertebrates is representative of all taxa in estuaries, but this does not seem to be true. For both algae and zooplankton (Fig. 4.12 d,e) there is a steady decrease in number of species present from sea to river, with no species minimum in the mid- or upper-estuary. Fish diversity is hard to determine due to their mobility, but appears to show a decrease in the mid-estuary area. However, a very different pattern is apparent when an alternative measure of diversity is used (see Chapters 1 and 14), in this case β-diversity, rather than the usual measures of α-diversity (Fig. 4.12f), demonstrating a peak in diversity in the upper estuary.

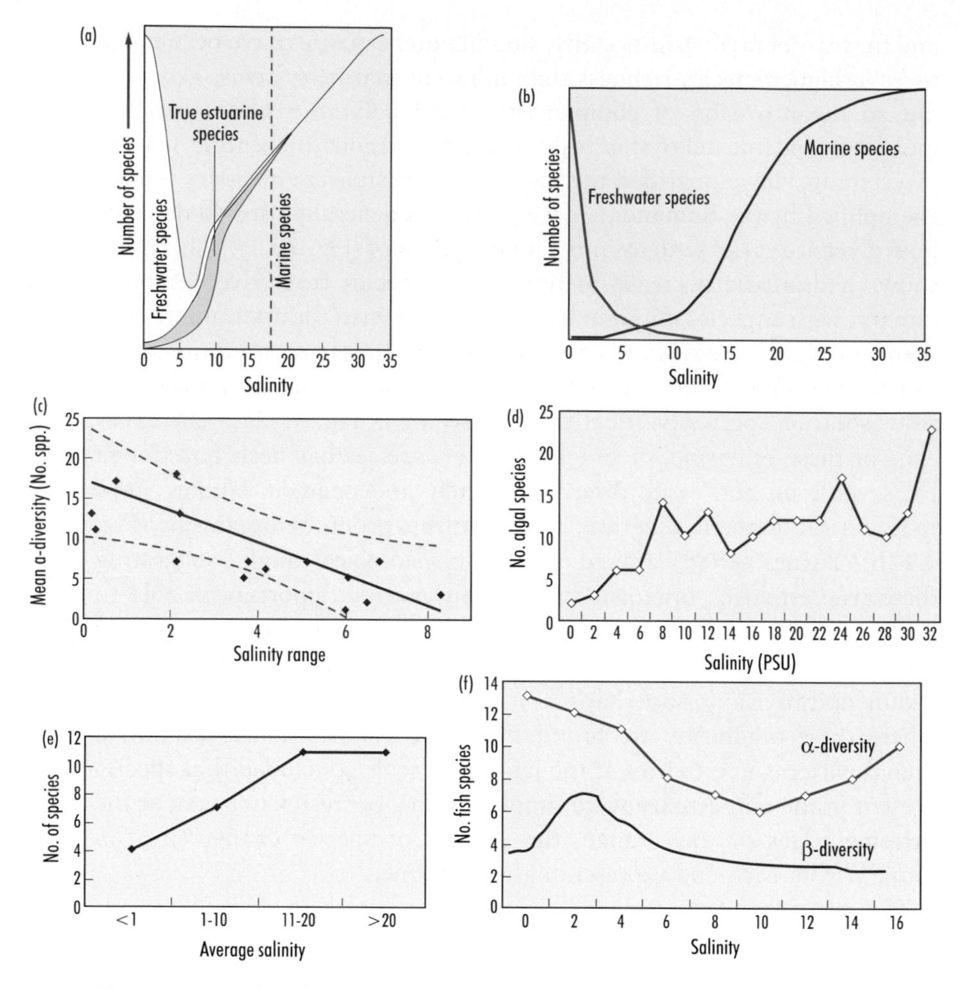

Fig. 4.12 Examples of diversity trends in estuaries. (a)–(c) Patterns of macroinvertebrate diversity: (a) original Remane diagram, (b) modified model removing estuarine species (see text), (c) linear model of diversity using salinity range. (d) Algal diversity in the Thames estuary. (e) Zooplankton diversity in the St Lawrence. (f) Fish diversity in Chesapeake Bay.

4.4.2 Larger-scale diversity trends

Anecdotal evidence has suggested that tropical estuaries are more diverse than their temperate equivalents, especially in the Indo-Pacific, so potentially a latitudinal diversity cline (Chapter 1) may exist in estuarine systems. If this is the case, it raises interesting questions about organism adaptation to fluctuating salinities. Empirical evidence for this is rather lacking, particularly for the key mid-estuary areas, and direct comparisons are often difficult due to the inherently different nature of temperate and tropical estuarine systems, particularly the overall salinity regime. Tropical estuaries, for example, often experience **hypersaline** conditions (see 4.5.2) due to intense evaporation, so for inner parts of these systems the salinity variability may be very different.

Direct comparisons of numbers of species found in mudflats across the globe are difficult due to variations in sampling methodology.

● Due to the huge level of variation in diversity within an estuary, between-estuary comparisons over large scales are difficult and sample location has to be carefully controlled to enable comparability.

Initial attempts to investigate the existence of large-scale diversity patterns have been made, however, in mid/outer estuary mudflats using a diversity index independent of sample size (Attrill et al. 2001). A significant increase in diversity was apparent at decreasing latitudes, supporting the idea that tropical estuaries are indeed more diverse than those at higher latitudes. Of the many explanations for the latitudinal diversity cline (Chapter 1), the most likely explanation is the impact of glaciation on high latitude systems (Box 4.1), many of which were effectively extinct until $<10\,000$ years ago. Tropical systems have therefore remained comparatively undisturbed for a much longer time period than temperate systems, and are thus much older systems which have had a longer **effective evolutionary time** to allow diversification. If tropical estuaries are found to be comparatively diverse, it is feasible that temperate estuaries are currently low diversity systems because they are geologically new rather than solely because of the impact of environmental variability.

● High latitude estuaries appear to have lower diversity than tropical ones, perhaps because they are new systems due to the impact of glaciation.

4.4.3 The role of biodiversity in ecosystem function

Two main questions describe what is perhaps one of the most important current ecological debates, namely whether biodiversity matters. Are high diversity ecosystems more stable? Is the functioning of an ecosystem affected by how many species are present? The comparatively low diversity to be found in mid-estuarine mudflats makes them excellent test systems for investigating the role of biodiversity in the functioning of an ecosystem. The estuarine assemblage can be manipulated and still be comparatively representative of the natural state, unlike experiments on grassland, rainforest, or coral reef systems for example, where such studies are only ever going to include a small fraction of extant species.

● Ecosystem functions include measures of primary productivity, decomposition rates, etc. The question is whether changes in biodiversity, without associated changes in biomass, influence these functions. Are all species necessary for the ecosystem to function?

Two main studies that involved estuarine invertebrates have been undertaken to address the relationship between diversity and ecosystem function. Emmerson et al. (2001) investigated diversity–function relations in a series of **mesocosms** measuring ammonia–nitrogen (NH_4–N) flux as an indicator or ecosystem function (a process essential for primary production) under varying treatments of biomass and diversity. Nitrogen flux increased with an increase in biomass, but relationships with diversity were idiosyncratic and varied between sites. Significant effects of species richness were apparent from a Swedish fjord, but in a Scottish estuary (Ythan) only species identity was important. In this case, the ragworm (*Nereis diversicolor*) had a disproportionate effect on NH_4–N production.

● Mesocosms are an intermediate-sized system, such as a dug-out pond or *in situ* enclosure, that can be replicated and manipulated to test both structural and functional parameters as a representative aquatic ecosystem.

However, one consistent result was that ecosystem function was much less variable and more predictable at higher diversities, supporting the idea that diverse systems are more stable. Bolam et al. (2002) undertook elaborate and extensive field experiment in Forth Estuary, Scotland, again manipulating the biomass and diversity of patches of mudflat and relating diversity and biomass to a set of indicators of the functioning of a detritus based ecosystem, such as organic content, nutrient fluxes, and community respiration. Neither changes in biomass nor diversity were found to significantly affect any of the ecosystem functions, with the exception of oxygen consumption, so provided no real evidence that diversity is related to the functioning of a mudflat community.

4.5 Other 'Brackish-water' Systems

In addition to estuaries, there are other aquatic systems that do not fit into truly marine or truly freshwater categories. Some of these, like estuaries, have salinity levels more dilute than seawater; others can have salt levels much higher than the sea. This section provides an overview of these other 'Brackish Water' systems, which can be categorized into three main groups: hyposaline seas (large bodies of water with salinity <34), hypersaline waters (coastal water bodies with salinity mostly >34), and lagoons (shallow coastal systems that exhibit a range of salinities).

● Hyposaline water has salinity less than full strength seawater; hypersaline water has salinity greater than seawater.

4.5.1 Hyposaline seas

From Scandinavia across through central Asia are several large, mainly enclosed bodies of water including the Baltic, Black, Caspian, and Aral Seas (Fig. 4.13a). None of these systems is freshwater, but have low levels of salinity and consequently many organisms of marine origin. The fauna of these seas has many similarities to estuaries, but crucially there are key differences between the two, exemplified by three examples.

The Baltic

The Baltic sea is a large (370 000 km^2) body of water surrounded by the countries of Scandinavia and northern Europe. The sea is generally very shallow (average depth 55 m), but with one basin dropping to 460 m. Crucially, there is only one narrow, shallow entrance linking the Baltic to the North Sea between Denmark and Sweden and restricting inflow of seawater. It takes 25–35 years to replenish the whole Baltic from the North Sea. Over 250 rivers flow into the Baltic, diluting the water retained within the sea. Generally salinity is <8, with a gradient apparent in salinity levels from Denmark to Finland. A large proportion of the northern Baltic is frozen every winter. However, despite the

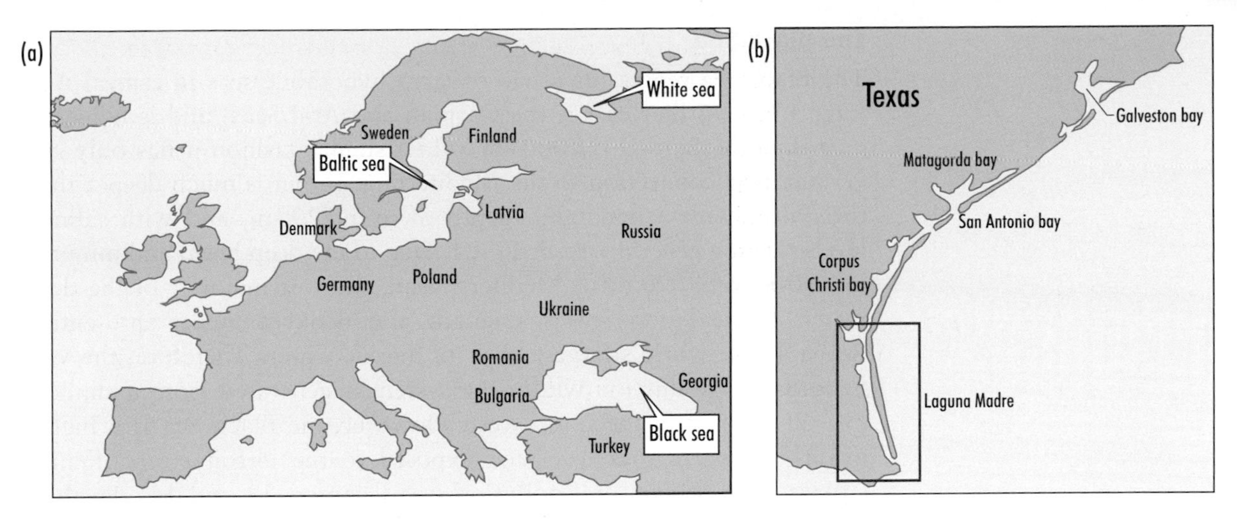

Fig. 4.13 (a) Map of Europe and Asia indicating the position of other 'brackish' water systems described in the text (b) Laguna Madre in Texas, U.S.

apparent salinity gradient, conditions in the Baltic are very different from those in an estuary as the Baltic Sea is practically tideless. Consequently, salinity levels at any one point do not vary to any great degree, resulting in a **stable salinity gradient** without the highly variable, and stressful, conditions apparent in estuaries.

● The Baltic has minimal tidal movement, so the salinity gradient present is stable, unlike the daily variation found in estuaries.

Some familiar marine groups, such as elasmobranchs and cephalopods, are absent from the main Baltic due to the consistently low salinity levels. For other organisms, particularly benthic species, their distribution in the Baltic tends to correspond much more closely to their theoretical **lethal salinity limit** than is the case in the fluctuating conditions of estuaries. Therefore, organisms are found in much lower salinities in the Baltic than in estuaries. Two outer-estuarine bivalves, for example, *Macoma balthica* and *Mytilus edulis* can be found over most of the Baltic down to salinities of 3 and 4 respectively. Due to the stability of the system and the extension of species ranges, in many parts of the Baltic there is an interesting mix of marine and freshwater taxa that is not found so clearly in estuaries. A fisherman on the east coast of Sweden, for example, could catch both pike (*Esox lucius*) and cod (*Gadus morhua*) from the same spot on the same day! Living in highly reduced salinities at the edge of an organism's range does have energetic costs, however. The extra energy required for survival at low salinities means that less energy is available for growth (the **scope for growth** is reduced); individuals of benthic species such as the clam *Mya arenaria* get smaller with a decrease in salinity up the Baltic. In summary therefore, the Baltic presents a similar set of species to those found in comparable estuaries, but the lack of tides and consequent stable salinity regime allows further penetration in relation to average salinity.

● Hyposaline seas, being low diversity, are very vulnerable to invasion by alien species.

The Black Sea

The Black Sea represents a trio of large hyposaline seas in central Asia (Fig. 4.13a, further east is the Caspian and Aral Seas) and is bigger in area than the Baltic (423 000 km^2). In a similar fashion it has only one very narrow connection to the Mediterranean, but is much deeper than the Baltic, with a maximum depth of over 2200 m, and with salinity levels ranging generally from 17–21. Due to the deep basin and minimal inflow of water from the Mediterranean, the vast majority of the deep water in the Black Sea is stagnant and deoxygenated, representing probably the world's largest body of anoxic water. Therefore, the vast majority of production within the system is generated from a shallow (30–60 m) shelf in the north-west. However, the Black Sea is a highly disturbed system that has been exposed to the introduction of alien species, over-fishing, and pollution (see Chapters 12 and 14). Furthermore, a history of inputs into the Black Sea from surrounding countries has exacerbated the deoxygenation of water in the basin and has caused the anoxic layer to rise in the water column, potentially threatening the survival of the shallow and productive north-western shelf.

● The Black Sea contains the world's largest body of anoxic water.

The White Sea

The White Sea (Fig. 4.13a) is a large (94 000 km^2) inlet in the Russian Arctic that presents apparently extremely harsh conditions for organisms. Water temperature varies dramatically over the year, the average Oct–May temperature being −1 °C, although the coldest surface temperature so far recorded was in the White Sea (−2.78 °C). During summer, however, temperatures can reach 20 °C. The sea surface and intertidal area are covered in ice from November to May. Generally, the salinity of the White Sea is 23–25, but following ice melt in summer it can drop to 5.

Despite these apparently hostile conditions, many typical estuarine species exist in the intertidal sediments, such as *Arenicola marina, Mya arenaria*, and *Macoma balthica*, and perhaps most surprisingly beds of seagrass (*Zostera marina*) develop during the summer.

4.5.2 Hypersaline waters

● Many tropical estuaries are often hypersaline, or have hypersaline reaches, during dry periods. Hypersaline waters do not include inland brines such as the Utah salt lakes which have no connection to the sea or a larger body of water.

Hypersaline waters are those with a salinity regime higher than that of seawater (i.e. >34). To achieve such concentrated saline conditions, the water body has to have an evaporation rate greater than the inflow of fresh water, so hypersaline waters are typically tropical in location, or situated in continental areas with extremely high summer temperatures. They also tend to be of a much smaller scale than hyposaline seas.

An example of a hypersaline water is the Sivash (others include the Suez Canal and Curaçao Lakes, West Indies). This is located on the

Crimea peninsula on the north coast of the Black Sea (Fig. 4.13a) and is a comparatively long, thin body of water isolated from the Sea of Azov (the north-eastern section of the greater Black Sea) by a land barrier. There is only a very small entrance to the Azov Sea, so water exchange is exceptionally limited. Salinity in the Sivash varies along a gradient from 11 at the connection with the hyposaline Azov Sea up to 132 at its most inland part. Tides in the area are minimal, so there is little daily fluctuation, but conditions vary seasonally due to the very cold winters and hot, dry summers experienced in the Crimea.

The conditions in the Sivash therefore present very different problems for organism colonization than estuaries or hyposaline seas. Organisms penetrating the Sivash originate from the hyposaline Azov Sea, and so will have to be able to tolerate both reduced and elevated salinity levels, their distribution up the Sivash being dependent on their **upper salinity tolerance**. However, many organisms that thrive in estuaries have also the ability to deal with increased salinity, examples in the Sivash being the flounder (*Platichthys flesus*) (occurs up to a salinity of 60), the cockle (*Cerastoderma edule*) (65), and the mullet (*Mugil cephalus*) (110). However, in the most hypersaline areas (>130) such organisms cannot survive, and here the exceptionally salty waters are colonized by a few specialist species, such as the brine shrimp (*Artemia salina*) and larvae of the midge *Chironomus salinus*.

● The Latin names of ultrahaline species indicate their salty habitat – note here the specific names '*salina*' and '*salinus*'.

The existence of hypersaline waters provides a salinity range from 0 to >130 over which organisms can be distributed and therefore extends the graphs of diversity trends developed for estuaries and hyposaline seas (Fig. 4.12). Generally, there is a decrease in diversity with an increase in salinity from 34 until the **ultrahaline** conditions are reached, which are inhabited by just a couple of species. The Remane diagram can be extended (Hedgpeth 1967; Fig. 4.12a) to demonstrate how the diversity trend over the full salinity range may be expressed (Fig. 4.14).

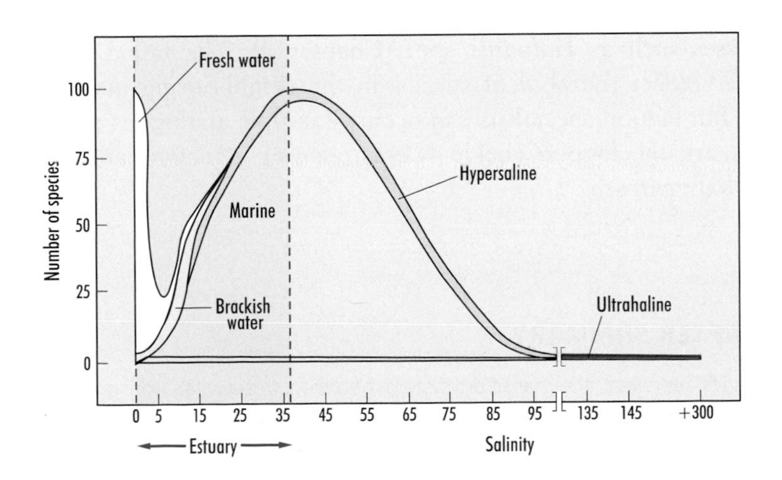

Fig. 4.14 Hypothetical diversity trend over the full salinity range including hypersaline waters.

4.5.3 Lagoons

Lagoons are comparatively small areas of sea that have been semi-isolated by the development of a barrier. Superficially, they appear similar to systems such as the Sivash, but generally lagoons are much smaller and in particular they are very shallow, generally only a few metres deep. Consequently, they rarely become stratified in the summer due mainly to wind action. Lagoons can form through the development of two main types of barrier. The deposition of sand can commence offshore, which eventually builds into a bar with no connection to the land (similar to bar-built estuaries), thus enclosing a shallow body of water. An example is the Laguna Madre in Texas, USA (Fig. 4.13b). Alternatively, barrier formation can extend out from the land (longshore barrier), eventually isolating the lagoon (e.g. the Vistula lagoon, Poland, Fig. 4.13a).

● Lagoons can vary in size considerable. The whole SE of the USA is a series of lagoons and bar-built estuaries extending 4500 km along the coast. Most are much smaller, the largest in the UK being the 14 km Fleet lagoon in Dorset, England.

Lagoons demonstrate a wide range of salinity regimes, from hyposaline (e.g. Vistula, 0–6) to hypersaline (e.g. Laguna Madre, 39–62). In rare cases a variable salinity gradient is apparent over very small scales within lagoons, an example being the Swanpool, Cornwall, UK. This is a small, generally hyposaline lagoon with depths <3 m. A low sill at the seaward end of the lagoon enables full-strength seawater to overflow at high tides. Due to its density, this seawater slides to the bottom of the lagoon creating a **halocline** with low-salinity water floating on seawater and a fast transition in terms of salinity from 4 to 34 (see also ROFIs in Chapter 7). In tropical areas, lagoons are often fringed by mangroves (see Chapter 9); reed beds, e.g. *Phragmites* are the dominant fringing plant in temperate areas. However, due to their size, shallow nature, and comparative stability, lagoons often differ markedly from estuaries and other larger brackish water systems by possessing a diverse submerged plant community, with macrophytes able to root on the lagoon floor. The plants present reflect the salinity of the lagoon, with pond species such as water lilies (e.g. *Nymphea*) present in the more freshwater systems while **halophytes** dominate saline lagoons, in particular seagrasses such as *Halodule* spp. (Chapter 9). The fauna of lagoons tends to reflect the pool of species in the neighbouring main body of water, but lagoon specialists can occur. Example analogous to estuarine species are the lagoon cockle (*Cerastoderma glaucum*) and the snail *Hydrobia ventrosa*.

● Halophytes are angiosperm plants that can tolerate increased salt levels in the sediment. These include seagrasses and saltmarsh plants.

CHAPTER SUMMARY

- Estuaries represent the transition system between freshwater and marine biomes, but most estuaries were only formed following the end of the last ice age 6–10 000 years ago.

- Conditions within tidal estuaries are exceptionally variable, particularly salinity, sediment type, oxygen and temperature. This variability results in harsh conditions that are amongst the most challenging for life in the marine environment.
- Estuarine systems include a mix of freshwater and marine organisms, few of which can tolerate the physico-chemical conditions present in the mid-estuary. Most successful organisms are excellent osmoregulators, which are capable of adjusting the balance between their internal body fluid salt concentrations and those of the external water. These species can be very abundant in estuarine mud.
- The distribution of organisms within an estuary is not just controlled by salinity. Many organisms require the presence of mud, or depend on cues from other environmental variables, such as temperature or freshwater flow, to determine where they are found.
- The amount of freshwater entering estuaries from the river has a major influence on the ecology of the system, affecting production, diversity and distribution of organisms.
- Estuaries are exceptionally important nursery grounds for many marine fish species, young fish exploiting the high food source and more favourable conditions for survival and growth.
- Estuarine food webs are powered by detritus from a range of marine, freshwater, terrestrial and estuarine sources, although it would appear that terrestrial carbon is relatively unimportant, despite vast amounts entering temperate estuaries in autumn and winter.
- Estuaries are exceptionally important habitats for birds, particularly those on migration. Four and half million shorebirds migrate to just the British Isles every year, where they exploit the high biomass of invertebrates in mudflats.
- Many inland seas, such as the Baltic and Black Sea, have conditions similar to estuaries, but generally do not have tides. In these systems the salinity gradient is much less variable and marine organisms tend to be found at lower salinities in these seas than in tidal estuaries.

FURTHER READING

Attrill and Rundle (2002) provide an overview of the ecotone/ecocline debate in estuaries. Emmerson et al. (2001) is a seminal study of the consistency of global patterns in estuarine biodiversity. Sutherland (1996) gives a detailed insight into the behaviourial interactions among and between avian predators of estuarine ecosystems.

- Attrill, M. J. and S. D. Rundle 2002. Ecotone or ecocline: ecological boundaries in estuaries. *Estuar. Coast. Shelf Sci* 55: 929–36.
- Emmerson M. C. et al. 2001. Consistent patterns and the idiosyncratic effects of biodiversity in marine ecosystems. *Nature* 411 (6833): 73–77.
- Sutherland, W. J. 1996. *From Individual Behaviour to Population Ecology*. Oxford University Press, Oxford.

Chapter 5

Rocky and Sandy Shores

CHAPTER SUMMARY

Human settlements abound on coastlines, a pattern of habitation that persists to the present day. Early marine biologists found the shore attractive for their science for similar reasons and the shore habitat is highly accessible. The biodiversity of the shore is exceptionally high compared to land, with all major taxonomic groups represented. Ecological research on shores has underpinned much of present-day marine ecology and has strongly influenced mainstream ecology. In addition, shores are of increasing concern to governmental policy-makers because of their recognized provision of goods and services for humans, such as coastal defence, recreation and fisheries products. Shorelines around the world are experiencing major impacts caused by human population pressure and there is a necessity to defend our economic and social investment in coastal development from the effects of accelerated sea-level rise. Understanding how shores function ecologically is therefore important not simply for intellectual reasons, but for practical purposes if their full value is to be maintained.

5.1 Introduction

● Shores are studied because they are accessible, taxonomically relatively simple and provide ecological goods and services to society (Preface).

The habitat created by the land meeting the sea is the most accessible part of the marine environment for humans. To explore deeper and further out to sea requires rafts, canoes or larger vessels and involves increased risks, while the shore itself, including adjacent shallow waters, offers rich resources that can be collected without great effort. It is therefore no surprise that early human settlements abound on coastlines, a pattern of habitation that persists to the present day.

Early marine biologists found the shore attractive for their science for similar reasons. The shore habitat is highly accessible (rubber boots or bare feet are the only equipment required). The biodiversity of the shore is exceptionally high compared to land, with all major taxonomic groups represented (even insects, if you look hard enough), and the fauna and flora are easily collected for study or manipulated experimentally. As a result, it is not surprising that ecological research on shores has underpinned much of marine ecology and has been the laboratory of preference for many mainstream ecologists.

In addition, shores are of increasing concern to governmental policy-makers because of their recognized provision of **goods and services** for humans, such as coastal defence, recreation, and fisheries products (Duarte 2000), and the current threats to these services (Brown & McLachlan 2002; Kennish 2002; Thompson et al. 2002). Shorelines around the world are experiencing major impacts caused by human population pressure and there is a necessity to defend our economic and social investment in coastal development from the effects of accelerated sea-level rise by spending many millions of dollars on coastal defence projects. Understanding how shores function ecologically is therefore important not simply for intellectual reasons, but for practical purposes if their economic value, as well as their aesthetic and cultural values are to be maintained.

● Shorelines around the world are experiencing increasing pressure from human developments.

5.2 What is the Shore?

Defining the shore is not as straightforward as it might seem. The early domination of shore ecology by north-west European and North American scientists who worked in macro-tidal areas has resulted in a very restricted perspective of a shore: the area between high and low water marks. As a result, shore ecology became synonymous with

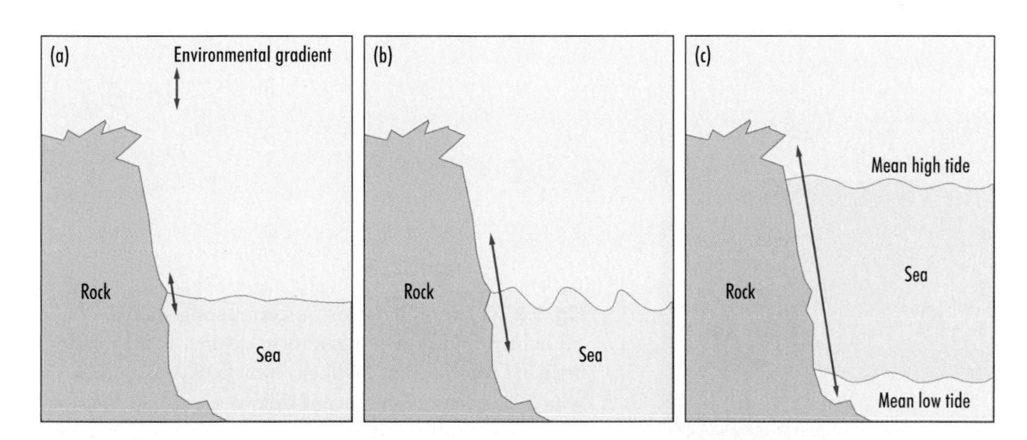

Fig. 5.1 The wet-dry gradient (a) that characterizes the shore is set up by the interface between water and air. Waves will extend this gradient up and down the shore (b), and tides (where present) further amplify the gradient (c). The effect of the tide is thus to greatly expand an existing gradient, tides do not create the gradient *per se*.

intertidal ecology. This was unfortunate for several reasons. First, many shores around the world do not experience significant tides and in such areas changes in air pressure will cover and uncover the shore to a far greater extent than tidal action (Fig. 5.1). Second, the distribution and abundance of shore biota were inevitably understood in the context of tidal rise and fall, yet tides *per se* cannot be responsible for these patterns (Figs 5.1 and 5.2). Third, the emphasis on tides as a controlling factor promoted the view that physical variables limited the distribution and abundance of shore organisms and thereby distracted attention away from the importance of biological processes. Fourth, the functional influence and extent of the shore may extend far into the terrestrial hinterland and down beyond the surf zone hundreds of metres offshore (Brown & McLachlan 1990). The influence of the shore extends even further if a larger scale perspective is taken to include the supply of

(a)

(b)

(c)

Fig. 5.2 Zonation patterns on contrasting shores. (a) lichens and brown algae form distinct bands on an artificial rock wall at Pwllheli, North Wales, an area which experiences significant tidal rise and fall of a few metres. (b) a typical shoreline in an essentially atidal area at Tjarno, Swedish west coast. (c) The angiosperms *Spartina* and *Salicornia* form characteristic zones at the edge of a salt marsh, Sylt, Germany.

larval stages from offshore water masses, the origins and end points of migratory taxa such as turtles and shorebirds, and the material linkages between shorelines of different types. Thus, as for other marine systems, defining 'the shore' is quite an arbitrary exercise and understanding patterns and processes within an area like the intertidal zone requires acknowledgement of a much larger scale system within which that limited area of habitat is set.

● Not all shores experience significant tides, yet they support a typical 'intertidal' fauna and flora.

● The functional extent of the shore extends well above and well below normally recognized limits.

In this chapter, the focus is very much on rocky and sandy shores. Other kinds of shore habitats (mangrove, estuary, tropical reef) are dealt with elsewhere in this book, but given the broader network in which rocky and sandy shores operate, reference to these other systems draws attention to the important ecological linkages that exist between them.

5.3 Environmental Gradients and the Shore

The shore is characterized by several environmental gradients (termed **ecoclines** elsewhere; Whittaker 1974). These gradients interact and intersect in quite complex ways to generate specific environmental conditions for shore organisms. All other things being equal, specific and predictable biological assemblages will be found at the intersections of these gradients according to their physical, competitive and physiological ability to occupy particular sections of the gradients, a phenomenon known as **zonation** (Figs 5.2 and 5.3).

● Different species have different physiological and competitive tolerances and hence occupy different sections of environmental gradients.

There are four main gradients on shores: wetness/dryness; exposure to wave action; substratum particle size; salinity.

5.3.1 Wetness/dryness

The wetness/dryness gradient is set up at the tension between water and air. The environment becomes progressively drier with distance from the water surface and is amplified (not created) by waves and tides (Fig. 5.1). Almost all of the plants and invertebrates encountered on rocky and sandy shores are marine aquatic in phylogenetic origin (Fig. 5.3) and the majority require access to the marine environment to complete their life history. Generally, species have different requirements (tolerances) and are able to live further or nearer to the water surface according to these tolerances. It does not follow that high shore organisms cannot withstand immersion in seawater: many appear to be restricted to high shore levels because of biological pressures (e.g. predation or competition) from species living at lower shore levels (5.4.1 and Chapter 1).

● Most shore species are marine in origin and need regular access to the sea.

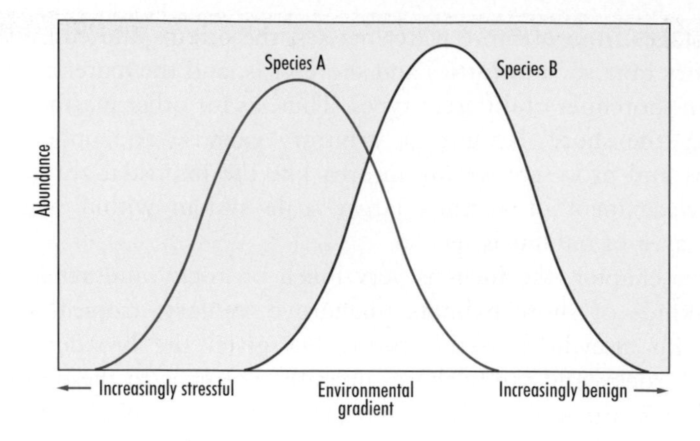

Fig. 5.3 Zonation of two species along a generalized environmental gradient running from stressful to benign conditions (with respect to the organisms). The two species A and B occupy different sections of the gradient according to their ability to tolerate physiological stress (left-hand tail of the distribution) or biotic interactions (right-hand tail of the distribution – see also 5.4). Note that for terrestrial taxa, such as saltmarsh plants, the sea is a stressful environment, while for marine taxa, it is the land that is unfavourable.

5.3.2 Exposure to wave action

A feature of all coastlines is that they experience wave action to varying degrees (Box 5.1). Waves are generated by the frictional drag of wind on the sea surface and the longer the distance over which the wind blows (termed **fetch**), the higher the waves. Waves may also arrive on the shore in the form of swells; these are the ghosts of large wind generated waves produced by storms far offshore. The forces expended when waves arrive at the shore will depend on how much energy is extracted by the immediate offshore substratum and any associated biological structures, such as kelp forests, reefs, or sandbanks. Thus, shallow sloping offshore areas will tend to dissipate the energy in the waves arriving at the shore, while steeper cliffs will experience a much greater physical impact of wave action. Because waves are wind generated, physically sheltered localities such as fjords, narrow estuaries and shores facing away from prevailing wind directions will have less wave action on average. Wave action, of course, will vary daily and seasonally. Many of the anatomical and behavioural attributes of shore organisms reflect attempts to minimize the risk of dislodgement by wave action and this allows different species to occupy different sections of the wave action gradient.

● Wave action is a major determinant of community structure and composition, as well as individual shape, form, and behaviour.

Shore ecologists use the terms '**wave action**' and '**exposure**' synonymously, and the latter should not be confused with exposure to air when shores are exposed by receding tides, more properly termed '**emersion**'. In addition, the concept of exposure to wave action is difficult to apply

Box 5.1 Wave formation

Waves are generated by friction from winds at the air/water interface. The length of fetch (the distance over which air moves unimpeded by a land mass), wind speed, direction and duration and the depth of water all affect the **period** and **height** of waves. Waves can continue to be propagated even without the influence of wind, this is known as **swell**, which decays gradually over time in the absence of wind. Coasts that are exposed to the ocean typically have swell with a long period. Given the right seabed conditions, these coasts produce some of the best waves sought by surfers. Although the surface features of waves are clear to see, **internal waves** occur beneath the surface, which decay towards the seabed. These internal waves rarely persist beyond a depth of 50 m. However, as waves pass into shallower areas the internal waves will create physical disturbance on the seabed, which will increase with decreasing water depth. This produces a gradient of wave disturbance on the sea floor that decreases with distance offshore and with depth.

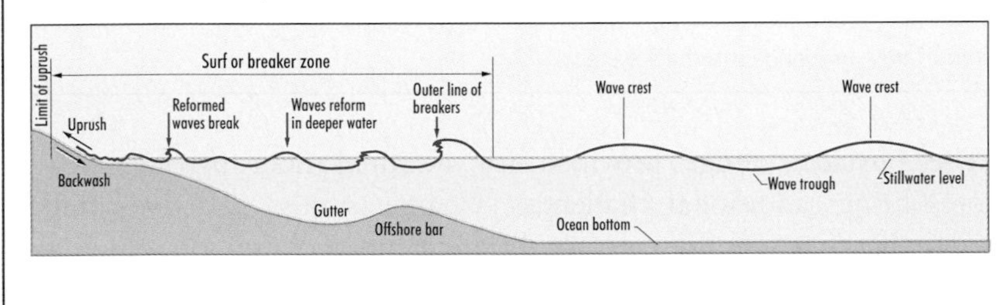

sensibly to sandy beaches and mudflats, because of the complex interactions between particle size, beach slope, and wave energy. Sediment shore ecologists thus prefer to describe a beach as having a particular **morphodynamic state** that occupies a point along a **reflective–dissipative** spectrum, rather than a wave exposure gradient (see 5.3.5).

● Morphodynamic state is preferred to the term 'wave exposure' for describing sandy beaches and mudflats.

5.3.3 Particle size

Shores can be ordered along a particle size gradient ranging from extremely large particles, such as a cliff or boulder beach, to those made up of very fine particles only a few microns in diameter (Fig. 5.4). Large particles provide a stable surface for attachment, and **epifauna** and **flora** dominate such shores. In contrast, finer particle sands are often too unstable to permit surface attachment and the fauna lives within the beach (the **infauna and meiofauna** (Box 5.2)), unless wave action is sufficiently low. Shores of intermediate particle size are very inhospitable for marine life because they are too unstable for surface dwellers and are comprised of particles that are too large for an infaunal life style (Fig. 5.4). Thus, different species are capable of living in different sections of the particle size gradient and many of them are specialized to cope with the unique conditions presented.

● The sizes of the particles that make up the shore have a huge effect on the kinds of organisms that can survive there.

Box 5.2 Macrofauna and meiofauna

In addition to the commonly encountered larger **infauna** of beaches and mudflats, such as clams, shrimps, and polychaete worms, a rich variety of tiny organisms live on and between the individual particles, including nematodes, harpacticoid copepods, gastrotrichs, archaic groups of polychaetes, kinorhynchs, and flatworms. These are grouped together as the **meiofauna**. Because of their small size (typically less than a millimetre) these taxa are not familiar to most marine biologists, let alone the non-expert. They occur in very large numbers, hundreds per 100 cm^2 of beach, and because of their small individual body size (only a few μg), they have high respiration rates per unit mass. This means that their productivity can be almost as high, and in many cases higher, than that of the macrofauna in the beach. Meiofaunal taxa also occur on rocky shores (Hicks 1985, Gibbons & Griffiths 1986) in association with the microhabitats provided by larger species, but they have been less well studied than those in sediments. Notwithstanding their small size and their taxonomic and identification challenges, meiofauna are an extremely rewarding group to work with, since so little is certain about their ecology and functional importance in marine intertidal systems.

The physical refuge provided for infaunal taxa by fine particle shores brings additional challenges. Coarse (sandy) and fine (muddy) particle beaches will present very different physical and microbiological environments, mainly due to their differences in surface area available for microbial activity and their capacity to retain water at low tide (Fig. 5.5).

5.3.4 Salinity

This gradient has been largely dealt with in Chapter 4 and only a summary is provided here for completeness. The salinity gradient is generated by the meeting of fresh water and seawater and the habitat where this occurs most obviously is the **estuary**. Here, river water with a very low concentration of ionic salts meets marine water with very high levels of such ions. The degree of physical mixing between seawater and fresh water can range from very little, with the less dense fresh water flowing over the top of the marine water, to complete mixing, where turbulence by large waves can result in similar salinities at the water surface and the seabed. The exact nature of the mixing and hence structure of the vertical and longitudinal salinity gradient will be variable for any one estuary, depending on the relative volumes of fresh and sea water, wind-driven physical turbulence that results in mixing of fresh and sea water and spring–neap tidal patterns. The distribution of the fauna and flora along this gradient will thus represent a response to the average conditions present. The majority of species found in estuaries are marine in origin, with few freshwater taxa penetrating far downstream.

● Interactions between fresh and sea water drive sediment distributions in estuaries.

Fig. 5.4 The environmental gradient of particle size, ranging from large rocks and cliffs to mud composed of grains only a few microns in diameter. (a) exposed cliff (Otago Peninsula, New Zealand), (b) sheltered cliff (Little Loch Broom, Scotland), (c) boulder shore (Isle of Wight, England), (d) shingle beach (Norfolk, England), (e) sandy beach (Clacton, England), (f) mangrove (Manakau, New Zealand), (g) seagrass bed (Moray Firth, Scotland), (h) Shore organisms can only live on the sides of the larger particles (epifauna and flora), and between the particles (infauna) on sandy beaches and mudflats (except in extremely sheltered habitats where some epifauna and flora reappear). Intermediate-sized particles that make up shingle and cobble beaches present a hostile environment for both types of organism, because they are too large to retain water and too mobile for surface attachment.

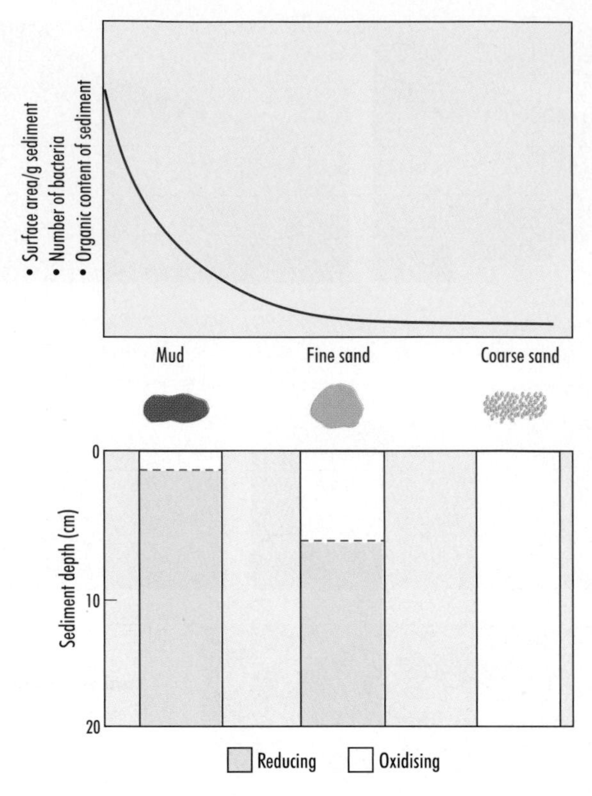

Fig. 5.5 The average particle size that makes up sandy beaches and mudflats has a profound effect of the physico-chemical conditions experienced by the infauna. Muds have a much greater surface area available for microbial activity by virtue of their small individual particle size. They also retain water better than sands and are usually waterlogged. The microbial activity reduces compounds such as iron and sulphate in the sediment creating anoxic conditions the extent of which is reflected in the depth at which the sediment changes colour from orangey-brown (oxidized iron particles) to dark brown or black (reduced iron sulphide).

5.3.5 Interactions between gradients and zonation patterns

● Many different gradients interact to produce conditions suitable for particular species and communities.

All the above gradients interact to generate particular conditions for life on shores. For instance, increasing wave action will amplify the wetness–dryness gradient, and thereby **uplift** biological zones with increasing wave exposure. Wave action and water movement will **sort** particles according to their mobility and in the process drive the overall beach environment towards a more reflective or a more dissipative morphodynamic state (Table 5.1). The interactions between salinity and particle size are complex and dealt with in Chapter 4.

The types of fauna and flora recorded at any location on a shore can be understood as the biological responses to the product of these interacting gradients, allowing for biogeography (Chapter 1). The overall

Table 5.1 General features of beaches at the extreme ends of the morphodynamic spectrum. Intermediate morphodynamic states have intermediate features. After Raffaelli and Hawkins (1996). Dissipative (a) and reflective (b) beaches on the Coromandal coast, New Zealand.

	Dissipative beaches	Reflective beaches
Sediments	Fine	Coarse
Waves	Large	Small
Slope	Shallow	Steep
Tidal range	High	Small
Wave period	Long	Short
Swash conditions	Benign	Harsh
Fauna	Rich	Impoverished

(a)

(b)

patterns of distribution and abundance are revealed as **zonation patterns**, which are most obvious on exposed macro-tidal rocky shores, and least obvious in sheltered sandy flats. Not surprisingly, we understand much better the processes maintaining rocky shore zonation than is the case for other shore habitats.

One of the most important features of rocky shore zonation patterns is the similarity that occurs worldwide. These were described most elegantly by Stephenson and Stephenson (1949, 1972), in a period when marine ecology was developmental and more descriptive. Similar, but not identical, types of fauna and flora occupy similar positions on shore gradients, independent of biogeographical region (Fig. 5.6). The significance of these **universal features of zonation** is that they imply similar and strong underlying structuring processes that operate on all rocky shores: only a limited range of body forms and phylogenies can cope with life on shores. Also, these consistencies in the major zoning species worldwide have permitted the invention of a spatial referencing framework for the unambiguous location (within a shore) of where particular studies were undertaken (Fig. 5.6). Thus, the term midlittoral (or **eulittoral**) would conjure the same mental image of a shore habitat for all rocky shore ecologists.

● Remarkably similar general patterns of zonation recur throughout the world.

Fig. 5.6 The universal zonation scheme proposed by Stephenson and Stephenson (1972). The main zones are where biota occur worldwide, with slight modifications according to biogeography, and is well illustrated by this rocky shore on the Otago Peninsula, New Zealand.

Similar schemes have been proposed for sandy shores (Fig. 5.7), and there is some evidence of universal zonation patterns for crustaceans (Dahl 1952). However, most general sandy shore zonation schemes are a reflection more of the physical environment of the beach, than its biology (Salvat 1964).

5.4 Causes of Zonation

Given the striking zonation patterns seen on rocky shores, and to a lesser extent on sandy shores and mudflats, it is not surprising that a major preoccupation of shore ecologists has been discovering the **determinants of zonation**. Earlier workers naturally assumed that the tides must be in some way responsible, given the intimate association between the so-called intertidal area and the twice-daily rise and fall of the tides.

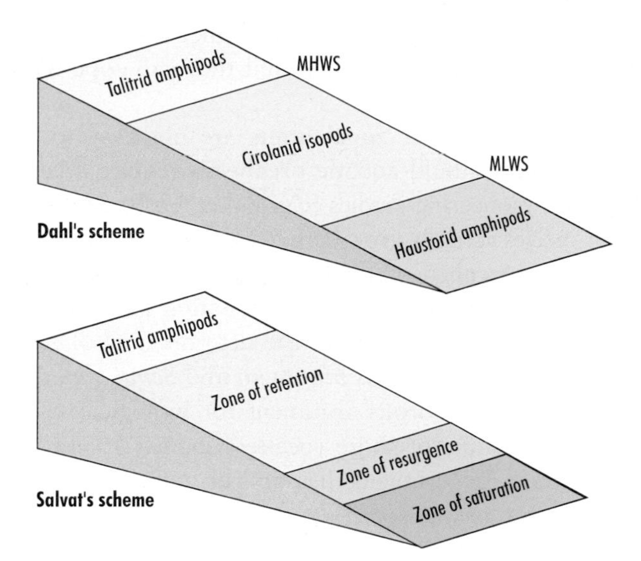

Fig. 5.7 'Universal' zonation schemes for sandy beaches provided by Dahl (1952) based on the occurrence of crustacean groups, and Salvat (1964) based largely on the degree to which water is retained by the sediment at low tide, producing visibly different zones on the shore.

Various schemes and theories were advanced, exemplified by the **critical tidal level** arguments put forward by Colman (1933), and later developed by Doty (1946) and others. It was argued that as one moved down the shore, there were certain regions where only small differences in shore level were characterized by large changes in the period of immersion, when averaged over a year. In other words, the immersion–emersion gradient was particularly steep at these points and thus critical for any species that decided to settle in these regions. Upper and lower distributional limits of several species seemed to coincide precisely with these critical levels. While plausible, indeed compelling, Underwood's (1978) re-examination of this theory revealed the basic science as thoroughly flawed, species distributional limits occurred haphazardly along the shore rather than in groups, and re-plotting of the tidal data did not support the idea of critical levels. Nevertheless, the concept lingers on in several marine ecology textbooks, perhaps because of a reluctance to take the tides out of intertidal ecology.

● Zonation patterns on shores cannot be explained by tidal rise and fall.

Given that the majority of species encountered on rocky and sandy shores are marine and aquatic in origin, we should not be surprised that the **upper zonal limits** of a large number of species have been shown to be associated with **physiological tolerance** to factors such as desiccation and thermal stress. In general, species found higher on the shore are more able to tolerate dryness and thermal stress, than lower shore species. However, it is noteworthy that many of these experiments have been focused on adult individuals. It is often the environmental conditions experienced by the **recruit** (larval) stages that more importantly determine the adult distributions, especially for sessile taxa such as barnacles and mussels. The juveniles of many marine snails recruit to

● Upper distributional limits of species are generally (but not always) set by their tolerance to physical factors.

lower shore levels, and later migrate to higher levels as they become larger and their physiology alters such that they can cope with different environmental conditions.

The determinants of **lower zonal limits** are not easy to attribute to physical factors. Why should aquatic organisms require a certain period of drying? Several pioneering studies (e.g. Baker 1909) demonstrated that high intertidal species actually grew better under a lower shore tidal regime and Stephenson and Stephenson (1949) suggested that competition and/or predation might be responsible for some zonation patterns. However, it was Connell's classic experiments on the determinants of zonation patterns in the barnacles *Chthamalus montagui* and *Semibalanus balanoides* that provided the most rigorous argument for biological factors setting lower distributional limits of shore species (Connell 1961a,b). The persuasiveness of these experiments lies in the controlled manipulation (removal in this case) of one species (*Semibalanus balanoides*) occupying the zone immediately below another potentially competitive species (*Chthamalus montagui*). Connell was able to show that in the absence of competition for space, a **limiting resource** on his rocky shore, the higher shore species could survive at lower shore levels than those at which it was normally found. This **manipulative approach** to understanding the determinants of zonation patterns has been emulated repeatedly by many workers. As a result, it is now something of a paradigm in rocky shore ecology that lower distributional limits of species occur as a result of biological factors, while physical factors set upper limits on distribution. However, there are exceptions, especially amongst the macroalgae (Hawkins & Hartnoll 1983). Thus, grazers may prevent the upshore extension of foliose seaweeds on some Australian shores (Underwood & Jernakoff 1981), red and brown seaweeds grows further upshore in the absence of limpets on the Isle on Man (Hawkins & Hartnoll 1985) and the green seaweed *Codium* may be partly limited by grazing from above (Ojeda & Santilices 1984). Notwithstanding these exceptions, an intriguing question posed by this paradigm is whether it is applicable to zonation along other kinds of gradients, particularly the wetness-dryness gradient on sandy shores, but also to salinity, exposure, and particle size gradients in other habitats.

● Competition and predation have been shown to be important determinants of lower distributional limits of species on many rocky shores.

There has been very little exploration of this question for the last three environmental gradients, but some work has been done on **zonation patterns on sandy shores**. One of the issues for sandy shore (and mudflat) ecologists is that the description of zonation patterns in these habitats requires the destruction of the medium in which the organisms live: quantitative sampling involves digging out volumes of sediment and separating the fauna, usually by sieving or elutriation. Nevertheless, this approach reveals zonation patterns that are usually not visible at the sediment surface (Fig. 5.8).

● Zonation patterns, as well as their causes, are much harder to detect on sandy beaches and mudflats.

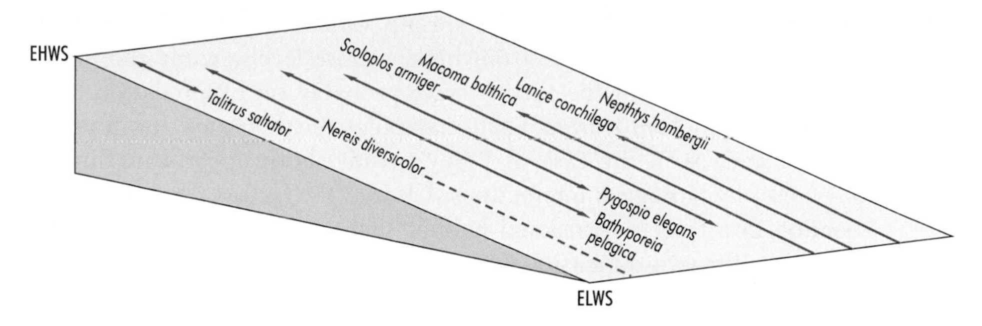

Fig. 5.8 Zonation on a sandy beach, Newburgh, Scotland. Only the most abundant species are represented for the sake of clarity. Data from Raffaelli et al. (1991).

Marine ecologists have sought to understand the role of individual species within sediment communities by systematically removing particular species and studying the consequences. Carrying out such species removal experiments in sandy beaches or mudflats is fraught with interpretational problems if the habitat has to be disrupted to such a severe extent. Just think how difficult it would be to remove 10 individuals of one species from a patch of sandy beach that contains 100 individuals comprising 20 species. A second problem arises in that, unlike the sessile fauna of rocky shores, the infauna is mobile and can easily move back into areas cleared in manipulative experiments. In other words, it is difficult to maintain the integrity of the experimental design of species removal experiments. Third, the beach fauna lives in a three-dimensional environment and displays zonation vertically as well as horizontally, and some species redistribute themselves at each high tide. Experiments aimed at removing potential competitors therefore have an additional layer of complexity. Fourth, it is not clear whether space is a limiting resource for which competition occurs in sandy beaches and mudflats, in the same way that it does on rocky shores. Certainly, the intensive **interference competition**, where individuals overgrow or crush one another, that is seen between space-occupying species on rocky shores is virtually absent in sediment assemblages, probably because individuals tend to be mobile (as opposed to the many attached biota found on rocky shores) and can simply move away from one another.

● Space is not such a severe constraint on sandy shores as the fauna are mobile and can relocate to avoid competition.

Despite these issues, several investigations have demonstrated the importance of physical factors in setting upper distributional limits, such as Petersen and Black's (1987, 1988) transplant experiments of the bivalves (*Circe* and *Placamen*) to higher shore levels where they grew more slowly and suffered higher mortality due to physical factors. The importance of biological factors in setting distributional limits is illustrated by Posey's (1986) study of a Californian beach. Here, the burrowing activities of the ghost shrimp *Callianassa* excluded a tube dwelling worm *Phoronopsis* from higher shore levels, while the lower distributional limits of the shrimp were probably set by predation by a fish, the sculpin *Leptocottus*. Such manipulative studies are few and far between and, as in the case of Posey's study, both upper zonal limits may be set by biological interactions. Clearly, the factors responsible for zonation in a three-dimensional habitat such as a sandy beach are not as clear as for two-dimensional rocky shores.

5.5 The Organization of Shore Communities

● Shore communities are organized by many different kinds of processes, such as predation and energy supply.

In common with all other biological assemblages, shore communities are organized by a combination of **top-down** (consumer-driven) and **bottom-up** (resource-driven) processes. Because many of the classic studies on top-down processes in food webs were first performed on rocky shores, there has been a tendency to generalize these results across all shores and even across all types of ecosystem. Notwithstanding the fact that some of these earlier studies would be unlikely to pass the present-day peer-review process (Box 5.3), they have made a major contribution to mainstream ecological theory. Before we discuss the relative importance of different processes in the dynamics of shore assemblages, it is worth rehearsing the features of shores that make them so amenable to studying these kinds of questions.

Box 5.3 Manipulative field experiments

The most convincing manipulative field experiments are those where the treatment and control plots are highly replicated to provide the necessary statistical power to detect an effect of manipulation. Such powerful designs will convince the most critical reviewer or editor. It is therefore somewhat ironic that the two rocky shore experiments which ecologists are completely persuaded by are fatally flawed in this respect. The first is Lodge's (1948) removal of the limpet *Patella* from a wide strip running down a shore on the Isle of Man and the subsequent bloom of algae in the manipulated area. The second is Paine's (1974) set of experiments on *Pisaster* on the Washington coast, where removal of the starfish saw a massive increase in mussels and a collapse in community structure (Fig 5.9). Neither study used a properly replicated experimental design, and there were essentially no controls (Raffaelli & Moller 2000). Yet the scientific community is persuaded by these experiments that *Patella* and *Pisaster* are keystone species. Why? Because (a) the effects were so dramatic, and (b) the resultant changes in abundance and distribution of algae and mussels were completely outside the 'norm' for similar rocky shore communities. However, Lodge and Paine were lucky to get away with it, the effects of predator removals are often much more subtle and a proper experimental design is advised to all would-be manipulators!

For further discussion, see Raffaelli and Moller (2000).

5.5.1 The role of field experiments

Of all the arguments that can be used to convince scientists that a particular viewpoint is correct, the experimental falsification of hypotheses has proven the most persuasive (i.e. the rejection of the null hypothesis). **Field experiments** have to be properly designed and analysed so that there is no ambiguity in their outcome (Underwood 1981, 1997; Hurlbert 1984). An important aspect of their design is that the experimental plots (areas within which the manipulation occurs) need to be replicated many times and dispersed appropriately over the study area (Chapter 14). This is often very difficult to achieve for many types of taxa and ecosystem (Raffaelli & Moller 2000), but not for shores. The small body size, high densities, and small-scale nature of spatial patterns of shore species means that many relatively small plots can be located within a small spatial extent. The three-dimensional variation within shore sediments is usually compressed into a layer no more than 20 cm deep. In addition, processes of interest tend to operate over relatively short ecological time scales (weeks to years), allowing them to be investigated within the traditional 3- to 5-year research grant period. Finally, of course, shores are highly accessible and enable sampling to occur with high precision; you can be sure that you collected your samples from within the experimental plot you intended to sample. These features mean that it is much easier to investigate certain types of processes on shores compared to other habitats, such as forests or the deep sea.

● Experimental demonstrations of interactions are only persuasive if executed correctly and unambiguously.

However, while shores are highly amenable to the experimental manipulative approach, not all questions can be satisfactorily addressed in this way, especially those questions concerned with processes that operate over large spatial and long temporal scales (Raffaelli & Moller 2000). Seductive as they are, field experiments are not the answer to every problem, although they are a powerful investigative tool.

Despite the experimental design issues that surround some of the earlier field experiments (Box 5.3), there is no doubt that top-down processes are of prime importance on some kinds of shore. Experiments involving the removal of suspected key consumers, such as predatory starfish, or whelks, and herbivorous sea urchins or limpets, have often revealed strong competitive interactions between species whose populations are regulated by the available spatial resource. Often a competitive dominant emerges in the absence of the consumer, with a resulting loss of inferior competitors and a lower overall community biodiversity (Fig. 5.9). While such experiments are dramatically effective in demonstrating the importance of top-down control by what have become termed **keystone predators**, their outcomes need to be placed in context (Pace et al. 1999).

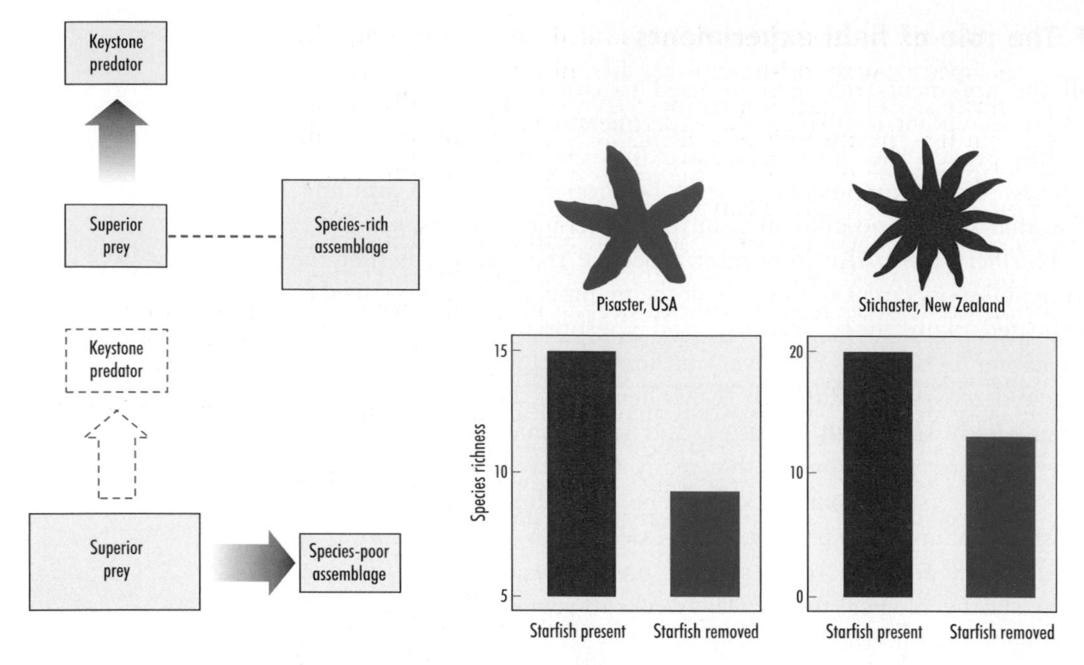

Fig. 5.9 The role of keystone predators in maintaining species richness. Left top: a strong controlling interaction between a competitively superior prey and its predator permits the coexistence of an associated module of other species. Left bottom: breaking the link releases the superior prey, which monopolizes the resources on which the module species depend. After Paine (1980). On the right are examples of how species richness is reduced in the absence of keystone starfish in the U.S. and New Zealand. After Raffaelli and Hawkins (1996).

5.5.2 Keystone predator or prey?

The keystone status of starfish, such as *Pisaster* (Fig. 5.9), or limpets, such as *Patella*, is not due to any inherent attributes of these species. Their status depends entirely on the response of their resource prey to reduced consumption, and in that sense the actual identity of the consumer is irrelevant. In other words, the factors that drive the competitive dynamics of the prey should be the focus of investigation. For instance, while *Pisaster* has been shown to be a keystone predator on exposed coastlines along the north-western USA, on other shores this is not the case. Similarly, the whelk *Morula* has an organizing role on some shores in New South Wales, but by no means all (Fairweather & Underwood 1991).

● A keystone species may simply be another brick in the wall in different circumstances.

One factor that determines whether a 'keystone effect' will be seen in the relatively simple communities of temperate systems (see below) is the recruitment dynamics of the dominant prey. This realization has led to the development of an extremely important, but at the same time incredibly challenging, area of shore ecology known as **supply-side ecology** (Gaines & Roughgarden 1985; Gaines et al. 1985; Lewin 1986). The significance of this area is best demonstrated by considering the dynamics of species with offshore early life history stages, such as barnacles, mussels, and algae. Recruitment of these taxa usually occurs every year, but it is the strength of a particular year's **cohort** of recruits that will determine the outcome of competition and hence community organization on the shore. Heavy recruitment by the dominant competitor will, in the absence of a consumer, lead to the exclusion of inferior competitors. However, if recruitment of the dominant competitor is poor, other species will not decline in the absence of the so-called keystone predator. These processes are also critical determinants of the likelihood of the successful colonization of species that have been introduced into new ecosystems through human activities.

● Recruitment processes operating far out at sea may profoundly affect the outcome of interactions on the shore.

The central question for understanding the dynamics of these communities is therefore 'what determines **recruitment strength**?' The answer rests in the ocean climate determinants of the offshore currents that carry the larvae or spores to the shore. If there are plenty of larvae in the currents and if these sweep the shore at the right time, then recruitment will be strong. In this respect it is unfortunate that there has been an emphasis on the keystone nature of the consumer, because the dynamics of these shores may in fact be driven by events hundreds of kilometres offshore. If this larger-scale perspective is taken, it could be argued that bottom-up, not top-down processes organize the shore.

Finally, in shore communities with a more complex set of predators, such as those found on many tropical shores (Menge & Lubchenco 1981; Menge et al. 1986) and in temperate estuaries and mudflats (Reise 1985; Raffaelli & Hall 1992), the removal of any one predator does not usually lead to the kinds of cascading effects documented for some temperate rocky shores, since the remaining species simply mop up any released prey resource. This effect is known as diffuse predation (Hixon 1991). Keystone predators are not therefore a characteristic of such shores.

5.5.3 Primary and secondary space

Top-down effects on shores are most obvious between consumers and those prey species that are **primary space limited**. That is, they compete aggressively for the rock surface. Such taxa include truly sessile groups, such as algae, barnacles, oysters, and sponges, as well as semi-sessile taxa, such as mussels. Interactions between these taxa are typically of the interference type, including overgrowth, crushing, smothering, and chemical warfare (algal and sponge allelochemicals). Taxa that are not sessile can move away from such interference competition, which explains why competitive exclusion, and hence keystone effects, are harder to demonstrate in the mobile species found in sandy shores and mudflats, as well as for rocky shore gastropods and amphipods. Indeed,

● The structure created by beds of primary-space occupiers, such as mussels, permits a high local biodiversity of associated secondary-space occupiers on rocky shores.

by keeping mussel densities low, *Pisaster* is partly responsible for maintaining a high diversity of primary space occupiers, but reduces biodiversity overall by reducing the habitat available for those secondary-space occupiers living in the highly complex mussel matrix, a system of extremely high biodiversity on rocky shores (Suchanek 1992).

5.5.4 Bottom-up processes

Despite a great deal of effort, much of it involving manipulative field experiments, it has been hard to demonstrate consistently an important role for top-down processes for sandy beach and mudflat communities (Raffaelli & Hawkins 1996). These systems often support a much larger range of consumers, including shorebirds, crustaceans, and fish, which occur in large numbers and have high energy demands (Table 5.2; Chapter 4). One might therefore expect top-down control in such systems, but this has been difficult to demonstrate experimentally. There are a number of factors that might be involved: low natural predator densities, poor experimental design, prey movement in and out of cages, inappropriately sized cages, insufficient time duration for the experiment, and, finally, a real absence of top-down control. There is no doubt that many experiments that have attempted to discern the importance of

Table 5.2 Examples of high consumption rates by shorebirds feeding on (a) macrophyte standing crop, and (b) invertebrate production. Data from Thayer et al. (1984) and Baird et al. (1985), respectively. Inset photographs: (a) seagrass bed Moray Firth, Scotland; (b) wading birds that are typical invertebrate feeders.

Prey	% consumed
(a) Macrophytes	
Zostera	30–75
Ruppia	20
Potamogeton	13
(b) Invertebrates	
Ythan estuary	36
Tees estuary	44
Langebaan lagoon	20

(a)

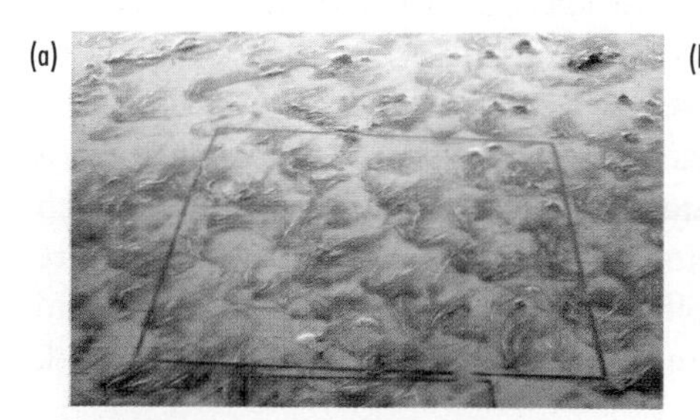

(b)

predation in these systems suffer from experimental design problems (Raffaelli & Moller 2000). Experiments in these habitats are typically short-term (weeks to months) and may be too short to reveal anything but the initial, transient dynamics, which may or may not be the same as the longer term behaviour of the system. This is important given that the effects of competitive interactions, for example for food resources, among prey species are likely to be subtle and long-term.

● The outcome of competitive interactions between mobile species may be much less dramatic than for sessile taxa.

Notwithstanding the above criticisms of manipulative field experiments in sediment systems, it would seem that bottom-up processes dominate, with predators limited by their prey, rather than vice versa. Many prey individuals remain inaccessible to predators, living out of reach for much of the time within the sediment and only a proportion of the prey standing stock (as opposed to production) may be predated. The dominance of bottom-up processes is perhaps not surprising given that sandy shores and mudflats tend to be net importers of organic matter from elsewhere, for example, upstream and along the coast in the case of estuaries, and from kelp beds in the case of many exposed sandy beaches.

● Most prey individuals in mudflats and beaches are unavailable to predators like shorebirds and fish, and their numbers are driven by the supply of organic matter and other food resources.

5.5.5 Disturbance and bioturbation

A feature of sediment systems, whether intertidal, sublittoral or deep-sea, is physical disturbance to the sediment fabric and alteration of the physico-chemical environment by the organisms themselves or external events such as storms and ice-scour. Species living within the sediment move through it, ingest and egest particles and draw oxygen-rich water down from the surface to depth (Rhoads 1974). This local-scale biological disturbance (**bioturbation**) can change the environment for other species, for instance by loosening and destabilizing the sediment fabric and making it more vulnerable to erosion by water movement. Alternatively, species that pump oxygenated water through burrows create a more favourable environment for other species (Fig. 5.10; Chapter 7).

In addition to changing the local environment for associated species, bioturbation will also alter the flux of nutrients between the sediment and the overlying water and in this respect the biodiversity of intertidal flats may have an overall impact on ecosystem functioning (Box 5.4). Many epibenthic predators, including flatfish, crabs, and shorebirds, disturb the sediment surface intensely during their feeding activities (Hall et al. 1993). While the resultant pits and surface features created tend to fill in with fine material and perhaps detritus, on most beaches the local effects are quickly erased by bedload transport (Fig. 5.10).

In summary, the picture that emerges from evaluations of top-down and bottom-up processes on shores is that a mixture of both is always present to differing degrees and this mixture will vary with the relative strength of consumer and resource recruitment, as well as the influence of wider-scale processes, such as ocean currents and catchment run-off.

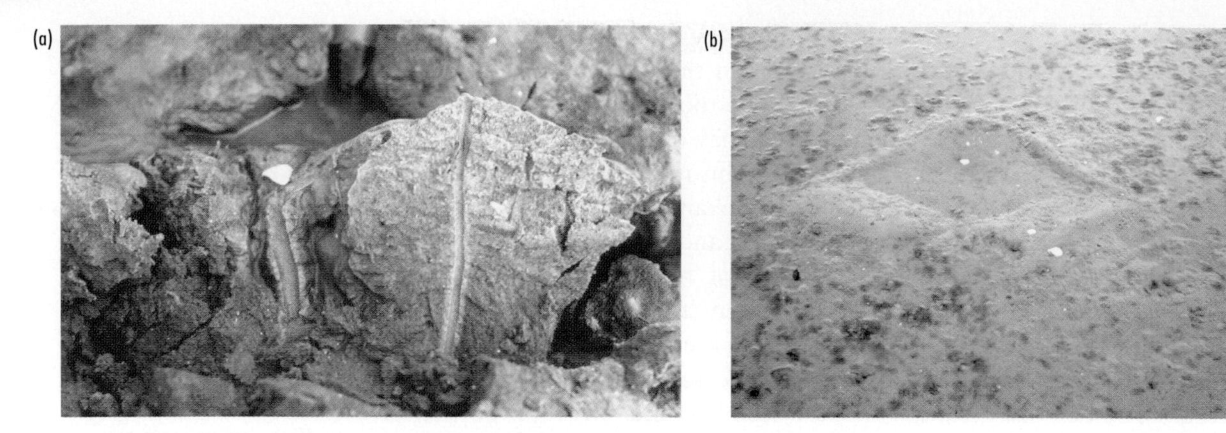

Fig. 5.10. (a) effect of bioturbation by burrowing ghost shrimp *Callianassa*, (Manakau, New Zealand), which piles up a large growing mound (seen here in section) by ejecting processed sediment through the top of its burrow entrance (Tamaki & Flach 2000). Note the oxygenated burrow walls caused by the animal pumping water from the surface (shown as light brown sediment lining burrow). (b) a depression or pit made by an eagle ray feeding at the sediment surface on small bivalves (Manakau, New Zealand). Such pits form significant structures on intertidal flats (Thrush et al. 1993).

The latter is a particularly important concept in the emerging field of coastal zone management in which the interactions between land and the adjacent marine areas are of fundamental importance.

5.6 The Shore Network

The previous section has stressed both the openness and connectivity of shores. Ecologists are usually required to delimit spatially (and temporally) their study areas in order to make sense of the complexity present, but the linear continuity of the coastline and its openness to the ocean necessitates a broader perspective if one is to truly understand how shores work. While conceptualizing a particular shore as a part of a larger network is relatively straightforward, quantifying those linkages will be a daunting task and one which is best suited to modelling as opposed to empirical approaches. However, there are a number of general features that can be described empirically.

● Shores are highly open systems, receiving and exchanging resources and propagules with each other and with offshore systems.

Coastal water movements have an incredibly significant impact upon the change and pattern in shore organisms, and are responsible for the transport of inorganic nutrients, organic material, sediment and its associated infauna, larvae, and spores within the water column, as well as smaller fish and crustacean consumers. Often, major shifts in community structure and composition on the shore can only be satisfactorily explained by considering near coast hydrodynamics. The irony is that

Box 5.4 Bioturbation in shore habitats

The bioturbatory activities of many infaunal species of beaches and mudflats affect the nutrient flux directly by changing the sediment physico-chemical environment, and indirectly through secondary impacts on the microbial community responsible for many chemical transformations (e.g. sulphate reducing bacteria). Experiments have been conducted to explore the significance of bioturbation for the release of nutrients from sediments, and in particular whether the presence of different species affect the flux (Raffaelli et al. 2003). It appears that a few species, such as the polychaete *Nereis (Hediste) diversicolor*, are extremely important, while others, such as the mud snail (*Hydrobia ulvae*), are less important in this respect. Furthermore, the importance of *Nereis* varies with near-bed flow, probably due to changes in the behaviour of the species in still and flowing conditions (Raffaelli et al. 2003). Finally, the more functional types (active burrowers, sediment reworkers, and oxygenators) present within the sediment, the greater the release of nutrients from the sediment. These experiments imply that the diversity of infauna present in sediments is important for maximizing this particular ecosystem process, but that functional group diversity is more important that species richness per se.

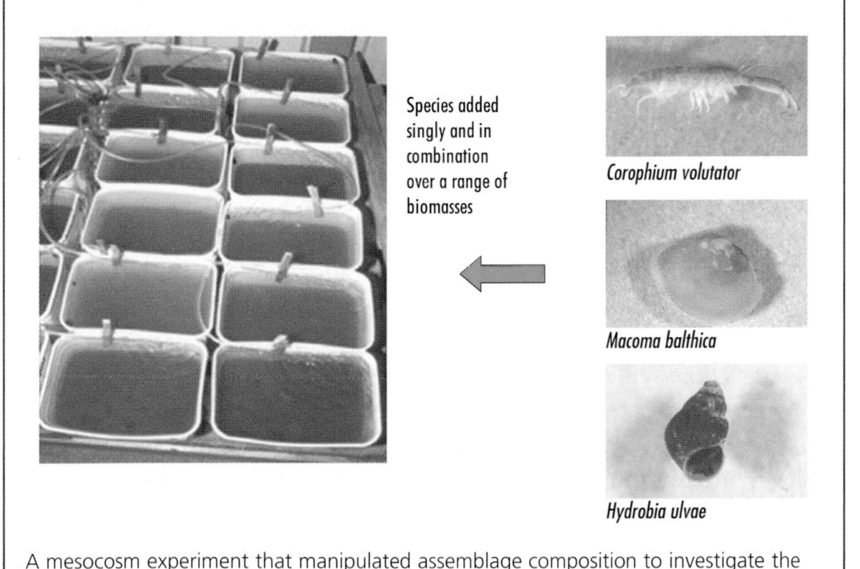

A mesocosm experiment that manipulated assemblage composition to investigate the ecosystem role of biota that perform different functional roles.

many shore ecologists are not trained in this area and that mainstream ecologists seeking to use the shore as a convenient laboratory for testing theory may not be aware of the importance of water movement. Our understanding of such physical processes is further hampered by the paucity of data routinely collected. Clearly, this is an aspect of shore ecology in need of urgent support and development.

● Often, major shifts in community structure and composition on the shore can only be satisfactorily explained by considering near coast hydrodynamics.

In addition to the physical transport of living and non-living material on- and offshore, and between different shore types, larger organisms can make purposeful migrations. These include larger fish, reptiles, birds, and mammals, which undertake such journeys for breeding, feeding, or to find a refuge from bad weather or predators. Their use of the shore can be considerable, especially for warm-bloodied taxa with high energy demands, such as shorebirds and marine mammals (Table 5.2), although, as argued in 5.5.5, cascading effects on the community are rarely detectable.

5.7 The Future of Rocky and Sandy Shores

The future of rocky and sandy shores depends on the temporal perspective taken. A series of key analyses looking at impacts and threats to 2025 (Brown & McLachlan 2002; Kennish 2002; Thompson et al. 2002) have identified a continued increase in many of the anthropogenic impacts that shores experience today, mainly because of projected increases in coastal populations worldwide. Looking further ahead to 2080, accelerated sea-level rise (ASLR) will undoubtedly impinge upon low lying sedimentary shores and estuaries, but also on rocky shores because of changes in currents and hence transport patterns, as well as changes in wave climate (Chapter 14).

● Accelerated Sea-level Rise, due to climate change, will have large-scale impacts on sandy beaches and mudflats over the next 50–100 years.

Beaches are dynamic physical entities, maintained by processes operating above and below the beach as commonly defined (Brown & McLachlan 1990), so that changes in sediment supply and wave climate are likely to impact greatly on beach dynamics. For sandy shores, ASLR may increase erosion, create a steeper beach profile and increase turbidity (Goss-Custard et al. 1990). With less organic matter retained in the beach there will be a lower biomass of infauna. These problems will be exacerbated if beaches are protected by hard engineering, as the available beach area will be progressively sandwiched between a rising sea level and an immovable structure, the so-called **coastal squeeze**. The net result of sea-level rise for many soft shores will therefore be a smaller, less productive beach, although other scenarios are possible depending on sedimentation patterns and the relative rise of the sea and land (Beukema 2002, Goss-Custard et al. 1990).

● Sea-level rise not only reduces intertidal area, but profoundly alters sediment distributions.

It is likely that rocky shores, at least those below high cliffs, will not be subject to coastal squeeze, since the biology can migrate upwards over time without impediment. However, both rocky and sandy shores in some regions could experience a more severe wave climate, due to an increase in storminess with climate change (IPCC 2001). Of most concern, perhaps, is the suspected relationship between sea level and the return time of **storm surges**, catastrophic flooding events which can remove entire beaches and mudflats. For instance, one estimate suggests that an increase in sea level of 0.5 m, well within the range predicted by

2080, can alter the return time of a 1.5 m storm surge from 1 in every 100 years to 1 in 10 years or less (IPCC 2001). How shore communities, both soft and hard, would respond to catastrophic disturbances at a frequency that approximates the lifespan of much of the shore biota is hard to predict. However, longer-lived species would have difficulty adapting to such conditions due to their lower fecundity and infrequent recruitment.

● The frequency of catastrophic wave action is likely to increase dramatically with rises in sea level.

CHAPTER SUMMARY

- Rocky and sandy shores are the most accessible parts of the marine environment and contain representatives of almost all the major classes of animals and plants.
- Rocky and sandy shores occur at either end of an environmental gradient of habitat particle size that ranges from very large (cliffs and boulders) to very small (individual sand grains). Species occupy sections of this gradient and the wetness/dryness (shore level) and wave action gradients and reveal zonation patterns that are most obvious across the shore level gradient.
- The distribution and abundance of these species are determined by their tolerances to physical factors, such as water movement and desiccation, and to biological factors, such as competition, bioturbation and predation.
- Rocky and sandy shores provide excellent laboratories for exploring mainstream ecological concepts. Much of the pioneering research using controlled experimental manipulations has been carried out on the seashore.
- Different kinds of shore are net exporters or importers of energy, especially detrital material. Understanding ecological functioning requires shores to be viewed as connected networks.
- Sandy and muddy shores are particularly vulnerable to climate change induced sea-level rise, because they are often prevented from transgressing inland.

FURTHER READING

Little and Kitching (2001) is an excellent concise book that deals with rocky shores, while Raffaelli and Hawkins (1996) consider in depth the development of intertidal ecology. Raffaelli and Moller (2000) critically evaluate the use and misuse of experimental approaches in ecology. Reise (1985) is a detailed consideration of the ecology of tidal flats.

- Little, C. and Kitching, J. A. 2001. *Biology of Rocky Shores*. Oxford University Press, Oxford.
- Raffaelli, D. and Hawkins, S. J. 1996. *Intertidal Ecology*. Chapman and Hall.
- Raffaelli, D. G. and Moller, H. 2000. Manipulative experiments in animal ecology – do they promise more than they can deliver? *Adv. Ecol. Res.* 30: 299–330.
- Reise, K. 1985. Tidal Flat Ecology. *An Experimental Approach to Species Interactions*. Springer-Verlag, Berlin.

Chapter 6

Pelagic Ecosystems

CHAPTER SUMMARY

Away from coastal boundaries and above the seabed, the pelagic environment encompasses the entire water column of the seas and oceans. The pelagic environment extends from the tropics to the polar regions, and from the sea surface to the abyssal depths, and is a highly heterogeneous and dynamic three-dimensional habitat. The pelagic is home to some of the most revered and reviled marine inhabitants, but great whales and jellyfish alike are subject to the consequences of pelagic ecosystem variability. Physical processes in the pelagic exert major control on biological activity, and lead to substantial geographic variability in production. Knowledge of biophysical interactions is essential for understanding ecological patterns and processes in the pelagic environment, and will be key for predicting changes there induced, for example, by climatic warming.

6.1 Introduction

The term **pelagic** means **'of the open sea'** and the pelagic realm is a largely open, unbounded environment in which the inhabitants have freedom, within physiological limits, to move in three dimensions. Contrary to the common perception of the sea as an unchanging, relentless expanse, the open ocean is an environment where variability is very much the norm. Patchiness in physical properties (e.g. temperature, salinity, turbidity), biological production, and biomass exists at a range of scales in space (centimetres to hundreds of kilometres) and time (minutes to decades). One of the key challenges to understanding open-ocean function lies in understanding the mechanisms that cause, and consequences of, this patchiness (Mackas & Tsuda 1999).

● The open ocean is an highly variable environment.

Despite the fact that much of the open ocean is remote from land, beyond the horizon for land-based observers, it has not escaped human impacts. For example, 90% of stocks of large pelagic fish such as tuna (Scombridae) and jacks (Carangidae) may have been removed by fishing (Myers & Worm 2003), and whole zooplankton communities have shifted their spatial distribution (Beaugrand et al. 2002) possibly in response to ocean warming, itself most likely caused by anthropogenically-released greenhouse gasses (IPCC 2001).

● Despite its apparent remoteness, the open ocean has been influenced strongly by human activities such as fishing and anthropogenic climate warming.

The pioneering studies of open-ocean ecology made from vessels with evocative names such as *Discovery, Challenger*, and *Atlantis* have been enhanced in recent years with observations from technologically advanced research platforms that include Earth-orbiting satellites (Bricaud et al. 1999) and unmanned **autonomous underwater vehicles** (AUVs) (Griffiths 2003). The aim of this chapter is to provide a synthesis of open-ocean ecosystem function and the factors that control it, insight into difficulties associated with sampling the heterogeneous pelagic realm, and some examples of ecological step-changes (**regime shifts**) in the global pelagic environment.

6.2 Definitions and Environmental Features

The **pelagic** realm spans the entirety of the water column, beginning at the sea surface and ending just above the seabed (the **benthic** realm; Chapters 7 and 8). The pelagic realm can be subdivided by total water depth and distance from shore. The **neritic** zone lies adjacent to shore, over continental shelves, and covers about 8% of the Earth's total sea area. Out beyond the continental shelf break, which is delimited typically by the 200 m depth contour, lies the vast, open **oceanic** zone (92% of the total sea area that covers 65% of the Earth's surface). There are numerous differences between the neritic and oceanic zones, differences that arise not least because of differences in proximity to land and consequent differences in nutrient and sediment loading in the water column (much of the sediment load in the seas and oceans is **terrigenous** and is delivered by rivers and estuaries; Chapter 4). Sailing out from a coastal port towards the open sea it is common to notice a transition from turbid to clear, blue waters. This transition is obvious not just from the deck of a ship but is visible from space. Indeed, interpretations of satellite remote sensed observations of ocean properties have to distinguish 'Case 1' waters, where the ocean colour is determined predominantly by algal pigments, from 'Case 2' waters where reflections from particulate matter dominate (Fig. 6.1) (Babin et al. 2003). One adaptive biological consequence of the difference in sediment and particulate loading between the oceanic and neritic zones can perhaps be seen in squid anatomy. Myopsin squid inhabit the neritic zone (for example *Loligo forbesi*, which is common in north-west European coastal waters, or *L. opalescens* from the west coast of the USA) and have a membrane across the eye that may serve to protect the eye from particulate irritants suspended in the water: in the open ocean myopsin

● The neritic and oceanic zones are very different, due in part to their differing distances from the coastline.

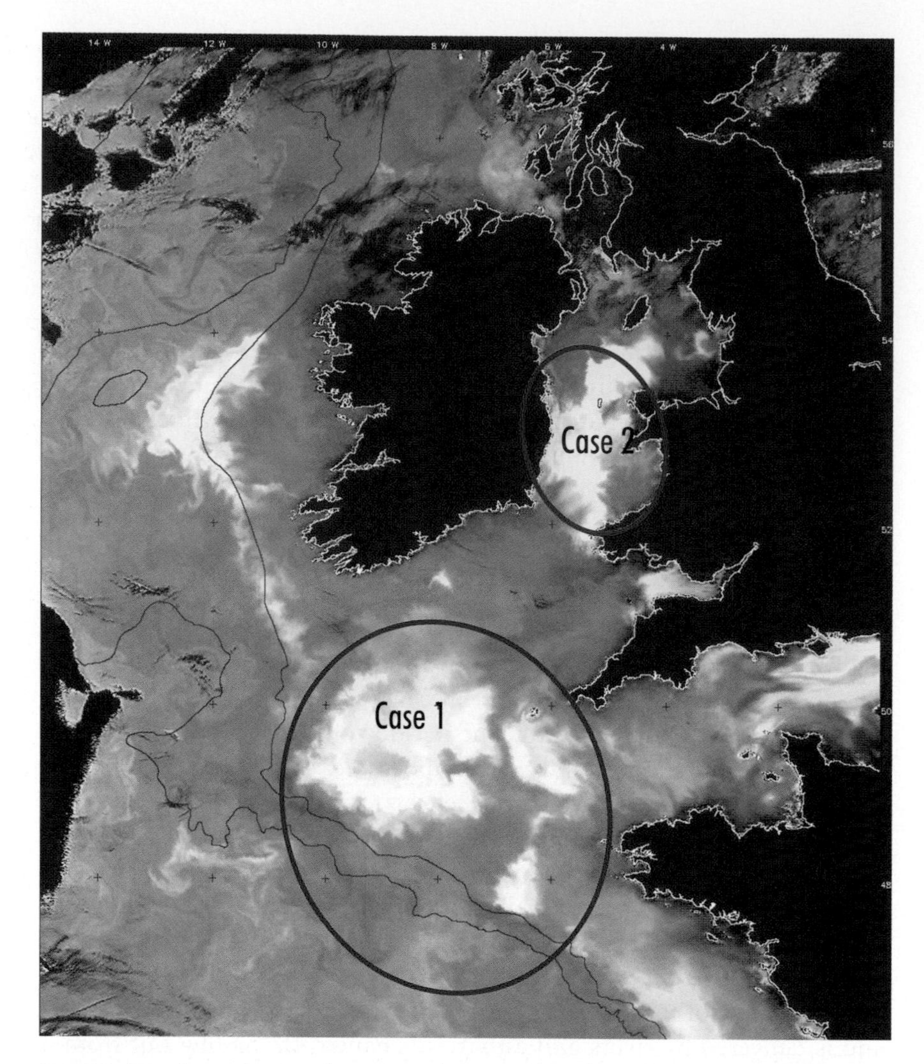

Fig. 6.1 An enhanced true-colour view of ocean colour from the SeaWiFS satellite (18 May 1998 1308 GMT) showing Case 1 and 2 waters around and to the south-west of the British Isles, North-West Europe. Satellite images were received by the NERC Dundee Satellite Receiving Station and processed by Peter Miller and Gavin Tilstone at the Plymouth Marine Laboratory (PML) Remote Sensing Group (**www.npm.ac.uk/rsdas/**). SeaWiFS data courtesy of the NASA SeaWiFS project and Orbital Sciences Corporation.

● The presence of a protective eye membrane in neritic squid may be an adaptation to heavy particulate loading in near-shore seas.

squid are replaced largely by members of the suborder Oegopsina (for example the European flying squid, *Todarodes sagittatus*) in which the membrane is absent, possibly because it is unnecessary.

The pelagic component of the open ocean can be divided further by depth. The upper surface of the ocean is known as the **neustic** zone and, in the tropics especially, is an habitat made harsh by exposure to high levels of ultraviolet radiation. Floating organisms inhabiting this zone typically have a blue colouration (Fig. 6.2) due to the presence of protective pigments that are able to reflect this damaging part of the light spectrum. The development of the ozone hole in the Earth's atmosphere has resulted in increased levels of UV radiation reaching the Earth's surface (the ozone layer acts as an UV shield), particularly in the southern hemisphere. In the Southern Ocean, Antarctic krill (*Euphausia superba*) may be particularly vulnerable to UV-induced DNA mutation because krill DNA is rich in thymine, which is the base that is most susceptible to UV radiation damage (Jarman et al. 1999). Since krill migrate away from the sea surface during daylight hours, however, their behaviour will probably serve to limit DNA damage, but UV damage to other species remains a distinct possibility.

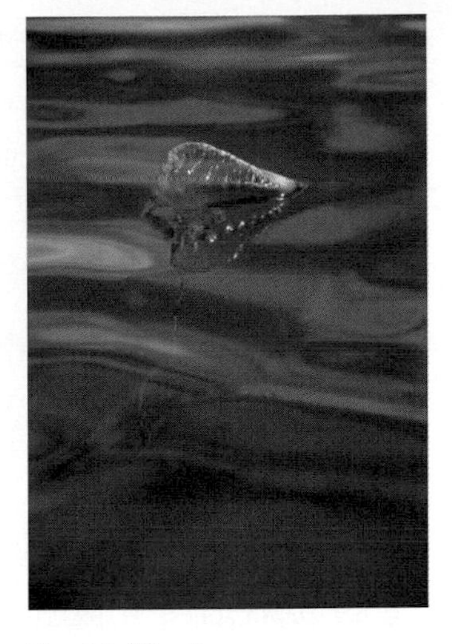

Fig. 6.2 The Portuguese man-o-war (*Physalia physalis*, a colonial Cnidarian) floats at the sea surface and has a blue colouration. (Photograph: joaoguaresma.com.)

Light also plays an important role in pelagic ecosystem function away from the neustic zone, both because it drives primary production (Chapter 2) and because it enables visual predation (predators that hunt using the sense of sight). In clear oceanic waters the threat from visual predators is increased because these predators can detect prey over greater ranges (Aksnes & Giske 1993). Shark attacks on humans often occur in turbid waters (Cliff 1991), possibly because under these conditions prey recognition is difficult and humans are mistaken for typical prey such as seals. The upper part of the water column into which light penetrates is called the **photic zone**. In clear tropical oceanic waters this zone may extend as deep as 200 m (much less in more turbid, temperate locations), although at this depth light intensity will usually be too low to drive photosynthesis (Chapter 2).

- Organisms that live in the neustic zone are particularly vulnerable to changes in UV radiation.

The upper 200 m of the water column is also known as the **epipelagic** (Fig. 6.3). Light at the red end of the spectrum is absorbed rapidly by seawater and does not penetrate far into the epipelagic zone (Chapter 2). Red colours are effectively invisible at depth, therefore, and many pelagic crustaceans adopt this colour as a means of camouflage against visual predators (Fig. 6.4). Below the epipelagic zone are, sequentially, the **mesopelagic** (200 m to 2000 m), the **bathypelagic** (2000 m to 4000 m), and the **abyssopelagic** (4000 m to 6000 m) zones. As depth increases, organisms in the pelagic environment are faced with increasing physiological challenges: pressure increases by 1 atmosphere for every 10 m increase in depth (Box 6.1; 1 atmosphere = $1\ \mathrm{kg\ cm^{-2}}$ or approximately the mass of a Mini on an area $25\ \mathrm{cm} \times 25\ \mathrm{cm}$), and in some locations oxygen-minima layers (Rogers 2000) arise at depth because oxygen is depleted by bacteria breaking down material sinking from the sea surface.

- Daily and seasonal changes in incident light intensity have profound effects on biological processes in the pelagic environment, driving vertical migrations and seasonal plankton blooms.

In general terms the total mass of biota per unit volume of seawater decreases with depth (Yamaguchi et al. 2002). This is because, with the exception of energy input by chemoautotrophic processes at hydrothermal vents (Chapter 8), biological processes in the deep sea are

- The total mass of biota per unit volume of seawater decreases with depth.

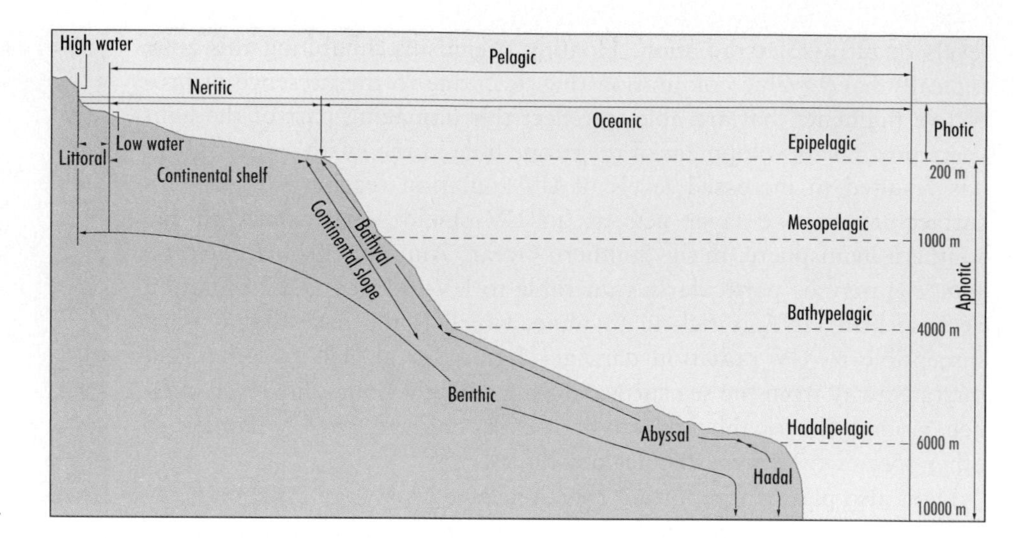

Fig. 6.3 Depth zones.

Fig. 6.4 A mesopelagic zooplankton/nekton sample showing the predominance of red-coloured crustaceans. Photograph: Andrew Brierley.

fuelled entirely by photosynthetically-generated organic matter that sinks from the illuminated surface region. As material sinks further from the surface it becomes distributed through an ever-increasing volume of water. This dilution effect means that as distance from the surface increases, food availability decreases, with the consequence that less animal biomass can be sustained at depth. In the deep pelagic zones,

Box 6.1 Seawater as a dense and viscous (sticky) medium

Seawater is denser and more viscous than air. The increased **density** offers advantages to some marine organisms, for example providing physical support that reduces the need for skeletal strength, but also presents challenges such as rapidly increasing pressure with depth. Seawater density varies as a function of the concentration of dissolved salts (**salinity**; salinity is reported without units since it is defined in terms of the ratio of the electrical conductivity of seawater to the electrical conductivity of a potassium chloride standard, see also Chapter 4) and temperature (warm water is less dense than cold water). The strength of cohesion, or stickiness, of a fluid is quantified by its **dynamic viscosity**. The dynamic viscosity of olive oil, for example, is 40 times that of seawater. The motion of a particle of a given size and velocity is more impeded through a medium of greater dynamic viscosity (think of marbles sinking through a bottle of water and a bottle of olive oil). For a fluid of a given dynamic viscosity (e.g. seawater) the continuity of motion of a particle is controlled by its velocity and size, and can be quantified by the **Reynolds number** (Re). For a given fluid, Re is simply ((particle velocity × particle size)/dynamic viscosity). Re is dimensionless because the units used in its calculation cancel out. Broadly speaking for small organisms moving at slow speeds Re is less than 1000 and seawater is 'sticky': ciliates, for example, stop the moment they cease swimming. For larger organisms travelling at higher speeds Re is greater than 1000: the inertia of a large pelagic fish, for example, enables it to continue gliding through the water even after it has ceased active swimming (see also Chapter 3).

energy efficiency is particularly important and animals adopt stealthy, sedentary lifestyles and use cunning mechanisms to ambush prey in darkness (Seibel et al. 2000).

● Energy efficiency is important in the deep pelagic due to the dispersed nature of potential food.

6.3 Pelagic Inhabitants: Consequences of Size

Pelagic organisms can be divided in to two categories on the basis of their locomotory prowess. **Plankton** are unable to counteract the influence of currents and drift passively in the horizontal plane (ocean currents often exceed 1 knot or *c.*0.5 m s^{-1}). Plankton can move vertically, adjusting their depth, and do so pronouncedly during **diel vertical migrations** at dawn and dusk (Fig. 6.5). Diel vertical migrations are a ubiquitous feature of pelagic ecosystems (Hays 2003) and are thought to be driven primarily by the trade off required to enable plankton to feed in the food-rich upper water column and yet to avoid the illuminated upper layer in daylight because of the increased risk of visual predation incurred there at that time (Tarling 2003). Some organisms, such as copepods and chaetognaths, complete their entire life cycle as plankton and are called **holoplankton**. Others, such as fish larvae and barnacle

● The word plankton is derived from the Greek word planao, which means 'to wander'

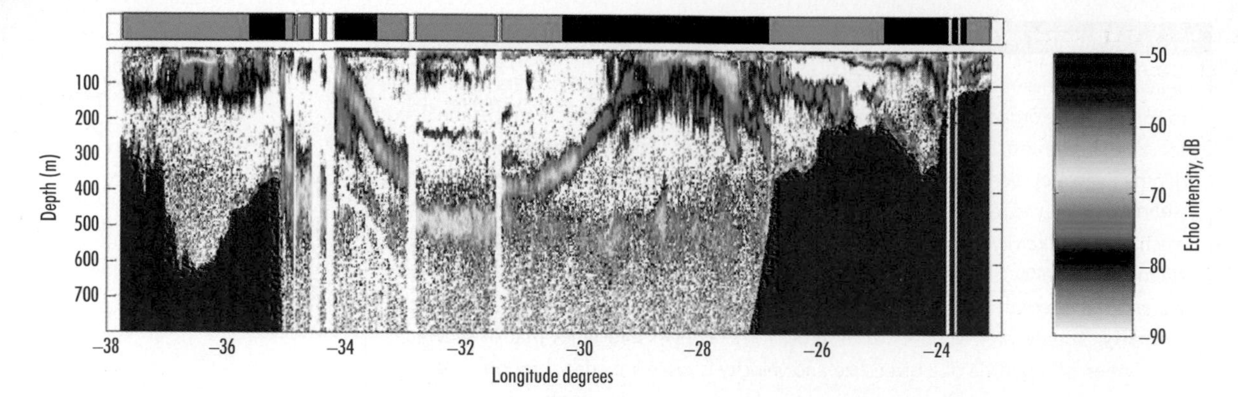

Fig. 6.5 An echogram showing the rapid ascent of zooplankton and nekton (as detected using a 38 kHz scientific echo-sounder) from depth to the near surface at dusk. This migration from 400 m to 50 m takes less than 2 hours. The upper bar indicates periods of day (light blue) and night (dark blue).

● Animals whose larvae have a planktic phase generally produce a relatively high number of eggs for their body mass.

larvae, spend only a part of their life cycle as plankton, either growing or settling out to the seabed as they age, and are called **meroplankton**. A **planktic** phase provides opportunities for dispersal and colonization, but is also a stage that is particularly vulnerable to predation (Pechenik 1999, see also Chapter 13).

Organisms that are capable of swimming to the extent that they can overcome currents are known as **nekton**. Inhabitants of the open ocean span several orders of magnitude of size, ranging from viruses, bacteria, and protozoa to large predators such as sharks and whales, which may reach many metres in length and body masses of several tonnes. Generally speaking, large organisms are **nektic** and smaller organisms are **planktic**: **micronekton** have intermediate swimming abilities and are of the order of 4 cm in length (for example large euphausiids). This size-related difference in mobility arises in part because of the interaction between size and viscosity (Box 6.1): seawater is essentially a 'sticky' medium for small organisms (Van Duren & Videler 2003) and a constant expenditure of energy is required to maintain their movement. There are exceptions, however, and the Arctic lion's mane jellyfish (*Cyanea arctica*), for example, may attain tentacle lengths of 40 m but is a passive, planktic drifter. The very smallest planktic organisms include viruses and bacteria (Azam & Worden 2004) and protozoa (Struder-Kypke & Montagnes 2002). The small size of these organisms belies their importance to pelagic ecosystem function. Dissolved organic carbon is taken up by bacteria, which are consumed by heterotrophic nanoflagellates and in turn by ciliates in the so-called **microbial loop** at the base of the food chain (Chapter 3). This loop recycles organic matter that is too small to be consumed by metazoan plankton, and the metazoans are able to prey upon ciliates. The microbial loop therefore fuels the pelagic food chain, and is especially important in oligotrophic waters (Lenz 2000).

● Very few pelagic organisms dismantle their prey before eating, a behaviour more common in benthic biota such as crabs.

As well as impacting mobility, organism size is also a major architect of pelagic food web structure. Pelagic organisms will typically consume food items whole and the size of item that an animal can consume is constrained by its mouth size such that predators are usually substantially larger than their prey (Cohen et al. 1993; Jennings & Warr 2003). This concept is well captured by Brueghel's picture *Big fish eat little fish*, in which small fish are tumbling out of the mouths of successively bigger fish (Fig. 6.6). A more trophically extensive example from the North Sea has unicellular algae such as diatoms (*c*.100 μm diameter), grazed by copepods (e.g. *Calanus finmarchicus, c*.3 mm length), which are in turn predated by herring (*Clupea harengus, c*.20 cm length), which might be consumed by gannets (*Sula bassana*, wing span *c*.180 cm). This strongly size-structured progression is in marked contrast to many terrestrial food chains where small predators (for example hyenas, body mass *c*.40 kg) may cooperate in social groups to take herbivores that are considerably larger (e.g. wildebeest, *c*.250 kg).

● In terrestrial systems, animals that forage in social groups can deal with prey items much larger than themselves.

Pelagic food webs are often far more complex than the simple four-linked-chain diatom-copepod-fish-bird example from the North Sea. An analysis of a 29-species food web for the Benguela ecosystem undertaken

Fig. 6.6 '*Big Fish Eat Little Fish*:' after, by Pieter Brueghel the Elder (1557) showing smaller fish tumbling from the mouths of bigger fish. The Metropolitan Museum of Art, Harris Brisbane Dick Fund, 1917. (17.3.859)

to determine if culling Cape fur seals (*Arctocephalus pusillus pusillus*) would increase hake (*Merluccius* spp.) biomass (Yodzis 2000) noted that there were over 28 million pathways from hake to seals! This complexity of food webs, and the fact that most levels can be controlled either from above (**top-down control**, e.g. by predation) or below (**bottom-up control**, e.g. by food limitation, Verity & Smetacek 1996) renders it very difficult to make predictions of the consequences of bioregulation. In general smaller mean predator : prey body size ratios are characteristic of more stable environments, and food chains are longer when mean predator : prey body size ratios are small (Jennings & Warr 2003). Systems that have shorter food chains are generally much more susceptible to **trophic cascade** effects (Chapter 7).

● The complexity of food webs means it is difficult (if not impossible) to predict the outcome of selective culling of higher predators on lower tropic levels.

6.4 Temporal and Spatial Variability in Pelagic Ecosystems

The open ocean is not homogenous. Interactions between physical and biological processes result in variability over a range of temporal and spatial scales, and patchiness is a key feature of pelagic ecosystems. The Stommel diagram (Fig. 6.7) illustrates the scales of variability that are inherent characteristics of pelagic ecosystems, from centimetres to thousands of kilometres and from seconds to millennia, and shows how variations in time and space are interlinked. It is important to appreciate the interplay of temporal and spatial scale, and it is a theme that recurs throughout this book. Phytoplankton are short-lived and are influenced by small-scale mixing processes (Martin 2003). The diel vertical migration of zooplankton at dawn and dusk is restricted to certain times of day but occurs everywhere (Pearre 2003). Fish, which are generally longer lived than zooplankton, are impacted by environmental variability over longer time scales: the Peruvian anchoveta (*Engraulis ringens*), for example, is influenced strongly by the El Niño Southern Oscillation that exhibits decadal scale variability (Chavez et al. 2003, see also Chapters 7, 12, and 14).

● Fish, which are generally longer lived than zooplankton, are impacted by environmental variability over longer time scales.

6.4.1 Physical processes contributing to temporal and spatial variability

The physical properties of the open ocean are heterogeneous by depth, position (latitude, longitude), and over time. In addition to the depth-related changes in light intensity and oxygen concentration already mentioned, vertical gradients in water temperature, density, and nutrient concentration may also exist. Solar heating warms the upper ocean leading to the development of a thermal gradient by depth. In some

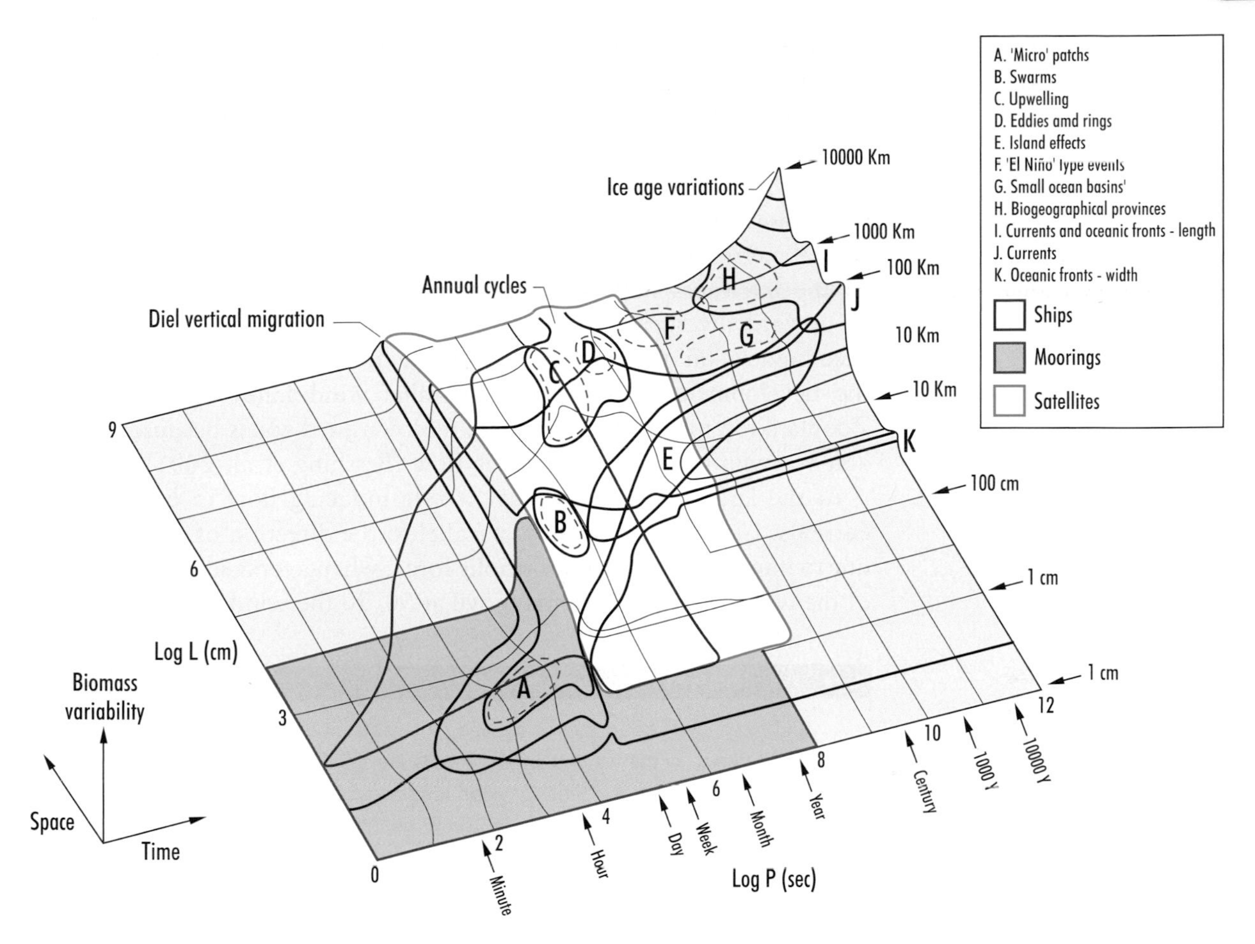

Fig. 6.7 The Stommel diagram, overlain to show the scales that can be sampled with various platforms, and features such as fronts. Modified from ICES Zooplankton Methodology Manual, Harris et al. (eds., 2000), page 36, Copyright 2000, with permission from Elsevier.

regions, combinations of tidal and wind-driven processes cause turbulence and mixing of heated surface water with cooler water below. However, in regions or seasons where winds are light and wave action slight, or in water that is too deep to be mixed completely by tidal flow (Chapter 7), pronounced vertical stratification can become established: warm surface waters then become effectively isolated from cooler, deeper waters by a **thermocline**. Vertical stratification is also promoted in situations where fresh water is introduced. Rain, river run-off, and ice-melt all introduce fresh water to the surface of the ocean. Low salinity waters are less dense than high salinity waters (Box 6.1) and stabilize the upper water column because more energy is required to mix low salinity waters downwards. The strong density gradient between the mixed, buoyant, low salinity surface waters and underlying high salinity waters is known as the **pycnocline**. The depth of the mixed layer,

● Where winds are light and wave action slight, or in water that is too deep to be mixed completely by tidal flow, vertical stratification can develop.

as bounded by the pycnocline or thermocline, will vary depending upon prevailing conditions (Chapters 2 and 7).

● Langmuir circulation can lead to the accumulation of zooplankton and flotsam, which forms visible wind lanes, at the sea surface.

As well as causing downward mixing, wind can lead to the upward transport of water from depth. Such wind-driven **upwelling** occurs over a range of scales. At the small scale, **Langmuir circulation** is generated as wind blows steadily across calm water, causing near-surface vortices several metres in diameter to develop parallel to the wind flow (Box 6.2). At the interfaces between neighbouring vortex cells, alternating lines of upward divergence and downward convergence develop. Flotsam accumulates on the surface above the downward zones, leading to the development of prominent, parallel **wind lanes** on the surface. Zooplankton may also accumulate in downward zones because they are able to swim upwards against the flow (Pershing et al. 2001).

At the large scale, wind plays a role inducing flow in most surface currents. Currents do not flow parallel to the direction of the wind but, due to interactions with the **Coriolis force**, when averaged over the whole of the water column, currents move at 90° to the wind. Movement is to

Box 6.2 Wind-driven circulation processes

As wind blows over the surface of the sea it generates waves and induces vertical and horizontal motion in the water. **Langmuir cells** are small-scale, parallel, helical vortices that are often apparent as a series of **wind lanes** on the sea surface running parallel to the direction of the wind. Vortices are usually not large enough to bring nutrients up from deep water beneath the pycnocline.

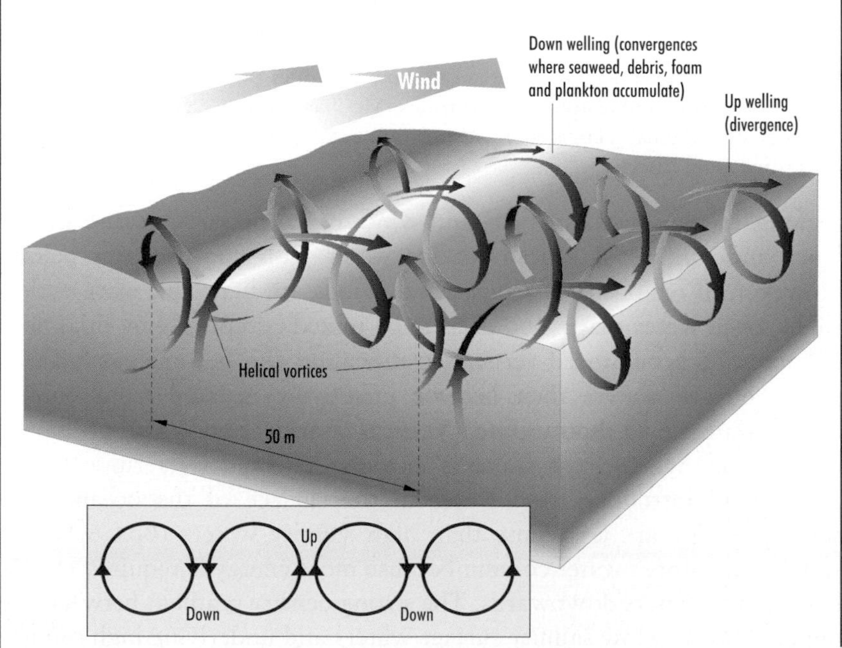

continues

BOX 6.2 continued

Large-scale ocean currents are also induced by wind, but occur at angles to the direction of the wind rather than parallel to it. This is because of the interaction between friction and the **Coriolis force**. The Coriolis force is the force experienced by a moving body of water due to the fact that the planet is rotating. The water column can be thought of as a series of horizontal layers. The upper layer at the sea surface is subject to wind friction (**wind stress**) at the top and water friction (**eddy viscosity**) at the bottom. Subsequent layers are impacted by friction with layers above and beneath. Slippage between layers result in an exponential decrease in current speed with depth until, below the **depth of frictional influence**, wind influence ceases. Cumulative impacts of Coriolis force result in an increasing angle of deviation away from the wind with depth. Current vectors in all layers form a spiral pattern known as an **Ekman spiral**. The averaged effect of the spiral is that the mean motion of the wind-driven (Ekman) layer is at right angles to the wind direction. Reprinted from Ocean Circulation, 2nd ed., the Open University Course Team 2001, pp. 42 and 68, Copyright 2001, with permission from Elsevier.

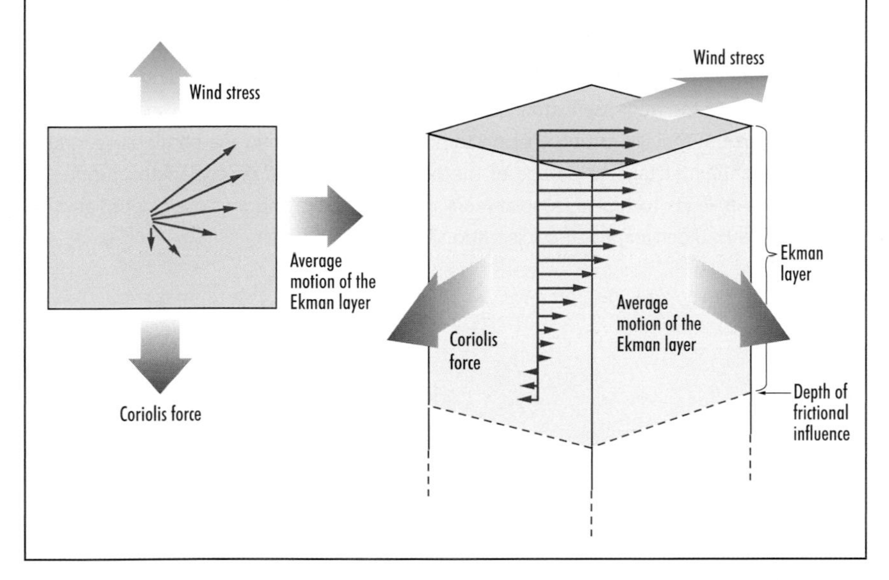

the right of the wind in the northern hemisphere and to the left in the south. This movement is called **Ekman transport** (Box 6.2). It contributes significantly to general ocean circulation and can result in pronounced **upwelling.** In the case of the Benguela Current, for example, and other southern hemisphere **eastern boundary currents**, the prevailing south-easterly winds blow alongshore and drive near-shore Ekman flow away from the coast. This in turn draws cold, nutrient-rich waters from depth to the surface at the coast (Carr & Kearns 2003). Changes in global wind patterns have the potential to affect upwelling and the ecosystems that are dependent upon them (Grantham et al. 2004).

Horizontal boundaries between water masses with different physical properties are known as **fronts** (Box 6.3). Fronts occur at a range of

Box 6.3 Front formation and elevated biological activity at fronts

In the same way that fronts on weather maps mark boundaries between different air masses, fronts in the sea are boundaries between dissimilar bodies of water. Fronts occur off estuaries at boundaries between fresh and salt water, in shelf seas between mixed and stratified waters, at continental shelf breaks adjacent to upwelling regions and, at the global scale, between major current systems. Fronts tend to be sites with higher biological activity that the surrounding water masses, often because nutrients are transported upwards into the stratified euphotic zone at fronts.

Tidal mixing fronts (also known as shelf sea fronts) occur between tidally mixed and stratified waters. They occur when the intensity of turbulent mixing caused by tidally induced flow over the seabed is sufficient to overcome the barrier to mixing caused by thermal stratification. In simple terms, this is a function of the strength of the tidal flow and water depth; a strong tidal flow will generate sufficient turbulence to completely mix shallow water. On the stratified side of the front, nutrient concentrations in the warm surface waters are depleted and the strong thermal gradient prevents nutrients from beneath being mixed upwards. Phytoplankton growth is therefore nutrient limited and low. On the well-mixed side of the front, although nutrients are not limited, phytoplankton are continually mixed down out of the illuminated surface layer and growth is light-limited. At the front itself stratification weakens sufficiently to enable some vertical nutrient flux but remains strong enough to hold phytoplankton in the photic zone long enough for them to take advantage of the nutrients. Increased phytoplankton production at fronts leads to higher zooplankton standing stocks and increased densities of predators and underlying benthos (see also Chapters 2 and 7).

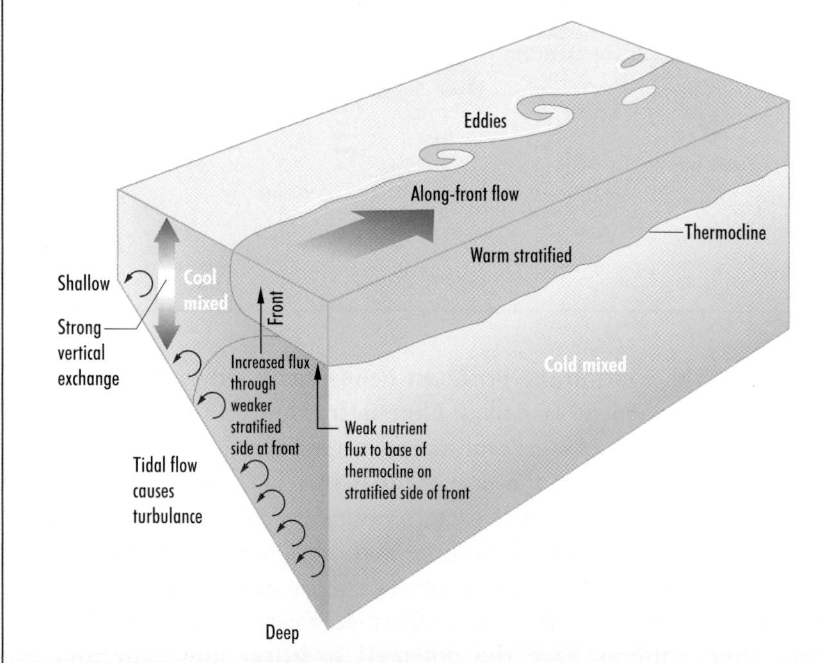

scales, from tidal mixing fronts that separate mixed and stratified waters in coastal seas (Hill et al. 1993) to major oceanographic boundaries such as the North Wall of the Gulf Stream and the Antarctic Polar Front (Taylor & Gangopadhyay 2001). High-velocity current jets associated with fronts can be important long-distance transport routes for many marine organisms: Antarctic krill (*Euphausia superba*) are transported widely throughout the Scotia Sea from breeding centres off the Antarctic Peninsula on frontal currents (Thorpe et al. 2004), and large oceanic squid such as *Illex illecebrosus* take advantage of currents for distribution and feeding (O'Dor 1992). Meanders in fronts can lead to columns of water (**core rings**) being shed from one side of the front to the other; the retroflection of the Agulhus current around the southern tip of Africa, for example, sheds warm core rings regularly into the south Atlantic (Garzoli et al. 1999) and is an important mechanism for trans-oceanic mixing.

● Most fronts are highly mobile, dynamic features that cannot be represented accurately by single, static lines on charts.

6.4.2. Consequences of temporal and spatial physical variability for pelagic primary productivity and biogeography

The depth to which water column mixing occurs, the **mixed layer depth**, has major implications for primary production in the open ocean because photosynthesis only takes place in illuminated surface waters. If the mixed layer is deep it is possible that phytoplankton will sink or be carried down below the **compensation depth** (Chapter 2), reducing net production. In temperate waters, primary production is minimal during winter when light levels and temperatures are low and the upper water column is thoroughly mixed by storm action and convection (Backhaus et al. 2003). Phytoplankton blooms do not commence until calmer, warmer weather in the spring leads to upper water column stratification. From this point on, phytoplankton cells are retained in the upper mixed layer, benefit from the increased illumination from the sun as it reaches higher angles in the sky, and grow and reproduce rapidly. Phytoplankton require nutrients such as phosphate, silicate, and nitrate to grow. As phytoplankton blooms develop, concentrations of these nutrients become depleted and the effective isolation of the mixed layer from the larger nutrient pool beneath means that nutrients are not replenished. In temperate waters, nutrient limitation may inhibit phytoplankton growth throughout the summer months. A second bloom may though occur at the onset of autumn when wind-driven mixing brings nutrient-rich waters from beneath the pycnocline up in to the illuminated surface layer (Diehl 2002).

● Although the development of stratification is a necessary precursor to bloom formation, the persistence of an upper layer that is effectively cut off by density and temperature gradients from the waters beneath can eventually inhibit phytoplankton growth.

● A phytoplankton maximum can develop at the pycnocline late in the season as this is the interface between nutrient-rich deep water and illuminated but nutrient-depleted surface stratified waters.

In regions of the world where upwelling is persistent throughout the year (Box 6.2) nutrients tend not to be limited and annual primary production levels are high. However, in El Niño years, changes in prevailing weather conditions reduce the usually strong upwelling off the coast of western South America and ensuing nutrient limitation has dramatic negative consequences for primary production and fisheries in coastal waters. Tropical ocean basins tend to be permanently stratified and primary production levels are low. As a consequence surface waters in these tropical regions lack particulate matter and are very clear: such waters are termed **oligotrophic**. Other areas of the world ocean have low phytoplankton biomass despite the presence of high nutrient concentrations. These **HNLC** (high nutrient, low chlorophyll) regions include the Southern Ocean and Equatorial Pacific, and hypotheses proposed to explain their existence include grazing pressure and absence of trace elements, particularly iron (Chapter 2).

● Iron is thought to be one of the key limiting trace elements in HNLC regions, and experimental iron fertilization in these locations has stimulated phytoplankton production (Boyd 2002).

Global variation in the pattern of annual primary production is strikingly clear in images of averaged chlorophyll concentration obtained through satellite imagery (see Fig. 2.3, Chapter 2). Regional coherence in the pattern of annual phytoplankton production has been used as one diagnostic feature in the hierarchical separation of the global ocean into distinct biogeographic **biomes** and **provinces** (Longhurst 1998). A biome is the largest coherent community unit that it is convenient to recognize, and Longhurst (1998) distinguishes four in the global ocean, which are characterized by the principal mechanisms driving their mixed layer depth: in the **Westerlies biome** local winds and irradiance force the mixed layer depth; in the **Trades biome** the mixed layer depth is influenced by large-scale ocean-circulation processes; in the **Polar biome** the presence of buoyant, fresh water from ice melt in spring constrains the mixed layer depth; and in the **Coastal biome** diverse processes including upwelling force the mixed layer depth. Within these biomes, 51 provinces are recognized (Chapter 1). Separation of the global ocean into provinces is very useful because it allows regional differences in physical oceanography to be used to gain understanding, and make predictions, of regional differences in ocean ecology.

● Knowledge that a particular area of ocean lies in a particular province enables predictions to be made regarding ecosystem function there, even if field data for the specific area are lacking.

6.4.3 Consequences for higher trophic levels of variability in primary production

Regions of the world's ocean with high primary productivity support richer pelagic communities, with higher total biomass, than do regions with low primary production. In fact there is a direct linear relationship between the magnitude of annual primary production and nekton (fish and squid) production (Sommer et al. 2002) (Fig. 6.8). This relationship is apparent in the distribution of global fish catches (Chapter 12): nutrient-rich shelf seas and regions with strong upwelling account for the vast majority of the world's commercial catch (Watson & Pauly

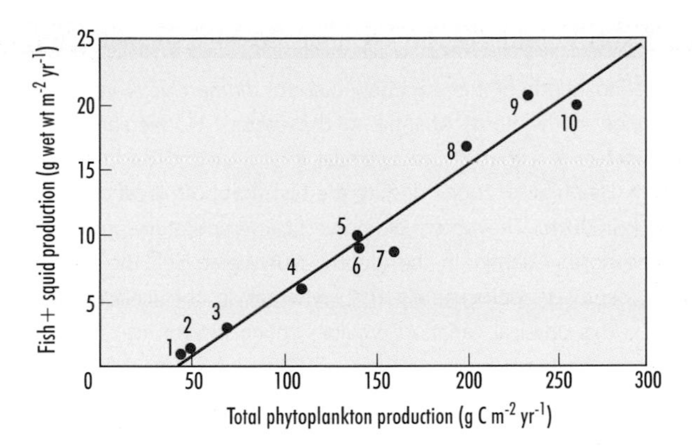

Fig. 6.8 The relationship between primary production and nekton production. Fish and squid production = (0.095 Phytoplankton production) – 3.73, $r^2 = 0.96$. 1 = Atlantic Ocean gyre centre, 2 = Atlantic ocean gyre boundaries, 3 = Hawaiian waters, 4 = Bothnian Sea, 5 = Gulf of Riga, 6 = Gulf of Finland, 7 = Baltic Sea, 8 = Nova Scotian shelf, 9 = Gulf of Maine, 10 = Mid-Atlantic bight. Redrawn from Iverson 1990. Copyright 2000 by the American Society of Limnology and Oceanography, Inc.

2001), whereas oligotrophic central open-ocean basins contribute little. Commercially important pelagic fish species do not consume phytoplankton directly but usually predate zooplankton and micronekton that are **primary consumers**. Understanding zooplankton ecology is therefore key to understanding fisheries production.

● GLOBEC (*glo*bal ocean ecosystem dynamics) **http://www.pml.ac.uk/globec** is a global research effort to understand interactions between primary consumers, higher trophic levels and fisheries.

Zooplankton blooms are only able to develop once phytoplankton biomass and production has become sufficient to sustain zooplankton grazing rates. In temperate waters, therefore, peaks in zooplankton biomass occur in spring and autumn slightly after the phytoplankton blooms. In high latitudes, where seasonality is extreme and the phytoplankton bloom is limited to a single spring/summer peak, some zooplankton species survive the dark, food impoverished winter months in deep water in a dormant state called **diapause**. Copedods including *Calanus finmarchicus* in the sub-Arctic north Atlantic and *Calanoides acutus* in the Southern Ocean build up large stores of lipids during summer feeding, a small proportion of which fuels their survival over winter. At the end of summer, growth and development are arrested and individuals sink, overwintering in a state of hibernation at depths between 500 m and 2000 m (Box 6.4). In late winter or early spring copepods emerge from diapause and migrate to the surface to spawn. Because of the short production season, the timing of reproduction is critical at high latitudes. By using lipid reserves accumulated in the

● *Calanus finmarchicus* survives food depleted winter months in the north Atlantic by drawing on lipid reserves. This also enables it to produce young in advance of the phytoplankton bloom in the following year.

Box 6.4 Diapause depth, water density, and lipid composition

The overwintering depth of the copepod *Calanus finmarchicus* varies throughout its distribution range in the north Atlantic. In the eastern Norwegian Sea, for example, *Calanus* overwinters at about 800 m whereas in the Iceland Basin overwintering is at around 1500 m (Heath et al. 2004), despite the fact that both locations have similar total water depths (c.2000 m). The water column vertical temperature profile varies markedly throughout the north Atlantic: in the eastern Norwegian Sea the temperature at the overwintering depth is approximately 0 °C whereas in the Iceland Basin it is much warmer (4 °C). This physical variation provides much insight into the variation of the overwintering depth. At the onset of diapause *Calanus* becomes physically inactive and sinks passively until it reaches the depth where it is neutrally buoyant. This is the depth at which the density of the copepod is the same as the density of the surrounding seawater, which itself is a function of ambient temperature and salinity. One of the major contributors to variation in density between copepods is lipid composition, such that density decreases as the proportions of lipid increases. Using knowledge of the temperature and salinity at the overwintering depth in several locations, and hence the density there, it has been possible to predict the proportion of lipid that should be expected in individuals overwintering at particular locations. A very good linear relationship has been found between predicted and actual values (Heath et al. 2004).

It is not yet clear what the main evolutionary drivers of variation in copepod lipid composition are. In order to survive the winter individuals need to descend below the depth of winter mixing, and will also have increased chances of survival overwinter if they descend below depths where predators can operate. In Norwegian fjords overwintering is shallower in fjords where visual predators are absent (Bagoien et al. 2001), but the winter distribution of predators in the open ocean is not yet known. Ironically, therefore, individuals that feed too successfully over summer and lay down excessive lipid reserves may be unable to sink to depths below predators; this would provide a strong selective pressure against over consumption and may be one reason why some copepods enter diapause early in the year when the phytoplankton bloom is still in full swing. Since current flow is different at different depths, the overwintering depth will also have a profound influence on the location at which individual *Calanus* surface after diapuse. In order to complete its life cycle successfully *Calanus* has to surface at a location where offspring can be spawned and hatch and eat, and descend to overwinter to complete the life cycle. It is likely that few depths will enable this to be achieved, providing another source of selection. It is clear that the life cycle of *Calanus* is tied very closely to its environment; understanding the environmental space-time dynamics will be vital to gaining full understanding of regional variability in *Calanus* abundance and consequences to higher trophic levels such as fisheries.

previous year to fuel reproduction, these copepods can spawn early, independent of the present year's phytoplankton bloom, and ensure that their young are in place to fully exploit the short livid phytoplankton bloom.

The timing of the phytoplankton bloom not only influences zooplankton secondary production but also places a significant control upon organisms that predate upon zooplankton. Developing fish larvae 'surf on the wave of zooplankton production' and the **match-mismatch hypothesis** (Cushing 1990) proposes that larval fish survival will be greatest in years when the period of plankton production overlaps most closely with the period of larval food demand. Satellite data that revealed between-year variations in the timing of the spring phytoplankton bloom support this hypothesis with regard to haddock (*Melanogrammus aeglefinus*) on the shelf east of Nova Scotia (Platt et al. 2003; Fig. 6.9). An index of larval haddock survival (size of age 1 year class divided by spawning stock biomass) showed that two exceptionally strong year classes occurred when the peak of the spring phytoplankton bloom was between 2 and 3 weeks earlier than the long-term average. An early bloom may result in lower larval mortality caused by starvation.

● Survival of larval fish is greatest when the overlap between the period of planktic food availability and food demand by fish is greatest.

Spatial as well as temporal coherence is required between production and consumption if high-biomass pelagic communities are to develop. At the large scale, it has been suggested that iron fertilization in regions of low primary productivity could enhance production up the food chain to zooplankton and fisheries, and that the increased photosynthesis could also lead to an increased drawdown of atmospheric carbon dioxide that may mitigate against climate change. Complex trophic interactions, however, make implementation of this far from straightforward (Buesseler & Boyd 2003; Gnanadesikan et al. 2003). At the smaller scale, patchiness is essential for maintaining ecosystem function. If phytoplankton and grazers were mixed homogenously then resource depletion would soon occur, whereas spatial segregation enables higher

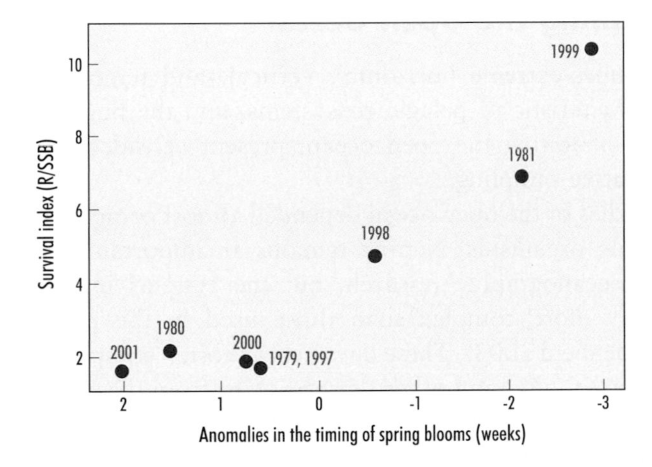

Fig. 6.9 Larval haddock survival (size of the year class at age 1 (R) divided by spawning stock biomass (SSB)) against deviation from the mean time of the annual peak in phytoplankton production. Timing of the peak of the spring phytoplankton bloom explains 89% of the variance in larval survival, providing strong support for the match-mismatch hypothesis. Redrawn from Platt et al. 2003 with permission from the author.

overall biomass to be maintained (Brentnall et al. 2003). Furthermore, if zooplankton did not aggregate in high densities then pelagic filter feeders such as basking sharks (*Cetorhinus maximus*) and baleen whales would not be able to survive on a diet of these organisms. Whales incur high energetic costs while foraging (Acevedo-Gutierrez et al. 2002) and it has been estimated, for example, that Right whales (*Eubalaena glacialis*) require copepod prey concentrations to exceed a minimum threshold of 4500 individuals m^{-3} of seawater just to balance the energy expended during feeding (Beardsley et al. 1996).

● If phytoplankton and grazers were mixed homogenously then resource depletion would soon occur, whereas spatial segregation enables higher overall biomass to be maintained.

Frontal regions tend to be characterized by increased primary production and to support particularly rich pelagic communities (Box 6.3). Fronts are therefore sites of intense feeding activity and are targeted by mobile predators including fish, squid, marine mammals, and birds (Durazo et al. 1998). The exact mechanisms by which predators locate prey in the wide expanse of the open ocean remain largely unknown and different cues are likely to be important at different spatial scales (Fauchald et al. 2000). In some ecosystems prey are located regularly in production 'hot spots' (Davoren et al. 2003), attracting larger predators in a predictable manner. Conservation measures aimed at reducing conflict between wildlife and fishers need to account for geographic variability such as this in ecosystem management plans. Fisheries for Antarctic krill, for example, may in future be required to operate outside the foraging areas of land-based central-place foragers during their breeding season (Constable & Nicol 2002).

● Conservation measures aimed at reducing conflict between wildlife and fishers need to take in to account the fact that predators forage at production hotspots that occur at oceanographic features such as fronts.

6.5 Sampling the Open Ocean

The sometimes-extreme horizontal, vertical, and temporal patchiness that is characteristic of pelagic ecosystems, and the huge size range of organisms inhabiting the open ocean, present considerable difficulties for quantitative sampling.

Early studies of the open ocean depended almost completely on nets to sample living organisms. Netting remains an important component of biological oceanographic research, but the systems in use today are considerably more complex than those used in the pioneering days (Wiebe & Benfield 2003). These days nets are often equipped with depth, temperature, salinity, and other sensors. Data from these sensors can be relayed to the ship in real time, either along conducting cables or via acoustic links, and enable nets to be placed accurately in the section of the water column of particular interest (Brierley et al. 1998). Optical particle counters (OPCs) can be used instead of nets to obtain estimates of zooplankton numerical density (Heath 1995), and photographic and video devices allow high-quality images of ocean inhabitants to be obtained

Box 6.5 Microlayers

Advances in the resolution of optical and acoustic sampling technology have lead to the discovery of widespread 'thin layers' of high biological activity in the ocean. These layers, which range in thickness from a few centimetres to a few metres, may be many kilometres in horizontal extent and may persist for several days. They contain densities of organisms several orders of magnitude higher than adjacent depth zones, and layers at different depths in the same area may contain distinct plankton assemblages. Thin layers produce microenvironments of physical, chemical, and biological parameters. Microlayers sometimes occur at the pycnocline as zooplankton forage on material that is suspended there, but may be deeper or shallower. Layers tend to occur in stratified water where current sheer is low. The species or populations that comprise each distinct thin layer probably aggregate in response to different sets of biological and/or physical processes. The existence and persistence of planktic thin layers generates great biological heterogeneity in the water column, and may go some way to explain the 'paradox of the plankton' in which high species diversity occurs in small, apparently homogeneous bodies of water.

(Benfield et al. 1996). Light is attenuated rapidly by seawater though, and visual sampling is often constrained by water clarity (Chapter 2). Sound, on the other hand, propagates very efficiently through seawater, as is testified by the long-range vocal communications of some whales. Scientific echo-sounders can be used to detect and quantify abundance of zooplankton and fish (Holliday & Pieper 1995), and ever-increasing sampling resolution is detecting biologically important features such as micro-layers (McManus et al. 2003), which are likely to be of very major importance to pelagic ecosystem function (Box 6.5).

● While modern imaging and acoustic technology has improved our ability to sample the ocean realm, we continue to be reliant upon often-rudimentary nets to obtain biological samples.

At the larger scale, organisms that must come to the sea surface to breath (e.g. seals, whales), or that forage over the sea surface (e.g. seabirds) can be counted at sea by observers on research vessels. Although much biological oceanographic data is still collected from ships, logistic constraints place restrictions on the amount of time that ships can spend at sea. In order to make longer-term observations, or observations over large extents of ocean, other sampling platforms or techniques are required. Moored instruments can be used to collect long time-series of data from spot locations (Schofield et al. 2002). Autonomous underwater vehicles (Box 6.6) have the potential to be able to operate in weather conditions that curtail sampling from ships, and in addition can work in environments that are impenetrable to ships. Earth-orbiting satellites are able to provide coverage of the entire surface of the global ocean on a weekly basis and deliver near-synoptic information on, for example, sea surface temperature, chlorophyll concentration, and frontal position (Miller 2004).

● Satellites can relay information from tags attached to air breathing animals such as whales and turtles, collecting data about their behaviour and patterns of movement in real time.

Satellites can also be used to track the movements of larger animals as they forage at sea over extended periods of time (Thompson et al. 2003).

Box 6.6 Autonomous Underwater Vehicles

Autonomous Underwater Vehicles (AUVs) are unmanned submersibles that can be programmed to navigate in three dimensions underwater. They can carry a variety of scientific instruments and are able to make measurements in parts of the ocean that are inaccessible, either physically or operationally, to conventional research platforms such as ships. The *Autosub* AUV, for example, has been equipped with a scientific echo-sounder and deployed on missions beneath Antarctic sea ice. There, it has made observations on the distribution of krill under ice, and of ice thickness, that were impossible to make using ice-breaking research vessels (Brierley et al. 2002). *Autosub* is among the largest of AUVs presently available to the scientific community, with an instrument payload capacity of 100 kg (weight in water). *Autosub* is 7 m long × 1 m in diameter, weighs 2400 kg, is powered by manganese alkali batteries, propeller-driven, has a range of about 800 km, and a maximum depth capability of 1600 m. Web link: **http://www.soc.soton.ac.uk/OED/index.php?page = as**. (Photograph: Andrew Brierley.)

Leatherback turtles (*Dermochelys coriacea*), for example, have been tracked in the north Atlantic (Hays et al. 2004). Leatherbacks are critically endangered, and a major source of mortality for them is capture by pelagic fisheries. Knowledge of Leatherback distribution and dive characteristics obtained via satellite telemetry could lead to the implementation of conservation measures designed to reduce the interaction of turtles with fisheries, and thus reduce by-catch.

The capacity of scientists to be able to collect data from the pelagic realm seems ever to be increasing. Plans are afoot to establish a series of permanent, automated ocean observatories that will be able to deliver multidisciplinary data continuously in real time, year on year. Although these systems will contribute enormously to our understanding of ocean ecosystem function, they will present new challenges in terms of extracting meaningful summaries from potentially overwhelming quantities of data.

6.6 Pelagic Fisheries

Fisheries for pelagic species have the potential to be among the most sustainable and least damaging to the environment. Shoaling species like the Atlantic mackerel (*Scomber scombrus*) and North Sea herring (*Clupea harengus*) form single-species aggregations and by-catch is minimal (Chapter 12). Indeed, at the time of writing, stocks of these two species seem to be bucking the global trend of decline and are thriving under good management and regulation. Myctophids, or lantern fish, are small mesopelagic fish that form a major component of oceanic deep scattering layers. They have been fished historically in the south-west Indian Ocean and in the south Atlantic, but fishing ceased in 1992 because of unfavourable economics and market-resistance, and myctophids are not presently under threat.

● Fisheries for pelagic species have the potential to be among the most sustainable and least damaging to the environment.

Planktivorous forage fish such as sardine and anchovy have vital ecosystem functions, particularly in upwelling zones, where they typify mid-trophic-level **wasp-waist** ecosystems (Cury et al. 2000). Abundance of these species can fluctuate wildly under variable environmental regimes and high fishing pressure and may result in major ecosystem changes. Fishing for large tropical pelagic fish including tuna and jacks has also had substantial impact. Analyses of long-lining data suggest that 90% of biomass of large pelagic fish may have been removed (Myers & Worm 2003). Open-ocean fisheries have tended to develop ahead of management procedures (maybe by as much as 15 years) and, in the case of pelagic long-line fisheries, it is possible that the estimated pre-exploitation biomass to which management processes are anchored are unrealistically low because they represent already-depleted stock levels. This 'missing baseline' presents particular difficulty for the long-term restoration of stocks because the size of the stock pre-exploitation remains unknown, and thus it is very difficult to say when or if restoration has been achieved. Not only has long-lining hit pelagic fish hard, but it has been and is still responsible for substantial declines in albatross populations. Albatrosses take baited long-line hooks as they are thrown from fishing vessels, drowning as the long-line sinks, and some populations are showing marked and continuing decline (Tuck et al. 2001). Seine netting for tuna also suffers from by-catch, particularly of dolphins, although, following international outcry, practices are in place to reduce the impact to levels that are now ecologically sustainable (Hall 1998).

● It is possible that the estimated pre-exploitation pelagic fish biomass to which management processes are anchored are unrealistically low because they represent already-depleted stock levels, the so called 'missing-baseline' effect.

● The whole issue of 'dolphin-safe' tuna remains the subject of debate and there are moves in the United States to ease legislation relating to the definition of the product.

Squid are fished on the open seas, mostly using hooked, coloured lures (jigs) at night to catch animals attracted to bright lights. The high intensity lights used to attract squid to jigs are so bright that they can be seen from satellites, and this has opened a new mechanism for potentially monitoring and managing open-ocean squid fisheries (Rodhouse et al. 2001). Robust management is particularly important for squid because they are short lived, often **semelparous** (spawn once and die), species and therefore vulnerable to over-fishing since there are few cohorts to provide a buffer from failure of any single generation. Squid also respond rapidly to changing oceanographic conditions and it is becoming increasingly clear that it is essential to understand interactions between squid and their ocean environment in order to predict inter-annual variations in recruitment. Recruitment of the squid *Illex argentinus* to the Falkland Islands fishery, for example, increases in years when water temperatures over the squid egg-hatching grounds are favourable (16 to 18 °C). In years when movement of the highly dynamic front between the Brazil and Falkland currents displaces waters of favourable temperature from over the hatching area, recruitment is reduced (Waluda et al. 2001).

● Semelparous animals such as squid are particularly vulnerable to over-fishing as their populations are not composed of multiple cohorts that provide a safety net against over-exploitation.

6.7 Regime Shifts in Pelagic Marine Ecosystems

Ecologists have long recognized that ecosystems may exist in 'multiple stable states'. In the oceans, conspicuous jumps from one state to another have become known as **regime shifts** (Scheffer et al. 2001). Shifts typically take less than one year to occur and regimes may persist for decades (Hare & Mantua 2000). Regime shifts may be driven by climatic changes, fishing pressure, or both, and may be manifest in parameters that measure physical and biological ecosystem state. In the north Pacific, statistically significant regime shifts in 1977 and 1989 are apparent in a composite index of 100 biological and physical time series including the Pacific Decadal Oscillation (PDO), zooplankton biomass estimates and salmon catches (Fig. 6.10).

● The 1988 regime shift in the North Sea may have been caused by a change in the North Atlantic Oscillation.

Regime shifts present major challenges for scientists attempting to manage fisheries. In the North Sea a regime shift in 1988 was evident from plankton time-series data from the Continuous Plankton Recorder (CPR) surveys (Reid et al. 2001). It has been suggested that this shift was caused by increasing flow of Atlantic waters into the North Sea, an increase that was correlated with a change in the North Atlantic Oscillation Index (NAOI). Recruitment of cod (*Gadus morhua*) in the North Sea has declined since the mid 1980s and it is possible that changes in the plankton following the regime shift have had a negative impact on the supply of food to young cod (bottom-up control) (Beaugrand et al. 2003). In the face of such possible environmental impacts on fisheries, it is clear that future attempts to manage fisheries will need to take environmental factors into account as well as data on fish population dynamics and catch levels. This realization has led to calls for the development of an holistic, **ecosystem approach** to fisheries management (Pitkitch et al. 2004; Chapters 12 and 15).

Although not necessarily a symptom of regime shift *per se*, jellyfish appear to have increased in prominence in many pelagic marine ecosystems worldwide in recent years (Mills 2001). Jellyfish blooms have occurred in the Bering Sea, the northern Benguela current and elsewhere,

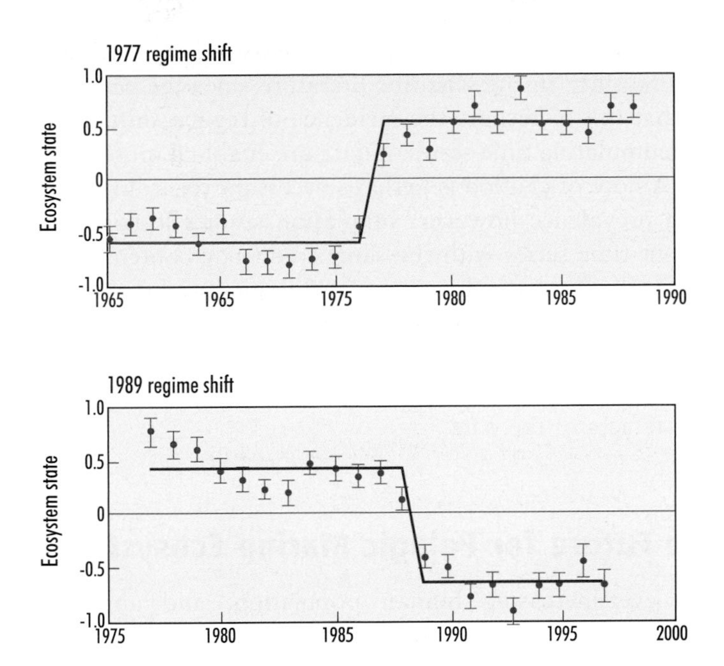

Fig. 6.10 Mean and standard error of a composite index of 31 physical and 69 biological parameters from the north Pacific between 1965 and 1997, showing significant step changes or 'regime shifts' in 1977 and 1989. The physical time series represent atmospheric and oceanic processes, while the biological time series all relate to oceanic species ranging from zooplankton to salmon and groundfish. Each of the time series was normalized before plotting and statistical analysis by subtracting the mean across both regimes and then dividing the data for each regime by the standard deviation for that regime. Standard errors for each year were computed as $s/\sqrt{n}$, where s is the standard deviation across all variables within a year and n is the number of time series used in the calculation (≤ 100). Redrawn from Scheffer et al. 2001 with Nature Publishing Group's copyright permission and permission from the author.

possibly in response to climate and fishing effects. Indeed, it has been suggested that jellyfish-dominated communities are the inevitable end point in pelagic ecosystems perturbed by fishing (Pauly & MacLean 2003). In the North Sea, correlations between the abundance of jellyfish and an Index describing the periodically fluctuating North Atlantic Oscillation (NAOI) have been detected (Lynam et al. 2004). Furthermore it seems as though the recruitment of herring (*Clupea harengus*) in the North Sea is adversely effected by high jellyfish abundance (although it is not clear as yet whether this is due to predation by jellyfish on herring eggs or larvae, competition between jellyfish and herring for zooplankton food, or both). Complex interactions between climate and jellyfish may therefore impact fish stocks, even in the absence of fishing, and could have major implications for the recovery of fish stocks even after any cessation of fishing.

● Jellyfish-dominated pelagic communities may be one consequence of overexploitation of pelagic fish stocks.

● The NAO is presently in a high phase, possibly restricting jellyfish abundance. If the NAO were to switch phase, climatic inhibition of jellyfish abundance may be relaxed and their numbers may increase with adverse consequences for fish stocks.

There has been an almost exponential rise in the incidence of the term 'regime shift' in the scientific literature since the early 1990s. It is possible that this is because the incidence of regime shifts is increasing, or that accumulating time-series of data are enabling more changes to be detected. A note of caution is perhaps necessary regarding this apparent increasing prevalence however: simulation studies looking at random, independent time series with the same frequency content as the Pacific Decadal Oscillation have shown that techniques used to identify 'regime shifts' may find them in noise. Detection of step-changes does not therefore necessarily provide evidence of processes leading to any meaningful regime shift (Rudnick & Davis 2003) since the step changes may be artefacts of the data.

6.8 The Future for Pelagic Marine Ecosystems

With an ever-increasing human population, and an ever-growing demand for food protein, it seems likely that pressure on the open ocean is likely to continue to grow. There is a history of fisheries advancing further from shore, into deeper and more distant waters, as conventional coastal resources are depleted and this looks set to continue. Fishing effort has already had major impacts on the global ocean. As traditional fish species are removed, fishing effort turns from these higher-trophic-level predators to smaller species. This phenomenon has become known as **fishing down the food web** (Pauly et al. 1998) and is ecologically unsustainable.

Humans are not just altering the open-ocean ecosystems by removing biomass but are also degrading it by addition. The incidence of waste in the ocean is increasing, with floating rubbish potentially distributing species far beyond their usual ranges, leading to alien colonizations of distant shores (Barnes 2002). Introductions of alien species in ballast water from cargo ships has also had devastating effects on pelagic ecosystems, such as the introduction of the ctenophore *Mnemiopsis leydii* to the Black Sea (Kideys 2002). This ctenophore, a native of the eastern USA, was predatory upon fish eggs and led to the collapse of the Black Sea anchovy fishery. Dumping CO_2 at sea in an attempt to reduce further increases in atmospheric concentrations is being investigated (Hunter 1999). As well as the addition of objects and organisms, human activity has also increased noise levels in the ocean. Low frequency noise from shipping, oil-exploration, and military activities may adversely impact cetacean communication and foraging (Croll et al. 2001) by masking the sounds these animals generate. Killer whales (*Orcinus orca*) in the waters of Washington State, USA, increase the lengths of their calls significantly (by about 15%) in the presence of whale-watcher

● The incidence of waste in the ocean is increasing, with floating rubbish potentially distributing species far beyond their usual ranges leading to alien colonizations of distant shores.

boat traffic, and probably do so in an attempt to overcome the noise generated by these boats that may mask their usual calls (Foote et al. 2004).

A recent forecast of the likely state of aquatic ecosystems in 2025 identified climate warming as the most significant single threat (Chapter 14), and climate changes have already had measurable impacts on sea ice extent and zooplankton distributions (Polunin, 2005). Perhaps the biggest climate-related threat to pelagic marine ecosystems arises from the possibility that increased warming and consequent freshening of the Arctic may switch off the north Atlantic current and hence perturb global ocean circulation (Rahmstorf 2002). It is probable that changes like this have happened multiple times during the Earth's history, and occurred over very short periods. If, as some models predict, this were to happen again in the near future, the consequences for the Earth's ecosystem and climate would be so severe that concern for the state of the pelagic realm would probably not be at the top of humanity's agenda.

CHAPTER SUMMARY

- The pelagic realm is highly heterogeneous, and production is patchy in both space and time. Generally production is higher closer to land, because of increased nutrient input (rivers, upwelling), and close to the surface because of light availability. There is a direct link between primary production and fisheries production.
- Organism size has a major bearing on mobility in the pelagic environment. Plankton are generally small (<10 mm long) and are unable to swim against currents and drift passively on them. Larger organisms (nekton) can move actively against currents. Plankton can however move vertically and undertake pronounced diel migrations.
- Pelagic food webs are size-structured: small organisms are consumed by a succession of larger grazers or predators. Most biomass occurs at the lowest trophic levels (grazers) and gradually decreases at increasingly higher trophic levels.
- Environmental heterogeneity and the large range of pelagic organism-size (from plankton to whales) presents a severe challenge for sampling the pelagic environment. Technological advances provide the means to collect ever-increasing quantities of data, but net sampling remains important for collection of biological material.
- Pelagic fish that form large single-species shoals should be amongst the most straightforward to manage and can be exploited with little risk of bycatch. Nevertheless, even pelagic species that inhabit remote locations far from land have been impacted severely by fishing.
- Pelagic ecosystems can suffer step-changes, shifting rapidly from one state to another. Such regime shifts may be due to impacts of climatic change, and have major implications for ecosystem management.

FURTHER READING

Longhurst (1998) provides an excellent description of the causes and consequences of geographic variability throughout the world's ocean. Mann and Lazier (1998) give a broad coverage of biological responses to physical processes in the ocean. A useful plankton atlas of the North Atlantic was published in Volume 278 of Marine Ecology Progress Series (2004). This provides a summary of Continuous Plankton Recorder (CPR) methods, and describes how this invaluable long-term record has become an important implement in our understanding of how pelagic ecosystems respond to global change. Steele (2004) provides a brief review of regime shifts and their definition. His article is the first article in a special issue dedicated to regime shifts.

- Anonymous 2004. Continuous plankton records: Plankton atlas of the North Atlantic Ocean (1958–1999). *Marine Ecology Progress Series* 278, Supplement Available on line at **http://www.int-res.com/abstracts/meps/CPRatlas/contents.html**
- Longhurst, A. R. 1998. *Ecological Geography of the Sea*. Academic Press, San Diego.
- Mann, K. H. & Lazier, J. R. N. 1996. *Dynamics of Marine Ecosystems: Biological-Physical Interactions in the Oceans*. Blackwell, Oxford.
- Steele, J. H. 2004. Regime shifts in the ocean: reconciling observations and theory. *Progress in Oceanography* 60: 135–41.

Chapter 7

Continental Shelf Seabed

CHAPTER SUMMARY

Continental shelves are the most heavily exploited and utilized areas of the world's oceans and support the greatest level of biological production. The ecology of the shallow shelf areas is strongly influenced by physical processes such as waves, tides, currents, erosion, and inputs of material from the adjacent land mass. These processes generate a great diversity of ecosystems and habitats at regional and local scales. The composition of the seabed and its associated biota are a direct reflection of the physical processes that act upon it and vary from mud sediments honeycombed with burrowing crustaceans and worms through species-impoverished mobile sands to bedrock encrusted with luxuriant growths of particulate feeders and algae. The benthos is a critical link in the transfer of organic material and nutrients from the water column above to the seabed. The flow of energy through food webs varies considerably among regions, with greater web complexity at mid to high latitudes. Systems with strong linkages among individual species are prone to trophic cascades and are more sensitive to the removal of key species leading to ecosystem regime shifts.

7.1 Introduction

The world's continental shelves contribute only 8% of the global sea surface. Nevertheless, their shallow nature means that most of the waters of the continental shelf fall within the euphotic zone (Chapters 2 and 6), which coupled with organic and mineral inputs via riverine discharges and strong physical mixing of the water column and seabed, make them among the most productive and economically important regions of the world's oceans (Costanza et al. 1997). In regions where the shelf is narrow, nutrients are additionally supplied through upwelling of water from the deep sea. The high productivity found on continental shelves fuels major global fisheries (Chapters 2, 6, and 12).

● Primary production in shelf seas fuels 90% of the world's fisheries (Pauly & Christensen 1995).

The abundant plankton and fish fauna found in these waters are a key food resource for populations of top predators such as seabirds and marine mammals. In addition to their status as hotspots of biological activity, the proximity to the coastline and shallow nature of shelf seas means that they have become the focus of intensive human activities and in addition receive agricultural and industrial contaminants from terrestrial run-off. Shipping, fishing, exploration for and extraction of hydrocarbons, mining of sediments and minerals, underwater cable laying, wind farm and tidal barrage development, offshore aquaculture and recreation are just some of the major activities that occur and impact upon the continental shelf sea environment. The ecology of the continental shelf is under ever-increasing levels of human usage, which has resulted in major changes in ecosystem structure in some localities.

● The intensive use of the coastal shelf increases the risk of environmental damage from activities such as over-fishing, eutrophication, mineral extraction, dumping of waste, and oil spill accidents.

7.2 Definitions and Environmental Features

The continental shelf extends from the extreme low water mark on the shoreline down to a depth of approximately 200 m and is termed **neritic** (Chapter 6). This region extends beyond the land from between nearly zero up to 1500 km offshore out to the **shelf break.** Beyond this point the continental shelf slopes down to the abyssal plain (Chapter 8). The shelf break is an area where biological and geological material from the continental shelf is supplied to the shelf slope, through a variety of processes such as the death of organisms or more dramatically via submarine mudslides. Only on extreme low water spring tides, or in the strandline wreckage in the upper shore that occurs after an onshore storm, are we able to observe unaided some of the organisms that live at the shallowest edge of the continental shelf. Generally the continental slope has a shallow gradient of approximately 1° except in regions where glacial activity has sculpted a more dramatic seabed. The shallow depth of the continental shelf and its position adjacent to the physical barrier of the land mass mean that it is strongly influenced by **physical forcing** processes such as glaciation events, currents, waves, the formation of fronts and water turbidity (Fig. 7.1). The interaction of these processes is influenced by shelf width and geographic disposition and is strongly related to consistent patterns in regional ecosystem structure.

● Physical processes perform a key role in the continental shelf environment due to its proximity to land, inputs of fresh water, seabed topography and shallow depth.

7.2.1 Influence of glaciation events

The physical and biological characteristics of the continental shelf habitat are strongly influenced by the geological composition of the seabed, much of which has resulted from past **glacial** events that have had a profound influence on coastal margins and shallow near-shore

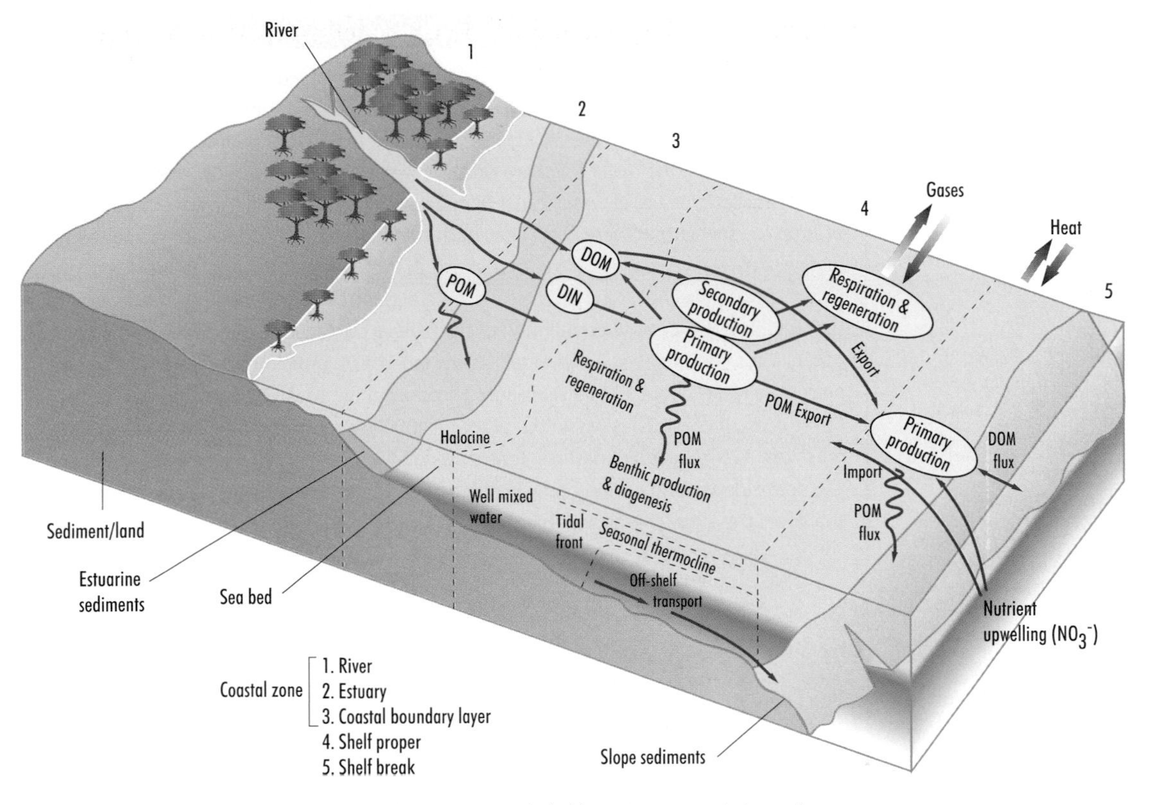

Fig. 7.1 Key processes that influence the continental shelf environment (adapted from Alongi 1998). POM = Particulate organic matter, DIN = Dissolved inorganic nitrogen, DOM = Dissolved organic matter.

seabed structure. The melting and formation of ice sheets causes the sea level to rise and fall respectively, while the Earth's crust is lowered under the weight of ice, but rebounds (rises) in its absence (Box 7.1). Thus areas hundreds of kilometres offshore may have been dry land at some point in recent geological history. This is reflected in the surface topography of the seabed where it still possible to see drowned river deltas (e.g. off the River Congo in Africa), while evidence of hominid butchery of large mammal bones has been found to the west of Florida in the Gulf of Mexico at a present-day depth of 40 m below sea level. Glacial deposits characterize much of the continental shelf bed in the higher latitudes and are typified by fields of boulders and gravel beds.

● The continental shelf is relatively young in geological terms and has undergone dramatic expansion and contraction as a result of glaciation events. Evidence for these events can be seen vividly on the seabed as drowned river plumes, boulder fields, and glacial scouring.

7.2.2 Importance of waves and flow

The physical effects of waves have important consequences for the ecology of the shallower areas of the shelf, and their effects reach down to a depth of 80 m on open Atlantic coasts during gale force conditions (Chapter 5). Clearly, the depth to which waves influence benthic ecology

Box 7.1 Rise and fall of the continental shelf edge

The coastal margins of State of Maine (USA) have left clear evidence of the rise and fall of coastal margin in response to glacial events. Land-level altered substantially from 14 000–10 000 years ago during the last glaciation event. After this period, global sea-level changes have had the major influence on the coastal landscape. About 14 000 years ago global (**eustatic**) sea level was about 110 m below its present level. However, the weight of local ice sheets that were over 1 km thick depressed the level of Earth's crust which meant that relative sea level was 70 m above that of the present day. Eustatic sea level was relatively low, but rising as the ice sheets began to melt. Ultimately, the release of the weight of the ice allowed rebound of the land to outpace the rate of the rising global ocean. Thus, local relative sea level fell (as the land rose), until it reached 55 m below the present sea level about 11 000 years ago. This caused the coastline to move well offshore of the present coast. Thus within the space of 4000 years the coastal margin varied by as much as 125 m with a coastline that expanded or contracted by tens to hundreds kilometres depending on the slope of the local continental shelf and coastal margin (the shallower the slope the greater the incursion or excursion by the sea).

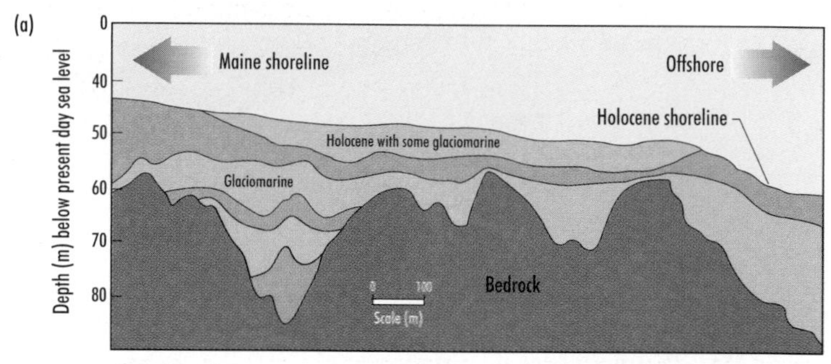

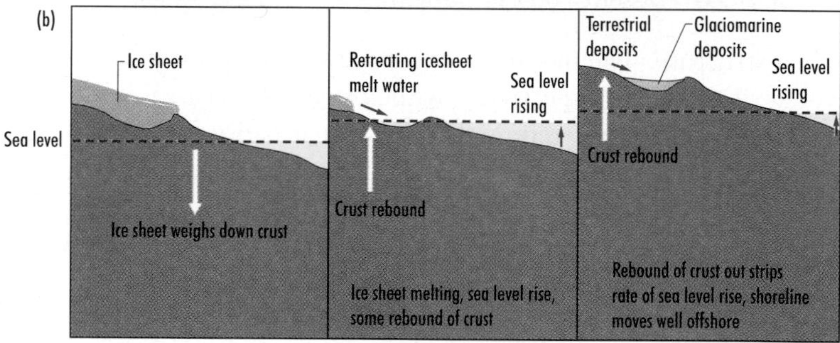

(a) Reflection of sound waves from the seabed strata at a depth of c.40 m offshore of Maine USA shows the successive deposition of glaciomarine material during various glaciation events. (b) A schematic image to show the various stages of vertical movement of the continental shelf in relation to glacial retreat and changes in sea level.

For more information on this specific example see **http://www.state.me.us/doc/nrimc/pubedinf/factsht/marine/sealevel.htm**.

will depend on the extent to which the coastline is sheltered from prevailing winds and the extent of fetch (Hiscock 1983). In areas where the physical force of waves causes sediment movement, wave action can be a major cause of mortality among benthic animals and has been shown to affect secondary production by limiting the body size of organisms that can survive in a highly energetic environment (Emerson 1989). Water movement generates currents and these affect both the shallow and deeper parts of the shelf. The phenomenon of flow is extremely important for the ecology of the continental shelf, as it affects the passive and active transport of organisms, their gametes and larvae, the rate of supply of food from the pelagic system to the seabed, and places upper physical constraints on the type of organisms that live in particular habitats. Thus close to the shoreline, the seabed is affected most strongly by both waves and currents, as we move further offshore the effects of waves reduce and the effects of currents begin to dominate the physical and biological processes on the seabed. This gradient is clearly reflected in the biomass of **sessile** filter and deposit feeding biota, which increase with distance away from shallow into deeper water where the physical stress associated with waves and currents decreases and the seabed is therefore less frequently subjected to physical disturbance.

- Fetch is the uninterrupted distance over which winds exert friction at the sea surface.

- The ecological importance of wave action diminishes with reducing fetch and distance from the shore as water depth increases.

- Animals that are either attached to a substratum or that move relatively little or infrequently are termed sessile.

The rising and falling water mass has important implications for organisms that live on the shore (Chapter 5 and 6). An equally important result of this oscillation is the flow or **current** that is generated as the water floods into or ebbs from restricted areas of coastline. Current is increased when a water mass moves through or around land bounded restrictions or across irregularities in the seabed topography. Typically straits and the narrow mouths of estuaries have some of the strongest tidal flows, reaching speeds of up to several metres per second. Headlands and bedrock protrusions from the seabed also present restrictions to the flow of water and lead to strong currents around their apex (Fig. 7.2).

As water moves over the seabed, the uneven seabed induces friction, which slows the immediate water column above, so seabed currents are rarely as fast as sea surface currents (except in areas of the deep sea). Currents and the associated bottom shear stress, influence food availability for benthic communities (Jenness & Duineveld 1985) and benthic secondary production (Warwick & Uncles 1980; Wildish & Peer 1983). High levels of shear stress cause scouring of the seabed and its biota while high currents velocities inhibit effective feeding activity. Biota that live in such environments tend to have characteristics or behaviours that enable them to cope with the extremes of physical stress. Typically, attached biota are highly flexible or encrusting, while mobile fauna will often seek shelter from currents within the sediment or in crevice

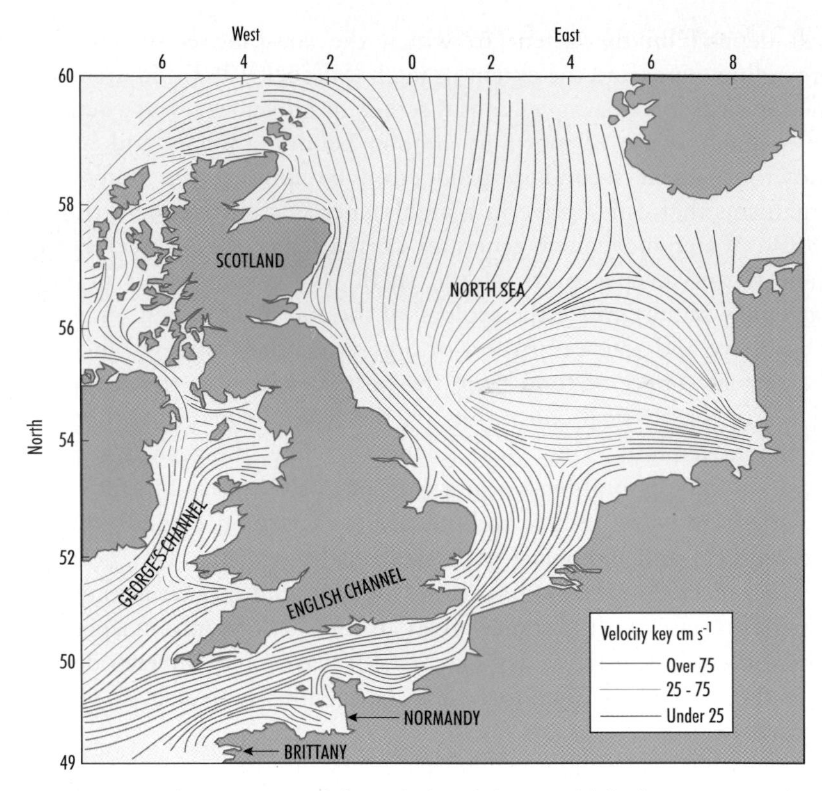

Fig. 7.2 Current velocity generated through the tidal rise and fall of water in a sea basin is exacerbated close to the coast where coastal morphology and restrictions increase the speed of flow. This can be seen clearly for the example of the north-western European shelf above, with strong mean surface velocities in the English Channel, the southern North Sea, the George's Channel. Current velocities are also increased around headlands such as off the North of Scotland and around the coast of Brittany and Normandy on the coast of France. The darker the blue lines the higher the current velocity. Adapted from Metcalfe et al. 2002.

habitats. At the other extreme, areas with reduced water movement at the seabed will not replenish the supply of food to the seabed, which then becomes a limiting factor for the growth of benthos. At a certain current velocity threshold, food particles transported from other areas will begin to sink to the seabed, where they become available as food to the benthos (Creutzberg 1984). The exact figure for the current velocity threshold will depend upon particle size and buoyancy. The latest research highlights how flow is strongly related to the rate of passive exchange of nutrients and oxygen between the top few millimetres of the sediment surface and the overlying water. A reduction in flow reduces the diversity of the surface sediment microbial assemblage, which in turn alters the physical properties of the sediment. Different microbes produce different types of extracellular products that coat sediment grains and hence affect properties such as shear (Biles et al. 2003).

● Flow velocity influences the rate of supply of particulate matter to the seabed, is linked to the rate of nutrient exchange at the sediment – water interface, and places upper constraints on the body size of biota that can live in a particular habitat.

7.2.3 Fronts and production

The development of **fronts** has important ecological implications in areas where water becomes **stratified** (Chapter 6). During winter periods, the water column of the continental shelf is held in a well-mixed state by high winds and wave action. This means that dissolved oxygen, carbon dioxide, and nutrients are relatively evenly distributed from the sea surface to the seabed. The water column remains well mixed throughout the year in areas where the seabed currents are strong and cause turbulent mixing, typically close to the coast in shallow water. In offshore waters and areas not strongly influenced by tidal mixing, periods of prolonged warm temperatures and calm conditions lead to stratification. This results in rapid depletion of oxygen at the seabed and nutrients in the surface layers of the water column. Frontal systems occur at a point where a stratified and mixed body of water meet. The resulting **density gradient** between the two bodies of water means that a flux of nutrients occurs from the mixed water mass to the nutrient-depleted upper stratified waters of the adjacent water mass. This fuels primary production to a level higher than in the surrounding waters (Chapter 6).

As full salinity seawater approaches the coastline it interacts with lower salinity water discharged as an estuarine plume. The difference in density between the two bodies of water sets up a **frontal system** due to the density gradient between the two bodies of water (the less dense estuarine water is more buoyant than full salinity seawater) and may deflect the estuarine water along the coastline in one direction due to the **Coriolis** force (Chapter 6). A good example of such a plume is the discharge from the River Rhine in the North Sea, which flows northwards along the Dutch coastline. These areas have become termed regions of freshwater influence or **ROFIs**.

● Elevated production of phytoplankton at fronts attracts both fish and their predators such as seabirds and marine mammals, and results in a greater supply of organic matter to the benthic fauna.

The elevated primary production at fronts fuels production in the benthos and in pelagic and demersal fisheries. As a result, fronts are often the foci of fisheries activities but also attract piscivorous and planktivorous seabirds and sea mammals, and there is evidence to suggest that fronts may be associated with above average by-catches of marine mammals such as the harbour porpoise (*Phocoena phocoena*) (McGlade & Metuzals 2000).

7.2.4 Light and turbidity

Water depth and **turbidity** are important determinants of the distribution of benthic algae in the shallow waters close to the coast. Areas of the seabed affected by estuarine plumes are generally severely light-limited due to the associated high load of **suspended sediment** and **phytodetritus**, which **attenuates** light in the first few metres of the water column (Chapter 2). Thus these areas of the seabed tend to be dominated by animals, and any algae are restricted to the very shallowest water adjacent to the shore. In contrast, coastal areas that are typically open to the ocean and have limited riverine discharge have much clearer waters and here a full range of algae can be found with clear **zonation** from the shallow to the deeper water of green (shallow), brown, and then red (deep) algal dominance. In these situations algae can be dominant in terms of their **biomass**. The width of the algal-dominated zone will depend on water clarity and slope of the seabed (the clearer the water and lower the slope the greater the area of seabed that will be suitable for algal growth).

● Algae are restricted to a narrow zone of near-shore shallow waters in regions where major riverine discharge and near-bed tidal resuspension of sediments increase the turbidity of the water column.

7.2.5 Regional ecosystem types

Basin-scale oceanographic processes vary with shelf seabed and coastal morphology (Warwick & Uncles 1980). These characteristics vary on a regional basis and are the primary determinants of regional shelf ecosystems. In the absence of the effects of fisheries, pollution, or the flux of anthropogenic materials, a relatively small number of physical factors regulate vegetation types or phytoplankton **biomes** that are modified by basin-scale circulation patterns and coastal morphology (above). The close coupling between physical processes and the biology of shelf seas enables them to be partitioned into broad regional ecosystem types, which Longhurst (1995) attributed to seven model systems. These systems are presented in Table 7.1 and show the link between regional physical processes, primary production and then the linkages to other compartments in continental shelf ecosystems (Chapters 1 and 6). While each of the ecosystems within each model displays differences in species composition, there is a high degree of similarity among the ecosystems at higher levels of organization (e.g. primary producers, secondary production processes, top predators).

● In the absence of human intervention through fishing, pollution or other forms of ecological disturbance, continental shelf ecosystems can be categorized under seven models according to the dominant physical processes and shared ecological features at higher levels of organization.

7.3 The Seabed Habitat and Biota

While it may be tempting to consider the continental shelf seabed habitat as rather two dimensional compared to the pelagic habitat above (Chapter 6), it is very much a three-dimensional habitat even though the third dimension, depth, is much more compressed than in pelagic systems. While in pelagic systems the depth dimension can exceed several thousands of metres in which biota can be found from the water's surface to just above the seabed of the deep sea. The seabed habitat is bounded by the depth to which biota can burrow and continue to live (usually no more than 2 m within the sediment) and the extent to which they penetrate into the waters above (30 to 50 m for some of the largest kelp genera such as *Macrocystis*) (Fig. 7.3). In some cases the biota are

Table 7.1 Seven models of continental shelf ecosystems giving their geographic locality, the principal primary and secondary production processes and other key components and characteristics at higher trophic levels (adapted from Longhurst 1995).

Model	Geographic location	Primary and secondary production	Higher trophic levels
Model 1 Polar irradiance-mediated production peak in regions permanently ice-covered	Eastern Siberian and Laptev Sea coasts of Siberia. Northeastern and Northern coasts of Greenland. Northern coasts of Canadian archipelago to west Beaufort Sea. Almost entire coast of Antarctica.	Productivity is light limited, with seasonal cycle symmetrical about local irradiance maximum. Latter corresponds with solar maximum or minimal snow cover. Below ice, phytoplankton productivity and zooplankton biomass are low, but abundant flora in the infiltration zone at the ice-seawater matrix. Underside of ice <2m thick may support dense growth of diatoms associated with abundant polychaetes, copepods and amphipods.	Benthic invertebrates are abundant and diverse providing food for abundant but low diversity populations of fish and squid. Large euphausiids (krill), characteristic of open-water regions, are replaced by small *Euphausia crystallophorias* and provide food for pelagic fish (*Pleurogramma* spp.) and crab-eater seals.
Model 2 Polar irradiance-mediated production peak in regions where ice-cover disperses partially or completely in summer, or only where broken pack-ice develops.	Coasts of Greenland. North America from Newfoundland to the Aleutians. Northern Asia from Finland to the Sea of Okhotsk. Short sectors of Antarctic coast in midsummer in eastern Ross Sea, to east of Ronne ice shelf and in Dumont d'Urville Sea.	Shallow polar halocline induces water column stability very early in open-water. Productivity is light limited, its seasonal cycle being symmetrical about the local irradiance maximum. Where pack-ice remains, conditions may resemble Model 1. After ice-melt, in open water, phytoplankton accumulates during the period when productivity increases, and then tracks its initial decline. Phytoplankton is dominated by diatoms, a subsurface chlorophyll maximum is often observed; in shoal water, significant biomass of benthic macroalgae develop. Planktonic herbivores are represented by abundant large copepods, euphausiids and salps, of which some species form swarms that support major stocks of baleen whales and seals.	Production exceeds consumption in the water column and supports rich and diverse macrobenthos, especially in boreal regions, where shelf areas uncovered by ice are much more extensive than around Antarctica. Low diversity fish fauna especially in the Antarctic, where small Notothenids dominate. The wider Arctic shelves support a greater diversity of Gadidae, Sebastidae, *Anarhichas* spp. Grey whale, walrus and bearded seal are boreal benthic feeders, having no austral equivalents.
Model 3 Canonical spring-autumn blooms of mid-latitude continental shelves	On mid-latitude continental shelves, under the influence of the global westerly winds. From Finland to Iberia and off the Mediterranean From Newfoundland to florida. Off Tasmania and southern Australia.	After winter mixing, a pulse of productivity and chlorophyll is induced by establishment of water column stability. Thereafter summer stratification is associated with relatively low productivity. Progressive mixing in autumn may induce renewed productivity fuelled by nutrients accumulated below the summer pycnocline. This sequence is modified by intermittent wind-induced coastal convergence and divergence, and persistent water column mixing in regions where tidal velocities exceed critical value. Effects of estuarine turbidity plumes may mask the canonical sequence, which weakens towards the equator. The balance between pico-autotrophs and larger algal cells is more equitable than in very high latitudes. In shoal water, especially at higher latitudes (Norway, Iceland, Newfoundland, Tasmania) there is significant autotrophic production in kelp beds. Small copepods dominate the inshore herbivorous plankton, larger species nearer the shelf edge, often over-wintering in deep water. Their seasonal cycle of abundance follows that of phytoplankton.	Most autotrophic production passes directly or indirectly through the macrobenthos, which is abundant, diverse, and characteristic of each sediment type. Diversity of fish fauna exceeds that in polar ecosystems (typically 200 species of >50 families). In boreal regions, major stocks of shoaling Clupeidae and Scombridae together with mainly demersal Gadidae, Percidae and Pleuronectidae. In much more restricted austral shelf regions, clupeids occur as in the north, together with a more-difficult-to-specify demersal fauna. Energy flow from pelagic invertebrates is mainly to clupeids and scombrids, from benthos mainly to demersal fish fauna.

Table 7.1 continued

Model	Geographic location	Primary and secondary production	Higher trophic levels
Model 4 Topography-forced summer production	Falklands shelf, the southern North Sea, the shelf of the Gulf of Alaska where tidal mixing consistently dominates the stability of the water mass, Temperate North Pacific where the oceanic permanent halocline passes inshore across the shelf, so constraining winter mixing to water above the pycnocline. New Zealand and similar locations where, topographically-forced upwelling sites in shallow water dominate the productivity sequence.	The seasonal productivity schedule, otherwise appropriate to Model 3, is instead forced by other factors (see adjacent column), differing regionally. Consequently, phytoplankton productivity tends to peak in mid-summer rather than in spring. In many respects the autotrophic biota, and energy flows, are broadly similar to those appropriate to Model 3 ecosystems.	In many respects the heterotrophic biota, and energy flows, are broadly similar to those appropriate to Model 3 ecosystems.
Model 5 Intermittent production at coastal divergences	The Atlantic from Iberia to Senegal and Gabon to Benguela. The Pacific from Oregon to Mexico and Peru to southern Chile. The Indian Ocean from Oman to Kenya.	These are the 'classical' coastal upwelling regions of eastern boundary current and other coasts, some of which occur at relatively low latitudes. The shelf is characteristically narrow, and the influence of river effluents minor. The equator-ward component of the Trade winds induces strong and persistent offshore Ekman drift, inducing upwelling of nutrient-rich deep water. This process is usually strongest during summer. Similarly, in the Indian Ocean the Southwest Monsoon forces offshore drift, principally off Somalia. Upwelling results in a rapid increase in primary production of phytoplankton, principally diatoms, and chlorophyll accumulation coincides with the duration of upwelling periods. The biota have low diversity and high collective biomass. Specialised invertebrate herbivores are large calanoid copepods (typically *Calanus* or *Calanoides*), euphausiids and filter-feeding anomuran crabs, each having life history tactics that take them into deep water, or the shallow sea-floor, during non-upwelling periods. Kelp forests reach their maximum development and generate significant autotrophic production and accumulation of biomass. There is heavy, intermittent settlement of large phytoplankton cells and faecal material to the sediments, frequently resulting in an oxygen deficit at certain depths on the narrow shelves.	Benthic consumers are abundant, but not diverse, and there is much physical export of organic material into deep basins on the shelf or across the shelf edge into deep water. Along the shelf edge, deep-water rockfish (*Sebastes*) are abundant, characteristic and diverse. Anchovies (*Engraulis, Cetengraulis*) are ubiquitous vertebrate herbivores, with in addition, sardines (*Sardinella longiceps*) in the Indian Ocean. Populations of predators, mostly pelagic clupeids (*Sardina, Sardinops*), mackerel (*Scomber*), hake (*Merluccius, Micromesistius*), sealions and piscivorous seabirds are characteristic of these regions.
Model 6 Small amplitude response to trade wind seasonality in regions with significant coastal river discharges which dominate over oceanic processes that would otherwise determine seasonal changes.	Amazon, Niger, Congo, Indus, Ganges, Irrawaddy, Mekong and others. These are wet tropical coasts, dominated by the effluent of a few major rivers or many smaller river systems. In the Atlantic, the Gulf of Guinea, the Guianas and northern Brazil. In the eastern Pacific, from Columbia to southern Mexico.In the Indo-Pacific, from the South China Sea to south-western India, including much of the Indonesian Archipelago and the northern coast of Australia.	Trade wind regimes force only weak seasonality in mixed layer depth, observed minor changes represent the geostrophic response of the pycnocline to seasonality in trade wind stress rather than mixing. The basin-wide slope of the thermocline has important consequences in the Pacific and Atlantic Oceans; to the west it lies deeper than the shelfedge, but in the east it is at mid-shelf level. In the east, therefore, the benthic regime has typical tropical character shorewards of this line, but more temperate characteristics in cool water seawards. On eastern Atlantic and Pacific coasts, the nutrient cline is perennially shallower than the photic depth, except during exceptional events, and vertically-integrated production rate is not normally light-limited. Everywhere, river discharges into the low salinity surface layer have strong seasonality, reflecting regimes of	Benthic community types conform to sediment types, and where these resemble those of cooler seas, members of the global suite of benthic infaunal communities occur (e.g. clams such as *Venus*). Inshore, organic-rich sediments may be extensive and support stocks of crustacea dominated by penaeid shrimps. These sediments are prone to resuspension by wave action in monsoon seasons.

Table 7.1 continued

Model	Geographic location	Primary and secondary production	Higher trophic levels
		wet and dry seasons, so that the seasonal schedule of primary production rate is governed by nutrient input from the land and possibly reduced irradiance due to prolonged heavy cloud cover during wet seasons. Autotrophic organisms are typically small cells, except in coastal blooms fuelled by river-borne nitrate, which are dominated by *Coscinodiscus* and other diatoms. The biomass of coastal subtidal macroalgae is not significant due to water turbidity. Consumers are numerically dominated by small copepods, but diatom blooms support large stocks of herbivorous clupeids (Atlantic – *Ethmalosa, Brevoortia*; Indo-Pacific – *Sardinella longiceps*). A large proportion of diatom material sinks to the seabed.	Fish fauna is diverse at all taxonomic levels and includes a higher proportion of pelagic species than in Models 3 and 4 at higher latitudes.
Model 7 Small amplitude response to trade wind seasonality in regions off dry coasts with relatively minor river discharges.	In the Atlantic, only the Caribbean. In the Indo-Pacific parts of the Arabian Sea, the Red Sea, north-east Australia and of the Indonesian archipelago.	Ecosystem of shallow seas off the coasts of the dry tropics, where river effluents are minimal. Many isolated islands and archipelagos in tropical seas are surrounded by reduced Model 7 ecosystems, of which the dominant characteristic is the development of coral reefs where topography permits, elsewhere unconsolidated sediments are dominated by carbonate sand. There is weak seasonality in mixed layer depth, and nutrient cline is usually shallower than the photic depth, except during exceptional events. Most primary production occurs in the benthos. Macroalgae (*Sargassum*), encrusting coralline green (*Halimedia*) and red algae, cyanophyte mats and sea-grass meadows dominate community production, in addition to the activity of symbiotic dinoflagellates within the tissues of many invertebrates: scleractinian corals, giant clams (*Tridacna*), coelenterates (alcyonians, anthozoans and scyphozoans), large ascidians and encrusting sponges. Nutrient sources and fluxes are various: advection, upwelling, vertical flux in fractured basement rocks, some terrestrial runoff. Nitrogen-fixing bacteria occur in the tissues of some corals, and cyanobacteria fix nitrogen within algal mats. Internal exchanges of nutrients within and between organisms is highly complex. Export from the benthic ecosystem, except in the form of carbonate eroded to sand and gravel factions, is a small but complex flux. Water clarity is high, and phytoplankton is dominated by the pico- and nano-fractions, as in oligotrophic ocean ecosystems. These are consumed by protists and small zooplankters themselves the prey of many filter- and tentacular-feeding polyps of corals and other coelenterates.	The macrobenthos associated with coral reef formations is exceptionally diverse at all taxonomic levels. On open sandy sediments, unencumbered with reefs, very high densities of filter-feeding crabs (*Pinnixa, Xenophthalmus*) are typical,together with other organisms, especially filter-feeding clams. Fish fauna is also diverse, both taxonomically and functionally. Parrot-fish (Scaridae) are among the most important herbivores, directly consuming coralline and other algal mats, and these may form a large-fraction of total fish biomass. An intense and complex network of trophic links between fish andbenthic invertebrates is characteristic of this ecosystem. This trophic complex supports a wide variety of large predators.

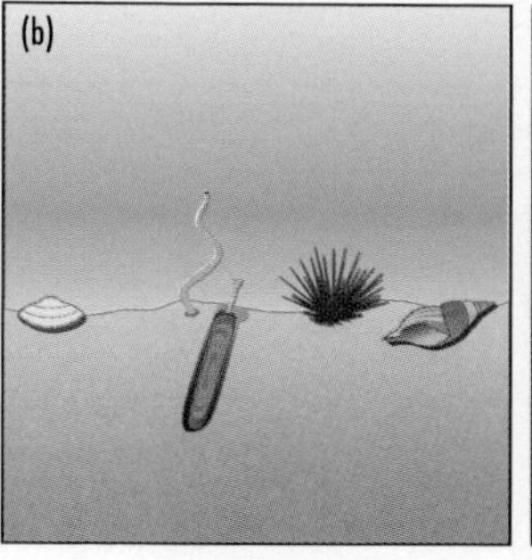

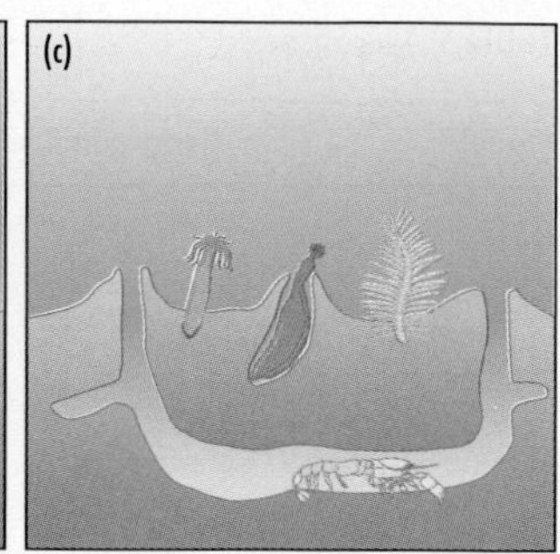

Fig. 7.3 The three-dimensional extent of the seabed habitat will vary considerably depending on the hardness and stability of the seabed. Rocky substrata provide a firm and stable anchorage for many animals and plants such as kelp, soft corals and sponges. The stability of this substratum means that often these animals are able to attain much larger body sizes that similar or the same species found in less stable habitats (a). Very few organisms can penetrate bedrock. Substrata that are composed of particles can vary considerably in stability from bedrock to fine muds and clays. Even the largest and heaviest particles can become mobile in certain sea conditions. Boulders provide interstitial spaces that act as refugia and provide a firm anchorage for epibiota. As particle size decreases, so the ability of animals to penetrate into the sediment increases until in the softest sediments it is limited only by the depth to which oxygenated water can be circulated into burrow systems. On mixed sediments (gravely shelly sands) the upper size of epibiota is constrained by near-bed currents (animals that grow too big will be sheared off the seabed) leading to adaptations such as a flattened body form (b). However softer sediments like mud tend to have a less diverse array of emergent epibiota and are usually limited to soft bodied animals like anemones and sea pens (c).

the key constituent of the habitat as in kelp forests, mussel and oyster reefs, maerl beds, corals and sponges. We will deal with specific examples of these habitats later in the chapter (7.7).

Body size affects both production processes and the degree to which biota are associated with particular habitats. The body size of animals living in the seabed is affected by burrowing ability and constraints on respiration due to oxygen exchange. The body size of attached biota that inhabit the surface of the seabed are constrained by physical process such as shear due to current velocity. The degree of mobility of the biota (i.e. their life mode), varies from those organisms that are firmly cemented to a substratum to anemones that are able to reposition themselves by a few mm per day to highly mobile crabs and fishes that migrate tens or hundreds of kilometres every year. The mobility of the biota affects their ability to respond to environmental change and places a constraint of the extent and range of habitat that can be utilized.

7.3.1 Body size and position

The major groupings of animals based on body size are **macrofauna**, **meiofauna**, and the **microbiota**. The meiofauna are usually defined as those organisms that would pass through a sieve with a mesh diameter of 0.5 mm but that would be retained on a mesh of 0.063 mm. The meiofauna are the least well studied, yet probably the most diverse, group of marine organisms that can be observed directly using light microscopy (Lambshead et al. 2001). Anything larger than this is termed the macrofauna. The small body size of meiofauna means that their rates of production (how much biomass is generated within a given time) is much higher than for the larger macrofauna. Accordingly, meiofaunal population responses to environmental and human disturbances can occur within days and weeks, which make them excellent indicators of environmental stress (e.g. pollution, physical disturbance). Most research has been devoted to the study of macrofauna due to their amenability to experimental and observational studies. Nevertheless, the contribution of the microbiota (dinoflagellates, diatoms, bacteria) to biomass and production in coastal seas is often under-represented, and in the tropics can considerably exceed that of meiofauna and macrofauna combined (Fig. 7.4). Body size and longevity is also an important determinant of the rate of recovery after a disturbance has occurred. Small animals recolonize areas of disturbance via larval recruitment on a scale of days or weeks whereas increasingly larger or slow growing biota can take many months or years to recolonize and attain their former biomass. The most extreme example of the latter would be slow growing deep water corals found on the continental slope edge or on seamounts (Chapter 8) or maerl (calcareous algae) beds (Hall-Spencer et al. 2000).

● Despite their contribution to benthic production and diversity, meiofauna and the microbiota are the least well-studied size classes of marine biota.

The biota can be categorized according to their relative position within or on the seabed. **Epibiota** are those emergent organisms that are anchored in or upon the substratum (e.g. corals, anemones, hydroids) or those free-living organisms that move about on the surface of the substratum (e.g. gastropods, starfish, crabs, fishes). Macroalgae are almost exclusively **epibenthic** and grow attached to the substratum. **Infaunal**

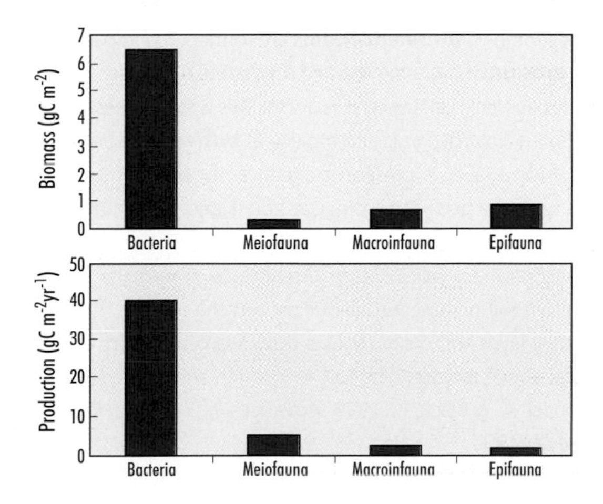

Fig. 7.4 Estimates of mean biomass and production of different size classes of benthic biota in the Gulf of Papua, northern Coral Sea (Alongi, D. M. & Robertson, A. I. 1995. Factors regulating beuthic food chains in tropical river deltas and adjacent shelf areas. *Geo-Marine Letters* 15: 145–152. Copyright Elsevier 1995.)

organisms live buried within the substratum, either entirely (such as free burrowing polychaete worms and heart urchins) or partially. Examples of partially buried infauna include burrowing shrimps, which build galleries within the substratum but emerge periodically to forage for food or mates or burrowing sea cucumbers that live within the sediment and extend feeding tentacles onto the sediment surface. In very shallow subtidal areas with good illumination at the seabed unicellular diatoms and dinoflagellates contribute to the **microbial community**, and along with cyanobacteria and bacteria, inhabit the interstitial spaces between sediment particles (Paterson & Black 1999) (Box 7.2).

● The microbiota have an important influence on the cohesive properties of the sediment through the production of extracellular organic secretions that modify flow and other important ecosystem functions such as nutrient exchange.

Box 7.2 Current flow over the seabed

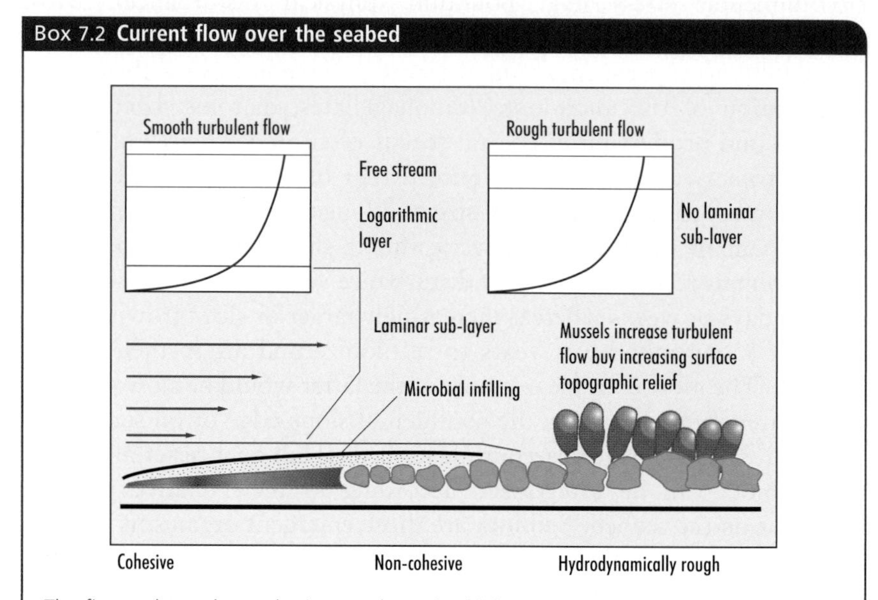

The figure above shows the impact that microbial organisms can have on flow over the seabed by infilling the gaps between sediment particles to create a 'smoother' surface. A rough bed will experience greater **stress** for the same overall flow than a smooth bed. This has important consequences for plants and animals living at the sediment surface. Under rough turbulent conditions, **turbulent eddies** are formed that impact the bed and increase the likelihood of **erosion** of the sediment and resident biota. Under such conditions it is not possible for the viscous laminar sub-layer to form. This is sometimes termed the **boundary layer**, which acts as a hydrodynamic and molecular **buffer zone** between the bed and the flow above. When this layer is present it significantly affects the flow of particles and nutrients to and from the bed. The biological significance of the physics becomes clearer when we consider the consequence of animal growth on the seabed. Mussels settling onto a smooth substratum will increase the surface complexity of the bed making it 'rougher'. This in turn will increase turbulent flow over the seabed. The turbulent flow breaks down the boundary layer and encourages a downward transport of phytoplankton from the layers of water above, encouraging further growth and an increase in the roughness of the bed. Paterson, D. A. & Black, K. 1999. *Advances in Ecological Research*, 29: 155–193. Copyright © Elsevier 1999.

continues

BOX 7.2 continued

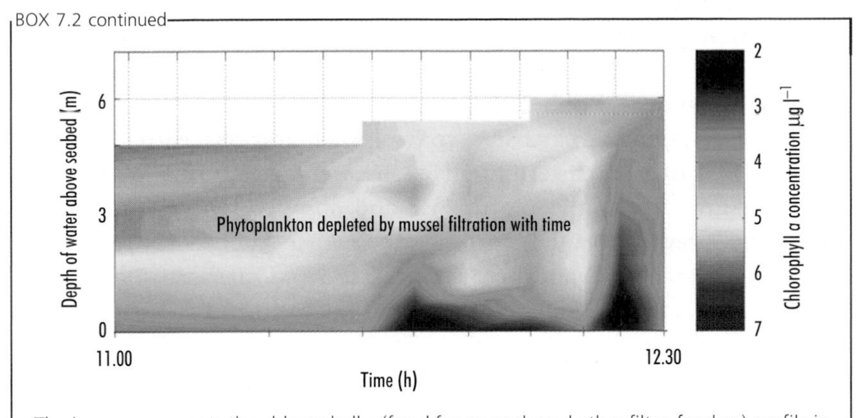

The image represents the chlorophyll *a* (food for mussels and other filter-feeders) profile in a body of water followed across a mussel bed over a 1.5 h period. Within 1.5 h nearly a third of the plankton was removed from the entire water column. Water height rises from 11.00 to 12.30 with the incoming tide. (Image: J. Gascoigne.)

7.3.2 Life mode and mobility of biota

The sessile components of the benthos are more closely associated with the composition and physical characteristics of the seabed and show seasonal growth followed by periods of winter dormancy at higher latitudes. This is true both of macroalgae and particularly of colonial animals such as hydroids. In contrast much of the epibenthos exhibits varying degrees of mobility from crawling gastropod molluscs and starfishes to highly mobile fish. Despite their apparently slow speed, starfishes are known to occur in dense feeding aggregations, which move slowly and systematically across the seabed, consuming animals in their path. Where these 'swarms' coincide with commercial shellfish beds they can cause severe economic losses for cultivators. Many crabs (e.g. *Cancer pagurus* and *Maja squinado*) move inshore in the spring and summer where mating occurs, often in dense aggregations, followed by a movement into deeper offshore water during the winter months to avoid wave action and severe decreases in near-shore water temperature.

● Sessile biota show seasonal patterns of growth and dormancy at high latitudes, while mobile epifauna exhibit varying degrees of inshore/offshore movement in response to water temperature. Inshore migrations of many decapods (crabs and lobsters) coincide with mating aggregations.

The distribution of bottom-dwelling fishes can be linked strongly to certain habitat types within regional seas, particularly when associated with specific habitats such as reefs or kelp beds. However, in regions where sediments predominate at mid to high latitudes fish species are often associated with a range of sedimentary habitats that may vary from fine sand to gravel as in plaice (*Pleuronectes platessa*) (Kaiser et al. 1999). Such variation in habitat use can be attributed to different behavioural characteristics at different life-history stages. Juvenile plaice are strongly selective of specific sediment grain sizes as this determines their ability to burrow into the substratum to evade predation, however adult plaice are released to some extent from predation pressure (apart from fishing) and utilize a wider range of habitats (Gibson & Robb 1992).

Some species make extensive spawning migrations (Box 7.3). For example, plaice in the eastern English Channel migrate 100s km to reach spawning grounds in the North Sea and show activity patterns that utilize the prevailing tidal currents to reduce the energy expended during the migration (Metcalfe & Arnold 1997). Other fish such as bass (*Dicentrachus labrax*) undertake seasonal migrations that track the rise and fall of seawater temperature as they require a minimum seawater temperature of 9 °C for ovary maturation to occur. Typically stocks move north from the coast of France and up through the English Channel and Irish Sea in late spring, returning south in the late autumn with falling water temperature.

● Fish show high habitat affinities when the habitat is structured as in reefs or kelp beds. However, fish associated with sedimentary habitats tend to be less closely associated with a specific habitat and may occur across a wide range of sediment types at different stages of their life history.

Box 7.3 Migration in continental shelf seas

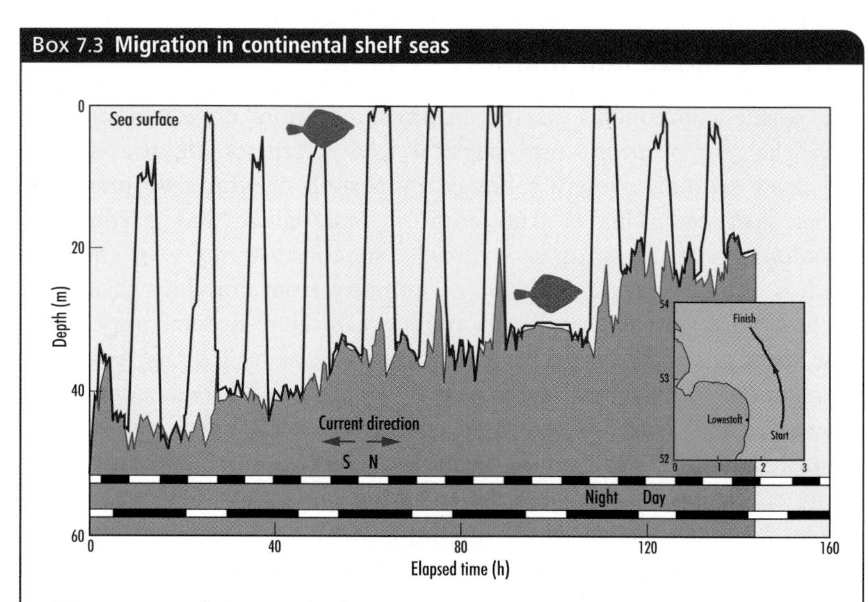

Plaice are normally bottom-dwelling fish that spend much of their time buried in the sediment and although they are widespread they show temporally stable patterns in distribution which is strongly associated with the environmental characteristics of the seabed habitat (Hinz et al. 2003). However, plaice also undertake extensive movements to their spawning grounds. The data above was obtained during a study in which plaice were tagged with data loggers that recorded the fishes' depth throughout the day. When the tidal stream was northerly towards the spawning grounds the fish is seen to swim up into the water column where it swims with the current. When the tide turns in the opposite direction, the fish descends to the seabed where it waits for the next turn of the tide. By adapting its behaviour to take advantage of the prevailing current, plaice are able to move several hundred km in less than a week (redrawn from Metcalfe et al. 2002).

7.4 Functional Roles of the Biota

As in all biological systems, the continental shelf seabed community has representatives that can be categorized according to what they consume (predators, scavengers, herbivores, filter feeders, suspension feeders) or what they do to the physical structure and processes within the habitat (bioturbators, eco-engineers).

7.4.1 Predators and scavengers

Most animals in continental shelf systems are highly flexible in terms of their feeding strategy. Nearly all **predators** will also **scavenge carrion**, for example fish (dab, *Limanda limanda*), starfish (*Asterias rubens*), decapods (*Cancer irroratus*), and gastropod snails (e.g. whelks, *Buccinum undatum*). Even **herbivorous** sea urchins (e.g. *Echinus esculentus*) and suspension feeders such as brittlestars (e.g. *Ophiura ophiura*, *Ophothrix fragilis*) are known to feed on carrion from time to time. These animals are known as **facultative** scavengers. There is some controversy as to the possible existence of **obligate** scavengers, i.e. animals that consume only carrion, but it appears possible considering the physiological energetic constraints. The most likely candidates would seem to be small (<6 mm long) lysianassid amphipods of the genus *Orchomene*, which appear to specialize in the consumption of crustacean carrion. This group of amphipods show many specialist adaptations to a scavenging lifestyle such as the ability to survive extended periods without food and to gorge on carrion such that their body size increases by up to 500%. However a convincing experimental demonstration of an obligate scavenging lifestyle remains elusive (Ruxton & Houston 2004).

- Many consumers at high latitudes exhibit multiple feeding modes and often resort to scavenging when carrion is available, these are facultative scavengers. Carrion is more likely to occur at high latitudes where physical processes are a significant source of natural mortality.

- The existence of a purely scavenging (obligate) lifestyle remains speculative in continental shelf systems. If they occur obligate scavengers are most likely to be found among the small-body-sized scavenging amphipod fauna.

7.4.2 Grazers

Grazing animals include herbivorous fishes such as some blennies (Blenniidae), sea breams (Sparidae), gastropod molluscs, and sea urchins, all of which play a key role in the maintenance of diversity within algal-dominated communities. **Herbivory** in fishes is much more prevalent towards lower latitudes, while at higher latitudes herbivores are primarily invertebrates. In systems in which a particular species or guild is the predominant grazer, they can have a **keystone** or **eco-engineering role** on the habitat through their consumption of certain types of algae. Grazers are not the only eco-engineers, as we will see a little later (7.5.1). Grazers also include **carnivores** such as nudibranch molluscs (sea slugs) that consume encrusting bryozoa, soft corals and sponges by scraping at the colonies with their radula. Some of these associations

● The development of complex ornamentation in gastropod shells is a strong evolutionary driver of cheliped morphology (claw design) in snail eating crabs and is a classic example of the 'evolutionary arms race'.

may be very specific, for example the sea slug *Tritonia hombergii* graze dead men's fingers (*Alcyonium digitatum*) and may be one of the few predators of this soft coral. Many of the external morphological features of bryozoa (spines) and the extracellular products of sponges and soft corals act as defence mechanisms against the predatory activities of sea slugs which has developed into an '**evolutionary arms race**' between predator and prey.

7.4.3 Particle feeders

● Biomass is sometimes calculated as the dry weight of living tissues (including shell material) minus the ash remaining after combustion, i.e. ash-free dry weight.

● Pseudofaeces are the means by which filter-feeding biota rid themselves of indigestible or rejected particles. These particles are bound in mucus secreted by gill tissues. Mucus production requires energy expenditure and is one reason why filter feeding ceases when suspended sediment loads become excessive.

Filter feeders such as oysters, mussels and clams extract phytoplankton from the water column and suspended matter from just above the seabed (**suspension feeding**). An individual animal of 1 g ash-free dry weight is estimated to filter 57 litres of water per day (Heip et al. 1995). Consequently, filter feeders have an important role in **bentho–pelagic** coupling as they process phytoplankton and suspended organic into faeces and pseudofaeces, which are deposited on the seabed. This material is rich in organic matter, which is processed by the microbial community and in turn feeds **suspension feeders** such as clams (e.g. *Mya truncata*), and bulk sediment processors such as irregular sea urchins (*Echinocardium cordatum*), sipunculans, and polychaetes. Bryozoa, hydrozoans, sponges, and anemones are particle feeders; although there is some evidence that the latter two animal types can also absorb dissolved organic matter through their body wall as a supplementary food source. These groups are more prevalent in deeper water than filter feeders that rely upon a supply of phytoplankton. Indeed, there is a strong negative relationship between depth and filter-feeder biomass, declining down to 50 m from 3 to 0.2 g carbon m^{-2} for the North Sea (Bryant et al. 1995).

7.4.4 Bioturbators

● Bioturbators have left their mark in fossilized sedimentary rocks. These are termed trace fossils.

Animals can have both a stabilizing and destabilizing role within sediments. **Palaeoecologists** have studied in great detail the manner in which live animals perturb sediment structures. This has given them insights regarding the likely agents of **trace fossils** that have recorded the passage of animals through or across sediment habitats. A seminal publication by Schafer (1972) describes in detail the different modes of sediment disruption that occur as a result of animal activities. These vary from the surface **bulldozers** such as irregular sea urchins and gastropods, to the **feeding pits** of starfish and elasmobranch fishes such as rays, to the burrow labyrinths and chambers created by burrowing crustacea (Box 7.4). The cumulative effects of these animal related sediment disturbances are known as **bioturbation**.

Bioturbators perform a key role in seabed systems as they enhance the passage of oxygenated water deeper into the sediment than it would

Box 7.4 Influence of fauna on sediment structure

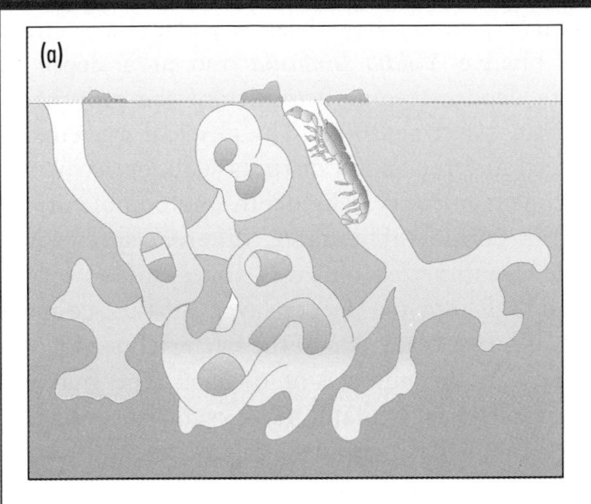

(a) Callianasid shrimps live in soft muddy/sand sediments and create intricate sub-sediment passageways. Apart from the burrow entrance there are other connections to the sediment surface that emerge within sediment mounds that act as chimneys on the seabed. These chimneys cause water to accelerate as it passes over the mound, which draws water within the chimney out. This causes the water within the burrow system to move through the passageways and thereby encourages new oxygenated water to be drawn into the burrow system. The shrimp store food and 'garden' microbial activity inside chambers.

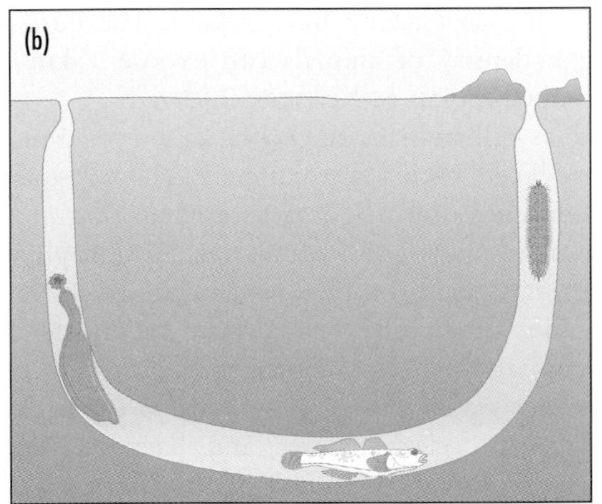

(b) Echiuran worms use their highly extendable proboscis to gather surface sediment matter at great distances (150 cm) from the entrance of their burrow. Once the sediment has passed through their gut it is expelled in a faecal mound. Echiuran burrows are often inhabited by lodgers such as blennies and scale worms.

(c) Feeding pits are excavated by starfish while trying to consume a burrowing sea urchin (Photograph: James Perrins).

otherwise penetrate by passive diffusion between sediment particles. Increases in surface sediment porosity have been found in association with the deposit-feeding bivalve *Yoldia limatula* and at a depth of 9–12 cm as a result of the deposit-feeding activities of the polychaete *Heteromastus filiformis* (Mulsow et al. 2002). The physical movements of these animals as they feed and reposition disrupts sediment structure and thereby increases sediment porosity. The depth and complexity of the burrow structures varies considerably among different species and can vary in depth from a few centimetres to metres (Jones & Jago 1993). These burrows can extend well into the deeper layers of the sediment that are characteristically coloured black due to the bacterial production of hydrogen sulphide (H_2S) in the absence of oxygen. The burrows themselves increase the surface area available for oxygen and nutrient exchange that encourages enhanced microbial activity on and in the burrow walls. For example, the brittlestar (*Amphiura filiformis*) occurs in densities as high as 700 m^{-2} in Galway Bay, Ireland. The burrows constructed by such a high density of animals can expose 1.4 m^2 of burrow wall m^{-2} of the seabed (Solan & Kennedy 2003).

● Bioturbators increase the complexity of the surface of the substratum through the creation of pits and mounds and enhance the exchange of oxygen and nutrients across burrow walls that occur to a depth of up to 2 m in the sediment.

Sediment conveyors such as callianassids and echiuran worms transfer surface sediments to deeper layers within the sediment, which results in a peak in chlorophyll *a* concentration that is greater than that at the sediment surface. Branch and Pringle (1987) found that chlorophyll *a* concentration was greatest at a depth of 15–25 cm in the presence of *Callianassa kraussi*. At this depth the sub-surface sediments are reworked by burrowing animals such as urchins (*Echinocardium cordatum*) and polychaetes such as *Scoloplos armiger*, which ingest sediment from which they digest organic matter and bacteria. These processes recycle minerals and nutrients when they are transferred to the surface of the seabed as sediment mounds and faecal pellets.

Sediment conveyors make a significant contribution to the resuspension of sediment in the overlying water column and influence seabed topography. Rowden et al. (1998) undertook both laboratory and field trials using the mud shrimp (*Callianassa subterranea*) and found that sediment reworking varied with temperature and hence season. They estimated an annual sediment turnover budget of 11 000 g (dry weight) $m^{-2} yr^{-1}$. Field observations at a site in the North Sea demonstrated that the sediment expelled by the mud shrimp formed unconsolidated volcano-like mounds, which significantly modify seabed surface topography. *Callianassa subterranea*'s maximum contribution to sediment resuspension results in a lateral sediment transport rate of 7 $kg\,m^{-1}\,month^{-1}$. Thus, mud shrimps not only rework sediment deep within the substratum but are also important determinants of the rate of sediment exchange between adjacent sedimentary habitats.

Echiuran worms are abundant bioturbators of mud sediments. However they are extremely difficult to study in laboratory studies due to their fragility, and they appear to be relatively immobile once established. Hughes et al. (1999) studied the rate of sediment ejection by the echiuran worm *Maxmuelleria lankesteri* in the field. They found that burrows and sediments persisted throughout the one-year study period and that worms ejected sediments year round. The mean ejection rate was 2748 g burrow^{-1} y^{-1}. The superficial sediment, on which *M. lankesteri* feeds, was very rich in organic matter, with monthly values of approximately 12–17% sediment dry weight. Ejection rate remained constant at labile fractions of organic matter below ~50%, but increased sharply above this threshold, so sediment ejection rate increased with increasing food quality. Burrows appeared to be static and individual ejecta mounds could persist for longer than a year. This is quite different to deep-sea echiura for which changes in burrow morphology are more frequent and probably relates to the higher quantity of labile organic matter on the continental shelf.

7.5 Food Webs in Shelf Systems

The study of **food webs** provides the framework to understand consequences of the interactions between species within a system, i.e. the pathway of consumption and the flow of energy from one trophic level to another. Several metrics of food webs provide insight into biomass partitioning and production in an ecosystem. Species richness and the number of species interactions (links between species) are among the most important of these metrics. Even for relatively simplified systems in which many species have been lumped together in general categories (e.g. bacteria) the linkages among different species or groupings is highly complex (Fig. 7.5).

7.5.1 Ecosystem resilience

The flexibility of feeding modes (including cannibalism) gives rise to great complexity in continental shelf systems (Link 2002) (Box 7.5). This complexity means that the removal of one or two or more species from the system is **unlikely** to bring about significant **cascade** effects at either higher or lower trophic levels, and is a general feature of marine foods webs (Dunne et al. 2004).

● Food web complexity confers ecosystem resilience unless major environmental changes (e.g. El Niño) or human intervention cause multiple species replacements or extirpations.

Ecosystem resilience can weaken considerably when environmental and human sources of stress act simultaneously and may lead to a **phase shift** in the ecosystem. A combination of chronic over-harvesting across a wide range of fish and shellfish species combined with decadal changes

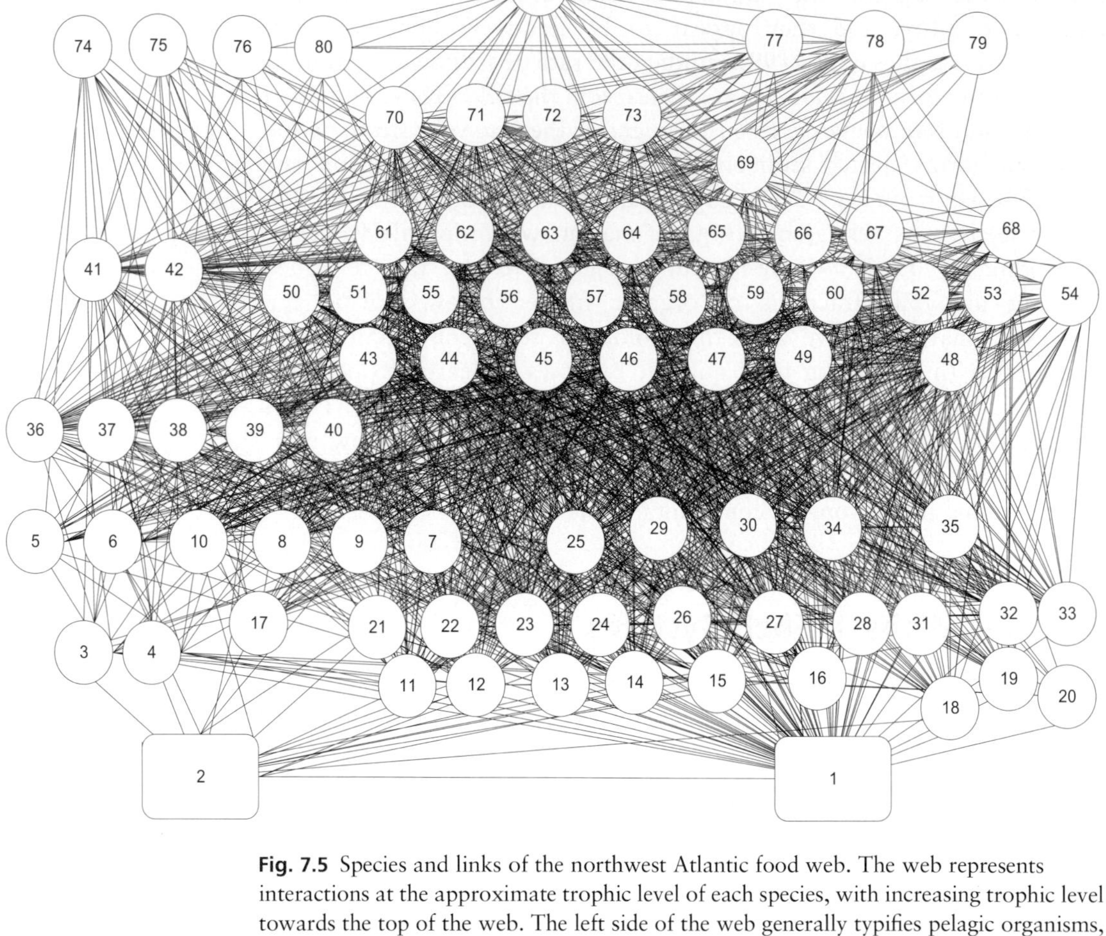

Fig. 7.5 Species and links of the northwest Atlantic food web. The web represents interactions at the approximate trophic level of each species, with increasing trophic level towards the top of the web. The left side of the web generally typifies pelagic organisms, and the right to middle represents more benthic/demersally oriented organisms. 1 = Detritus. 2 = Phytoplankton. 3 = *Calanus* sp. 4 = other copepods. 5 = Ctenophores. 6 = Chaetognatha (i.e., arrow worms). 7 = Jellyfish. 8 = Euphasiids. 9 = *Crangon* sp. 10 = Mysids. 11 = Pandalids. 12 = Other Decapods. 13 = Gammarids. 14 = Hyperiids. 15 = Caprellids. 16 = Isopods. 17 = Pteropods. 18 = Cumaceans. 19 = Mantis Shrimps. 20 = Tunicates. 21 = Porifera. 22 = Cancer crabs. 23 = other crabs. 24 = Lobster. 25 = Hydroids. 26 = Corals, Anenomes. 27 = Polychaetes. 28 = other worms. 29 = Starfish. 30 = Brittlestars. 31 = Sea cucumbers. 32 = Scallops. 33 = Clams, Mussels. 34 = Snails. 35 = Urchins. 36 = Sand lance. 37 = Atlantic Herring. 38 = Alewife. 39 = Atlantic Mackerel. 40 = Butterfish. 41 = Loligo. 42 = Illex. 43 = Pollock. 44 = Silver Hake. 45 = Spotted Hake. 46 = White Hake. 47 = Red Hake. 48 = Atlantic Cod. 49 = Haddock. 50 = Sea Raven. 51 = Longhorn Sculpin. 52 = Little Skate. 53 = Winter Skate. 54 = Thorny Skate. 55 = Ocean Pout. 56 = Cusk. 57 = Wolfish. 58 = Cunner. 59 = Sea robins. 60 = Redfish. 61 = Yellowtail Flounder. 62 = Windowpane Flounder. 63 = Summer Flounder. 64 = Witch Flounder. 65 = Four-spot Flounder. 66 = Winter Flounder. 67 = American Plaice. 68 = American Halibut. 69 = Smooth Dogfish. 70 = Spiny Dogfish. 71 = Goosefish. 72 = Weakfish. 73 = Bluefish. 74 = Baleen Whales. 75 = Toothed Whales, Porpoises. 76 = Seals. 77 = Migratory Scombrids. 78 = Migratory Sharks. 79 = Migratory Billfish. 80 = Birds. 81 = Humans. (from Link 2002).

Box 7.5 Complex feeding interactions

The flexibility of feeding modes in shelf systems makes food web analyses extremely complicated, but has given rise to some fascinating behavioural interactions between predators and their prey. In the North Altantic, the starfish (*Leptasterias polaris*) is a voracious predator of the whelk *Buccinum undatum*. When resting, whelks remain buried in the seabed to reduce the chance of detection by predators. *Leptasterias* are extra-oral feeders, such that once they have begun to digest their prey they are less likely to be able to catch and consume anything else. Extra-oral digestion can be messy, and amino acids and adenosine triphosphate (a potent attractant of scavenging invertebrates) leak into the surrounding water. Despite the normal risk of being eaten, whelks will approach feeding starfish and are able to consume some the prey that is in the process of being digested (Rochette et al. 2001). This behaviour is known as **kleptoparasitism**. Rochette et al. (2001) found that this behaviour was more prevalent in larger female whelks just before the time they were due to lay their eggs. Hence the whelks were bolder when their energy demands were greatest. After their eggs had been laid, female whelks were far less likely to approach or feed in the presence of predatory starfish.

● Extraoral digestion is a process whereby starfishes extrude their stomach onto their prey which is thereby enveloped and covered in digestive enzymes. Not all starfishes use this technique and many ingest their prey whole whereupon digestion occurrs intraorally.

(a)

(b)

(a) A Buried whelk *Buccinum undatum* with its siphon protruding above the sandy substratum and (b) a starfish *Asterias rubens* succumbs a whelk using extra-oral digestion. Hermit crabs (*Pagurus bernhardus*) are attracted by the molecular stimulants released during extra-oral digestion (Photographs: James Perrins).

● Organisms within food webs are assigned to a particularly trophic level according to their relationship with other components (i.e. their role as predators and prey) in the food web, hence bacteria and other primary producers are found at the bottom of the food web at the lowest trophic levels and apex predators (e.g. seals) at the top, and hence highest trophic levels.

in water temperature and stratification (which affects primary production), has caused the decoupling of the benthic–pelagic system on the eastern Scotian Shelf off Nova Scotia. Severe declines in demersal (bottom-dwelling) fish productivity have coincided with compensatory increases in pelagic fish biomass (Choi et al. 2004). What we have seen is a switch from a system dominated by benthic biota and bottom-feeding fishes like cod and haddock to one dominated by zooplankton and planktivorous fishes. However, it is important to note that phase shifts also occur in response to strong environmental forcing (e.g. El Niño events) without any interference from humans (Grantham et al. 2004).

7.5.2 Trophic cascades

● Trophic cascades are likely to occur when the linkages between species are strong and the linkages occur between species assigned to major trophic levels.

Scientists have looked hard for evidence of strong species interactions in both terrestrial and aquatic food webs with limited success (Table 7.2). In situations where strong linkages occur among species, several rules of thumb seem to apply. Some examples of cascades are given in Table 7.2. In all cases, none of the cascades involve more than three interactions

Table 7.2 Examples of trophic cascades from marine, freshwater and terrestrial systems. (adapted from Pace et al. 1999).

Ecosystem	Cascade	Evidence	Effects	References
Marine				
Oceanic	Salmon-zooplankton-phytoplankton	Ten-year time series	Increased phytoplankton when salmon abundant	Shiomoto et al. 1997
Coastal	Whales-otter-urchins-kelp	Long-term data and behavioural studies	Predation by whales on otters leads to urchin population expansion and increased grazing of kelp	Estes et al. 1998
Intertidal	Birds-urchins-macroalgae	Manipulation of bird predation on urchins	Algal cover greatly increased when urchins reduced	Wooton 1995
Fresh water				
Streams	Fish-invertebrates-periphyton	1° and 2° production affected by predation of invertebrate populations	Annual 1° production affected by 6 fold difference	Huryn 1998
Shallow lake	Fish-zooplankton-phytoplankton	Observations of lakes under clear and turbid conditions	Reductions in fish abundance led to shift in zooplankton size-structure with consequent effects on phytoplankton	Jeppesen et al. 1998
Terrestrial				
Meadow	Lizards-grasshopper-plants	Observations and experimental studies	Grasshopper density mediated by lizards, plant biomass decline with decreasing lizard predation	Chase 1998
Tropical forest	Beetles-ants-insects-piper plants	Observations and manipulation of beetle density in enclosures	Beetles predate ants that remove herbivorous insects that consume plants. More foliage consumed in presence of beetles.	Letourneau & Dyer 1998
Boreal forest	Wolves-moose-balsam fir	Study of moose grazing over 30 years	Population cycles of wolves, moose and balsam fir	McLaren & Peterson 1994

between different species, i.e. they are simple systems even though they may be embedded within highly diverse and complex systems. These embedded mini-systems are perhaps to be expected in high diversity ecosystems that are characterized by more specialized feeding interactions (e.g. coral reefs). The interactions that occur within the cascade are between organisms from major trophic levels within a system, e.g. **predator – herbivore – primary producer**, few involve intermediate trophic levels. In contrast, the food web of the Benguela upwelling system has at least seven interactions between the bottom and the top of the system. Key predators or herbivores in **trophic cascades** are usually the dominant organisms at their trophic level and often these can be **ecosystem engineering** biota. In the case of a significant decrease in the population of a predator or herbivore, there are few others to take their place, which means that linkage between one trophic level and another is strong. This contrasts sharply with many of the more 'open' marine ecosystems where there may be three or more predators that exert similar levels of predation on one or more species (e.g. birds, cetaceans and seals all eat the same pelagic fishes in the Benguela upwelling system) (Yodzis 1998). Cannibalism is rare within species involved in systems prone to trophic cascades in sharp contrast to many marine ecosystems in which cannibalism is common, for example North Atlantic cod (*Gadus morhua*) and hake (*Merluccius* spp.) are both good examples of fish that exert considerable density-dependent predation pressure on their own species.

● Ecosystem engineering biota are those that due to their abundance and feeding or other activities exert a strong influence on the structure of the ecosystem, e.g. urchins grazing kelp.

7.5.3 Hunting, fishing, and trophic cascades

Understanding the characteristics of communities in which trophic cascades are likely to occur is critical for effective conservation management (Chapter 15). On the west coast of North America, the effective conservation of **sea otter** (*Enhydra lutris*) may have helped to conserve kelp beds that provide important habitats for many fish and invertebrates. Past hunting of sea otters depleted their populations so severely that they no longer exerted significant predatory control over urchin populations. Under normal circumstances urchins consume algal material at the base of the kelp canopy, but with an increase in their population the urchins began to deplete kelp beds through grazing pressure. This excessive grazing resulted in **urchin barren** grounds, where the bedrock was colonized by small **epilithic** algal species and urchins.

● Once established, urchin barren grounds are persistent unless predation pressure increases, the urchins are physically removed or mortality increases due to disease.

Urchin barren grounds represent a phase-shift that has been repeated in other systems around the globe, notably on coral reefs. Urchins are capable of surviving on minimal energy intake for extended periods of time. Once urchin barrens are established, only urchin disease or an increase in predation can possibly reverse the shift to the former state. In the late twentieth century, otter populations have grown following hunting bans and improved conservation, so they have exerted in turn more predation pressure on the urchins. This has allowed kelp beds to establish on areas of barren ground. Clearly, these effects can be confounded by many other factors such as oceanographic effects (e.g. temperature and current patterns) on urchin recruitment, and kelp growth and survival, which has been negatively affected by pollution and over-harvesting in the past decade.

In general terms, the effective conservation of otters has helped to conserve the kelp beds (Estes & Duggins 1995). Or so we thought, until Estes et al. (1998) reported annual declines of 25% year on year for the population of sea otter at Adak Island, Alaska. Severe declines in the population of Steller sea lions (*Eumetopias jubatus*) appear to have triggered a switch in the predatory behaviour of killer whales (*Orcinus orca*) in the Aleutian Islands archipelago. The otter population living between Kiska and Seguam Islands, (an area spanning about 700 km) totalled about 53 000 sea otters prior to the early 1990s but was reduced to 13 000 some six years later. Given that a single killer whale can easily eat more than 1800 otters per year, it is entirely feasible that whale predation may have been one cause of the decline. The decline in sea lions is either related to cyclic environmental changes in species of forage fishes, or due to over-fishing of the latter, or both of these effects acting together.

● Indirect effects of human activities or environmental changes that occur elsewhere can have indirect effects on species linkages. This may explain the sudden increase in sea otter predation by killer whales as a result of a decline in the population of their normal prey, sea lions.

7.6 Characterization of Seabed Communities

Lists of species that comprise a community are useful at one level in that they provide an immediate idea of the main characteristics of the benthos in a particular area and they are the starting point for understanding the composition of food webs (above). However, an over-emphasis on single species perhaps misses more interesting general structuring patterns that might be observed across a range of communities that have different components at the species level. As we have seen, most communities contain a range of taxa that can be assigned to different trophic levels or functional groups within the community. Box 7.6 highlights the similar **functional role** performed by different species within distinct assemblages. A broader categorization of community constituents may be more ecologically meaningful, particularly when one is interested in examining long-term changes in community structure. Over long periods of time (years) within a community, one species may be replaced by another, but if their functional role is similar then the functional integrity of the community should be maintained. This situation may occur with rising global sea temperatures. For example, as high latitude boreal species begin to retreat further north they will be replaced by colonizing species from lower latitudes that may fulfil

● An over-obsession with single species approaches to monitoring may miss more relevant patterns at higher levels of organization within communities.

Box 7.6 Functional consistency between different assemblages

Two community types as described by Thorson (1971). Although we would avoid using the specific 'labels' or names for these assemblages, both contain different species, but in each case it is possible to identify different species that perform similar ecological functions. This consistency in assemblage structure is ecologically much more important than the original aim of Thorson's descriptions which attempted to categorize multiple assemblages as specific types. Thus, it would appear that general rules can be applied to the composition of soft-bottom assemblages across a variety of systems. The distribution of these functional types within assemblages can alter in response to external forcing factors such as organic pollution or physical stress.

Macoma Community	Function	*Abra* Community
Mya arenaria Bivalve	Surface suspension feeder Deep bioturbator	*Mya truncata* Bivalve
Arenicola marina Polychaete	Sediment processor Sub-surface bioturbator	*Lagis koreni* Polychaete
Spionid Polychaete	Suspension and surface deposit feeder	*Abra alba* Bivalve
Macoma balthica Bivalve	Surface deposit feeder	*Ophiura* sp. Echinoderm
Cardium edule Bivalve	Suspension feeder Shallow bioturbator	*Phaxus pellucidus* Echinoderm
Pandalus sp. Crustacean	Carnivore Scavenger Surface bioturbator	*Buccinum undatum* Gastropod

a similar role. This has important implications for how we monitor the composition of coastal communities through time (Chapter 15).

Marine benthic ecologists have typically tried to characterize benthic assemblages in terms of their main (most abundant) constituent species or those species that occur on a regular basis in consecutive samples taken from the same area (often the latter is more important than the former). In many ways it is similar to describing terrestrial systems in terms of oak woodland, bog, or heather moorland. The notion of distinct assemblages that could be assigned labels was pioneered separately by Petersen (1914) in the North Sea off Denmark, and later by Jones (1950) and Thorson (1957). Petersen gave titles or names to certain types of benthic assemblage that were encountered, usually based on the more conspicuous bivalves or echinoderms that were typical of these assemblages, e.g. *Abra* community, *Brissopsis* community, shallow *Venus* community. Jones (1950) used a slightly different labelling approach, which was based mainly on a description of the location and sediment association of the community, e.g. boreal offshore muddy-sand association. These approaches have influenced a great many subsequent studies that have often attempted to pigeon-hole the assemblages encountered in the terms used by Petersen, Jones, and Thorson, hence it is not unusual to read papers that contain phrases such as 'similar to the *Abra* community as described by Petersen'. However, it is clear that benthic communities are highly varied at different scales in space and time, hence attempting to attach descriptions to assemblages that have only a partial relationship to the limited range of assemblages of biota studied by Petersen, Jones and Thorson is somewhat meaningless.

● There is an almost unavoidable tendency for ecologists to attempt to give names to particular assemblages of species. It is questionable whether this approach is useful given the infinite number of possible variants created by the substitution of one species for another.

The characterization of biological communities on the basis of habitat characteristics and indicator species provides a means of describing specific **biotopes** that then become a reference point against which to measure temporal change or consistency in community structure. The biotope classification scheme used by conservation agencies in the UK is highly advanced and a useful source of ecological and habitat information (Fig. 7.6). This system can be linked to the European Union Nature Information System (EUNIS) so that it is comparable at a wider-scale. The problem with such characterizations is that they are potentially infinite as the multivariate community statistics on which they are based are so sensitive that it only requires a slight deviation in community composition or habitat descriptor (e.g. pebbles versus pebbles with sand) to initiate yet another biotope description. New combinations or intermediate communities are conceivably limitless – and the more we sample the same marine environment the more variants we will encounter. Global temperature changes are likely to lead to an increase in the number of new variants of biotopes recorded as warm water species invade colder latitudes.

● What Keith Hiscock, Programme Director of the Marine Life Information Network thinks of the utility of biotope classification: 'It is more accurate to indicate that survey data can usually be matched to a particular biotope (enabling like-with-like comparisons of species richness, extent etc.) but, inevitably, some survey data or results of data analysis will be impossible or very difficult or questionable to match to a particular biotope'.

Erect sponges, Eunicella verrucosa and Pentapora foliacea on slightly tide-swept moderately exposed circalittoral rock. MCR.ErSEun

Bedrock with the sea fan *Eunicella verrucosa* and the soft coral *Alcyonium glomeratum* amongst a hydroid turf.
Image width ca 1.0 m.
Image: Rohan Holt

Recorded and expected MCR.ErSEun distribution for Britain and Ireland

Key Information researched by: Angus Jackson & Dr Keith Hiscock **This information is not refereed.**

Distribution of biotope in Britain and Ireland

Recorded in southern England from Dorset westwards, in the Isles of Scilly and on the north coasts of Devon and Cornwall, Lundy and Skomer. One record from Bardsey Island in north Wales (without *Eunicella verrucosa*) and also from several locations on the west coast of Ireland including Donegal and Bantry Bays.

Description of biotope

For a full description of this biotope including characterizing species, distribution, survey information and references visit http://www.jncc.gov.uk/mermaid/biotopes/643.htm

Mainly found on exposed and moderately exposed rock, in slight tidal currents and often relatively silty, with a rich variety of species typically including branching and cup sponges, the sea fan *Eunicella verrucosa* and the ross coral *Pentapora foliacea*. Typically a bryozoan turf of *Cellaria* spp. and *Bugula* spp. is present amongst the larger species (see the biotope CR.Bug). The branching sponges *Axinella dissimilis*, *Stelligera* spp. and *Raspailia* spp. are typically present, with cup sponges *Axinella infundibuliformis* and *Phakellia ventilabrum* found in some cases. *Alcyonium glomeratum* and *Parerythropodium coralloides* (now *Alcyonium hibernicum*) may also be present and short vertical faces sometimes have the star anemone *Parazoanthus axinellae* and/or *Parazoanthus anguicomus*. There are numerous examples of sites with lots of branching and cup sponges where sea fans have not been found (but are often known to be present within the same geographical area); some of these are included in MCR.ErSPbolSH. *Diazona violacea* is also often recorded in this biotope although it occurs in MCR.ErSSwi also. There are a few instances of *Swiftia pallida* being found at the same sites (in SW Ireland) as *Eunicella verrucosa*. Where this biotope occurs on more open coast (e.g. SW Britain and W Ireland) the cotton spinner sea cucumber *Holothuria forskali* is often present.

Fig. 7.6 An example of database information for a circalittoral biotope as described by the Marine Life Information Network for Britain and Ireland (MarLIN). The biotope classification scheme is linked directly to the EUNIS classification system at certain levels within its hierarchical structure. The biotope description is composed of a combination of physical habitat and biological constituents. For more information on biotope classification and images of biotopes and ecological information see **http://www.marlin.ac.uk**. Photograph: Rohan Holt.

7.7 Specific Habitats

The seabed habitat is composed of the inert non-biological and **biogenic** material (e.g. broken shell material) and the biota that live in association with it. In addition to the prevailing water column conditions overhead, the inert elements of the habitat dictate the basis of what can, or cannot, live within or upon that seabed. This normally depends upon the interaction between seabed hardness and stability. The physical nature of the seabed ranges from bedrock, boulder fields, cobble pavements, gravel lags, coarse mixed sediment, sands, and muds. Excluding bedrock, the physical size of the individual particles that make up these habitats vary in diameter from micrometres up to several metres (see also Chapter 5). At present, sediments dominate the seabed habitat of the continental shelf and vary according to their tectonic history the quantity and quality of riverine inputs of sediment and the transport by waves and currents. On the inner continental shelf at depths of less than 65 m, muds and sands are the most common substratum (37% and 47% respectively) while hard substrata are relatively rare (6%) and the remainder are made up of reefs and shell debris (Hall 2002). For the purposes of the next sections we will deal with the general categories of substrata as hard (bedrock to cobble), soft (gravel to mud), and biogenic (habitats formed by living organism) habitats. Coral reefs, mangroves, and seagrasses are the subject of separate consideration in later chapters.

● Sediments dominate the present-day continental shelf seabed, bedrock and biogenic reef habitats are relatively uncommon.

7.7.1 Hard habitats

Hard substrata, though rare on the continental shelf, are probably the most familiar habitats to the general public (aside from coral reefs) due to their photogenic biota. Bedrock is inherently the most stable substratum (unless subject to seismic activity), followed by decreasing sizes of boulders and cobbles. All of these hard substrata can be found in a gradient of physical conditions that range from high to low stress depending upon their exposure to physical processes such as near-bed flow, wave action, and scour by water-borne sediments and glaciers. Thus it is possible to find bedrock rich in luxuriant epibiota in areas with clear water and moderate current, or to have bedrock that is species poor due to high near-bed currents or severe wave action. Hard substrata provide a secure anchorage for sessile biota many of which are either macroalgae, encrusting calcareous algae, particle or filter feeders. As for rocky shore habitats, competition for space is intense and some organism such as soft corals, anemones, or sea squirts may come to dominate a given patch of the habitat.

● Hard substrata are occupied by encrusting sessile biota, hence competition for space is a key community structuring process.

Hard substrata are particularly important for algae that require a secure substratum on which they can settle and then develop a holdfast. Most macroalgae would soon be washed away by currents and waves if they were anchored in soft sediments. Some smaller algae such as *Enteromorpha* spp. are able to attach securely to small pebbles or even the shells of bivalve molluscs, but these occur in relatively low-energy environments. At higher latitudes, macroalgae are confined to a relatively narrow coastal band of shallow water in which light is attenuated rapidly due to the load of suspended matter in the water column. At mid to low latitudes in regions of upwelling, the biomass of macroalgae is greatest due to water clarity and the enhanced supply of nutrients from deeper oceanic water. At lower latitudes, the water is clear enough to support photosynthesis by epilithic encrusting algae down to mid shelf depths (75–95 m).

● Hard substrata are a fundamental requirement for the establishment of many macroalgae and enables giant kelp to attain a length of up to 40 m.

Kelps are the subject of fisheries in northern Europe, the USA, and Chile, and are restricted to mid to high latitudes (Steneck et al. 2002). Their economic importance and sensitivity to ecological regime shifts, harvesting and other forms of human disturbance mean that harvesting of this resource requires careful management to ensure sustainable use (Box 7.7). In Chile, management of the harvesting of kelp resources has occurred for at least a century. The 'parcela' system occurs in two main regions where fishers extract the kelp *Durvillaea antarctica*. Fishers are organized in syndicates in which each participant is allocated the harvesting rights to a number of boulders on which the kelp grow. Off the coast of California, giant kelp (*Macrocystis pyrifera*) are managed with the use of marine protected areas. This species can grow at rates of up to 50 cm d^{-1}. The stipe and fronds of these algae are buoyed by gas-filled pneumatocysts that support the stipe and ensure that the fronds remain close to the sea surface where irradiance is greatest. The high rate of primary production and retention of kelp detritus in giant kelp forests supports highly diverse communities with over 200 species of macrobiota of which 36% may be highly associated with the kelp forest (Dayton 1985; Steneck et al. 2002).

Rock reef and boulders with high surface relief will offer a number of different microhabitats. The sides and overhangs of reefs and boulders tend to be dominated by particle feeders such as sponges, bryozoa, and sea squirts. Many of the sessile biota exhibit adaptation to life in changing water currents. Flow over polyps and feeding tentacles will affect the rate of delivery of food particles. At low flow rates, the delivery rate of particles may be insufficient; hence feeding may cease because it is not energetically profitable. When flow is too strong, feeding may not be physically possible as particles evade capture or entrapment and the increased drag incurred may elevate the risk of shear or body damage, hence feeding structures are withdrawn. Many of these organisms also exhibit diurnal patterns in feeding behaviour that coincide with movements of zooplankton in the water column.

● Sessile particle feeders modify their feeding activity in relation to flow velocity by cessation of feeding when the delivery rate of particles is too low or when current velocity might incur physical damage as a result of shear.

The crevices and interstices associated with reef, boulder, and cobble habitat permit exchange of well-oxygenated water and provide ideal

Box 7.7 Distribution and structuring processes in kelp communities

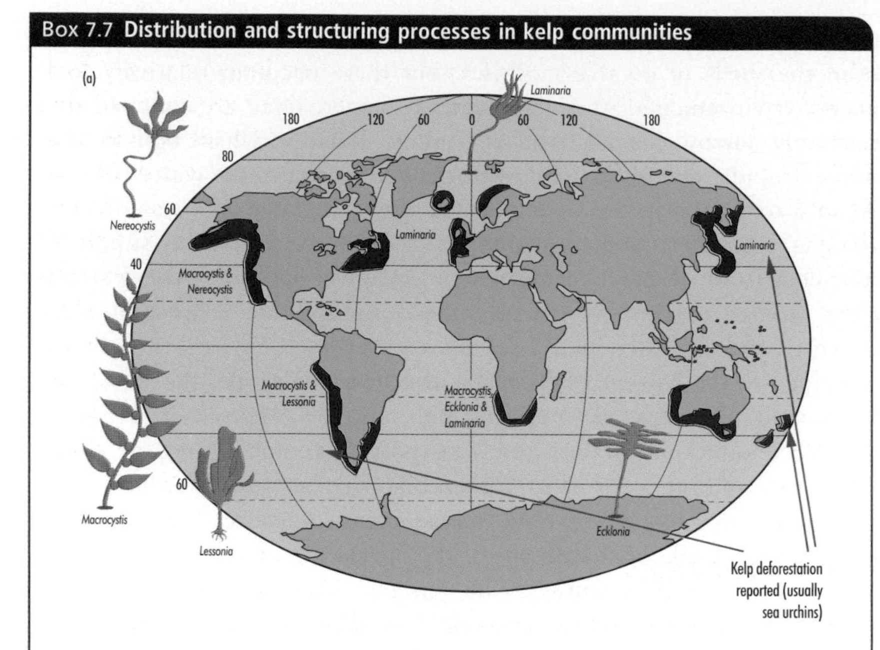

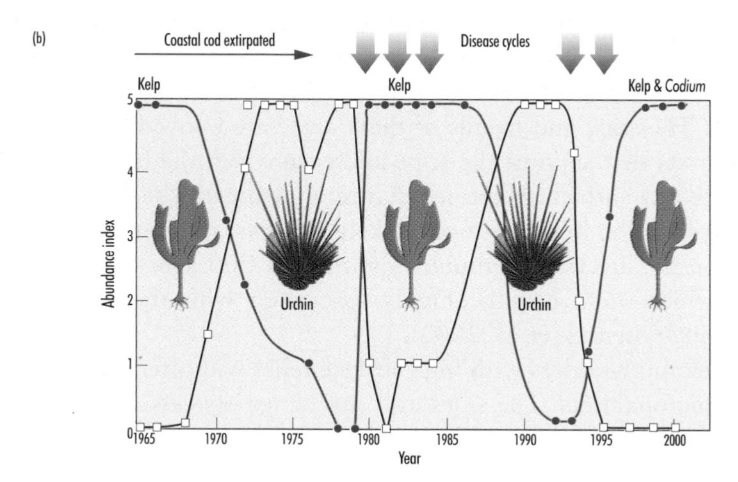

(a) The distribution of dominant kelp genera around the world (from Raffaelli & Hawkins 1996). The global distribution of kelps is physiologically limited by light at high latitudes and by nutrients, warm water and competition from macrophytes at low latitudes (Steneck et al. 2002). (b) Urchins *Strongylocentrotus droebachiensis* are major herbivores of kelp. Over-fishing and the subsequent extirpation of cod from the inshore waters of Nova Scotia released sea urchins from predator control leading to a sequence of phase shifts between sea urchin and kelp dominated communities. Only the intervention of sea urchin disease increased natural mortality sufficiently to permit the periodic re-establishment of the kelp in this system (see also 7.5.3) (Adapted from Steneck et al. 2002).

refugia both from strong near-bed flow and predators. Many commercially important fishes utilize boulder reefs in their juvenile stages and experimental and field observations demonstrate the importance of the protection from predators provided by this habitat (Auster & Langton 1999). The provision of suitable cryptic habitat is a critical constraint of commercially important taxa such as lobsters *Homarus gammarus*, which compete aggressively for the best refuges, causing displacement of individuals when there is a shortage of suitable habitat. Small lobsters generally remain within a relatively restricted area of the seabed moving a distance of <4 km during a 30-month period, while larger individuals can move as much as 45 km in the same time (Smith et al. 2001). Fishery enhancement programmes have sought to increase available habitat through the use of artificial reefs made from a variety of materials including concrete blocks, tyres, and even scrapped streetcars.

● The availability of suitable crevices can lead to strong density dependence for animals such as lobsters.

7.7.2 Soft habitats

Soft-sediment characteristics mirror the hydrodynamic processes and topography of the seabed. Substrata with particle sizes from gravels down to coarse sands are closely determined by physical processes (water flow and wave effects) and as a result are subject to frequent physical disturbance. However as the influence of physical processes decrease, finer sediments predominate and biological and chemical processes begin to have an important influence on the physical properties of the sediment properties (Fig. 7.7). Not surprisingly, you only tend to find finer sediments in less physically perturbed environments i.e. in sheltered shallow

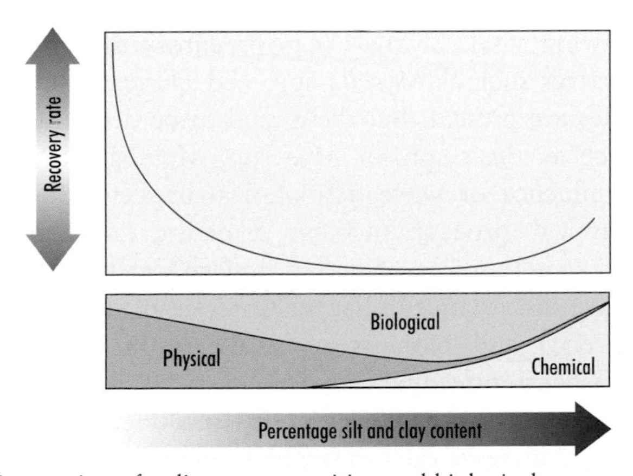

Fig. 7.7 The interaction of sediment composition and biological recovery rate in disturbed sediments. In coarse sediments with a low percentage of silt and clay physical processes are dominant. As the percentage of silt and clay content increases so the contribution of biological processes such as microbial activity become more important. Chemical attraction is most important in the finest sediments. Biological recovery rate is slowest in those sediments that are influenced by a mixture of physical, biological and chemical processes. Adapted from Dernie et al. (2003).

bays, fjords or beneath gyres or areas of deeper water with minimal current. The stability of sedimentary habitats depends upon the interaction between physical, biological and chemical processes.

Geologists and ecologists typically describe soft sediments in terms of their particle-size composition. This necessitates the **disaggregation** of particles within the sample. As a result we turn 'living' sediment into one that has lost most of its biological properties, i.e. the structural effects of organic molecules and microbial films that coat the surface of individual particles. During this process other crucial information about the nature of the habitat is lost such that we gain only a partial impression of the habitat. For example, many of the fine silt and clay particles (<0.063 mm diameter) are elongated or flat with high surface-to-volume ratios and are attracted to each other via **Van der Waals forces**. In contrast, larger particles are generally spherical and exert no significant inter-particle attraction. Particle-size analysis tells us nothing about the packing of the components of the sediment and hence we are unable to infer to what extent interstitial spaces exist between particles and hence the **porosity** (and hence water content) of the sediment. Sediments with high porosity will have a greater exchange of oxygenated water, which will increase the depth within the substratum that free-living animals can exist. Sediment porosity also affects how easily organisms can burrow into and through the sediment.

● Particle-size analysis of sediment habitats destroys the physical properties conferred by chemical activity and biological processes. Techniques that measure the property of sediment *in situ* would give a better indication of properties relevant to benthic communities.

Near-shore sediments less than 30 m deep are often highly perturbed by wave action. As a result, emergent sessile epifauna are often absent while highly mobile, robust, scavengers are often dominant (e.g. swimming crabs, hermit crabs, starfishes, whelks). Much of the infaunal assemblage is characterized by highly mobile short-lived polychaetes and rapidly burrowing small bivalves (e.g. predatory worms *Glycera* spp. and small bivalves such as *Mysella* spp. and *Donax* spp.). Some larger biomass species are present, but these tend to be deep (20 cm or more) burrowers such as *Ensis* spp. or *Mya* spp. Moving into deeper water beyond the influence of wave action, near-bed currents become the dominant physical process and can generate considerable habitat diversity over spatial scales of a few hundred metres (Fig. 7.8). The absence of wave disturbance in deeper water coincides with an increase in the occurrence and biomass of sessile biota such as hydroids, bryozoans, suspension-feeding bivalves such as scallops, and tube-building polychaetes such as *Chaetopterus* spp (Fig. 7.9).

● Soft-sediment habitats can vary considerably over distances of <100 m. The habitat variability is driven by physical processes such as near-bed currents.

Emergent biota increase the topographic complexity of the seabed and are often associated with their own diverse assemblage of biota. For example, Haines and Maurer (1980) found that the tube complexes of serpulid worms were closely associated with 54 species in contrast to the surrounding sediments that contained 107 species. The patterns of species association seem more strongly related to the architecture of the structure rather than the identity of the tube-building species. The tube complexes of both the serpulid *Hydroides dianthus* and *Pomatocerus*

● Emergent sessile biota can support their own micro-community of high diversity per unit volume of habitat. In addition to seabed features such as sand waves, emergent fauna provide shelter or a source of prey for many demersal fishes.

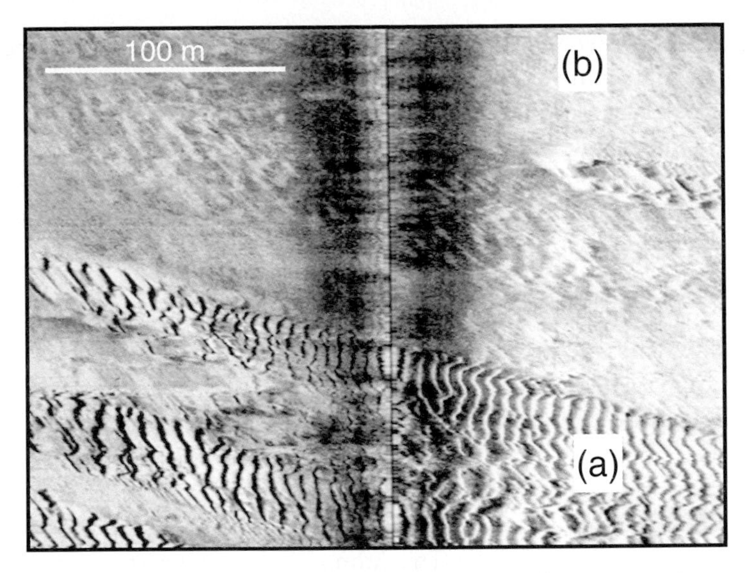

Fig. 7.8 Near-bed currents can generate considerable habitat heterogeneity in near-coast subtidal systems. This image is derived from acoustic data of the seabed showing a strip of seabed approximately 200 m wide. Two distinctly different sediment habitats are apparent, (a) mobile megaripple systems which are subject to daily physical stress and erosion that are dominated by small-body-sized fauna, and (b) more stable shelly gravel sand sediments that are typified by filter and suspension feeding communities.

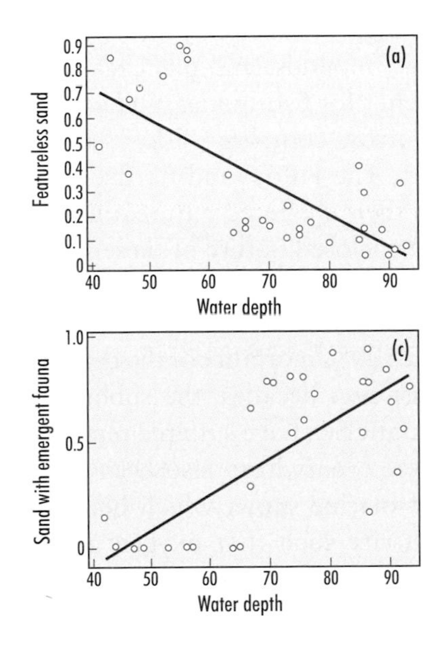

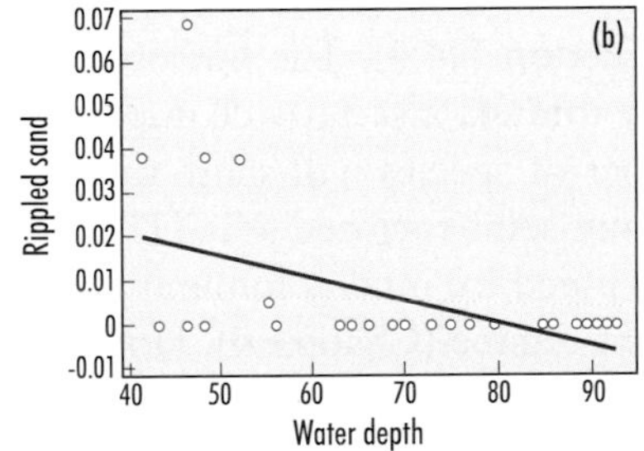

Fig. 7.9 The changes in soft-sediment characteristics with increasing depth as measured from photographic images of the seabed. Featureless sand habitat (a) that is frequently rippled (b) is typical of relatively shallow water depths (40 to 60 m). Emergent fauna were increasingly evident as sandy substrata occurred in water deeper than 60 m (c). (Lindholm et al. 2004).

triqueter had similar associates: *Nereis succinea, Lepidonotus squamatus, Polydora ligni, Eumida sanguinea, Syllis gracilis* (Kaiser et al. 1999b). Thus, similar structures on the seabed would appear to attract comparable groupings of fauna.

The additional structural complexity afforded by emergent fauna and bedforms has important implications for habitat use by more mobile species. Bottom-dwelling fishes such as silver hake (*Merluccius bilinearis*) use small-scale seabed features such as sand ripples to shelter from currents or as cover from where they ambush prey (Auster et al. 2003). The high abundance of prey associated with complex seabed structures is an important food source for many bottom-dwelling fishes as revealed by analysis of their diets. In softer mud sediments cerianthid anemones and sea pens (*Virgularia* spp.) are among the few emergent fauna that exist in this habitat (Fig. 7.10).

In deeper water (>50 m deep), in basins or in shallow sheltered areas, finer sediments such as silts and muds are able to settle out due to reduced physical forcing. The epifauna tends to be sparse in such areas with few sessile emergent species (mainly anemones and sea pens) and low abundances of mobile scavengers such as starfishes and hermit crabs. The fauna is typified by burrowing megafauna that shape the surface of the seabed with burrow entrances and mounds of excavated sediment and faeces (7.4.4). The fauna is dominated by crustaceans, typically callianassids, with commercial fisheries for Norway lobster (*Nephrops norvegicus*) and hyperbenthic pandalid shrimps. Echiuran worms are deposit feeders that use a highly extensible proboscis to feed on surface organic matter at distances up to 1 m or more from their burrow entrance. These worms graze the surface sediment sequentially in a circular fashion such that the surface microbial community has sufficient time to become replenished through production processes (see also section 7.4.4). The burrow entrances, mounds, and pits excavated by the infauna also provide important shelter for fish species, and often a number of species will share the same burrow complex.

● Mud sediment habitats are typified by limited emergent sessile fauna and a low diversity of mobile epibenthos.

Many semi-enclosed seas (The Adriatic, The Baltic) and fjordic areas are typified by mud communities due to their sheltered nature and low current regime (Chapter 4). However the enclosed nature of these water bodies encourages strong stratification of the water column in the summer. The combined effects of eutrophication and stratification can lead to periodic anoxic events at the seabed with a 100% mortality of the benthos (Chapter 14). For example, over the past two decades, the sublittoral benthic communities of the Northern Adriatic Sea have suffered repeated large-scale mortalities (Justic 1991). These events have also been associated with an excessive development of **marine snow**, which blankets benthic biota and elevates microbial activity such that oxygen at the seabed is rapidly depleted. The benthic communities in the Northern Adriatic are composed largely of sessile, epibenthic filter- and suspension-feeding organisms. They are long-lived and typically aggregated into so-called multi-species clumps, which is an unusual feature for muddy sediments (Box 7.8). The filter-feeding activity of these multi-species clumps is critical for benthic–pelagic coupling in this system. The longevity of the species involved mean that repeated anoxic events do not leave sufficient time for these associations to recruit and become reestablished.

● Soft-sediment communities in semi-enclosed seas areas are subject to periodic anoxic events that can lead to 100% mortality of the macrobenthos.

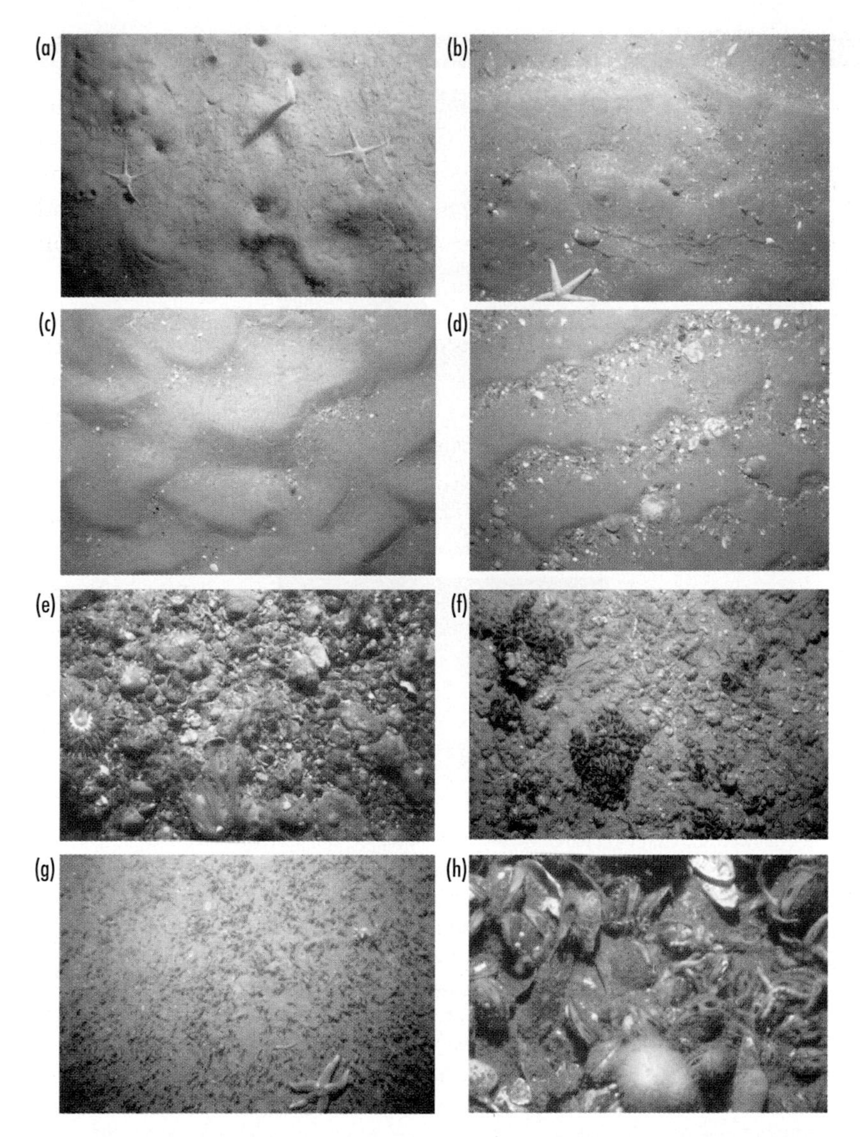

Fig. 7.10 A selection of sedimentary habitats. (a) A fine muddy habitat with brittlestars and a sea pen, burrow entrances and mounds indicate the presence of a high number of bioturbators in this habitat. (b) A sandy mud habitat with evidence of some reworking of seabed sediments by bottom currents. A surface dwelling sea slug *Scaphander* sp. leaves a distinctive surface furrow just above the starfish. (c) A mobile sandy habitat formed by near-bed tidal currents, which is typified by small infaunal organisms and an impoverished epifaunal community. (d) A typical mixed sediment with a veneer of sand overlying a coarse gravel and shell sediment. (e) A stable mixed sediment dominated by cobbles and gravel. This stable substratum is typically colonized by large attached filter feeders such as hydrozoans and bryozoans and anemones. (f) A cobble and mixed sediment habitat dominated by the slipper limpet (*Crepidula fornicata*) with clumps of blue mussels (*Mytilus edulis*). (g) A muddy sand habitat stabilized by a lawn of tube-building polychaetes *Lanice concheliga*. These habitats provide rich feeding grounds for flatfishes. (h) A biogenic habitat dominated by horse mussels showing the deposition of silt and organic matter among the shell matrix. This community is typified by a high abundance of filter and deposit feeding epibiota and has a high diversity of associated species. Photographs © E. Ivor S. Rees.

Box 7.8 Multi-species clumps in the Northern Adriatic

(a)

(b)

(c)

(a) Characteristic high-biomass community consisting of sponges, ascidians, and brittlestars at 25 m depth in the Gulf of Trieste. The organisms are typically aggregated into so-called multi-species clumps. Marine snow events are often associated with hypoxia-related mortalities. In an advanced stage, enormous cloud-like mucus aggregates eventually sink to the sea floor, further accelerating the collapse of benthic communities. (b) Dead multi-species clump covered with marine snow during an oxygen crisis. Death of the sponge or ascidian core of such aggregations also kills the associated fauna, for example crustaceans and polychaetes. (c) The first recolonization phase is characterized by more rapid-growing, opportunistic forms, here by tubeworms and the ascidian *Ciona*. They bear little resemblance to the typical multi-species aggregations formerly found here. Knowledge about such successions provides valuable information on past collapses and on the resilience of these communities. Photograph A: Kurt Fedra; photographs B–C: Michael Stachowitsch.

7.7.3 Biogenic reefs

● Reef-building fauna can exert strong grazing pressure on primary production in the water column through filter feeding. The removal or degradation of such structures through harvesting or pollution has led to large-scale ecosystem changes particularly in the zooplankton community.

Biogenic habitats or reefs are structures created by aggregations of organisms, which may or may not rise from the seabed, that form a discrete community from the surrounding habitat. Biogenic reefs can be composed entirely of the reef-building organisms or accumulations of biota, organic, and inorganic material. Such habitats are formed by bivalves (e.g. oysters, mussels), polychaetes (e.g. *Sabellaria* spp.), corals, and sponges. Locally, biogenic reefs can contribute to significant proportions of the seabed habitat (Lenihan & Peterson 1998; Cranfield et al. 1999). For example, extensive oyster beds occurred in Chesapeake Bay on the eastern US coast and in the Foveaux Strait, southern New Zealand (*Crassostrea virginica* and *Tiostrea chilensis* respectively). However, in both localities, long-term harvesting has severely reduced the seabed coverage of oyster reefs.

In the case of Chesapeake Bay it was estimated that the oysters filtered the entire volume of the water body every day. With such a large turnover of water volume it is not difficult to understand why the ecology of water bodies such as Chesapeake Bay might be changed by reducing oyster reef biomass. These reefs act as a sink for phytoplankton production and transfer this biomass from the pelagic to the benthic system. In addition to their role in transferring energy to the seabed, biogenic structures greatly contribute to marine habitat complexity by increasing the three-dimensional relief of seabed topography and often have a nursery function for juvenile fishes and crustacea. In the case of fauna that provide a hard surface such as bivalves and tubeworms, the reef structure may provides a settlement surface for epibiota such as algae, soft corals, sponges, tunicates, hydroids, and bryozoa, which further contribute to the processing of water borne particulate matter. Reef structures represent irregular features on the seabed and consequently affect flow around and over them. Erosion due to currents will be greatly reduced in the interstices of the reef matrix, which creates the right conditions for deposition of faeces and other organic matter that fuels production of the associated benthos and microbial communities (Lenihan & Peterson 1998).

As biogenic reefs are constructed primarily of living organisms, they are particularly vulnerable to physical disturbance, fishing or pollution effects associated with eutrophication. The initial establishment of reefs depends upon the chance settlement of a cohort of recruits coupled with favourable environmental conditions. Hence restoring degraded reefs is fraught with problems and prone to failure. Restoration is further complicated when the reef-building organisms are slow growing as in the case of the calcareous algae known as maerl. Maerl grows at a rate of about $1\,mm\,y^{-1}$ into twiglets and branches that interlock to form a living sediment matrix. This open matrix has large interstitial spaces that permit deep penetration of oxygenated water into the maerl substratum. Maerl is also associated with the reef-building bivalve *Limaria hians,* which constructs 'nests' of shell debris and other material attached to its byssus threads and can form beds at densities of more than 700 individuals m^{-2} within the maerl matrix (Hall-Spencer & Moore 2000). This bivalve is scarce in any other type of habitat and hence is highly vulnerable to environmental impacts that might adversely affect the maerl habitat.

● Some reef-building biota can be many hundreds of years old and hence they are particularly vulnerable to any form of disturbance.

● CHAPTER SUMMARY

- Continental shelf systems account for approximately 8% of marine habitats yet they are the site of much of the ocean's global primary production that ultimately fuels major world fisheries.

- Shelf systems are relatively young on a geological scale with large changes in the extent and position of the coastal margin occurring during the last glaciation event. Drowned river deltas, glacial lag deposits and glacial scour are features of the current continental shelf seabed.
- Tidal currents generate turbulent flow as they move across the shallow seabed areas of the shelf close to the coastline. This turbulence generates mixing throughout the water column and prevents thermal stratification and enhances the flow of material to the seabed and hence benthic production. Further offshore in deeper water, stratification occurs and leads to the development of fronts at the interface with mixed water masses.
- The world's continental shelf systems can be categorised into regional ecosystem types according to the prevalent environmental conditions that impinge upon them and characterise the timing of the spring/summer phytoplankton bloom that fuels shelf food webs.
- The flora and fauna of the continental shelf seabed system can be subdivided according to their functional role (e.g. predators, herbivores) and ecosystem function (e.g. eco-system engineers, sediment processors, habitat formers). Large-scale population changes in key species can lead to effects on other trophic levels in some relatively simple systems.
- Early marine ecologists attempted to categorise particular assemblages of species that occur on continental shelves. However, it is questionable whether this approach is useful given the infinite number of possible variants created by the substitution of one species for another and the likely large-scale changes we will see with current trends in global ocean warming.
- Seabed habitats can be categorised according to the characteristics of the substratum (hard or soft-sediment). The physical characteristics of soft-sediment habitats are indicative of the physical energy that affects the seabed, with mud occurring in low energy environments, while coarse sands occur in tidally swept areas exposed to wave action. Some habitats are composed of living biota such as kelp forests, maerl beds and oyster reefs that are critical links with their associated communities of organisms.

FURTHER READING

Dayton (1994) discussed the issues of scale and stability in hard bottom marine communities, while Snelgrove and Butman (1994) review the debate on the relationship between animal assemblages and soft sediment habitats. Hall (2002) assesses the major impacts that affect present day shelf systems.

- Dayton, P. 1994. Community landscape: scale and stability in hard bottom marine communities. In P. Giller, A. Hildrew, and D. Raffaelli (eds) *Aquatic Ecology: Scale, Pattern and Process*. Blackwell Scientific Publications, Oxford, pp. 289–332.
- Hall, S. J. 2002. The continental shelf benthic ecosystem: current status, agents for change and future prospects. *Environmental Conservation* 29: 350–74.
- Snelgrove, P. and C. Butman. 1994. Animal-sediment relationships revisited: cause versus effect. *Oceanography and Marine Biology: An Annual Review* 32: 111–77.

Chapter 8

The Deep Sea

CHAPTER SUMMARY

The deep-sea floor represents the largest, yet least-known, habitat on earth and ranges from the edge of the continental shelf at 200 m down to the abyssal plain 5 km below the surface, with trenches plunging to 10 km in places. The environment is remarkably constant across the ocean floor: cold, dark water overlying soft, deep mud. While the high hydrostatic pressure is the most obvious physical feature of the deep, it is food supply from the surface that is the limiting factor for life on the abyssal plain. In temperate areas, the food input can be seasonal, providing cues for reproductive cycles. Due to the lack of food, the community of animals in the deep sea is of much lower abundance than in shallow waters, but some organisms can grow considerably larger than their shallow water relatives, such as 25 cm single-celled organisms! Potentially there are millions of species inhabiting the deep-sea benthos; the main groups of large, mobile organisms are echinoderms, decapod crustaceans, and fish, many of which show remarkable adaptations to deep-sea life. Recent exploration using submersibles has revealed exciting 'island' habitats in the deep sea with a level of production and diversity higher than the surrounding environments. These include hydrothermal vent communities, which are powered by production from chemosynthetic bacteria and are practically independent of sunlight. Vents have a unique, remarkable, and productive associated fauna, including huge vestimentiferan worms that have no mouth or gut, but rely on vast populations of bacteria within their bodies, and the Pompeii worm that lives in water temperatures above 60 °C – the most thermotolerant animal so far discovered.

8.1 Introduction

The deep ocean represents the largest habitat on the planet, yet remains the least known; humans have landed on the moon, but due to conditions inhospitable to humans, no one has yet set foot on the deep ocean floor. The kilometres of water separating us from the abyssal floor prevent much remote sensing of the bottom features. As a result, even the topography of the world's seabed is not fully documented, and the life that is found there even less well documented. Therefore, unlike other habitats in the marine realm, the study of deep-sea organisms has been exceptionally difficult and our knowledge of how the deep sea functions has, until recently, had to be pieced together from samples representing a tiny fraction of the ocean floor. Only the development of submersibles in recent decades has led to a more vibrant view of deep-sea ecology and enabled the discovery of some of the most exciting and fascinating habitats on earth. However, the dark, cold depths of the abyss still generate a sense of the unknown that has resulted in varied descriptions of deep-sea environment from a completely lifeless realm to one occupied by strange monsters, epitomized by books such as John Wyndham's *The Kraken Wakes*. The truth lies somewhere in between, and the aim of this chapter is to detail the environment that exists on the ocean floor, the problems facing organisms living there, what these organisms are, and how they are distributed; and to explore some of the astonishing 'island' habitats that have recently been discovered within the deep sea.

● The extreme environment of the deep sea make it one of the planet's most hostile habitats for exploration.

8.2 Definitions and Environmental Features

8.2.1 What is the deep sea?

The definition of 'the deep sea' is relatively vague, but generally represents all the water and seabed beneath the edge of the continental shelf (Fig. 8.1, this area being termed **neritic**, Chapter 7). Generally, this boundary tends to be at a depth of about 200 m, so the deep sea can be regarded as any water or benthic habitat situated below 200 m. The deep-sea benthos can be divided into separate depth regions that refer to the general topography of the ocean floor that is generated as a consequence of plate tectonics. At the edge of the continental shelf is the **shelf break**, where the gradient of the floor increases down the **continental slope**. This can be exceptionally steep at many continental margins

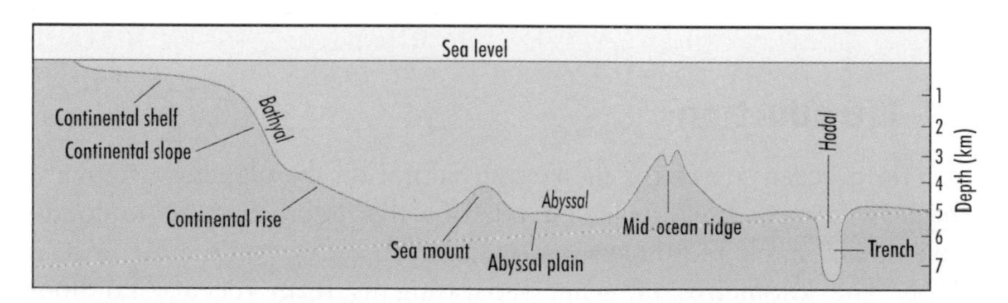

Fig. 8.1 A hypothetical cross section of an ocean basin indicating the main topographical features of the deep sea and terms used for areas of the deep-sea floor. Vertical scale is much exaggerated.

(e.g. off the south-west coast of the British Isles) and if visible would be among the most impressive natural features on earth. In such areas, the slope can extend from 200 to 4000 m, but more often the gradient is less dramatic, and the steep slope gives way to a gentler **continental rise** between 2000 to 5000 m. At around 4000 to 5000 m, the ocean floor is reached and extends over the ocean basins at depths generally around 5000 m (the **abyssal plain**). The benthic area associated with the gradient from continental shelf to abyss is known as the **bathyal** region (though some definitions end this region at a depth of about 2 km), the area on the sea floor is termed the **abyssal** region.

● The deep sea is generally considered to be the ocean below 200 m, though most of the ocean floor is about 5 km below the surface.

Early attempts to map the ocean floor relied on extensive extrapolation from the few sonar transects that were available. For example, the famous three-dimensional maps of the ocean floor produced in the late 1960s (published in *National Geographic* magazine) relied on relatively few transects across the ocean (Kunzig 2000), the topography between them was inferred by an artist's impression. Consequently, these early diagrams illustrated the abyssal plain as generally a huge, flat expanse – an image that has persisted in the minds of both professionals and amateurs. However, more recent detailed work in certain parts of the world has revealed the abyss to be far less uniform, with many small topographical features resulting in 'rolling hills' rather than plains. Additionally, such smaller-scale studies have revealed the existence of many **seamounts**, which rise from the ocean floor but do not break the surface (Fig. 8.1). In original maps that extrapolated from known points it is clear that such features would be missed, but these have recently been found to host fascinating and diverse communities. Other features of the ocean floor are much larger and well known and coincide with areas of generation or subduction (one crust moving under another) of the oceanic crust. The abyssal plain does not extend across the whole ocean basin, but is interrupted by a long chain of mountains known as the **mid-ocean ridge** (Fig. 8.1, Chapter 1). The ridge is the site of the formation of ocean crust, which spreads out evenly on each side, and therefore can also be the site of intense volcanic activity that results in **hydrothermal vents** (8.5). **Trenches** in the ocean floor (Fig. 8.1) occur where plates meet and the ocean crust buckles and deepens as it moves beneath an adjacent plate. These trenches represent the deepest parts of the sea and can extend down to 10 000 m. The deepest point of all is the Mariana Trench off the Philippines (10 912 m). The benthic area within a trench (a depth of between 6 000 to 10 000 m) is known as the **hadal** region.

● Continental plates are created at mid-ocean ridges such as the long-chain of mountains in the mid-Atlantic. The plates spreading out from the ridge and observed symmetry of magnetic patterns each side of the ridge partly led to the theory of continental drift.

8.2.2 Problems with sampling the deep sea

Despite being the world's largest habitat, the ecology of the deep-sea floor is the least understood due to inherent problems of sampling

(Chapters 6 and 14). Traditionally, our knowledge of the organisms of the deep has been pieced together from the contents of comparatively few a remote trawls, grabs, and dredges that have been hauled up from the depths. Piecing together the ecosystem on the abyssal floor, or even how animals live their lives, from such samples is extremely difficult (if not impossible) and involves considerable guesswork. To comprehend how difficult such an exercise is, sampling the deep sea has been likened to sampling the rainforest by flying a plane above the clouds, pulling a net through the trees and constructing the system from the contents of the net. This would clearly be ridiculous, but it is how deep-sea biologists have had to work. In recent years our means of investigating the deep sea has improved dramatically, first by the use of photography and particularly by the development of manned submersibles. For the first time scientists can actually see what the organisms are like in their natural surroundings.

● The total area of quantitative mud samples taken so far from the deep sea is only about 500 square metres – perhaps a millionth of the total seabed area (see **www.sams.ac.uk**).

● Most deep-sea trawls depressingly contain rubbish, such as tin cans generally from waste thrown overboard ships. Particularly below major shipping lanes, trawls return large amounts of *clinker* – the burned coal from steamships. Even the abyssal plain is not immune to our influence.

However, even taking a trawl from the sea floor is exceptionally difficult and requires a scale of sampling unnecessary for the rest of the marine realm. To pull a trawl or dredge along the abyssal seabed at a depth of 5000 m would require nearly 11 000 m of wire; considering the oceanic conditions involved a large ship is required. Additionally, knowledge of the location of the trawl is vital to know it is actually on the seabed, so sophisticated acoustic devices are required. Deep-sea sampling is therefore very expensive and prone to failure, even for the collection of the most basic of samples; simply placing and retrieving a sampling device on the deep sea floor takes hours rather than minutes.

Fig. 8.2 Photographs of organisms on the deep-sea floor. (a) A red crab *Geryon trispinosus* reacting to the approach of a photosled. (b) A 'bathysnack' photo of the spider crab *Neolithodes* attracted to bait. (Copyright Southampton Oceanography Centre)

As well as traditional gear, much of our knowledge of life on the ocean seabed, particularly detail of megafaunal animals, is from photography. Two main types of camera can be employed. First, cameras with open shutters can be mounted on sampling gear such as benthic sleds; the film-winding mechanism is connected to the flash unit. The camera therefore records the seabed before the sampling gear moves over it (Fig. 8.2a). Alternatively, stationary photographic gear can be dropped to seabed – basically a frame with a camera mounted on it known as a **bathysnap**. Photographs are taken at certain intervals (e.g. 1 hour, 1 day) and the gear is left on the seabed often for months. The camera is retrieved by transmitting an acoustic signal to the seabed, which releases weights attached to the camera mount, and the bathysnap floats to the surface to be retrieved. Such series of images have given us important information on the dynamics of the deep-sea bed. Cameras (still or video) can also be deployed with bait attached to the frame (e.g. a fish carcass) in order to attract and photograph scavenging animals (Fig. 8.2b). Such pieces of sampling gear are fondly known as **bathysnacks!**

The most expensive, but enlightening, method of sampling the deep sea is the use of manned submersibles such as *Alvin, Shinkai 6500*, and *Nautile*. Certainly our understanding of how the deep-sea system functions has progressed in leaps since the extensive use of submersibles for observation, sampling, and even experimentation, with scientists now able to set out experiments on the ocean floor and record the response of organisms in real time. Determining the ecology of deep-sea vents, for example, would never have been possible by using remote sampling techniques.

● For more details on *Alvin*, the first submersible to be used extensively to explore the deep sea, go to **www.whoi.edu/marops/**.

8.2.3 What is the environment like on the abyssal plain?

The deep abyss is clearly very different from all other marine habitats, but how is that reflected in the physical and chemical environmental parameters that influence organisms? Other chapters have demonstrated that single variables can have a large influence on the distribution of organisms within a system (exposure to air on intertidal shores, salinity in estuaries). Therefore, it is important to review the conditions in the deep-sea floor (at 5000 m) to determine which parameters potentially influence the organisms found there (Gage & Tyler 1991). Some parameters vary little below 2 000 m; for example, salinity is relatively constant at 34.8 ± 0.3.

8.2.4 Light

The abyssal plain is very different from practically all other marine habitats in that no light penetrates the deep ocean to these depths; the abyss is in permanent darkness. Consequently, photosynthesis is not possible, so no living plants exist in the deep sea resulting in a food web dependent entirely on energy from detritus and carrion originating from the systems above (with the exception of bacterial chemosynthesis in hydrothermal vents, see 8.5.2). However, many deep-sea organisms have evolved methods of generating light using light-emitting bacteria contained in special cells (**bioluminescence**, Box 8.1). This light generation in the darkness has many uses, such as finding mates of the same species, finding food, and predator avoidance. Scientists in the first submersibles entering these dark depths were amazed at the extent of the bioluminescence as organisms emitted light on contact with the submersible.

● While bioluminescence has traditionally been associated with bacteria, it is now clear that some groups of marine animals (particularly cnidarians) can generate their own light through the production of photoproteins.

● The deep-sea water column can be divided up in terms of light penetration. The top layers (euphotic) have enough light for photosynthesis. Down to about 1000 m there is still enough light for some vision (dysphotic zone), but below this there is no light at all (aphotic).

8.2.5 Temperature

The temperature of deep-sea water is low and constant. Below about 2000 m, temperature generally ranges between only −1 °C to 4 °C, with the majority of the abyssal water at around 2 °C. As organisms only ever

Box 8.1 Bioluminescence

Bioluminescence is the light produced by a chemical reaction within an organism. Perhaps the most familiar bioluminescent organisms are fireflies (actually beetles), but light production is comparatively rare on land and is much more a feature of the marine environment where it is a relatively common feature. Practically all marine taxonomic groups from dinoflagellates to fish have members that can bioluminesce (producing the light themselves or using bacteria housed in light organs or **photophores**), particularly below the euphotic zone in the ocean – for example, approximately 90% of animals are bioluminescent in the mesopelagic zone. Two main chemicals are required to produce bioluminescence: luciferin creates the light, but requires luciferase, which catalyses the oxidation of luciferin resulting in light. Most of the light is blue-green in colour (generally in the range 440–479 nm) as blue light travels furthest in water and most marine organisms are sensitive to blue light. Notable exceptions are deep-sea malacosteid fish (loosejaws), which have an unusual ability to produce red light (Haddock et al. 2004). This is achieved through the use of filters and a fluorescent pigment inside the photophore, which takes the energy and re-emits it as red light. Generally, fish cannot see red

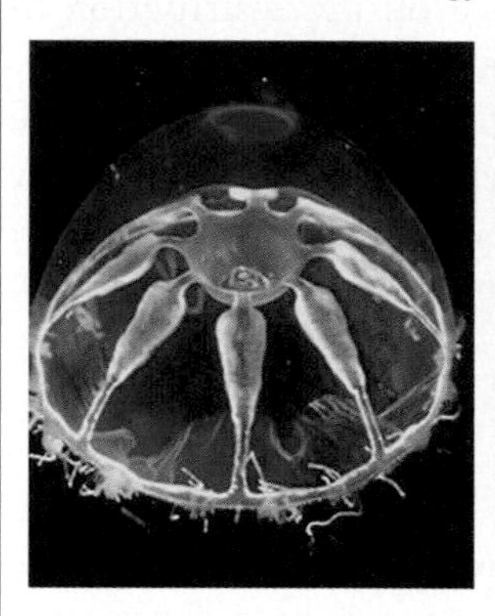

light, but loosejaws (e.g. *Aristostomias*) have extra photoreceptive pigments, allowing light to be detected in this range. This predatory fish can therefore emit light that only it can detect and allows red-coloured prey to be illuminated.

Images from **http://www.lifesci.ucsb.edu/~biolum/**. Copyright Steven Haddock. See website for more information of bioluminescence.

encounter cold water, many are killed by the temperature shock as they are brought to the surface in samples. The minimum recorded temperature is −1.9 °C beneath Antarctica, where cold water sinks. However, there are two particular deep-sea areas that have unusually high temperatures. The Mediterranean Sea has a very deep basin and water at 4000 m here has been recorded at temperatures of 14 °C, while water at 2000 m in the Red Sea can be as high as 21.5 °C.

8.2.6 Dissolved oxygen

The oxygen content of the water is relatively constant across the abyssal plain, with a concentration *c*.5 mg l^{-1} below 2000 m. Under normal oceanic conditions, the oxygen minimum tends to be just below the **euphotic layer** (400 to 500 m). Some areas of the deep sea have notably low oxygen levels, due to specific local conditions, such as the Santa Catalina Basin off California, which consequently appears to have a lower faunal diversity than comparable nearby regions such as the San Diego Trough.

8.2.7 Bottom currents

Levels of oxygen, and other variables, are kept comparatively constant because the deep-sea floor is constantly refreshed by a range of bottom currents, which can provide environmental cues for organisms (e.g. to direct the action of feeding appendages, migration, etc.). As well as directly measuring flow, we can see that currents exist from photographs of, for example, sea anemones whose tentacles are trailing in a uniform direction. Three main types of current exist in the deep sea. First, in some parts of the ocean, tidal currents can reach down as far as the base of the continental slope (e.g. Bay of Biscay, Europe) and can even form familiar ripple patterns on the deep-sea floor. Tidal currents will change direction, but generally the bottom currents are unidirectional and come from two sources. Second, the **oceanic conveyor** (Chapters 6 and 11) is vitally important for the movement of water and provides constant, unidirectional currents in the deep sea. For example, cold dense water sinks at the Antarctic and moves northwards across the Atlantic above the seabed. Third, the **Coriolis current** arises from the rotation of the earth and the relative movement of water in respect to the seabed. As the earth spins, the majority of the water in the oceans remains stationary, resulting in oceanic currents (Chapters 6 and 7). However, at the ocean floor, a small layer of water (the viscous sub-layer) moves with the earth whereas the water further up the column moves at a relatively slower speed. The sea floor is therefore moving beneath the water column, resulting in a current. However, due to the scale of the earth movement, the impression is of the water moving over the seabed.

Fig. 8.3 Before and after. The impact of hydrostatic pressure at 3000 m on a polystyrene cup. All the air in bubbles within the polystyrene has been squeezed out, shrinking the cup size.

8.2.8 Hydrostatic pressure

The most notable, and predictable, feature of the deep sea is the immense hydrostatic pressure due to the weight of overlying water, the single most limiting factor for humans. Pressure increases by 1 atmosphere every 10 m, so on the abyssal plain it is over 500 atmospheres (cf. 4 atmospheres on a very deep scuba-dive). At the bases of trenches it is even greater. An average-sized person standing on the bottom of the Mariana trench would experience pressure equivalent to holding up

48 jumbo jets! One consequence of pressure, as every diver knows, is that it compresses gases, including the swim bladders of fish. The impact on gas is illustrated by attaching a polystyrene cup to the outside of submersibles (Fig. 8.3), the cup shrinking to a much smaller size by the impact of pressure squeezing air from the material matrix. High pressure can also slow down the rate of enzyme catalysis.

● Even small amount of gas in the stomach can expand dramatically on raising a fish to the surface, with messy consequences! Fish stomachs can often protrude from their mouths and the eyes can look bulbous.

However, it has to be remembered that although the pressure seems severe to us, organisms that have evolved to live in the abyss do not experience any fluctuations in pressure, or the other variables discussed above, and so are not necessarily subject to environmental stress from such factors. Organisms adapt well to living at extremes; it is variability in the environment that is particularly hard to deal with (Chapters 4 and 5). There is one physical factor that potentially does pose problems for organisms living on the abyssal sea floor: sediment.

8.2.9 Sediment

● The problems of dealing with fine, soft sediment represent the only real physical variability for deep-sea organisms. Organisms have developed a range of strategies to remain above the mud surface, such as possessing long stilts, long stalks of sessile animals, climbing on other organisms, and flotation methods.

In parallel with the mid-estuary, the sediment across the vast majority of the ocean floor is soft, fine mud, though at a much greater scale. This soft sediment represents the world's largest single habitat due to the consistency of the physico-chemical conditions. The mud on the abyssal plain has been deposited over millennia, and consequently in some areas can be up to 1000 m thick, so benthic organisms have to develop strategies in order to remain above the sediment. As this parameter is probably the single physical variable affecting organism distribution, it is worth looking at the sediment in more detail.

Deep-sea sediments can be divided into two clear classes: clay particles, inorganic sediments found mainly under oligotrophic surface waters such as mid-ocean gyres; and **biogenic oozes**, sediments found below productive surface waters and containing >30% biogenic skeletal material. The composition of this biogenic material can vary, depending on the dominant taxa in the surface plankton and two subclasses of biogenic ooze can be identified. **Siliceous oozes** are silicon based, and are therefore made up from the skeletons of two main groups: diatoms (whose skeletons are known as frustules) and radiolarians. Due to the dominance of diatoms in many ocean regions, siliceous oozes are found beneath diatom-productive surface waters such as the sub-Antarctic and central-west Pacific. **Calcareous (or Foraminiferan) oozes** are sediments based on calcium carbonate ($CaCO_3$) and are composed of the skeletons of two groups of plankton with calcareous skeletons, namely the Foraminifera and coccolithophores. As foraminiferans are the more widespread of these two groups, the sediments are commonly named after them. Coccolithophores can, however, form massive blooms in surface waters (Chapters 2 and 6). Calcareous oozes tend to be found in shallower water than siliceous oozes as the hydrostatic pressure at great depth forces $CaCO_3$ into solution.

The composition of deep-sea sediments is therefore relatively predictable and depends on the productivity and species composition of the surface waters. This fall of material from the euphotic region of the ocean to the deep-sea floor has major implications for deep-sea life, as we see later (8.3), and is unlike the situation on the continental shelf (Chapter 7). Hard substrata are very uncommon in the deep sea, which has consequences for many sessile groups that in shallow waters tend only to be found on reefs. Examples of exposed rock, or other hard material, include hydrothermal vents (8.5), seamounts, steep slopes where sediment cannot settle (such as trench walls or mid-ocean ridges), beds of manganese nodules, and whale skeletons. On death, the bodies of whales fall to the ocean floor, where scavengers quickly remove the tissues and leave the skeleton. The bones appear to remain for a long period of time, forming an important 'island' habitat for a high diversity of organisms (Fig. 8.4, Baco & Smith 2003), including those generally found only on hydrothermal vents. It is possible that whale skeletons provide an important stepping stone for the dispersal of vent organisms (and others) as the bones contain sulphurous compounds necessary for chemosynthesis (8.5.5). The removal of whales during the intensive whaling period may therefore have had a more indirect impact by restricting the fall of whale carcasses, and thus skeletons, to the deep sea.

● Coccolithophores also bloom in coastal areas and turn the water a chalky bright blue colour similar to an alpine river.

● It has been estimated that whale skeletons may be on average only 9 km apart across the deep-sea floor. This distribution may have been modified by the whaling industry, both by removing whales and in dumping flensed skeletons regularly in certain areas.

Fig. 8.4 Hagfish on whale skeleton. Copyright NOAA.

In conclusion, the environmental variables discussed here do not vary greatly and therefore do not have as dramatic a controlling influence in the deep sea as in other marine systems such as estuaries and rocky shores. To discover what may be the major limiting factor for abyssal organisms we have to look at biological variables.

8.3 Food Supply to the Deep Sea

8.3.1 Production in the deep sea

As light is absent on the abyssal plain, no photosynthesis occurs. The ocean floor is therefore one of the few major habitats in the world where living plants are absent and thus the deep sea does not have a direct input from the primary production underlying the majority of the world's ecosystems. A different form of primary production does occur in some isolated parts of the deep sea, bacteria associated with hydrothermal vents (8.5) using sulphur compounds to fuel chemosynthesis (Box 8.2). However, the organisms occupying the vast majority of the ocean floor have no direct input from primary production, but instead have to rely on the input of organic material from the euphotic layer of the ocean, mainly in the form of **particulate organic matter** (POM). The deep-sea bed is therefore an entirely **allochthonous** system. Only about 1% to 3% of the surface net primary production reaches the abyssal seabed (Gage & Tyler 1991) and is the food supply for the whole

● **Allochthonous** systems rely on the input of organic material from outside the system, compared with **autochthonous** systems that mainly generate their own production.

Box 8.2 Chemosynthesis at hydrothermal vents

Microbiologists first uncovered chemosynthesis in the late nineteenth century, but until the discovery of hydrothermal vents it was considered to play no significant role in the photosynthesis-dominated carbon cycle of the earth (van Dover 2000). *Chemosynthesis* is a significant microbial process involved in seawater sulphur cycling. Organic matter is consumed by sulphate-reducing bacteria (generally in the absence of oxygen, such as in anoxic sediments), converting sulphates to sulphides. This product can then be subject to microbial oxygenation, resulting in the generation of organic compounds:

$$S^{2-} \text{(sulphide)} + CO_2 + O_2 + H_2O \rightarrow SO_4^{2-} + [CH_2O] \text{ (organic material)}$$

However, when this process occurs away from vents there is no additional organic material formed, since organic carbon is being oxidized to generate the sulphide in the first place. At vents, new supplies of sulphide are being generated from the geochemical interaction of seawater and heat, allowing a net gain of organic compounds and accumulation of new biomass. The process was termed 'chemosynthesis' to give a direct comparison with 'photosynthesis', but should be more thoroughly described as 'chemo-autolithotrophy' (van Dover 2000). If this seems a mouthful, animals (such as those at vents) utilizing chemosynthesis are termed '**chemoheteroorganothrophs**'.

deep-sea system away from hydrothermal areas. Therefore, food supply is the major limiting factor for deep-sea organisms and is the main reason there is a comparatively low abundance and biomass of animals on the ocean floor.

8.3.2 Dissolved organic matter (DOM)

While the most important food input is from POM, DOM may have an important role in providing carbon to sediment-dwelling organisms. Measurements of interstitial water have shown that DOM in deep-sea sediment can be 10 times that of the overlying water and it is suspected that some large animals (such as vestimentiferan worms and some polychaetes) derive a significant proportion of their carbon from DOM sources. However, DOM is created primarily from the metabolic processes of metazoans, bacterial action and decay, so it is primarily a reworking of carbon that has been derived from external sources.

8.3.3 Large food falls

The whole bodies of dead animals and large fractions of plants can sink intact to the ocean floor and provide a food source for a suite of scavenging organisms such as amphipods, brittlestars, and fish (in particular hagfish and rat-tails, which may be attracted to feed on the amphipods). Fish, whale and squid carcasses rapidly attract a range of large scavenging animals, as witnessed by baited cameras (Jones et al. 1998), and it is clear that such animal food falls are the prime resources for a range of deep-sea animals. However, the appearance of a carcass on the ocean floor is an unpredictable and sporadic event, so scavengers have to be able to respond to, and make the most of, any food fall that occurs. In particular, these organisms may have to endure long periods between meals and so some scavengers have become particularly adept at gorging. The giant amphipod *Eurythenes* (8.4.11) may be able to eat up to 75% of its body weight (Hargrave et al. 1994), enough to sustain a mature female for over a year.

● Hagfish are related to lampreys and represent the 'primitive' group of eel-like fish known as the Agnatha. They have no true jaws, just a circular mouth lined with teeth.

While such amphipod species appear to have adapted to feed on carrion, animal carcasses do not form the sole food source for the majority of fish species that appear when bait is provided, but provide important energy subsidies that are rapidly exploited. Initially it was proposed that some fish that quickly appear on bait, such as rat-tails, would demonstrate a 'sit and wait' strategy, as this would be more energy efficient than constant foraging. However, when radio transmitters were placed in bait (Priede et al. 1991) it was found that these fish are active foragers and the food fall is dispersed over great distances as faecal deposits.

Large food falls of plant material also provide an important carbon supply to the deep-sea floor. This can be from the shallow water marine environment, in the form of algae and seagrass, or plants that have been washed from terrestrial regions, in particular woody debris. Fruit such as mangos have even been recovered from the deep-sea floor. Terrestrially derived material is often comparatively unimportant in coastal marine systems (e.g. estuaries, see Chapter 4), despite its large biomass, due primarily to the lack of marine organisms able to process the cellulose and lignin present in terrestrial matter. In some deep-sea areas wood can be commonly encountered, such as in trenches and basins in the Caribbean, and seems to be a key source of carbon for some species.

- Large inputs of terrestrial material to the deep sea can be seasonal, even in tropical areas, associated with yearly events such as monsoons or hurricanes.

In particular, wood is processed by boring bivalves (Xylophagainae), which convert it to faecal pellets, making it available to other organisms. It appears that such species have symbiotic bacteria in their gills that can digest cellulose (Distel & Roberts 1997), although they lack the dense populations of gut bacteria found in termites, for example. It is an indication of the importance of woody debris in the deep sea that this feeding strategy is apparent. Experiments set up by submersibles using wood structures for settlement, etc., have been ruined by deep-sea animals eating the wood, while wood has also been suggested as an important stepping stone for the dispersal of vent organisms (Distel et al. 2000).

Both terrestrial plant material, and floating seaweed such as *Sargassum*, can provide a seasonal input to the deep sea associated with storm seasons such as the monsoons. The discovery of such seasonal inputs has major consequences for deep-sea animals and has altered our perception of the deep sea as a never-changing, seasonless system (8.3.5). This is much more apparent when the main input of food material to the deep sea is considered.

8.3.4 Particulate organic matter (POM) from the euphotic zone

The importance, and temporal patterns, of comparatively amorphous POM falling to the seabed from the ocean surface layers was quantified in the 1970s and 1980s using a combination of sediment traps and long-term photographic records (Billett et al. 1983; Lampitt 1985). The results of these studies provided one of the most important discoveries in marine biology that has changed the way we view the deep sea. Some dead, or dying, material from the plankton falls through the euphotic zone without being recycled (most organic matter is degraded in the water column) and sinks to the deep sea. This material can be categorized into three main groups: faecal pellets (primarily from copepods and includes partly digested matter and bacteria), moults (the hard parts of the plankton, including dead intact small organisms), and in particular amorphous aggregates. This latter category has been termed 'marine snow' as it is flocculent organic material that falls through the water column to the seabed and is from a range of sources, in particular gelatinous animals (such as salps), bacteria and the remains of diatoms. The aggregates are also comparatively rich in protein, trace metals, carbohydrates and lipids. An important feature of marine snow is that as it falls it 'scavenges' other particles, and so the size of each aggregate increases with depth. This has the effect of speeding up the sinking rate, so that the organic matter reaches the sea floor much more quickly than individual particles and can avoid decomposition in the water column. This input of POM is the prime supply of carbon to the deep-sea benthic ecosystem.

● The 'marine snow' falling from the euphotic zone is also termed *phytodetritus*.

8.3.5 Timing of food inputs to the deep-sea benthos

Primary production in the euphotic zone of temperate and polar areas is highly seasonal (Chapter 2), with the major temperate bloom occurring in spring. Therefore, the majority of phytodetritus sinking out of the upper layers of the ocean occurs during these blooms. Due to the aggregation of particles, sinking rates for most material is around $100\,m\,d^{-1}$ (rather than perhaps $2\,m\,d^{-1}$ for a large individual phytoplankton), so organic material would take between 1 to 2 months to reach the abyssal sea floor. Using primarily long-term photographic records of one portion of the deep-sea floor, the pattern of phytodetritus arrival has been determined. In the temperate Porcupine Sea Bight, minimal amounts of organic matter can be seen on the sea floor during May and June. However, during late June the floor starts to become covered in dark phytodetrital 'fluff', which builds up rapidly during July until it is covering practically the whole seabed (Lampitt 1985; Fig. 8.5). The inputs of material begin to decrease and the

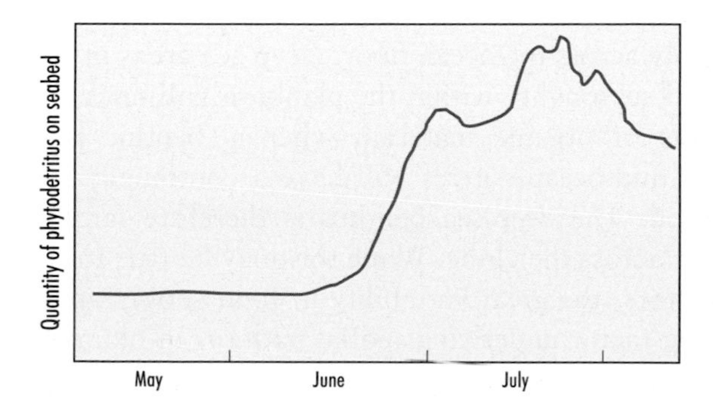

Fig. 8.5 Seasonal deposition of phytodetritus onto the ocean floor (4 km depth) of the Porcupine Sea Bight.

● Sediment traps can also be set up in the deep sea to record settling material. Like Bathysnaps they are deployed on the seabed and a series of 'cups' extend out of the gear for set time periods to collect material.

organisms on the seabed rapidly respond to the food supply, the detritus being quickly removed into the deep-sea food web. By mid-August much of the seabed is again clear. Very little material is not taken up by either microbes or metazoans, and so does not become incorporated into the sediment. Animals feeding on the sediment (e.g. holothurians) therefore have to process vast quantities in order extract sufficient organic material.

The input of food therefore provides a strong, regular and predictable seasonal signal for deep-sea organisms in what is otherwise a seasonless, unchanging environment. This has allowed the biology of deep-sea organisms to be viewed under a different light, with the potential for life strategies and breeding cycles to have developed in order to correspond to, and maximize the use of, this seasonal input of food. Several species appear to have *r*-selected characteristics (thought unlikely in the deep sea) and have populations that directly respond to the input of detritus. These tend to be small, opportunistic organisms such as foraminiferans, nematodes and copepods, foraminiferans such as *Alabaminella* becoming the dominant species after the arrival of the phytodetritus (Smart & Gooday 1997). Work on organisms in the Porcupine Abyssal Plain, and other areas, in recent years has demonstrated that many species, such as isopods, bivalves and echinoderms, demonstrate seasonal maturation of ovaries and egg production. Five species of echinoderm were found to show remarkable reproductive synchrony, all producing small plankton-feeding larvae in January or February, the larvae reaching surface waters in time for the spring phytoplankton bloom. Such reproductive strategies do still remain unusual in the context of the whole deep sea and reflect evolutionary developments in areas where such strong seasonal signals exist. For much of the rest of the ocean floor, megafaunal organisms reproduce throughout most of the year and tend to produce large eggs with a lot of yolk, indicating more direct or abbreviated development.

● ***r*-selected** species have generally fast generations times, short lifespans and mature early. They are commonly found in disturbed or rapidly changing areas and survive in these conditions as successive generations. The alternative (***k*-selected**) are long-lived, but competitive, organisms that rely on stable conditions to survive and reproduce. Such stability is clearly a feature of the deep sea.

A further consequence of the increase in sinking rates of material is that comparatively large amounts of food arrive on the ocean floor at specific times during the year, rather than a continual low input of food. This also means that there is a vast difference in the amount and timing of food supply across the ocean floor. Deep-sea areas in tropical regions with minimal seasonality within the plankton will not experience large, seasonal falls of organic material, whereas benthic regions beneath oligotrophic mid-oceanic gyres will have a continually low amount of available food. The deep-sea benthos is therefore far from being the same habitat across the globe. While this may be true for other physico-chemical factors, the great variability in food supply appears to be the prime forcing factor underlying spatial patterns in biomass, abundance, diversity and life histories.

8.4 The Organisms of the Deep Sea

8.4.1 General patterns of organism change from shallow water to the deep sea

One of the great gradients in the marine realm is the depth gradient from coastal waters, across the shelf and down the continental slope to the abyssal plain. Providing information on how assemblages of organism change along this cline clearly requires samples at all depths. This can be problematical as in many areas the continental slope is very steep and hard, if not impossible, to sample. Therefore, our knowledge of species patterns as we move from the shallows to the deep sea has been built mainly from a few geographical areas where the slope is shallow enough to allow detailed sampling over the whole depth range. One such area is the Porcupine Sea Bight to the south-west of Ireland. Samples from trawls, and other devices, can be pooled to enable general patterns to be explored and some very clear trends have become apparent. These have been supported by samples taken from the other side of the Atlantic.

● The scale and steepness of the continental slope can be dramatic in places. South of the Porcupine Sea Bight, the slope drops 2000 m over a few kilometres – a sloping cliff higher than those of the grand canyon.

8.4.2 Abundance and biomass of assemblages

The number of individual organisms in any given area, and their total biomass, shows a sharp decline as you move from shallow to deep water (Fig. 8.6a,b; Rowe et al. 1982). There are many more organisms per unit area in shallow water (Chapter 7) than there are in the deep sea; the abyssal plain is characterized by having a very sparse fauna. These trends have been related to overall food supply, and its quality. Lowest levels of biomass are found in abyssal regions under waters of low productivity, whereas coastal biomass is boosted by terrestrial run-off and high coastal productivity, which will diminish with increasing distance from land. An exception to this pattern can be deep trenches relatively close to land, which act as traps for coastal sediment and detritus and consequently can often have a standing biomass much higher than would be predicted from the trend in Fig. 8.6b.

8.4.3 Size of individual organisms

The trend in individual organism size shows a less consistent pattern and varies between different sized organisms. Generally, for the macrobenthos there is a decrease in mean size with depth, supported by trends in the meiofauna (Gage & Tyler 1991). This also has been related to decreasing food concentration. However, some larger organisms do not demonstrate such a body size decrease, even quite the opposite as exemplified by Figure 8.6c, which plots individual fish body size against

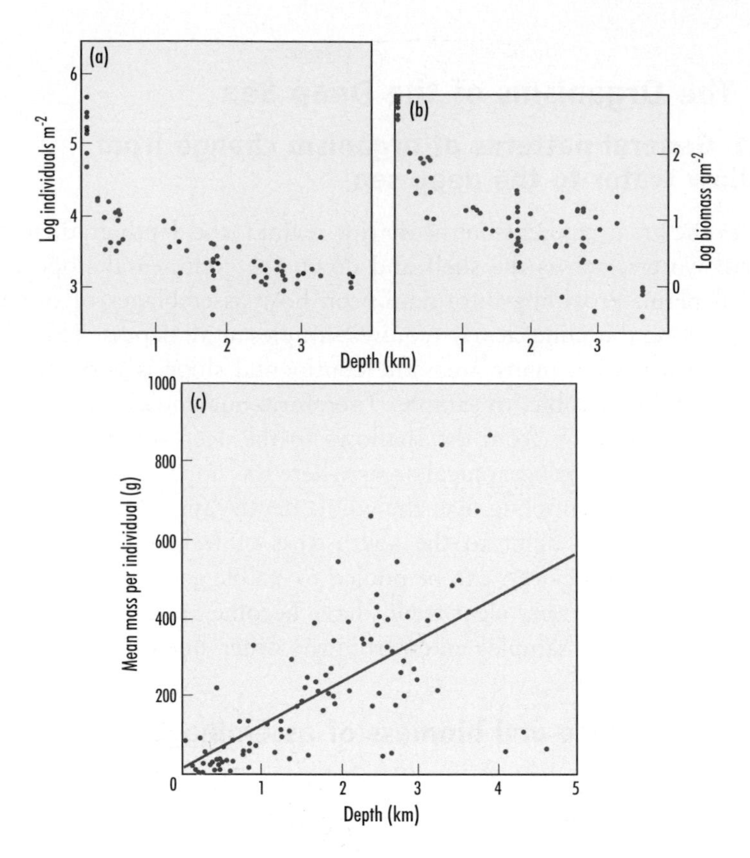

Fig. 8.6 Depth-related trends in (a) abundance of organisms, (b) total organism biomass, and (c) size of individual fish off New England.

● Large deep-sea fish are thought to be very old as their growth rates are slow in the cold water. The orange roughy, for example, found on seamounts can live 120–130 years and does not reach maturity until the age of 23–29. This species is therefore exceptionally vulnerable to over-fishing (Koslow 1997).

● Some early workers such as Forbes concluded that the deep sea was **azoic** = 'no animals', i.e. lifeless.

depth: fish on average get bigger as you move into the deep sea. This has become extreme in some deep-sea groups, which demonstrate **gigantism**: giant organisms are one of the major features of the deep sea (8.4.11).

8.4.4 Diversity trends

Our impression of diversity in the deep sea has changed dramatically over the period of its study. In the nineteenth century, Edward Forbes suggested that an 'azoic zone' existed below 600 m, following a series of dredges in the Aegean Sea. It just happened that this is a particularly sparse region of the deep sea, and with further sampling it was clear that life existed in the depths, but the arguments as to how many species are present in the deep sea still continue. We tend to think of coastal regions as teeming with species, whereas the dark depths of the abyss have comparatively little life. This image perhaps arises from the trend in abundance (Fig. 8.6a), and having fewer individual organisms clearly will affect the number of species present in any given area (Chapter 1 and 14). However, perhaps a more suitable measure of diversity is how many species are encountered when studying a certain number of individuals, which will control for changes in density. Work from the late

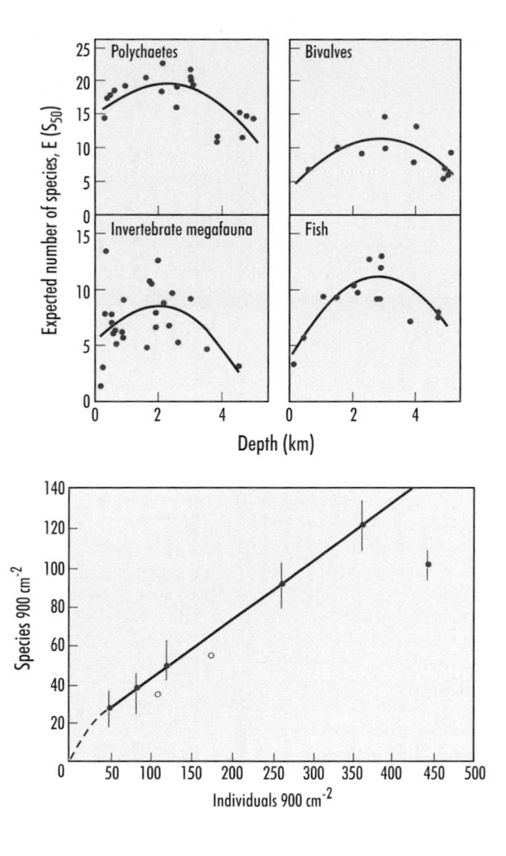

Fig. 8.7 Diversity patterns in the deep sea. (a) An example of the number of species expected for each 50 individuals encountered, indicating a peak in diversity around 2000 m. (b) The relationship between number of individuals in a 900 cm^2 sample and the number of species in the sample, demonstrating a remarkably similar ratio regardless of depth.

1960s and into the 1980s suggested that such patterns of diversity do not necessarily demonstrate a linear relationship with depth (Fig. 8.7a), as perhaps would be expected, but a peak in diversity at around 2000 m depth (Rex 1981). Potentially this relationship may be still affected by the declining density with depth, and clearly there is a need to increase sampling in the deepest areas to clarify this pattern. When quantitative samples from a range of deep-sea sites are collated there is a consistent ratio of number of species and number of individuals in a specified area, regardless of depth (Fig. 8.7b, Grassle 1989). Additionally, abyssal sites can vary considerably in their diversity, even those comparatively close geographically (see 8.2.6). A distinct latitudinal diversity trend (see Chapter 1) is also apparent in the deep sea for several groups (Fig. 8.8), although there is evidence that the southern hemisphere is more diverse than deep-sea areas in the northern hemisphere (Rex et al. 1993). The trend in diversity with depth therefore remains to be confirmed, perhaps due to the comparatively small number of samples we have managed to obtain from the deep sea. However, what is clear is that overall the deep sea is far more diverse than early workers could possibly have imagined (see section 8.4.6).

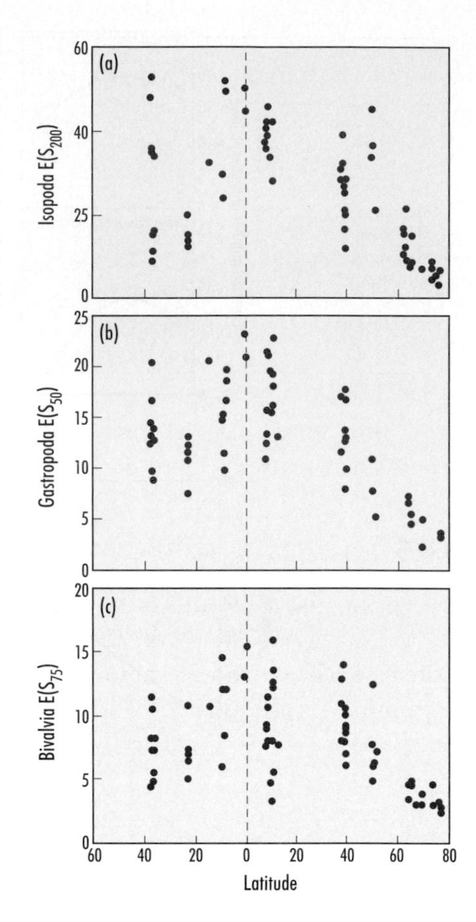

Fig. 8.8 Latitudinal diversity trends in three groups of invertebrate for deep-sea sites suggesting tropical deep-sea areas are more diverse than those at higher latitudes, particularly in the northern hemisphere.

8.4.5 Trends in species composition

Most large environmental gradients in the marine environment demonstrate a distinct sequential distribution of organisms, such as for rocky shores (Chapter 5) and estuaries (Chapter 4). Closely related organisms have specific environmental requirements and so fill a **realized niche** that has minimal overlap with similar potentially competing species. Perhaps surprisingly, given the huge spatial scale, a sequential distribution of similar species is also apparent along the depth gradient from coast to abyss. An example is for squat lobsters (galatheid crustaceans) in the Porcupine Sea Bight (Fig. 8.9), where species from two closely related genera each exist across comparatively narrow depth ranges. Where overlap does occur (e.g. *Munidopsis rostrata* and *M. bermudezi*), the species appear to be functionally separated, for example, each member of the pair has one of two different chela morphologies (Fig. 8.9). In the case of the example pair of species, *M. bermudezi* has spade-shaped chelae, while those of *M. rostrata* are spear-shaped.

● An organism's **realized niche** represents the range of conditions over which the species exists in its habitat. This contrasts with the **fundamental niche**, which is the range of conditions over which the species could survive if other factors, such as competition, were not involved.

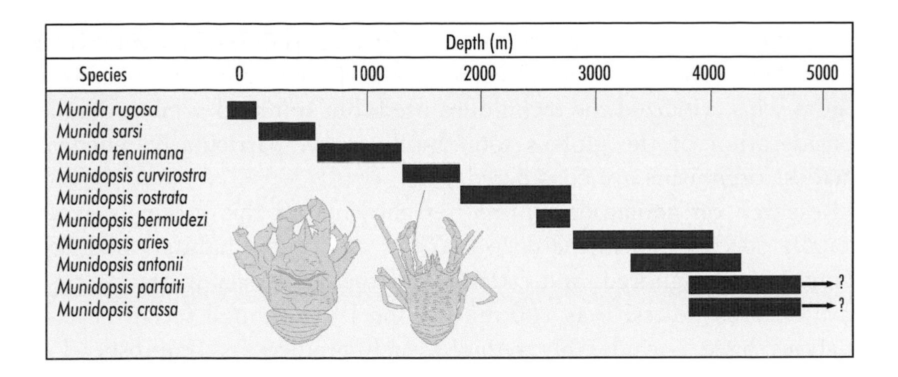

Fig. 8.9 Bathymetric distribution of squat lobster species in the Porcupine Sea Bight, illustrating a clear sequential distribution. Species illustrated are *M. bermudezi* (left) and *M. rostrata* (right).

How this morphological difference relates to feeding, and thus niche separation, is however unclear.

8.4.6 How many species are there in the deep sea?

One of the great debates in ecology is how many species actually exist on the earth. Much of the debate has centred on diverse terrestrial systems such as tropical rainforests (with their huge insect diversity), with inputs to the debate from coastal marine areas such as coral reefs. As the deep sea was generally regarded as species poor, the world's largest habitat had not been considered as being important in terms of global diversity until comparatively recently. The vast majority of work on the deep sea has concentrated on the larger animals, but as more and more samples are processed from the sediment it is becoming clear that the diversity of small organisms in the deep sea is very high indeed, most new mud samples that are thoroughly studied reveal species new to science. However, the problem arises with estimating the number of species in the whole ocean basin from the tiny fraction of the seabed that we have sampled. Attempts to undertake the latter extrapolation have launched the deep sea into the midst of the 'How many species?' debate.

● The diversity of microbes has only recently been approached using new techniques that amplify ribosomal RNA in samples. rRNA is a constitutive molecule in bacteria, and sequences can be determined in order to assess the number of species present, even if the bacteria cannot be cultured.

The World Resources Institute **http://www.wri.org** states that approximately 1.4 million species are known to exist on the planet, most of these are insects, especially beetles (Chapter 1). Estimates in the early 1990s suggested a total of 3 to 4 million species were present on earth, mainly in the terrestrial environment, but attention was drawn to deep-sea diversity following a study by Fred Grassle & Nancy Maciolek of Woods Hole Oceanographic Institute (Grassle & Maciolek 1992). They had taken a large number of replicate samples from along a deep-sea depth contour off New England, USA and had found a remarkably high number of species new to science, in particular polychaetes. Additionally, new species continued to be encountered with additional samples. The pattern of species capture with increasing sample area was assessed and used to estimate further species capture rates across the deep-sea basin, with a conservative conclusion that there were probably at least

10 million species in the deep sea. This caused much controversy, particularly among terrestrial ecologists and some coastal marine ecologists who criticized the techniques used, but initiated a more detailed consideration of the globe's total biodiversity, particularly when the smallest organisms are considered.

Research on nematode worms has shown that this group is exceptionally species rich in the deep sea, but as in most habitats had been comparatively ignored, and estimates of their diversity in the deep sea have been as excessive as 100 million species, although recent detailed analyses have considerably reduced such predictions (Lambshead & Boucher 2003). Recently, it has also become evident that there is a vast diversity of bacteria and viruses in the world's oceans (e.g. Beja et al. 2002), a large undefined number of parasite species, plus perhaps a much-underestimated insect diversity in the terrestrial biome. Biodiversity estimates are therefore constantly under revision. The only real certainty is that we do not know how many species there are, and estimates now range from 10 million to over 111 million species in total. It is unlikely we will ever be much more accurate, particularly when the majority of the earth surface remains unsampled.

8.4.7 The main groups of organism on the deep-sea bed

We are very familiar with the main groups of larger marine organism dominant in shallow waters, but comparatively few taxonomic groups of animal have proliferated on the deep-ocean floor and are most regularly caught in samples. Some taxa that are comparatively innocuous or species poor in coastal regions are dominant in the deep sea. The assemblage composition of the abyssal plain, even at the level of phylum, will therefore show marked differences from shallow water.

● The term meiofauna is generally used to represent those organisms living in the interstitial spaces between sediment particles, the size division generally representing the sieve method used. Macrofauna, in contrast, move through the sediment by pushing the grains aside, rather than travelling between them.

From the comparatively few samples fully analysed, the taxonomic composition of smaller organisms in the mud (termed macrofauna (>0.5 mm) and meiofauna (<0.5 mm)) are similar to equivalent sediments in shallow waters. The macrofauna are dominated by polychaetes, bivalve molluscs, and pericarid crustaceans (e.g. amphipods, cumaceans), while the main meiofauna constituents are nematodes, copepods, and Foraminifera. It is the larger megafauna that show the differences, organisms that are big enough to show up in photographs and to be caught in trawls. These can be divided into three groups.

8.4.8 Sessile megafauna

● Sessile organisms settle as larvae and then can not move after settlement, e.g. barnacles, sponges, most bryozoans.

Sessile organisms are generally uncommon in soft sediment environments due to the lack of suitable hard substratum for attachment. However, two comparatively unfamiliar groups of cnidarian that are found in coastal areas are much more successful in the deep sea. Both are members of the Anthozoa (a group containing corals and anemones), the sea fans (gorgonians), and the sea pens (pennatulids). These organisms are anchored to the sediment by means of a stalk, with the feeding and reproductive zooids held well above the fine mud. Both groups are often found in comparatively large beds and seem to be characteristic of relatively high-energy deep-sea regions. A further group of deep-sea anthozoans are the extensive hard coral reefs, whose size and importance has only recently been appreciated with the use of video technology.

Deep-water corals occur worldwide, often on seamounts; extensive reefs are present between 800–1300 m off west Ireland, for example. One *Lophelia pertusa* reef on the Sula Ridge off Norway was found to be over 14 km in length and rose 30 m above the bed in some places. The reefs are also extremely old, recent pieces of skeleton dated using ^{14}C were found to be at least 4550 years old! Due to their complexity compared with the muddy bottom, deep-sea coral reefs also have a diverse associated community, including fish, and thus have been popular areas for fishing. Only recently has it become apparent that trawling has done extensive damage to many of these ancient deep-sea reefs (Hall-Spencer et al. 2002). The only other group of sessile organisms that are commonly encountered in the deep sea are the sponges (Porifera). For example, beds of soccer-ball-sized glass sponges (*Pheronema*) can be found below 1000 m in the Porcupine Sea Bight and their main opening (**osculum**) is regularly used as ready-made, sediment-free burrows by animals such as *Munida* squat lobsters.

● The rare ^{14}C isotope of carbon is radioactive and thus decays. The rate of decay is known, so ^{14}C levels are used to determine how old a sample is – known as carbon dating.

8.4.9 Sedentary megafauna

The group of deep-sea organisms that we probably know least about are the large animals living below the mud surface. While our grabs and dredges can sample the smaller and shallow-living fauna, the only real information we have on anything larger or deeper is from photographs, either of part of the organism or more tantalizingly marks left on the mud surface (termed **Lebensspuren** – German for 'life-traces'). Another group of organisms that is not commonly encountered in shallow water, but appears to be a major constituent of the deep-sea infauna is the Echiura (spoon worms). These live permanently in burrows, but extend a feeding proboscis across the mud surface to collect depositing detritus. Occasionally the proboscis has been caught in photos and the extended part has been measured at over 50 cm long, suggesting an animal size of at least a metre. Echiuran worms also leave a distinctive, rosette-like feeding trace and examples of these have been photographed clearly made by organisms bigger than 1 m. It is likely that many unusual

● Sedentary – organisms that tend to stay in one place, but could move if required. Examples include sea anemones, large burrowing worms such as *Arenicola*, and students.

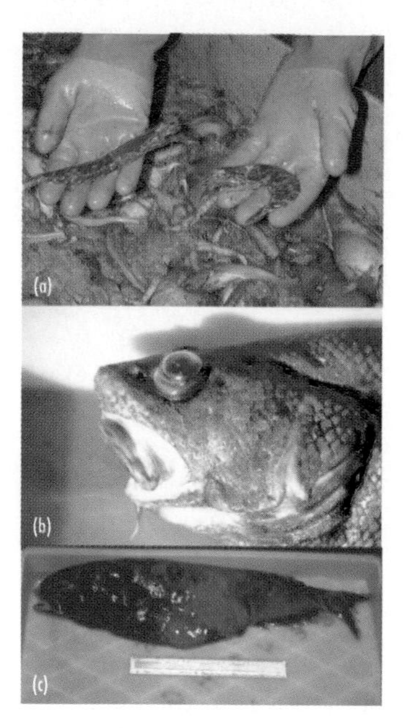

Fig. 8.10 (a) Examples of deep-sea prawns, highlighting their use of red coloration for 'camouflage'. (b) Head of the rat-tail *Nematomurus*. (c) The 'jellyfish' *Conocavara*.

● Deposit feeding sea cucumbers have to process vast amounts of sediment to obtain sufficient food. A single individual can pass 100 g of sediment through the gut every day.

and fascinating large animals live below the deep-sea sediment surface, but unless our technology develops dramatically they will remain unknown.

8.4.10 Mobile megafauna

The group of deep-sea organisms we know most about are the large animals that move about on, or just above, the mud surface and are caught in both photographs and trawls. Many taxonomic groups are represented in the mobile megafauna, but three are particularly dominant and characteristic of the deep-sea fauna: echinoderms, decapod crustaceans and fish.

Echinoderms

The deep sea could reasonably be termed the kingdom of the echinoderm as this phylum is far more prevalent than in most shallow regions. All classes of echinoderm are well represented in the deep-sea benthic fauna, but two groups are particularly widespread and diverse: the brittlestars (Ophiuroidea) and sea cucumbers (Holothuroidea). The holothuroids in particular are remarkably diverse and common in the deep sea, and show a fantastic range of morphological developments to life in the deep ocean, from swimming ability due to the development of 'fins' (e.g. *Enypniastes*) to highly modified oral tentacles (e.g. *Peniagone*) that are used as hand-like structures to collect sediment particles. In some food-rich parts of the deep sea, sea cucumbers can also be abundant forming '**herds**' that move together across the seabed. Examples include *Kolga hyalina* and *Scotoplanes globosa*, whose suction-like feeding method has been termed 'vacuum cleaning' as they suck up sediment and its inhabitants. Such organisms, together with burrowers, can have a large **bioturbating** effect and are highly important in structuring and altering the sediment, and in potentially influencing small-scale diversity patterns (Widdicombe et al. 2000).

Decapod crustaceans

Crabs, prawns and lobsters are a familiar component of practically all coastal habitats, but like echinoderms the most common large crustaceans on the deep-sea floor belong to a less familiar group. Below the edge of the continental shelf, the dominant group of walking (**reptant**) crustaceans are the squat lobsters – anomuran crustaceans related to hermit crabs (Fig. 8.9). Crabs are comparatively rare in the deep sea and seem to be mainly a shallow water group, with a couple of exceptions. The red crabs (*Geryon* and *Chaceon*) can extend down to below 1000 m in many parts of the world and are often the object of commercial fisheries. In the Porcupine Sea Bight, the red crab species is *Geryon trispinosus* (Fig. 8.2a), which appears to undertake an **ontogenetic** depth-related migration as it passes through its life stages. Young crabs are present only in deeper water, while the largest breeding adults are in depths at the top limit of the deep sea (2–300 m). It would appear that the larvae settle out at depth and through their life migrate up the continental slope to shallow-water breeding grounds (Attrill et al. 1990). It is possible that the absence of small crabs in the shallow areas is due to competition from the comparatively high densities of *Munida sarsi* squat lobsters present at these depths (Fig. 8.9). Below 1000 to 2000 m there are very few crab species to be found, with the amazing exception of a couple of species (e.g. *Cyanograea*) that exist on the ocean floor around hydrothermal vents (8.5.3).

● **Ontogenetic** – a factor that changes as an organism progresses through different developmental stages. Examples are ontogenetic shifts in diet, and changes in the shape or size of certain body parts, such as crab chelae. The noun is *ontogeny*.

The other main group of deep-sea decapod crustaceans are the prawns, which tend to swim above the seabed (**natant**). Many of these are large compared with shallow-water species, and are often a bright red colour (Fig. 8.10a). This would be an unusual anti-predator mechanism in coastal regions, but is a common strategy in the deep sea and related to the relative attenuation of light. Red light does not penetrate far into the water column, and so in the gloomy (**dysphotic**) regions of the deep sea where some light still penetrates, the red wavelength is absent. Therefore, red organisms will be 'invisible' as there is no red light to reflect off them. A fascinating predator–prey arms race has evolved in the deep sea, with some species of deep-sea fish possessing light organs that produce red light – a torch to reveal red prey in the dark! (Box 8.1)

● **Natant** organisms mainly swim, whereas **reptant** organisms mainly crawl on the seabed.

Fish

Perhaps the most evocative images of the deep sea are the weird, alien-like fish with huge teeth, bulging eyes, or enormous mouths, which have been trawled up from the deep and feature in many a documentary series. However, most of these species inhabit the featureless open-water habitat and few groups of deep-sea fish are actually adapted to a true benthic lifestyle. These are mainly rays, hagfish, some flatfish and angler-fish, a group that has diversified in the deep sea and generally employ some form of bioluminescence as a lure. Most fish found on the abyssal floor are known as **benthopelagic**, swimming just above the seabed but not resting on it. A diverse group of fish comprise this assemblage, but there appears to be a common, elongated body form regardless of taxa that is the ideal shape for deep-sea life. This is exemplified by the rat-tails (Macrouridae, Fig. 8.10b), with long dorsal and anal fins supported by many fin rays and practically no tail fin. In cold water with comparatively low energy inputs, fast movement may not be practical, so it is likely that most benthopelagic fish move slowly and continually across the bottom.

● There are several terms for fish (and other organisms) living just above the seabed, such as demersal, benthopelagic, hyperbenthic, and even proximo-benthic.

Rat-tails, such as the roundnose grenadier (*Coryphaenoides rupestris*) are a successful, diverse and comparatively abundant taxonomic group within the benthopelagic fishes and have supported commercial fisheries, particularly by Russian fleets (Randall & Farrell 1997). A major problem for benthopelagic fish is the ability to remain above the sediment. Hydrodynamic lift, as used by many coastal fishes, requires relatively fast movement and thus a large amount of energy, so is not a suitable strategy in the deep sea. To achieve neutral buoyancy, deep-sea fish either use buoyancy devices or reduce tissue density. As with shallow fish species, many still utilize swim bladders to achieve buoyancy, despite the problems of gas compression under the immense hydrostatic pressure – fish with gas-filled swim bladders have been found to a maximum depth of 5000 to 7000 m where the energetic costs of maintenance probably become too great.

● Rat-tails are members of the same major taxonomic group as cod and haddock (the Gadiformes).

An alternative buoyancy option comparatively unaffected by pressure is to use lipids instead of air in a swim bladder. Lipid accumulation in organs such as the liver is used for buoyancy by shark species, but some fish such as the orange roughy (*Hoplostethus atlanticus*) and the coelacanth (*Latimeria chalumnae*) have wax esters making up more than 90% of the swim bladder fat to boost buoyancy (Randall & Farrell 1997), while others (e.g. *Antimora rostrata*) have fat-filled swim bladders, mainly cholesterol and phospholipids. An alternative buoyancy method is to reduce body density (particularly skeletal and muscle density) by converting tissues to lighter substances such as water, gelatinous material or cartilage. Such fish with watery or gelatinous tissues (e.g. *Acanthonus*, *Conocovara*, Fig. 8.10c) are only slightly denser than water and true jellyfish!

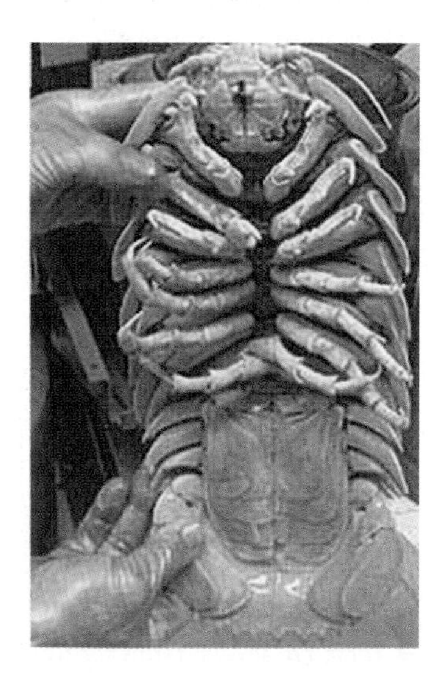

Fig. 8.11 The giant deep-sea isopod *Bathynomus giganteus*. Photograph: National Museum of Natural History. Copyright Smithsonian Institution.

8.4.11 Gigantism in the deep sea

A feature of the deep sea is that some groups that are classified as macrofauna in coastal waters have deep-sea species that are so large they can be caught in trawls and considered megafauna (Gage & Tyler 1991). Giant amphipod and isopod crustaceans are features of the deep-sea benthos (and also Antarctica, Chapter 11) and among the major scavenging organisms. Examples include the massive amphipod *Eurythenes gryllus*, which can grow to 140 mm, and the giant swimming isopod *Bathynomus giganteus* (Fig. 8.11), which can grow to 0.5 m in length and is commercially fished! Sea spiders (pycnogonids) also have giant representatives in the deep sea: *Colossendeis colossea*, which has a leg-span up to 0.5 m, its long legs acting as stilts that help to keep the body above the sediment. However, perhaps the most remarkable example of gigantism occurs in a group known as the **xenophyophores** (Levin 1994), originally thought to be related to sponges but now known to be giant single-celled protozoans (their own phylum within the Protoctista) that are so large they appear in photographs of the seabed and are bigger than most macrofaunal metazoans. Their form is varied, from spherical to disc-shaped and can have a branching network below the mud surface. The largest, the leaf-shaped *Stannophyllum*, is 0.25 m in diameter and 1 mm thick! In some areas of the deep sea, such as the south Pacific, xenophyophores are the dominant taxon and can comprise 97% of the biomass.

Why do giant taxa exist in the deep sea? Several theories have been produced, including links to the metabolic effects of great hydrostatic pressure, delayed onset of sexual maturity and indeterminate growth – organisms live a very long time and continue growing. Other ideas relate to the method of food capture, and perhaps have a clearer evolutionary route. For example, giant scavenging amphipods could be an adaptation to a foraging strategy where high motility is needed to locate sparely dispersed food. However, we are still not clear how, or why, animals often living in extremely oligotrophic areas become so large.

● The best-known example of a deep-sea giant is the giant squid *Architeuthis*. The maximum size of this species remains unclear, the largest caught specimen (1878) measuring nearly 17 m long. Sailors' observations and sucker scars on sperm whales suggest squid lengths >50 m, but it has also been pointed out that such scars may increase in size as the whale grows. However, there appears to be an even bigger species, the 'colossal' squid *Mesonychoteuthis hamiltoni*, a specimen of which was caught in 2003 from 2000 m off New Zealand (**http://www.tonmo.com/science/public/giantsquidfacts.php**).

8.5 Hydrothermal Vents – Islands in the Deep Sea

Thirty years ago, practically all deep-sea ecology was focused on the comparatively featureless deep-sea bed and our impressions of deep-sea life were therefore determined by this image. However, discoveries over the last couple of decades due to the use of submersibles and more intensive acoustic mapping, have demonstrated that the ocean basin is not a continuous, featureless floor but within these plains are a series of other features that have exciting, unique and often diverse communities: islands in a sea of mud. Most are based on unusual outcrops of hard substrate that have been formed by different means, and include sea-mounts (De Forges et al. 2000) and deep-sea coral reefs (Hall-Spencer et al. 2002). However, the most dramatic and unusual island communities are associated with hydrothermal vents.

● The first vent was observed in 1977 on the Galapagos Rift (Pacific) from the submersible *Alvin* by a team including Bob Ballard – the man behind the rediscovery of the *Titanic*. The system was found by 'chance' when investigating temperature anomalies in surface waters.

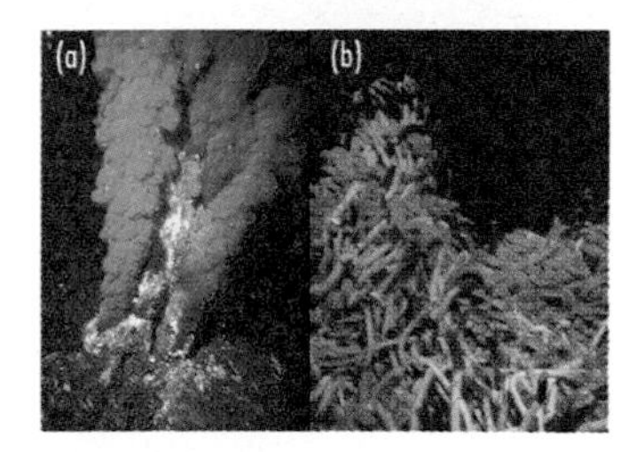

Fig. 8.12 (a) A black smoker. (b) A dense aggregation of vetimentiferan tubeworms (*Riftia*) near a hydrothermal vent. (Copyright: NOAA)

8.5.1 What and where are hydrothermal vents?

Hydrothermal vents are associated with parts of the ocean floor that exhibit high levels of tectonic activity, such as the spreading axes of plate formation (mid-ocean ridges), and vent clusters have been given evocative names such as 'Rose Garden' and 'Snake Pit'. Within such regions, hot magma chambers occur near the seabed and heat up water that has permeated into the ocean floor. This superheated water erupts back out of the seabed carrying with it a rich cloud of minerals such as sulphides, methane, manganese and many other trace metals. The vent therefore often appears as smoke emerging from a chimney formed by mineral deposits (Fig. 8.12a). The hottest, and most dramatic, type of vent are therefore termed 'black smokers' due to the colour of the plume, and water temperatures within the plume can be up to 350 °C. 'White smokers' also exist, which tend to be a little cooler, and in some areas hot water escapes through cracks and crevices in the seabed to form diffuse vents with water temperatures much lower than the smokers. Hot emissions from smokers are rapidly mixed in the cold deep sea so most vent animals generally live in water temperatures close to the ambient of 2 °C (the plume water also contains no oxygen), though there are astonishing exceptions such as the Pompeii worm (*Alvinella pompejana*, Fig. 8.13), which forms burrows on the vent chimneys (Cary et al. 1998). Water temperatures within the worm burrows measured by *Alvin* were found to average 68 °C, with frequent spikes up to 81 °C. Pompeii worms emerge from burrows to feed on filamentous bacteria and have been known to survive short exposure to 105 °C. *Alvinella* is the most thermotolerant eukaryotic organism known.

● Water temperatures above 100 °C are possible due to the high hydrostatic pressure that prevents water boiling.

● Until the discovery of hydrothermal vents, chemosynthesis was thought to play no significant role in a photosynthesis-dominated world.

8.5.2 Production at hydrothermal vents

Hydrothermal vents are exceptional among deep-sea islands, and vastly different from the rest of the deep sea, in having a huge biomass of associated organisms (Fig. 8.12b). Clearly, alternative methods of food supply are sustaining these communities and this production is autochthonous and related to the supply of reduced compounds (particularly sulphur compounds) emerging from the vent. Vent assemblages are sustained by primary production generated by bacteria through **chemosynthesis** (Box 8.2) and considered of little importance until recently, when vents were investigated.

Fig. 8.13 The Pompeii worm (*Alvinella popejana*) – the most thermotolerant multicellular organism on earth? (Copyright: Macmillan.)

Chemoautotrophic bacteria are present in hydrothermal fluid and tend to be members of the most ancient Archaea. These can tolerate exceptionally high temperatures and are therefore either known as **hyperthermophiles** (80–115 °C) or the most extreme **superthermophiles** (>115 °C). Bacteria are also free living in the vent environment, such as the filamentous bacteria fed upon by *Alvinella*, and thus provide a continually replaced food supply for vent animals. However, the comparatively large biomass associated with vents is not primarily explained by production of free-living bacteria, but through chemosynthetic, symbiotic bacteria that live within the vent invertebrates. A group of animals unique to vent environments has therefore evolved that relies upon a symbiotic relationship with bacteria.

8.5.3 Classic vent animals

Probably the most striking examples of this dependence on symbiotic organisms are the large **vestimentiferan** tubeworms characteristic of vent communities in the Pacific (Fig. 8.12b). The 1 to 2 m long worms, such as *Riftia pachyptila*, live in white tubes attached to the vent surrounds (15 to 20 °C) with a red tentacular plume extending from the tube, which can be retracted quickly when disturbed. While superficially the worms look like tube-dwelling polychaetes feeding on particulate material, they have two strange anatomical features: they completely lack a mouth and digestive system and they have a specialized organ (**trophosome**) that houses chemosynthetic bacteria deep within the animal's body; bacteria compose up 50% of the weight of the worm. *Riftia* relies on bacteria for its organic carbon supply, the plume of tentacles is used to uptake other nutrients from the surrounding environment.

● What are vestimentiferans? Previously they were classified with another unusual worm group, the Pogonophora, but have now been given a phylum of their own (Vestimentifera).

Other key components of the vent fauna are large, white bivalve molluscs, which form extensive clusters in cracks and crevices around smokers and other vents. These are typified by two genera, the giant clam (*Calyptogena*), which can grow up to 26 cm, and the vent mussel (*Bathymodiolus*). These molluscs also house symbiotic bacteria (in cell vacuoles rather than a specific organ), which supply the majority of the bivalve's organic carbon needs. *Calyptogena* takes in ambient CO_2-rich water over the gills through siphons, while its foot extends into the crevice where it lives in order to exploit warm sulphide-rich vent fluids.

The other major group present on vents are decapod crustaceans, which scavenge off other vent animals rather than having endosymbiotic bacteria. Squat lobsters are common, but Pacific vents also have associated crab species (e.g. *Bythograea thermydron*) that are key members of the vent food web but comparatively uncommon anywhere else in the deep sea due, perhaps, to the planktonic larval strategy generally adopted by the group. How this species has colonized and adapted to life on vents is therefore an interesting question, but it would appear that larval stages of the crab remain in the vicinity of the vent. Like shallow-water crabs, large numbers of eggs are produced, but first-stage zoea of *Bythograea* have been captured in plankton tows of bottom water over vents rather than at the surface thousands of metres above. However, the most remarkable feature is the megalopa (settling larval stage), which is enormous compared with shallow water crabs (5–10 cm carapace length) and common in both water overlying vents and within clumps of *Riftia* where they appear to take refuge.

Vestimentiferans are classic features of Pacific vents, but in the mid-Atlantic (whose vent systems have a very different fauna to the Pacific), the dominant animal is often a species of shrimp (e.g. *Rimicaris exoculata*), which can form huge swarms around hydrothermal vents.

● Atlantic vent systems have a very different, and generally less diverse, vent community to vents in the Pacific. This suggests the two basins have been physically isolated over much of the history of vent development.

● Huge swarms of copepods can also be found in the water directly above hydrothermal vents.

These shrimps are characterized by large, paired dorsal eyes that can extend halfway down the midline of the shrimp; the shrimps do not possess normal eyes and eyestalks. It has been proposed that *Rimicaris* can detect thermal radiation emitted from hot vent fluids, thus enabling the shrimp to navigate around the vent without being cooked.

While the tubeworms, bivalves and decapod crustaceans are the most prominent groups associated with vents, most other taxa are also represented, particularly sessile and sedentary animals such as sponges and anemones. Few fish species are found only in vent habitats, an exception being the well-named *Thermarces*, which has aspects of its biochemistry adapted for high temperatures as well as high pressure. The fish appears to feed on the abundant swarms of amphipod crustaceans (*Halice*) that can reach densities of $>1000\ l^{-1}$ above lower-temperature vents!

The exact composition of vent assemblages and the relative position of species at a vent seem closely related to the chemical speciation within the vent water (Luther et al. 2001). Significant differences in oxygen, iron and sulphur speciation are correlated with the distribution of taxa in different microhabitats.

8.5.4 Other types of vent and seep

● A large deep-sea oil 'lake' exists on the sea floor in the northern Gulf of Mexico. The lake is ringed by deep-sea bivalves, which can cluster so closely that individuals get pushed into the lake – with fatal consequences.

Not all chemosynthesis-based communities in the deep sea rely on hot water vents and distinct assemblages have been discovered associated with other conditions where a suitable concentrated source of nutrients is available. Most of these are termed **cold seeps**, for example where oil leaks onto the sea floor from underground reservoirs. These communities are often dominated by the bivalves that have symbiotic relationships with chemosynthetic bacteria. The deepest chemosynthesis-based community so far discovered is at 7326 m in the Japan trench (Pacific) and is dominated by the bivalve *Maorithyas hadalis* (Fujikura et al. 1999). At this site, as in other trenches, the assemblage is sustained by chemosynthesis using the high sulphide content of the sediment associated with the geologic fault.

8.5.5 Dispersal and gene flow in vent organisms

Most vent organisms are uniquely associated with vents or seeps, and a feature of vent fauna is the number of endemics, many with ancient lineages. The persistence of ancient taxa has led to suggestions that vents have been comparatively immune to major planetary extinction events, as they are only indirectly dependent on the sun. However, hydrothermal vents are comparatively temporary in nature as the earth's crust will move over the magma chamber and the vent will 'die'. Observation of the Rose Garden vents near the Galapagos Islands since discovery in 1979 has suggested some degree of succession as the vent matures and its chemical nature changes (Hessler et al. 1988; Van Dover 2000). *Riftia* seems associated with comparatively new vents, but declines in relation to decreases in sulphide concentration. The bivalves, however, persist and even expand their populations as vestimentiferans disappear, suggesting possible competition between mussels and tubeworms. New vents will clearly emerge through the same geological process, but one of the most interesting aspects of vent ecology is how organisms disperse between vents and how new vents become colonized. The majority of vent organisms appear to have larvae with abbreviated development (**lecithotrophic**), so will be colonizing nearby vents. Surveys in the Pacific have suggested vents are generally <10 km apart but there is up to 100 km between vent fields, presenting dispersal problems between these strings of islands for short-lived larvae of endemic vent species (Van Dover 2000). If dispersal, and thus gene flow, is limited then there would be extensive genetic differentiation between vent fields. Two different models of gene flow are apparent in vent taxa (Vrijenhoek et al. 1998; Van Dover 2000). *Riftia* fits a 'stepping stone' model with most gene exchange occurring between neighbouring populations and gene flow declining with distance (Fig. 8.14a), whereas patterns of gene flow are different for *Bathymodiolus* (Fig. 8.14b) where migration rate appears unrelated to distance. This is termed the 'island' model suggesting long-distance dispersal and mixing of larvae within a 'migrant pool'. The vent mussel appears to be unusual in this respect, though

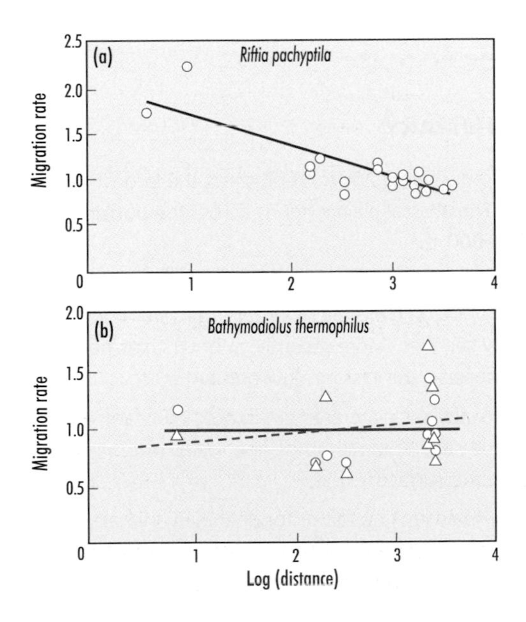

Fig. 8.14 Gene flow models in vent organisms. (a) *Riftia*, the isolation-by-distance model, with migration rate decreasing with distance. (b) *Bathymodiolus*, the island model where migration rate is constant, regardless of distance.

its apparent wide dispersion may be also influenced by its ability to colonize a range of seep and vent types.

Despite the general lack of dispersal ability, vent organisms appear to successfully colonize the majority of vents. The puzzle of how poorly dispersing larvae can travel large distances has focused on the supply of potential stepping stones between vents that provide the suitable nutrient supply for chemosynthesis to occur. Whale skeletons provide such a resource (H_2S is produced by bacterial decomposition of lipids), so larvae could disperse over several generations (Smith & Baco 2003). Conversely, it has also been suggested that some groups of organisms may have originally invaded vents from biological material such as whale bone and wood, analysis of such material from the deep sea revealing small mussel species (e.g. *Idas washingtonia*) in the same subfamily as *Bathymodiolus*, which appear to have preceded vent specialization within this lineage (Distel et al. 2000). These small bivalves possess chemosynthetic endosymbionts, utilizing sulphide produced by wood decomposition, so it is feasible that wood (and bone) were vectors that originally transported mussels to vents.

● It has been calculated that most deep-sea vent larvae could reach successive whale skeletons. If the dispersal stage lasts 30 days, a direct movement of 300 m d^{-1} would disperse larvae to the next skeleton site, assuming the average distance between skeletons of 9 km. Most bottom currents are faster than this.

While vent taxa appear to disperse effectively, through a variety of strategies, it would appear that some deep-sea islands remain uncolonized. Recently, a new vent field was discovered in the mid-Atlantic (named 'Lost City'), which had steep-sided black smokers and dense microbial communities, yet none of the large invertebrates associated with other Atlantic vent systems (Kelley et al. 2001). This new field lies 15 km off the main axis of the mid-Atlantic ridge and clearly is in a position where the main dispersive currents have not, as yet, brought settling larvae of vent organisms.

CHAPTER SUMMARY

- The deep ocean floor (below 200 m) represents the largest habitat on the planet, but the least known. The abyssal plain extends across the bottom of much of the ocean at depths of 4000–5000 m.
- Sampling the deep-sea floor is exceptionally difficult and is an expensive undertaking. As a result, much of our knowledge of larger organisms comes from photography and comparatively few samples. More recently, submersibles have revolutionised how we can sample and observe this distant environment.
- Conditions on the abyssal plain are remarkably constant, generally being dark and cold (2 °C) with the seabed made up of fine, deep mud. Hydrostatic pressure is over 500 times that at the surface.
- Food supply is the main limiting factor for deep-sea animals, as the food web relies on the fall of organic material from the surface layers, mainly as 'marine snow'. This fall is seasonal in temperate regions, giving regular annual cues to benthic organisms.

- Only 1–3% of surface production reaches the ocean floor, so animals on the abyssal plain are in low abundance and total biomass. However, individual animals can grow very large, with giant members of groups such as sea spiders, amphipods and xenophyophores (giant single-celled protozoans) a feature of the deep sea.
- There is much controversy over how many species live in the deep-sea, with estimates as excessive as 100 million! We will never know the actual number, but it is possible that the majority of species on earth live in the deep sea.
- Many deep-sea organisms produce light, known as bioluminescence. This can be for a range of functions, such as attracting prey, deterring predators, spotting prey in the dark and advertising for mates.
- In the late 1970s and early 1980s the use of submersibles assisted the discovery of an entirely new ecosystem in the deep-sea – hydrothermal vents. The sulphur in these vent systems is used by bacteria to generate production through chemosynthesis – primary production without the need for sunlight. Many vent animals, such as giant tube worms known as vestimentiferans, have vast numbers of these bacteria living symbiotically in their bodies and providing them with a food supply.
- How vent animals disperse continues to puzzle marine ecologists. Some species may use stepping stones on the ocean floor to reach new vents. Whale carcasses could be very important for vent animal dispersal as their bones contain sulphur.

FURTHER READING

For excellent, detailed accounts of deep-sea biology, see Gage and Tyler (1991) and Herring (2002). Van Dover 2000 provides a detailed account of hydrothermal vent ecology.

- Gage, J. D., Tyler, P. A. 1991. *Deep-Sea Biology: a Natural History of Organisms at the Deep-Sea Floor*. Cambridge University Press, Cambridge.
- Herring, P. 2002. *The Biology of the Deep Ocean*. Oxford University Press, Oxford.
- Van Dover, C. L. 2000. *The Ecology of Deep-Sea Hydrothermal Vents*. Princeton University Press, Princeton, NJ.

Chapter 9

Mangrove Forests and Seagrass Meadows

CHAPTER SUMMARY

Mangrove forests and seagrass meadows represent two of the most valuable marine habitats in the world, rivalled only by coral reefs for their importance in providing a high level of productivity and physical structure that supports a considerable biodiversity of associated animals. However, both systems are remarkable in that they are based on higher plants more generally associated with terrestrial systems, so mangrove trees and seagrasses demonstrate fascinating physical, biological, and life-cycle adaptations that enable survival in the marine realm. Mangrove forests host a unique mix of marine and terrestrial animals, including insects, birds, fish, and particularly crabs, whose activities heavily influence the productivity of mangrove forests. Species such as mudskippers demonstrate incredible levels of adaptation to life in this intertidal habitat. Seagrass beds provide a structural habitat on generally featureless soft-sediment bottoms, and so are utilized by a wide and diverse range of fish and invertebrates, providing a physical home, food supply and shelter from predation. Seagrass meadows are also vital grazing areas for large vertebrates, such as turtles and sea cows. Both mangroves and seagrasses provide a range of functions that influence the wider coastal ecosystem and have value for humans, so their fragmentation and loss has consequences far beyond the organisms that live associated with these plants.

9.1 Introduction

A fundamental, and obvious, difference between the ecology of terrestrial and marine systems is the taxonomy of the main primary producers that power the respective food webs. Production within the ocean environment (Chapter 2) is mostly generated by small, often microscopic algae, supplemented in coastal waters by larger brown, red and green macroalgae attached to the seabed and intertidal zone. Generally, standing biomass is low (on a global scale), although production itself can be very high. With comparatively few exceptions (as we will see), marine animals do not generally rely on plants for shelter, giving the impression that the sea is an animal-dominated rather than plant-dominated system (think of coral reefs for example). The terrestrial environment, in contrast, is dominated by large primary producers such as trees, grasses and herbaceous flowering plants (**angiosperms**), which are comparatively recent in evolutionary terms, terrestrial angiosperms, for example, only appearing extensively in the fossil record approximately 100 million years ago. Wherever conditions allow, the land is covered by plants and so the majority of land animals utilize plants as a structural habitat.

● The very oldest angiosperm fossils are about 130 million years old, but extensive radiation and diversity did not occur until the late Cretaceous, 90 to 100 million years ago.

In a similar manner to insects, which dominate the terrestrial animal fauna, few terrestrial vascular plants have been able to exploit marine conditions. In fact, only one group is able to live fully beneath the sea: the seagrasses. These angiosperms can grow, flower, and reproduce all within the subtidal environment and in parts of the world form vast meadows in coastal habitats. This results in a very unusual habitat, much more similar to those on land than those generally found in the sea, and therefore providing an important production source and home for many associated marine organisms.

Two other important coastal systems are created by higher plants: mangrove forests and saltmarshes. While these assemblages of plants cannot tolerate continual immersion in seawater, they are highly adapted to dealing with the problems faced by plants on the fringes of the marine environment. Consequently they too can cover huge areas of the land/sea interface and provide a key role both in terms of the functioning of coastal ecosystems and the provision of a structural habit for a diverse range of species. The ecology of saltmarshes is essentially terrestrial in nature (when compared with other marine systems); while the plants do demonstrate tolerance to salinity, their biology is not inherently tied to the sea (e.g. for reproduction). Similarly, the majority of associated species on saltmarshes are terrestrial, although they do provide important foraging grounds for marine fish at high tide. Mangroves, on the other hand, are closely connected with marine conditions, and have a diverse and important associated fauna. Together with coral reefs and seagrass meadows, mangroves are recognized by UNEP as the most important marine habitats. This chapter will therefore discuss how two plant groups (seagrasses and mangroves) are able to survive in the marine environment, and, in particular, investigate their relationship with a diverse and often unusual range of animals, which associate themselves with these plant-based habitats.

● For more details on saltmarsh ecology, see Adam (1993).

● For an excellent, more detailed text on mangrove ecology, see Hogarth (1999).

Fig. 9.1 Examples of types of mangroves: (a) riverine mangrove, (b) tide-dominated (or fringing) mangrove, (c) fringing mangrove roots with associated community of animals. Copyright: a. US Fish & Wildlife Service, b. NOAA. c. Martin Attrill.

● *Nypa* pollen and fruits have fossilized well, due to the conditions in mangrove forests, providing information on the evolution of palms and climate change. *Nypa* fruits from the Eocene (60 mya) have been found in the UK!

9.2 Mangrove Forests

9.2.1 What is a mangrove?

Mangroves are woody trees or shrubs that flourish at the sea/land interface in tropical estuaries and inlets. They can be subdivided into two categories: **true mangroves**, which only occur here, and **mangrove associates**, which can also be found elsewhere, e.g. rainforest. An individual tree is termed a mangrove, while the whole forest habitat is a **mangal** (or simply 'mangrove forest'). There are three basic forms of mangrove, depending on shore morphology and sediment, salinity regime of the surrounding water, and the relative tidal and freshwater influence (Fig. 9.1). **Riverine mangroves** form where there is a low tidal range and the dominance of freshwater flow, such as the deltas of major rivers. Most of the large areas of mangrove forest in Asia are riverine mangroves. **Tide-dominated mangroves** are fully intertidal, often in full-strength seawater and are subject to high-wave action. These are also termed '**fringing**' mangroves and are often the pioneer species that first colonize the intertidal mudflats. **Basin mangroves** occur to the landward side of fringing mangroves where there are lower tidal currents and wave action, but where salinity can be highly variable due to evaporation and rainfall.

Mangroves are taxonomically diverse (Tomlinson 1986), with representatives in mangals of 16 very different plant families. This suggests that the mangrove habit has evolved separately at least 16 times. Two main families dominate the world's mangrove flora, however, in terms of both diversity and ecological role. The first is the Avicenniaceae represented by eight species of the genus *Avicennia* (the black mangrove), the second the Rhizophoraceae which includes the red mangrove (*Rhizophora*, 8 species) together with three other genera (*Bruguiera, Ceriops* and *Kandelia*). Other important families include the Combretaceae (which includes the white mangrove, *Laguncularia*), the Sonneratiaceae (important in Australia and Southeast Asia, 5 species) and also one species of palm (Palmae), *Nypa*, demonstrating the width of mangrove taxonomic range.

9.2.2 Where are mangroves found?

Mangroves are strictly tropical, and their distribution across the globe coincides very closely with that of coral reefs, but they are also found in tropical regions where local conditions do not favour reef development, such as the Amazon region of South America. Distribution is generally confined to the 20 °C isotherm either side of the equator, modified by either warm currents extending their distribution (e.g. east Australia) or cold currents extending into the tropics (e.g. west South America). Diversity is highest in the Indo-West Pacific (IWP) and declines dramatically away from this region to a minimum in the Caribbean/West Atlantic (Ellison et al. 1999). This is thought to be due to increased local diversification following continental drift (**vicariance hypothesis**) rather than the IWP being the centre of origin of mangrove taxa and from where they subsequently dispersed to other parts of the world. Mangroves develop wherever shallow-sloping, generally muddy shores are available within this region, especially lagoons, estuaries and river deltas.

● A vicariance event is the formation of a barrier to genetic exchange that causes separation of related taxa (e.g. continental drift, glaciation). Diversity is then due to separate evolution of species rather than dispersal from a centre of origin.

Within a mangrove forest, there is a clear pattern of distribution of species with distance away from the sea. The fringing mangrove species are found at the sea edge, and across most of the world these pioneer trees tend to be species of *Rhizophora*. Other true mangrove families inhabit conditions behind these fringing species; in Florida for example, *Avicennia* and *Laguncularia* are located in the mid-swamp, *Laguncularia* preferring natural topographic highpoints where sediment is less waterlogged. Behind these true mangroves are distributed a range of mangrove associates, generally forest plants that are able to tolerate the conditions found in the high swamp, but cannot deal with the excessively harsh environment inhabited by the true mangroves. So why do mangroves show such a distribution pattern in relation to distance from the sea and why is it so difficult for terrestrial trees to inhabit the coastal fringe? Two main factors limit plant distribution: increased salt levels and waterlogged sediment, although other more minor factors (such as soil nutrient levels) can also have an influence.

9.2.3 How mangroves deal with living in a marine environment

When considering how terrestrial organisms penetrate the intertidal zone, we have to invert our usual perception about the stresses affecting marine organisms living in this habitat. On a soft-sediment shore, high salinity levels and water being maintained within the sediment are crucial to the survival of a marine organism, but both cause extreme problems for terrestrial plants used to living in dry soil with minimum salt content.

Waterlogged sediment

The thick, waterlogged mud present in mangrove forests presents a key problem for trees, the sediment is exceptionally low in oxygen and tree roots need oxygen to respire and function. Mangroves have developed three main morphological adaptations to enable their roots to obtain oxygen despite growing in anoxic conditions (Box 9.1). All involve root structures that exit the sediment enabling air to be taken in though

Box 9.1 Mangrove root structures

Different mangrove families have markedly different strategies enabling them to get oxygen to roots within anoxic sediment. *Rhizophora* produces aerial roots (A), which leave the tree up to 2 m from the ground and then penetrate the soil, giving the tree support. The aerial section is rich in lenticles to take in air. *Rhizophora* can also produce aerial branches to aid oxygen uptake. *Laguncularia* roots periodically break the soil surface during growth, producing '**knees**' above the sediment surface (B) through which air is taken up. *Avicennia* roots have a feature symbolic of mangrove forests, with vertical tubes (pneumatophores) emerging every 15–30 cm from horizontal roots (C). The tip of each tube has abundant lenticles for air uptake. A 2 to 3 m high *Avicennia* can have up to 10 000 pneumatophores,

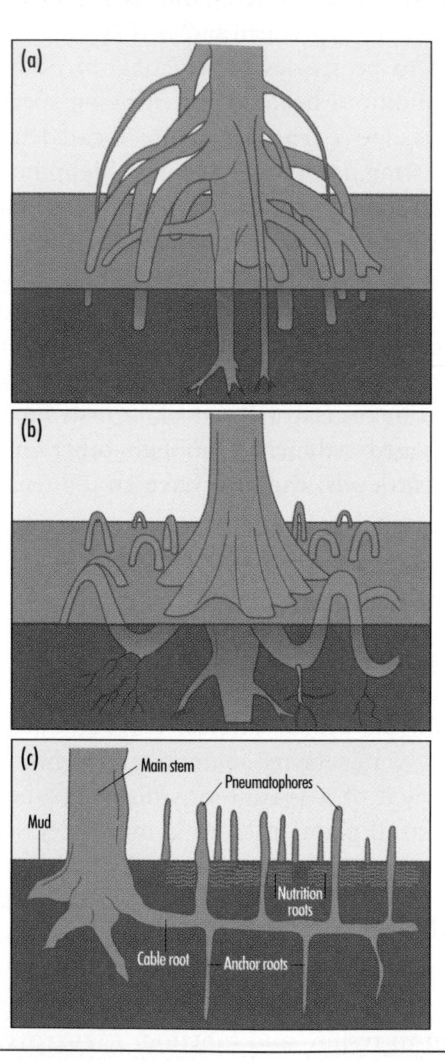

special pores (called **lenticles**) and keep the section of root within the sediment supplied with oxygen.

Dealing with increased salt levels

The presence of salt results in two main problems for trees living in marine conditions. First, the uptake of salt disrupts cellular mechanisms and is fatal for most plants. This can be dealt with by either exclusion of salt by roots, tolerance of salt in tissue (this is much higher in mangroves than other plants), or secretion of excess salt through the bark or by shedding leaves. Second, and perhaps most importantly, salt water in sediment reduces the osmotic difference between the root and the sediment making it difficult to take up water (Hogarth 1999). *Rhizophora* only takes water from the top 50 cm of soil (fresh water is less dense than seawater and hence occurs in the upper layers of soil in these systems) and it is thought that *Avicennia* excludes 90–95% salt at root surface (salt can be excreted, though we are not sure exactly how it is excluded). Due to this osmotic difference, mangroves take up water much less easily than other plants and so need a much greater proportional root biomass to achieve this (the additional growth needed for roots therefore restricting energy being put into vertical growth and reproduction). Within a mangrove forest, trees inhabiting higher salinity soils have a larger below : above ground biomass ratio than those in more freshwater conditions, highlighting the efforts required to survive in marine sediments (Saintilan 1997, Fig. 9.2a).

● It is easy to observe how mangroves excrete salt by shedding leaves. Lick a leaf and it will taste salty!

● Mangroves have much greater root biomass than other trees. In Australia, mangrove root biomass was estimated as 125 tonnes per hectare ($t\,ha^{-1}$), compared with 90 $t\,ha^{-1}$ for large eucalyptus and 32 $t\,ha^{-1}$ for *Acacia* (Snowden et al. 2000).

The difficulty in taking up water has additional important consequences. Plants tend to regulate temperature by transpiration, allowing water to be evaporated from leaves to cool the plant. Mangroves cannot afford to cool leaves primarily by this route as water is a limiting resource, but to achieve maximum photosynthesis leaves need to held perpendicular to the sun – which also maximizes temperature increase! *Rhizophora* holds leaves at angles to the sun depending on how exposed the leaf is. Those at the edge of the tree, in full sun, are held at 75°, thus preventing overheating, while shaded leaves further back into the tree are fully exposed to the sun direction (0°).

Salt tolerance also varies between species within the same genus, which relates to their natural distribution within a mangrove forest. For example, in Australia, *Sonneratia alba* has a higher salinity tolerance than *S. lanceolata* and is therefore found closer to the sea. However, *S. lanceolata* outcompetes *S. alba* at lower salinities as it can grow taller (Ball & Pidsley 1995; Fig. 9.2b).

9.2.4 How do mangroves reproduce?

Mangroves all produce flowers which require pollinating. A large range of flower size and method of pollination exists reflecting mangrove

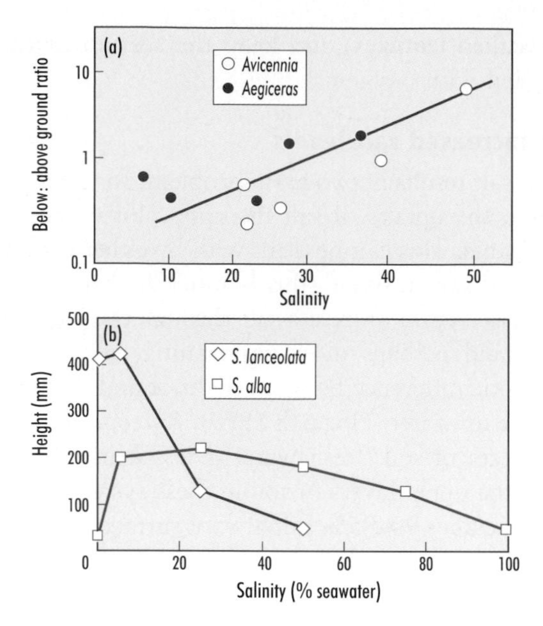

Fig. 9.2 Dealing with salt levels. (a) The cost of living in high salinity sediments – mangroves require proportionally larger root biomass to enable water uptake when salinity is high. (b) Comparative distribution along a salinity gradient of two closely related *Sonneratia* species, their natural distribution relating to a combination of salt tolerance and competitive ability.

Fig. 9.3 The propagules of *Rhizophora*. Photograph: Martin Attrill.

● **Vivipary** is the process of giving birth to live young and is generally associated with animals. In plants, it refers to the seeds germinating on the plant instead of falling.

taxonomic diversity, from the small wind-pollinated flowers of *Rhizophora* to the large, beautiful blooms of *Sonneratia* and *Bruguiera*, which are pollinated by bats, moths, birds and butterflies.

However, the key problem facing mangroves living in marine sediments is dispersal of resulting seeds, which would need to tolerate inundation and high salt levels, or be dispersed by the sea. Most mangroves demonstrate a degree of **vivipary**; following pollination the growing embryo remains on the parent plant (up to several months). The young plant therefore does not leave the adult as a seed or fruit, but as a fully developed seedling known as a **propagule**. The most advanced vivipary is shown by *Rhizophora* (Fig. 9.3) – seedlings in Southeast Asia have been recorded up to 1 m long. Mangroves put a huge amount of energy into propagule production; *Avicennia* in Costa Rica can produce more than 2 000 000 propagules ha^{-1} yr^{-1}, representing 10 to 40% of net primary production (Smith 1992).

Rhizophora seedlings drop from the parent plant directly into the water. Initially they float horizontally, turning vertically after about a month when roots develop (by 40 days). The roots drag on the bottom and the seedling can get stranded horizontally, or sink, where they erect themselves vertically after rooting. If not rooted after 30 days, propagules retain buoyancy and are viable for >1 year, although survivorship is species-specific and dependent on propagule weight.

Mangroves provide an incredibly complex structural habitat, both above the water line and below it through the root system. The assemblage of organisms associated with mangroves is therefore a unique mixture of animals that colonize the mangrove forest from both the land and the sea, resulting in a dynamic and diverse associated community.

● Why are mangroves tropical? The combination of three key components has a high energy cost: being a tree, tolerating salt, and coping with waterlogged soil. Temperate regions would not allow photosynthesis all year round and so the tree could not gain enough energy to survive.

9.2.5 Terrestrial organisms associated with mangroves

Non-woody plants are not common in the marine section of mangrove forest, due to the same pressures of salinity and waterlogged soil that affect the mangroves themselves. Epiphytic species that can avoid the sediment are the most successful, particularly those belonging to two groups, the bromeliads and the mistletoes. Insects, however, are successful, as anyone who has visited a mangrove swamp will know. Ants, termites and mosquitoes are particularly abundant (Hogarth 1999), with some mosquitoes moving onto unusual hosts; *Aëdes pembaensis*, for example, has been recorded feeding off mudskippers at low tide. This species also has an unusual life cycle, laying eggs on the pincers of crabs, after which the larvae develop in the crabs' burrows. The major ecological impact of insects, however, is their grazing activity on the leaves of the mangroves, damaging the leaf and thus reducing the potential of the tree for photosynthesis, growth and reproduction. The seasonal emergence of caterpillars (e.g. the moth *Nophopterix*) can have particularly dramatic impacts. Grazing can also have a controlling effect on the successful settlement and growth of new propagules. Sousa et al. (2003) noted that settling *Rhizophora* propagules were only successful in clear gaps within the existing forest (e.g. those caused by lightning strike). While offering better growth conditions, refuge from predation by the beetle *Coccotrypes rhizophorae* was just as important, with seedling mortality increasing with distance from the gap into the forest.

● Crocodiles are frequently associated with mangroves, *Crocodylus acutus* for example having special salt glands to help deal with high salinity conditions.

A range of terrestrial vertebrates has also become specialized for life in mangroves, including species of reptile and amphibian that have developed exceptionally high tolerance to high salinity water. One example is the crab-eating frog (*Rana cancrivora*), a group not generally associated with marine systems, which can survive as both tadpole and adult due to exceptional osmoregulatory abilities. The most diverse and obvious group of vertebrates in mangroves, however, is the birds, which use the mangrove for nesting, feeding and as a roost at high tide. Many mangrove systems around the world are protected specifically due to the high numbers of these high-profile organisms they support. The bird fauna is a mix between waterfowl and terrestrial species that feed off insects, etc. within the mangrove canopy.

The complex nature of the mangrove ecosystem allows extensive **niche differentiation** within the canopy bird fauna. Noske (1995) studied the feeding guilds of these birds in Malaysia, recording 17 species from four contrasting feeding guilds (nectarivores, aerial hawkers, bark foragers, foliage insectivores). Within these guilds, species avoided direct competition through small-scale spatial variation in their distributions. For example, two species of woodpecker were each restricted to a different tree species (*Avicennia* and *Sonneratia*), and thus separated within the forest due to differential distribution of the mangroves. Within the same tree, two species of foliage insectivores (flyeater, *Gerygone sulphurea*, and ashy tailorbird, *Orthotomus ruficeps*) were spatially separated by height, the flyeater only forages below 4 m in the canopy and the tailorbird in the canopy top.

● Hutchinson (1957) defined a **niche** as an 'n-dimensional hypervolume' representing a space of all environmental factors that affect the welfare of a species. Niche overlap can therefore potentially result in competition, niche differentiation allowing species to coexist.

Mangrove birds can be key species for the conservation of mangrove habitats, but full details of their ecology is necessary in order to plan suitable management strategies. A good case study is that of the scarlet ibis (*Eudocimus ruber*, Fig. 9.4) within the Caroni Swamp in Trinidad. The ibis forages within the channels of the swamp and returns to roost in the tops of trees at dusk – a spectacular sight when thousands of birds return at once. Caroni Swamp comprises over 5600 ha of mainly state-owned habitat, principally mangrove, and is fed by four rivers. The population peaks in autumn/winter (up to 10 000 birds), most birds then migrate to South America leaving behind a small breeding population (Bildstein 1990). Despite protection of their mangrove habitat, this is a reversal of the situation in the 1950s and 1960s when the majority of birds stayed to breed, so the decline in breeding numbers was somewhat

Fig. 9.4 The scarlet ibis, *Eudocimus ruber*, an inhabitant of mangroves in the Caroni swamp, Trinidad. (Photographer: Mike Lane, rspb-images.com.)

of a puzzle. Non-breeding birds feed off marine prey, such as fiddler crabs (*Uca*) and polychaetes (*Nereis*), but it was noted that nestlings raised on such salty prey did not develop properly, breeding adults switching to foraging in freshwater areas inland.

Despite protection of the mangrove, much of the freshwater wetland in Trinidad had been drained for rice production, removing the ibis's feeding grounds. As a result the birds began to migrate to the Orinoco delta to breed where freshwater habitat was still available. A further twist to the story was uncovered when high levels of mutation (chlorophyll deficiency) were recorded in *Rhizophora* in parts of Caroni swamp (Klekowski et al. 1999), traced to high levels of mercury within the sediment under the trees, particularly those where scarlet ibis were nesting. Samples of feathers from ibises and other birds in the roost revealed that the ibis had exceptionally high levels of mercury in their feathers (41 p.p.m. compared with 6 p.p.m. for other species). The high level of mercury in feathers was traced to the ibis's new foraging grounds in the Orinoco, which are key gold mining areas. The mining operations release high levels of toxic metals into the watershed. Mercury is accumulated in feathers (Chapter 14), so when the ibis returned to Trinidad for winter to moult, the feathers contaminated the soil and impacted the mangroves. Many birds demonstrate a remarkable tolerance to metal contamination, being able to 'dump' metals in feathers and then shed them, removing the toxin.

9.2.6 Marine organisms associated with mangroves

Examples from the vast majority of marine taxa can be found associated with mangroves or in the surrounding sediment; the **sessile** fauna attached to submerged roots is particularly diverse, the roots providing a hard substratum for attachment that otherwise is restricted in a soft-sediment environment (Farnsworth & Ellison 1996). Despite domination by long-lived groups such as sponges and ascidians, the mangrove root community in Florida was found to be incredibly variable, and the assemblage composition on individual roots changed dramatically over short time periods (1 to 2 months, Bingham & Young 1995). This was thought to be related to the intensity of **stochastic** perturbations, such as physical disturbance from strong tidal flow and predation causing large fluctuations in the abundance of certain species. This disturbance prevented competitive processes from producing a stable, equilibrium community.

● Reminder: **sessile** organisms are those that are fixed in place so they cannot move, such as barnacles and sea squirts.

● Stochastic means involving a random or chance variable. For example, a severe storm is a stochastic perturbation.

Fewer marine species are associated with mangrove trees above the water line, except for gastropod molluscs, the majority of which are detritivores or graze off epiphytic algae. The most abundant mangrove snails are members of the genus *Littoraria* and, in a similar process to mangrove birds, different species show distinct niche separation within the mangrove. For example in Papua New Guinea, *L. pallescens* lives on the mangrove leaves (where it grazes on algae rather than the leaf itself), *L. intermedia* inhabits tree bark in freshwater creeks, while *L. scabra* is also associated with bark, but on seaward trees (Hogarth 1999). Few marine snails are actually able to feed on the mangrove leaves themselves, an exception being *Terebralia palustris*. This species demonstrates an **ontogenetic** change in diet, the young feeding off detritus, whereas the adults graze directly on mangrove leaves once their **radula** has metamorphosed to be able to penetrate the leaf surface.

● **Ontogenetic** means relating to the development of an organism. Thus ontogenetic changes are those that occur over the developmental life cycle (e.g. a larva will have different characteristics to an adult, such as an ontogenetic change in feeding).

The classic marine organisms generally associated with mangroves tend to inhabit burrows within the surrounding mud and, unlike those in temperate mudflats for example, are much more active at low tide. Two key groups will be considered further here: mudskippers and crabs. Mudskippers (Fig. 9.5) are unusual fish related to gobies, represented by several long-named genera (e.g. *Periophthalmus, Boleophthalmus, Scartelaos*) and are exceptional among fishes in their amphibious behaviour. They live in water-filled burrows, emerging as the tide drops, often following the tide in and out (Ikebe & Oishi 1996) to forage, 'walking' across the mud surface using modified pelvic and anal fins. Most are omnivores, but some species specialize (e.g. *Periophthalmodon* feeds on crabs). Mudskippers are highly physiologically adapted to this amphibious life cycle, but do have to return periodically to their burrows (Box 9.2). However, they are remarkably tolerant to desiccation; *Periophthalmus cantonensis*, for example, can survive 2.5 days out of water, while *P. sobrinus* spends 90% of time out of water and can lose 20% of its body mass yet survive (Gordon et al. 1978).

Fig. 9.5 Mudskippers: mangrove fish remarkably adapted to life out of water. Here *Perioptbalmadon* is gulping air to take it down into its burrow. (Photographs: Toru Takita and Atsushi: Ishimatsu.

Crabs are perhaps the most important group of mangrove fauna. They are particularly abundant and diverse in this habitat, have a major impact on sediment dynamics through their burrowing activity, and are the main processors of mangrove detritus. The action of crabs has a major influence on the whole functioning of the mangrove system (as we shall see) and so they are an example of an **ecosystem engineer**. Two main families of crab are dominant in mangroves: the Grapsidae (e.g. *Sesarma*) and the Ocypodidae (e.g. fiddler crabs such as *Uca*). The grapsids have the most important role in terms of the functioning of the mangrove ecosystem, so we will look further at how they influence forest productivity.

● An **ecosystem engineer** is an organism that directly or indirectly modulates the availability of resources to other species. It is able to modify, maintain, and create habitats (see Lawton & Jones 1995 and Chapter 7).

Mangrove leaves are exceptionally unpalatable for invertebrates due to high proportions of carbon exacerbated by high tannin levels. Generally it is assumed that a C:N ratio of <17 is required to be profitable to invertebrates (i.e. there is enough nitrogen in the substance to be worth the effort), but ratios in mangrove litter can be 70 to 100 for *Rhizophora* (Lee 1998). Consequently, the majority of carbon produced by the mangrove forest is liable to be exported away from the system as leaves, or

Box 9.2 How mudskippers aerate their burrows

Mudskippers' burrows are constructed in highly anoxic sediment and measurements within burrows have shown that the water in deeper chambers contains practically no oxygen at all. It has therefore been somewhat of a puzzle as to how mudskippers cope with these inhospitable conditions, particularly as eggs are laid and reared in apparently hypoxic water. Ishimatsu et al. (1998) noted that walking near *Periophthalmodon* burrows caused bubbles of gas to be released up the main shaft. Further observations revealed how mudskippers oxygenated their living quarters – before entering the burrow the fish fills their mouth cavity with air and transports this down the burrow, often making return trips. Air released in the breeding chamber becomes trapped providing an oxygen reservoir for the fish and developing embryos that are often laid on the roof of chambers, where air will accumulate.

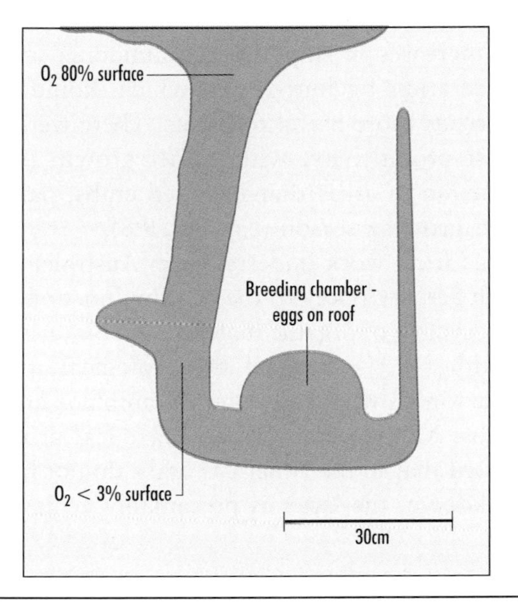

the nutrients become locked up due to slow microbial decomposition. *Sesarma* crabs, however, have the rare ability to utilize mangrove leaf litter as a prime food source, despite its apparent unpalatability, but the digestive mechanisms for this important feeding habit have yet to be elucidated. A few species, such as *S. messa*, are even able to feed directly off fresh leaf material; *S. leptosoma* has even been observed making daily vertical migrations up mangrove trees to graze on live leaves (Vannini et al. 1995). Grapsid crabs also store leaves within burrows, so fulfil an important function in retaining carbon material within the mangrove system. Robertson (1986) concluded that *S. messa* populations could remove or store up to 28% of all *Rhizophora* litter, with major implications for the reduction of organic matter export from mangrove forests and thus the flow of energy into coastal systems (Lee 1998).

Grapsid crabs also have a major ecological impact through their burrowing activity (**bioturbation**), which can influence the chemical make-up of the sediment and subsequent forest productivity. This is exemplified by a study in Queensland, Australia (Smith et al. 1991), which investigated the impact of removing crabs from mangrove areas. The study included three treatments: a removal area (R) where all crabs were trapped and removed, a disturbance area (D) where activity was undertaken but crabs not removed (to see if the physical disturbance of the experiment had an effect, termed **procedural control**), and control areas (C), which were left alone. Levels of soil chemicals were measured and forest production assessed by the fall of **stipules** (for growth) and the number of mature propagules on trees (for reproductive output). Burrowing was found to have significant effects on the sediment chemistry, with increases in sulphide and ammonia in exclusion plots due to lack of aeration by burrowing, which would normally allow oxidation to nutrients more useful to plants. There were also significant impacts on forest productivity, with higher growth (stipule fall) and propagule production in areas that included crabs, particularly during the main summer growing season (e.g. Fig. 9.6).

● A **stipule** is the outgrowth of the leaf base, which generally are shed as the leaf grows. Therefore, measuring the number of stipules falling gives an idea as to the rate of leaf growth.

The conclusions from work undertaken in Australia and SE Asia are unequivocal: crab activity is key to the healthy functioning of mangrove forests through recycling of organic material and bioturbation. However, McIvor and Smith (1995) assessed the ecological roles of crabs in Florida mangrove forests where the family Xanthidae are more common than grapsid crabs. Xanthid crabs do not have leaf processing abilities, so it was concluded that in the Americas crabs do not have a significant role in leaf breakdown, the latter is presumably achieved primarily by microbial action.

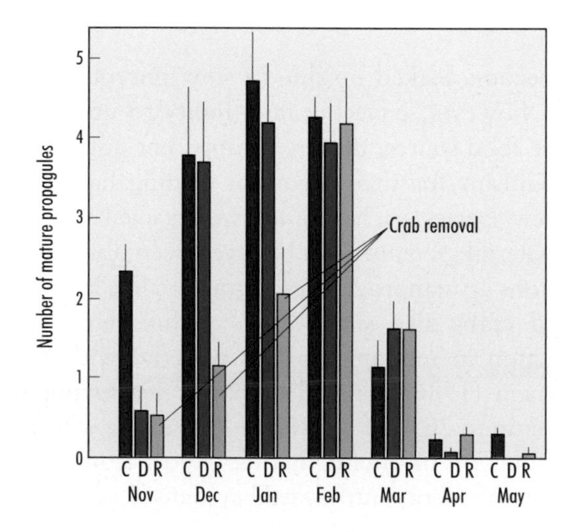

Fig. 9.6 The results of crab removal experiments on mangrove forest productivity. The production of new propagules was greater in areas that included crabs due to their bioturbating activities (Smith et al. 1991).

9.2.7 The wider role of mangrove ecosystems

Mangrove forests as a whole provide a range of valuable functions that influence surrounding coastal systems, and impact on human activities such as fisheries (Chapter 12). Two of the most important are their potential role in nutrient input and energy flux between mangrove and marine systems, and the provision of a nursery for fish species that recruit to local fisheries and coral reefs.

● The products and functions of an ecosystem that are useful to humans are often referred to as 'goods and services' (Preface).

Despite the efforts of crabs, much detrital material from mangroves is exported out of the swamp to surrounding systems, particularly from fringing mangroves (Robertson et al. 1992), providing a large potential source of food. The relative value of this material to marine organisms within adjacent systems is inconsistent, due to the general unpalatibility of the tree leaves to marine animals. The use of stable isotopes (Chapter 4.3.2) has proved valuable in assessing how important mangrove detritus is compared with other carbon sources. Schwamborn et al. (2002) investigated the uptake of different carbon sources by crustacean larvae in a tropical estuary to which large amount of mangrove detritus was exported. They found that the contribution of mangrove carbon to larval nutrition to be negligible, with only the zoeae of porcelain crabs having a significant proportion of mangrove carbon in their tissues. A similar, though perhaps even more unexpected, result was found in the tropical Embley River estuary in Australia (Loneragan et al. 1997). Stable isotope analysis was utilized to investigate whether seagrass- or mangrove-derived carbon were important carbon sources for prawn species in the outer estuary. Organic matter in the sediment was found to correspond to the main local source, but this pattern was not reflected in the tissues of the prawns. While prawns that inhabited seagrass beds did seem to rely completely on seagrass-derived material, mangrove-dwelling prawns did not depend on carbon from mangroves, but utilized either algae or seagrass material.

Mangrove leaf litter does not even seem to be valuable in nutrient-poor habitats, and Lee (1999) demonstrated that additions of mangrove detritus to sandy substrata did not enhance the marine benthic community, perhaps due to the high associated tannin levels. It therefore appears that carbon from mangroves is rarely directly utilized by marine organisms, the carbon probably entering the food web following bacterial decomposition and recycling. However, mangroves do have a clear physical role influencing the flux of material to coastal systems by providing a buffer for land run-off, e.g. from storms. In particular, mangroves reduce the amount of sediment washed into the coastal region, which would otherwise impact adjacent reef systems (Hogarth 1999); where mangroves have been removed (e.g. for shrimp farming) there are potential consequences for the health and survival of local reef systems.

● Tannins, among many other chemicals, have been produced by plants in an 'arms race' with insect grazers. Such chemicals are unpalatable, so help protect the plant. Tannins are responsible for the astringent taste of unripe fruit and red wine.

Mangroves do have a clear positive role in the provision of recruits to local fisheries and, in particular, coral reef systems. This conclusion has been somewhat anecdotal (Baran & Hambrey 1998) due to the presence of juvenile coral reef fish in mangrove systems, but Mumby et al. (2004) have now unequivocally demonstrated the value of mangroves to coral reef fish populations. Working in Belize, they assessed the assemblage of fish present on replicate reefs with or without a rich mangrove stand nearby. There were marked differences in the reef fish communities between the two sets of sites, and mangrove extent was the dominant factor that affected the fish assemblages. In particular, the biomass of fish species known to use lagoons or mangroves as juveniles was enhanced in reefs near to mangrove-rich areas. Mumby et al. (2004) concluded that the main reason for these results was that, once fry are large enough to leave seagrass habitats, mangroves provide a refuge for juveniles from predators and a plentiful food supply that increases juvenile survivorship and recruitment to coral reefs. They noted that the largest herbivorous fish in the Atlantic, *Scarus guacamaia*, which depends on mangroves, has become locally extinct since mangrove removal. It is clear that conservation policies targeted at protecting coral reef habitats also need to consider nearby mangrove systems.

9.3 Seagrass Meadows

9.3.1 What are seagrasses?

● For a detailed account of seagrass ecology, see Hemminga and Duarte (2000).

Seagrasses are the only truly marine angiosperms, generally growing in soft sediments in shallow coastal waters. Below ground there is a network of rhizomes and roots (Fig. 9.7), the rhizomes spread horizontally joining individual plants, while the roots extend vertically into the sediment. The root-rhizome network of the Mediterranean *Posidonia oceanica* can build up over hundreds of years, forming a peat-like '**matte**' several metres thick (Borg et al. 2002). Above ground the seagrass is made up of discrete shoots that emerge from the rhizome and comprise several strap-like, laminate leaves that grow from the base of the shoot (the leaf **sheath**), although the genus *Halophile* has rounded leaves. The number and height of leaves varies considerably between species, the longest leaves (over 6 m) belonging to the Japanese species *Zostera caulescens* (Aioi et al. 1998). In suitable conditions (see 9.3.2), seagrasses can form extensive meadows covering large areas of seabed (e.g. the seagrass bed within the Spencer gulf near Adelaide, Australia, covers over 4000 km^2, Keuskamp 2004) with densities of shoots >1200 m^{-2} (e.g. *Posidonia oceanica*, Borg et al. 2002), creating one of the most important and extensive subtidal marine habitats in the world.

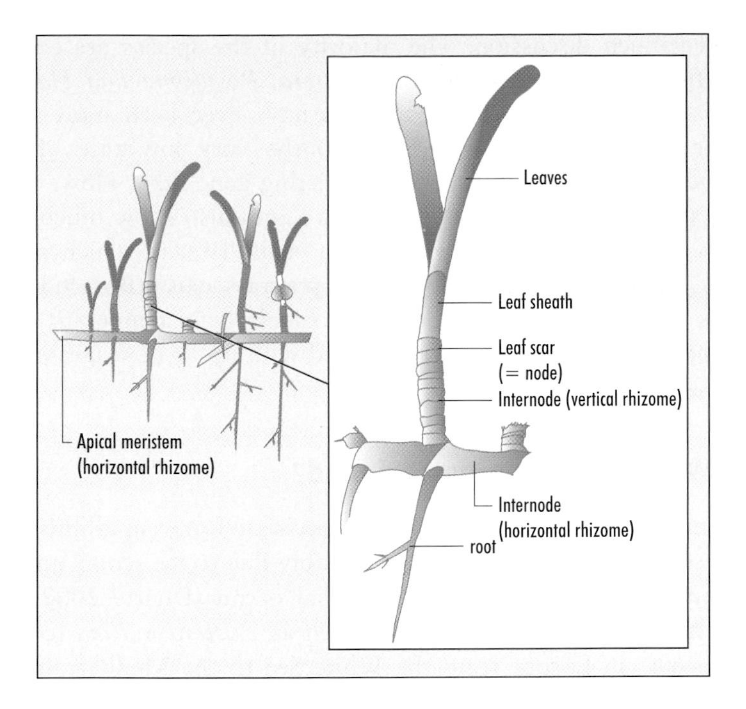

Fig. 9.7 Schematic diagram representing the main structural features of a seagrass.

The taxonomy and origin of seagrasses has been subject to much debate, but seagrasses are clearly **polyphyletic** and generally assigned to two families, Potamogetonaceae and Hydrocharitaceae, neither of which is related to the grasses familiar on land. Fossil evidence suggests that seagrasses first appeared in the marine environment around 100 million years ago, the oldest Cretaceous fossils include the genus *Posidonia*, but their ancestors are uncertain. Two candidates have been put forward: coastal plants (e.g. saltmarshes, mangroves) and freshwater **hydrophytes**. Some seagrasses have lignified stems and two genera are viviparous, linking them to mangroves, while freshwater plants show many of the adaptations required to survive in the marine environment, such as basal meristems and **lacunar** gas transport systems (Hemminga & Duarte 2000). Certainly it seems likely that the seagrass habit has evolved several times over the last 100 million years (perhaps from both ancestral routes), with the genera *Phyllospadix* and *Enhalus* appearing significantly later than other seagrass groups (Larkum & den Hartog 1989). However they have evolved, seagrasses possess three key attributes that enable them, uniquely, to colonize the marine environment (Hemminga & Duarte 2000): (a) leaves with **sheaves** adapted to high-energy environments; (b) **hydrophilous**, and thus submarine, pollination; and (c) extensive **lacunar** systems that enable transport of oxygen to the below-ground structures in anoxic sediments.

● **Polyphyletic** – a useful grouping of organisms that has more than one evolutionary root form. For example, 'winged vertebrates' would include birds and bats, but each has very different origins.

● Hydrophyte – a true plant that has fully adapted to live in water. Hence also hydrophilous.

Diversity is surprisingly low, with only *c*.50 species represented worldwide (Hemminga & Duarte 2000), although this figure has generated much discussion. The majority of the species are contained within three of the oldest genera: *Zostera, Posidonia and Halophila*. There is little evidence that seagrasses have ever been more diverse, a major reason perhaps being due to the very low rate of sexual reproduction (9.3.3) and dispersal restricting gene flow. However, seagrasses are exceptionally **plastic** in nature, and also show much genetic diversity even within meadows (Reusch et al. 1999a), which means it remains unclear how many species of seagrass exist (Box 9.3). Most seagrass meadows are monospecific, particularly in temperate regions, but tropical meadows have been reported with up to 12 seagrass species present (Duarte 2001).

9.3.2 Where are seagrasses found?

Unlike mangroves and saltmarshes, seagrass are found in all the world's coastal seas except for Antarctica (probably due to ice scour) and cover approximately 0.1 to 0.2% of the global ocean (Duarte 2002). Some species have extensive distributions, such as *Zostera marina* (eelgrass), which occurs in Europe from the White Sea to the Mediterranean, on both coasts of North America and the Northwest Pacific (Green & Short 2003). Other species are more restricted. *Posidonia oceanica* is the dominant seagrass across the Mediterranean, but is not found elsewhere, whereas some seagrasses have a very narrow range indeed; *P. kirkmanii* occurs only in a small area off SW Australia. The peak of diversity occurs in Malaysia, and seagrass species richness declines with distance along major currents from this point (Mukai 1993), resulting in

Box 9.3 Diversity of seagrass: how many species are there?

Seagrasses are morphologically **plastic**, a term that describes species that appear very different depending on the environment to which they are exposed. As classification is generally based on morphological features (e.g. shape of leaf margins), there is around a 20% uncertainty within the seagrass world as to how many species of seagrass actually exist, with most arguments focused within the three main genera (*Halophila, Zostera, Posidonia*). Molecular taxonomy (e.g. the use of DNA markers) has helped the confusion to some degree and tended to reduce the number of existing species defined using morphological features. For example, intertidal *Zostera marina* looks different from subtidal plants and in the UK is still considered a separate species (*Z. angustifolia*), although genetic evidence from mainland Europe suggests it is just a plastic form of *Z. marina*. Outside the UK the species is not recognized. However, the use of genetics has raised further problems, and has drawn into doubt the identity of some genera. For example, the genetic distance between *Heterozostera* and *Zostera* is similar to that between species of *Zostera*, suggesting that one genus should include both these taxa (Hemminga & Duarte 2000). Whatever the final total of species, it is clear that seagrasses are a very low diversity group.

suggestions that this may be centre of origin for seagrasses (Hemminga & Duarte 2000).

Within their biogeographical range, much of the coast is devoid of seagrass due to unsuitable habitat conditions; so individual seagrass beds can potentially be separated by large distances of inhospitable coast. The vast majority of seagrasses require a soft substratum that will enable the penetration of roots; although some species can grow on rocks (e.g. *Phyllospadix, Posidonia*) this is unusual and most only develop on sand or mud.

● Species of *Phyllospadix* are known as surf grasses and unusually are found on rocky shores. They produce significantly more root hairs than the closely related *Zostera*, these providing extra attachment.

However, some features of soft sediments make plant growth impossible, particularly highly mobile or exposed sediments that can result in burial of colonizing seagrasses, or those with very high inputs of organic matter resulting in reduced, anoxic sediment conditions. Paradoxically, the presence of a seagrass meadow on soft sediment can increase the sediment organic content, from both the seagrass production itself and the capture of other detrital material. Seagrass beds can also change the particle size distribution of the sediment, the baffling effect of seagrass blades slowing water currents and enhancing the deposition of fine sediment particles. Seagrasses therefore have a very complex relationship with the sediment, which we are only starting to untangle. Seagrass is also limited by the light levels, which must be above the threshold where photosynthesis is still possible. As a result, the maximum depth where seagrass meadows are found is related to their **compensation point.** This depth is controlled by the clarity of the water, so in comparatively turbid temperate regions seagrass beds are found at depths shallower than in clear tropical or Mediterranean water. *Zostera marina* in Europe is generally found above 6 m depth, whereas *Posidonia oceanica* in the clear waters off the island of Malta has been recorded as far down as 40 m (Borg et al. 2004). High turbidity water can therefore be a limiting factor for seagrass colonization and growth.

While the lower level of a seagrass bed is set by light levels (unless limited by habitat availability, e.g. the presence of coral reefs in the tropics), factors controlling the upper limits are less well studied (Hemminga & Duarte 2000). Several seagrass species, such as *Zostera marina* and *noltii, Phyllospadix* and *Halophila* spp., can form extensive intertidal meadows. Desiccation can be minimized by the structure of the bed, dense continuous seagrass trapping water beneath the flat-lying leaves. Ultraviolet damage is also thought to be a major factor preventing intertidal survival, particularly of subtidal species, while in some regions upward extension may be prevented by physical factors such as wave exposure, ice scour, or the lack of suitable substratum.

● The depletion of the ozone layer primarily through our use of CFCs has allowed increased fluxes of UV radiation, in particular the damaging UVb. There is much speculation about its effects on plants, but concerns include inhibition of photosynthesis and the cost of tissue repair and production of blocking compounds (see also Chapter 11).

Many seagrasses show a tolerance to a wide salinity range (for example *Z. marina* occurs in full-strength seawater and also down to a salinity of 5 in the Baltic), and seagrasses often are a major feature of estuaries and hypersaline lagoons (Chapter 4), although the majority of species perform optimally under fully marine conditions. However, success in estuaries and similar coastal systems may be limited by high levels of plant nutrients in the water, which is becoming an increasing problem through agricultural run-off (Chapter 14). Such high levels of, for example, nitrate and ammonium can directly affect the growth of seagrass (Short et al. 1995), but it also influences the competitive balance between the seagrass productivity and that of algae associated with the seagrass meadow, represented by an **epiphytic** assemblage growing on the leaves or by macroalgae that grow alongside, or among, the seagrass meadow.

● Epiphytes are organisms growing on the surface of plants, but not deriving nutrition from them. Generally the term is used for other plants (including algae), but sessile animals such as bryozoans can also be termed epiphytes.

Williams and Ruckelshaus (1993) demonstrated that *Zostera marina* showed a saturation-type response to increasing nitrogen levels; above sediment ammonium concentrations of 100 μmol l^{-1} the seagrass was unable to increase growth rates. Once over such a threshold, algal growth may continue, resulting in a shift in dominance within the seagrass bed (Short et al. 1995). This can have detrimental results for the seagrass, in particular due to light limitation imposed by the algae reducing seagrass productivity and, ultimately, survival. Hauxwell et al. (2001, 2003) demonstrated that eelgrass (*Z. marina*) was now absent or disappearing in all Waquoit Bay estuaries (Massachusetts) due to excessive algal growth (Fig. 9.8), except those that received the lowest land-derived nitrogen loads.

Introduced algae are also posing a severe threat to seagrass coverage in Europe. The alga ***Caulerpa taxifolia*** has been accidentally introduced

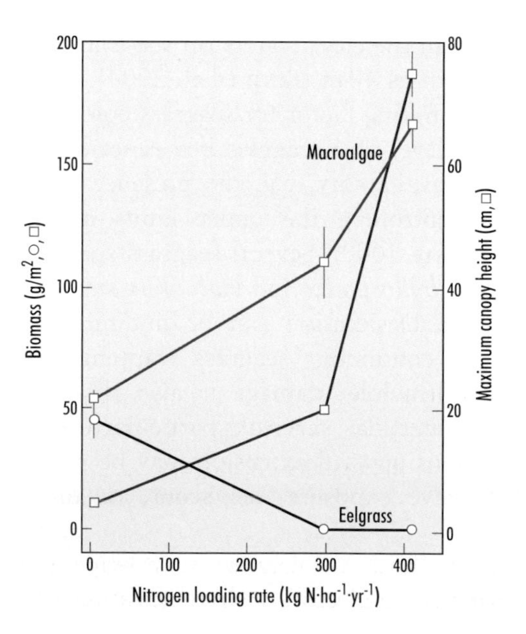

Fig. 9.8 The relationship between the growth of seagrass and macroalgae at increasing levels of nitrogen loading, indicating algal biomass increases relative to seagrass under increasing nutrient conditions.

to the Mediterranean from southern France and is spreading rapidly, smothering the extensive natural *Posidonia oceanica* beds with consequent loss of large areas of seagrass coverage. While not so dramatic, the Japanese seaweed ***Sargassum muticum*** poses a threat to native *Zostera* beds in Northwest Europe. While direct competition under natural conditions does not appear to occur, physical damage to seagrass beds (e.g. from boat anchors) can allow the colonization of *Sargassum*, and once established it appears that seagrass cannot reclaim the space. Large beds of *Zostera* in NW France have been lost to *Sargassum* by this mechanism (Givernaud et al. 1991).

● *Caulerpa* was accidentally introduced to the Mediterranean in 1984 from the Monaco aquarium. By 2000 it covered 131 km^2 of bed along 191 km of coastline (Meinesz et al. 2001) and could be regarded as one of the most serious invasive species in marine systems.

Changes to the global cover of seagrass have raised much recent concern, Green and Short (2003) stating that 15% of seagrass worldwide had been lost over the ten-year period up to 2003 due mainly to a combination of human activities. However, seagrass within the northern hemisphere, particularly *Z. marina*, was devastated during the 1930s by a mysterious wasting disease that killed massive areas of seagrass and resulted in the loss of many beds in Europe and North America. The culprit was identified as a slime mould *Labyrinthula zosterae*, causing brown spots to appear on the leaves, which spread to the shoot within weeks. Many of the surviving beds were located within estuaries, from where dispersal and recolonization could occur. It has been suggested that the salinity tolerance of the seagrass was greater than that of the slime mould, allowing some beds to survive in reduced salinity refugia (Durako et al. 2003).

● Slime moulds are rather mysterious organisms that have proved hard to classify. They were originally thought to be fungi, but are now included in the Protozoa.

9.3.3 Reproduction and growth of seagrasses

Seagrasses can reproduce both sexually and, rarely, asexually (e.g. through fragments of drifting rhizome), the prominence of sexual reproduction within beds varying widely across species and geographical location. The majority of seagrass genera are **dioecious**, which is relatively rare in terrestrial angiosperms and has been suggested as a way of avoiding self-fertilization (Hemminga & Duarte 2000).

● **Dioecious** – having separate male and female plants.

Flowers tend to be produced seasonally, even in tropical species where it often coincides with very high spring tides, and is generally controlled by temperature (Ramage & Schiel 1998), which facilitates simultaneous flowering of seagrasses across wide areas. However, only a small proportion of a bed will flower at once (<10%), and there is great variability between years, making flowering a comparatively rare event in seagrass beds with minimal allocation of carbon to this process by the seagrass (in direct comparison to mangroves). There is very little evidence, for example, of flowering in northern European subtidal *Zostera marina* beds, resulting in the existence of very old clones in some areas. One extensive single seagrass clone in the Baltic is thought to be

● The single genotype of *Zostera* in the Baltic covers a total area of 6400 m^2 and weighs approximately 7000 kg.

>1000 years old (Reusch et al. 1999b) and represents the largest known marine plant! In contrast, flowering occurs relatively commonly within intertidal *Zostera* beds, such as those in Northern Ireland. It appears that disturbance and stress enhance seagrass flowering, and thus sexual reproduction, compared with more stable subtidal beds, which favour vegetative growth.

Practically all seagrass species have **hydrophilous** pollination, a development within seagrasses that allows survival in marine systems, pollen being released into the water column to fertilize female flowers (Hemminga & Duarte 2000). The resulting seeds vary in their dispersal ability; some species have negatively buoyant seeds (e.g. *Halodule* spp.) that will not travel far, while others possess structures that enhance buoyancy (e.g. the seed coatings of *Zostera*). However, evidence suggests they do not disperse far from source, 80% of *Zostera* seeds from a Chesapeake Bay population remaining within 5 m of their source (Orth et al. 1994). The seedlings themselves may also be a dispersive stage, while in the two viviparous genera (*Amphibolis* and *Thalassodendron*), the seedlings develop attached to the mother plant. However, it would appear that the survival of both seeds and seedlings is very low indeed, the probability that any given shoot will successfully establish a new seagrass **genet** is <0.00001 (Hemminga & Duarte 2000). Sexual reproduction is therefore exceptionally inefficient within seagrass, resulting in clonal propagation as the major method of seagrass meadow survival. The consequence is ancient beds (as outlined earlier) and highlights how potentially vulnerable existing seagrass beds are; recolonization is not straightforward.

Established seagrass patches grow through the lateral extension of the below-ground rhizomes into uninhabited surrounding soft sediment (as long as the conditions of that sediment favour seagrass growth), eventually producing a new shoot unit, or **ramet**. Growth rates vary considerably between species but tend to be related to the size of the seagrass, with a negative relationship between rhizome diameter and horizontal extension rate (Duarte 1991, Fig. 9.9a). As a result, the smallest species such as *Halophila ovalis* can spread up to $5\,m\,y^{-1}$ and are often regarded as pioneer species in multi-species meadows. Rhizomes can also branch, allowing a two-dimensional colonization of new sediment.

● The time between the development of two seagrass units is known as the **plastochrone interval** and is a useful measure to compare growth rates between species and under different environmental conditions.

Leaf growth can be highly seasonal in temperate species, relating to temperature and especially light levels. A study on *Zostera marina* on the west coast of the USA exemplifies such seasonal patterns of production (Nelson & Waaland 1996, Fig. 9.9b), peak summer values being 6.5 times those in winter. Unlike the beds themselves, leaves do not live particularly long, with a lifespan generally shorter than land plants (Hemminga et al. 1999). Leaf age also varies considerably between

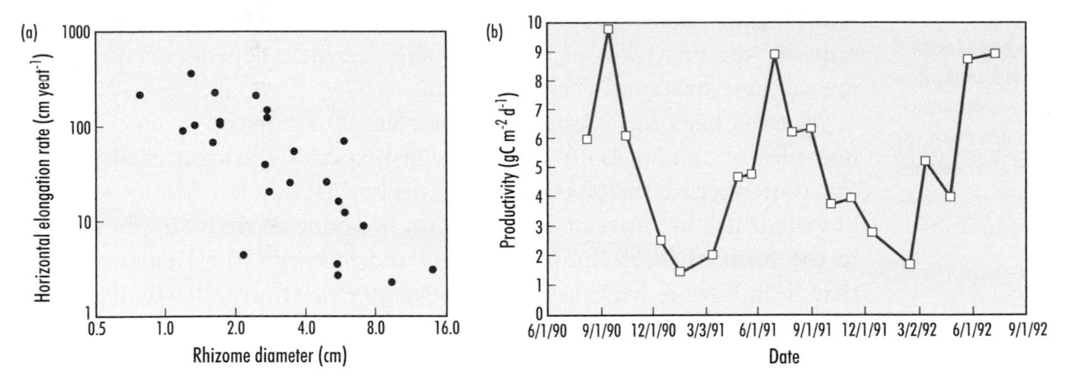

Fig. 9.9 (a) The growth of seagrass rhizomes of different sized seagrass species – smaller species with thinner rhizomes grow faster. (b) Seasonality in seagrass productivity – *Zostera marina* off Washington State, USA.

species (up to a year for *Posidonia oceanica*), but most will shed leaves through the growing season (e.g. *Zostera* mean leaf lifespan is <100 days), resulting in large amounts of plant organic matter entering the coastal ecosystem; in dense seagrass areas (e.g. parts of the Mediterranean) much of this litter can wash up on beaches.

● In parts of Greece, for example, seagrass detritus on beaches is harvested by local people for use as an agricultural fertilizer and soil improver.

9.3.4 Factors structuring the assemblages associated with seagrass

Seagrass meadows make available a high level of physical structure within what is usually a comparatively homogenous, featureless subtidal habitat. Additionally they are highly productive systems (from both the seagrass and the associated algae) and so provide potential food and shelter for a wide range of organisms. Many studies have demonstrated that seagrass beds have a richer associated community than surrounding soft sediments, including invertebrates living in the sediment and on seagrass blades (Connolly 1997, Lee et al. 2001), larger mobile invertebrates (e.g. crabs, cephalopods), and certain fish that shelter within the meadow (Jackson et al. 2001). This is perhaps not surprising as seagrass clearly provides a complex structure within which animals can hide from predators, and also provides completely new habitats (e.g. the leaves) that will host species not found in soft sediment. Lee et al. (2001) highlighted this, by demonstrating that all invertebrates recorded living in soft sediments off Hong Kong were also present in adjacent seagrass beds, but 48 additional species were only found associated with seagrass. However, the structural complexity of the habitat may not be the only factor influencing this boost in species diversity. Edgar (1999) undertook some elegant experiments with artificial seagrass and found that a much richer community developed in treatments that also included seagrass detritus than those

● Some fish are only found in a certain habitat, such as seagrass beds. These are termed obligate inhabitants, and would disappear with loss of habitat. Other fish use the bed out of 'choice' at certain times of the day. These are known as facultative inhabitats.

● A method used to assess relative predation levels in seagrass beds is **tethering**. Prey items (e.g. prawns) are attached to lines and placed within different parts of the seagrass bed. Comparative losses to predation are then recorded (e.g. Hovel & Lipcius, 2001).

with simply artificial leaves. He concluded that small invertebrates required the provision of food and showed little dependence on solely the seagrass structural characteristics.

Seagrass beds have an important role as a nursery ground for the juveniles of commercially important fish species (Jackson et al. 2001). For some species, the physical habitat is key as a shelter from predation, but other fish are attracted to seagrass beds due to their supply of food in the form of invertebrates; Jenkins and Hamer (2001) demonstrated that King George whiting (*Sillaginodes punctata*) juveniles tended to be associated with density of their prey (small crustaceans) as much as seagrass habitat. Clearly, seagrass beds are important due to a combination of shelter and food supply.

Similarly, seagrass beds may attract larger predators too, preying on the small fish and larger invertebrates sheltering in the bed. Predation pressure has been suggested to be a major force structuring the assemblages found within seagrass beds. For example, denser parts of seagrass beds (i.e. those with more shoots per metre) have been shown to provide more shelter for prey items (Edgar & Robertson 1992). Predators may be expected to forage less efficiently in dense, thick seagrass, thus providing a refuge for smaller animals. A further factor that may influence the success of predators is distance from the edge of a bed, the presumption being that the further in the bed a prey organism is, the harder it is for predators to reach it. This has been demonstrated in a range of experiments where survival of prey in the field has been measured, Bologna and Heck (1999), for example, reporting that bay scallops (*Argopecten irradians*) living along edges of beds suffered a higher predation than those within the meadows. However, these scallops on the edge also seemed to grow quicker, so there may be a trade-off between survival and growth.

Additionally, there is a direct relationship between the structural complexity of a meadow and the associated organisms, which also can explain their distribution within a bed. As a habitat becomes more architecturally complex, it is expected that more niches will become available allowing higher numbers of species to be supported. Such a relationship has been observed with coral reef fish and cactus-dwelling insects, for example. Similarly, there have been clear examples within seagrass beds of measures of complexity being related to diversity. Webster et al. (1998) demonstrated a positive relationship between shoot density and associated invertebrates, while several classic studies have positively related seagrass biomass to the number of species recorded (e.g. Heck & Orth 1980). Increasing seagrass biomass has been seen as an analogue for seagrass complexity, but Attrill et al. (2000) raised an alternative explanation that high seagrass biomass simply provides a greater leaf area (due to the two-dimensional nature of the leaves) and so the relationship is a species-area effect, with more species recorded because a greater area has been sampled. Attrill et al. (2000) highlighted this by demonstrating a close relationship between diversity and seagrass biomass, but not with alternative measures of structural complexity (Fig. 9.10). Whatever the reason, thicker, healthier beds do seem to provide a more favourable habitat for increased diversity.

The structure of a seagrass meadow itself can also affect the assemblage of organisms associated with it. A key concern in seagrass (and terrestrial) ecology is the fragmentation of beds (habitat) into smaller patches. Patchy beds can be naturally generated by specific environmental conditions, but much fragmentation of continuous beds is due to human activities, such as damage caused by boat anchors, propellers or moorings (Fig. 9.11). **Habitat fragmentation** has a range of consequences that impact organisms living within that habitat. The overall area of the seagrass can be reduced, which may impact big species requiring a comparatively large territory.

● The relative importance of fragmentation for conservation is a major issue. Is one large area better than several small areas? This has become known as the SLOSS debate: Single Large Or Several Small reserves.

Similarly, discrete patches of seagrass can form islands that are separated from the next patch by bare sediment, potentially isolating organisms from the main population. Crossing this sediment therefore presents a risk, and corridors of vegetation are very important for the movement of species such as crabs (Micheli & Peterson 1999). Fragmentation also increases the **edge effect** within seagrass habitat, a patchy bed having a much larger edge : area ratio than continuous beds. Previously we saw how predators are potentially much more successful at the edges of beds, so increasing the amount of edge can increase overall predation pressure. Fragmentation can also change water flow and sediment deposition, and ultimately physical conditions within a bed.

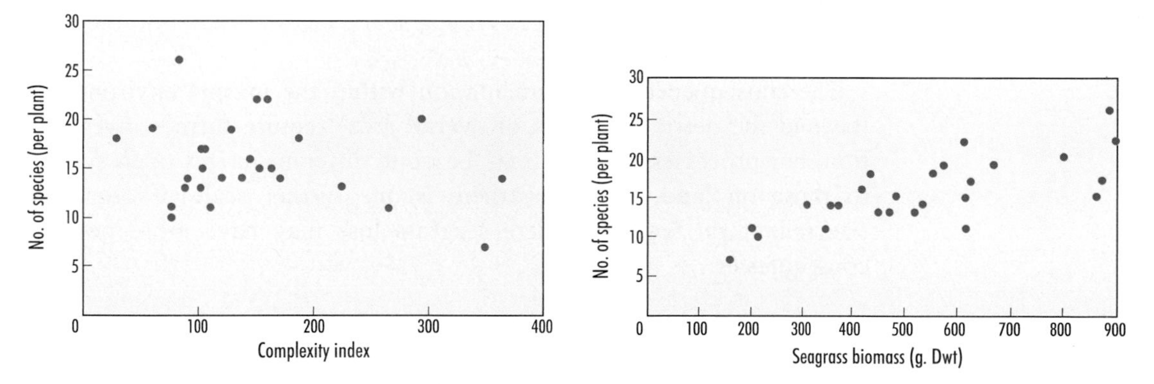

Fig. 9.10 The relationship between invertebrate diversity and seagrass architecture. Attrill et al. (2000) demonstrated that structural complexity was not related to associated diversity, but there was a relationship with seagrass biomass. They suggested this was due to increasing biomass providing a larger area to sample, and so more species were encountered.

Fig. 9.11 *Zostera marina* bed in Jersey, English Channel. The clear patches are gaps in the seagrass meadow caused by the physical impact of mooring chains used by the yachts. (Photograph: Emma Jackson/Paul Tucker.)

While the detrimental impact of fragmentation on land is comparatively well accepted, evidence from seagrass beds is more equivocal. Bowden et al. (2001) demonstrated higher diversity of sediment-dwelling invertebrates in large *Zostera marina* patches than small ones, but McNeill and Fairweather (1993) found that several small patches had a higher overall diversity than continuous patches of the same area. Studies into the impact of patch size on predation pressure have also produced contrasting results. Irlandi et al. (1999), for example, worked on the role of fragmentation on the survival of the same species as Bologna and Heck (1999), but found no consistent patch size effect.

● The sea is an open system as there are rarely any barriers at a localized scale to organism dispersal through the water between habitat 'islands', unlike between two forest areas, for example.

The consequences of fragmentation within the marine environment (beyond the detrimental loss of overall area) require further investigation, but processes involved may be quite different in this open system to those on land. However, there is one further scale of organism size reliant on seagrass where seagrass loss may have more notable consequences.

9.3.5 Seagrass as a food supply for large grazers

The majority of small grazers, such as gastropod molluscs and amphipods, that inhabit seagrass beds generally feed on the epiphytic algae rather than directly grazing the seagrass leaves. Such grazers therefore potentially have a role in limiting the growth of such algae within seagrass beds, and thus could influence the productivity of the seagrass itself by preventing algal overgrowth and shading (Williams & Ruckelshaus 1993). Additionally, invertebrate grazers may be key in reducing the impact of increasing nutrients within coastal systems, which may favour growth by algae over seagrass (9.3.2), although experiments are so far inconclusive (e.g. Keuskamp 2004). The only invertebrates that demonstrate a major grazing impact on the seagrasses themselves are sea urchins, such as *Lytechinus variegatus*, which can occur in huge numbers (up to 360 m^{-2}) within seagrass beds off the east coast of North America. The urchin has been demonstrated to seriously overgraze beds in Florida, removing (for example) nearly 0.8 km^2 of meadow in 8 months (Rose et al. 1999).

The major direct grazers of seagrass are, however, marine vertebrates, in particular two groups: the turtles and the sea cows (Sirenia). Both groups have representative species that feed primarily of seagrass and can have major impacts on the growth, productivity and structure of seagrass meadows. The green turtle (*Chelonia mydas*) grazes primarily on *Thalassia testudinum* (hence its common name, turtle grass), particularly in the Caribbean and adjacent tropical regions (Fig. 9.12).

● The green turtle is actually brown, so its name may appear somewhat inappropriate. However, it was named after the colour of its fatty flesh. The turtles were hunted primarily to make turtle soup.

Young turtles are pelagic omnivores, but once they reach 20 to 35 cm in size they begin benthic foraging, preferably on seagrass, though they will graze on algae where seagrass is not available (Bjorndal 1980). Adults demonstrate a diel feeding pattern, resting during the night (e.g.

Fig. 9.12 *Thalassia testudinum* is known as turtle grass. (Photograph: Tom Peschak.)

on coral reefs) and foraging during the daytime. Their specific foraging strategy depends on the seagrass density. In high-density areas turtles are selective, avoiding older leaves or those covered in epiphytes and consuming the young leaves or leaf tips, which are more nutritious and have lower lignin content. Turtles are seen to return to the same feeding area in order to graze newly grown vegetation. This continued grazing stresses the plant, resulting in reduced leaf production, so eventually these feeding areas are abandoned. In beds of low density, turtles are less selective and forage more widely.

Four extant species of Sirenia exist, three species of manatee (Fig. 9.13), and the dugong (*Dugong dugon*). Manatees are not seagrass specialists; while they feed off seagrass when in salt water, manatees spend much time in fresh water where they forage on submerged vegetation. For example, water hyacinths are a staple food for the West Indian manatee (*Trichechus manatus*) in many Florida rivers. The dugong occurs in the Indo-West Pacific region and is more strictly marine, with seagrass forming the main diet. Unlike manatees, dugongs can feed in large herds of 100 to 200 individuals, particularly in the extensive seagrass beds off northern Australia. Unlike turtles, dugongs often take the whole seagrass plant, roots and all, leaving distinctive feeding trails through the seagrass bed (Preen 1995). In dugong foraging areas off Australia, the seagrass beds are composed of two main species: *Zostera capricorni* is the dominant large species interspersed with patches of the fast-growing pioneer *Halophila* spp. Dugongs prefer to feed on *Halophila* as it has a high nitrogen content and low proportion of fibre, and as a pioneer species *Halophila* is the first to recolonize the disturbed areas left by dugong feeding. In this way, grazing by herds of dugongs alters the species composition of the seagrass meadow, stimulating the growth of *Halophila*, and herds have been observed grazing the same

Fig 9.13 Dugong feed on seagrass and generate disturbance to the seabed. Dugong and manatee use of habitat close to the shore in shallow water makes them vulnerable to injury by boat propellers. (Photograph: Oxford Scientific/photolibrary.)

location for up to a month as they crop the new growth of *Halophila* within their grazing trails. This activity has been termed **cultivation grazing**, as the dugongs' feeding results in a greater proportion of their favoured food supply.

● As well as four existing species of sea cow, a further species (the 10 m long Stellers sea cow, *Hydrodamalis gigas*) was hunted to extinction by 1768 – within 30 years of its discovery in the arctic waters of the Bering Strait.

9.3.6 The wider role of seagrass meadows

As well as providing a habitat maintaining high levels of biodiversity, a nursery ground for fisheries species and providing food for large endangered grazers, seagrass beds perform other critical functions that make them valuable to both coastal ecosystems and humans (Green & Short 2003).

For a marine ecosystem, seagrass beds have exceptionally high biomass and productivity, Duarte and Chiscano (1999) estimated an average production of 1012 g dry weight $m^{-2} y^{-1}$. This is higher than other primary producers, such as macroalgae (365 g dry weight $m^{-2} y^{-1}$) and phytoplankton (128 g dry weight $m^{-2} y^{-1}$). Seagrass beds therefore provide large amounts of carbon for input into coastal systems, supporting food webs and commercially important species such as prawns (see Loneragan et al. (1997) in 9.2.7 for example). The high leaf biomass produced by seagrass beds is also harvested by humans for a range of uses, such as packing material, fibre for use in mat weaving and even seagrass furniture and storage boxes (Green & Short 2003).

● For comparison with seagrass productivity: savannah = 900 g $m^{-2} y^{-1}$, boreal forests = 800 g $m^{-2} y^{-1}$, lakes and streams = 250 g $m^{-2} y^{-1}$. Rainforests, however, produce 2200 g $m^{-2} y^{-1}$ (Whittaker 1975).

Seagrass beds also provide key ecological services. The root-rhizome system enhances sediment stabilization and thus prevents erosion, while the foliage slows water currents through their baffling effect, encouraging sediment to settle and preventing resuspension. Extensive seagrass beds therefore are stabilizing features within the coastal landscape, and provide a natural form of coastal protection. In the low-lying Wadden Sea (Netherlands), seagrass debris was traditionally used to make dykes, and restoration of *Zostera marina* beds through transplantation is being investigated as a natural barrier to protect the coast (van Katwijk 2003). Perhaps the greatest value of seagrass beds is, however, more indirect through water purification and nutrient cycling (Green & Short 2003). Through sedimentation processes and the active uptake of nutrients into the seagrass meadow ecosystem, large seagrass beds can be effective in removing nutrients from the water column and trapping them for a comparatively long time in leaf litter; most algal-dominated or planktonic systems have a much quicker turnover of nutrients. Seagrasses can therefore help mitigate problems of eutrophication (9.3.2) and even bind organic pollutants. Similarly, seagrasses can help drawdown and remove carbon dioxide and play some role in the amelioration of climate change, particularly species such as *Posidonia oceanica* whose root-rhizome matte can persist for hundreds of years. While their overall impact compared with phytoplankton in the world's oceans will be low, their high productivity gives them a disproportionate influence in coastal systems (Green & Short 2003).

● The matte of *Posidonia* can be roughly aged due to known growth rates. It has therefore been used to date old wrecks in the Mediterranean!

Seagrass beds therefore provide a range of goods and services of benefit to coastal ecosystems and humans. Costanza et al. (1997) attempted to put an economic value on the world's ecosystems relating to these services, suggesting seagrass/algae beds are worth \$19 004 ha^{-1} y^{-1}, mainly due to their nutrient cycling role. Green and Short (2003) therefore estimated the global value of seagrasses to be \$3.8 trillion (i.e. $\$3.8 \times 10^{12}$), but pointed out that this does not represent the total worth of the ecosystem and is not a purchase value.

CHAPTER SUMMARY

- Mangroves and seagrasses are both 'true plants' and their aggregations represent two of the most valuable marine habitats in the world, being highly productive and with a very high associated biodiversity.
- Mangroves are found mainly in the tropics and are woody trees that can flourish at the land/sea interface. Through morphological and physiological adaptations, mangroves can deal with their roots being in waterlogged, anoxic sediment and tolerate the high levels of salt, surviving in conditions that would be fatal for most plants.
- A dynamic mix of terrestrial and marine organisms share mangrove forests. They are important sites for many bird species, including species of high conservation status such as scarlet ibis.
- Crabs are the most important marine group of organisms associated with mangroves, their burrowing activity and consumption of leaf litter being important for carbon cycling and in turning over the sediment (which affects the production of mangroves).
- Mangroves have an important wider role, exporting carbon to surrounding areas, protecting the coast from erosion and providing a nursery area for many fish species, including those from coral reefs.
- Seagrasses are the only angiosperms that can survive fully submerged in the marine environment. They are found across the world's coastal seas and can form huge meadows covering 1000s of km^2.
- Meadows mainly form in shallow subtidal soft sediments in clear water, but can extend into the intertidal zone or even grow on rocks. In the clear waters of the Mediterranean, *Posidonia* beds can grow down to 40 m.
- Seagrass leaves form a substratum for the settlement of a diverse epiphyte community. The relationship between seagrass and algae is complex, and can be altered by increased nutrient levels that favour the algae, outcompeting the seagrass.
- The physically complex nature of seagrass beds, compared to surrounding bare sand, results in meadows having a high associated biodiversity of marine animals (invertebrates and fish). The diversity and abundance of animals is associated directly

with the amount and complexity of seagrass that is present, from the density of shoots to the overall areal cover of seagrass.

- The global cover of seagrasses has been reduced dramatically (15% from 1993–2003) as they are sensitive to changes in light levels, nutrients and human mechanical disturbance.

FURTHER READING

Hogarth (1999) and Green and Short (2003) provide comprehensive examinations of mangrove and seagrass systems respectively. While Duarte (2002) examines the future of seagrass systems in the face of increasing human pressure on the marine ecosystem.

- Duarte, C. M. 2002. The future of seagrass meadows. *Environ Conserv* 29: 192–206.
- Green, E. P. & Short, F. T. 2003. *World Atlas of Seagrasses*. University of California Press.
- Hogarth, P. J. 1999. *The Biology of Mangroves*. Oxford University Press, Oxford.

Chapter 10

Coral Reefs

CHAPTER SUMMARY

The first sight of a coral reef has inspired many to seek a career in marine biology. The bands of bright colour fringing the coasts of tropical islands and highlighting small coral atolls in an otherwise deep blue ocean are a marvel. These reefs can be hundreds of metres thick, and yet they have been built by a thin veneer of living coral tissue. Species diversity on reefs can equal that in rainforests, but reefs are probably more accessible because you can float silently above them observing the inhabitants. Indeed, reef research has burgeoned with the advent of widely available scuba-diving gear in the 1950s. For millions of people in the tropics, reefs are not only a source of fascination, but their main source of food, building materials, and income. The activities of burgeoning human populations are threatening many reefs, and although reefs can be highly productive, reef growth is a transient process, and easily upset by changes in climate, sedimentation, fishing and pollution.

10.1 Introduction

Coral reefs support some of the most diverse and productive communities in the marine environment. Living corals create limestone formations that may be thousands of kilometres long or hundreds of metres deep. While the small polyps that form living coral are best viewed under a microscope, limestone coral reefs are clearly visible from space. Coral reefs constitute a shallow, productive, and brightly illuminated ecosystem, supporting an amazing diversity of plant and animal species. In many areas, by providing protection from wave energy, they also help foster ecological oases such as of mangroves and seagrass beds (Chapter 9), in otherwise deep and oligotrophic oceans.

This chapter describes the global distribution and typology of coral reefs, the biology of reef building corals, and the factors that influence their growth and reproduction. We also look at the productivity of reefs and the biology and diversity of animals supported by this production. Despite their vast size, reefs are among the most sensitive of marine habitats to human disturbance; they are in fact the marine ecosystem most threatened with anthropogenic degradation. We consider the roles of climate change, fishing, and pollution in driving this degradation.

10.2 Reef Development and Distribution

From the seafarer's perspective, reefs are rocks close to the sea surface that can damage a ship's hull. However, coral reefs are distinct because they are biogenic, or deposited by living organisms (Veron 2000). Growing corals and calcareous algae deposit carbonate and this can form vast limestone structures that raise the living reef high above the surrounding seabed. Thus, the Great Barrier Reef in Australia extends for 2000 km and has an area of 48 000 km^2 and Enewetak Atoll in the tropical Pacific is over 1300 m thick. While the reef at Enewetak Atoll is probably 50 million years old, most of the limestone that forms the structure of the Great Barrier Reef was deposited in the last 500 000 years and most modern reef growth dates only from the last 10 000 years, during the present interglacial period (Holocene) (Chapter 1).

● Coral reefs are biogenic – built by living organisms.

● Tiny coral polyps can build massive coral reefs.

Most reefs are found within a band 30° north or south of the equator (Fig. 10.1). The extent of shallow water coral reefs worldwide is 284 300 km^2 (Table 10.1); this is only around 3% of the total tropical continental shelf area but is inordinately rich in terms of biodiversity.

● See **www.wcmc.org** for global maps of coral reef distributions.

There are two broad groups of corals, reef-building or **hermatypic** corals and the non reef-building **ahermatypic** corals. Hermatypic corals are largely confined to the tropics, while ahermatypic corals are found worldwide. The growth of hermatypic corals and other reef-builders, and hence the distribution of coral reefs, is strongly influenced by the physical environment.

● Reef-building corals are known as hermatypic.

The latitudinal span of coral reefs corresponds with a temperature range of 18 to 36 °C, with optimum reef development occurring at 26 to 28 °C. Temperature helps explain the absence of reefs from much of the tropical coasts of the Americas (Fig. 10.1) and West Africa, where cool upwelling water (Chapters 6 and 7) does not support coral growth.

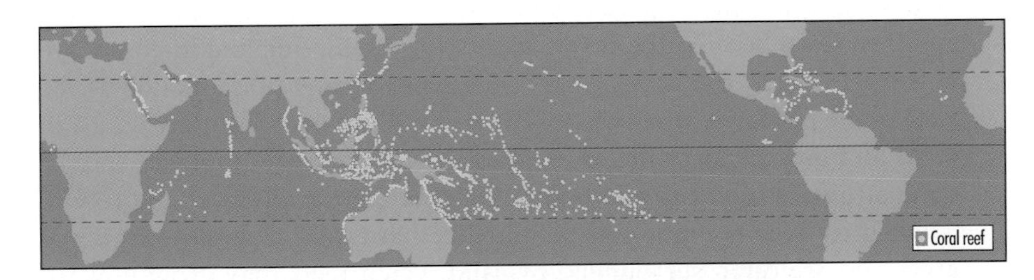

Fig. 10.1 Global distribution of coral reefs. After Spalding et al. (2001).

Table 10.1 The global distribution of warm water coral reefs. After Spalding et al. (2001).

Region	Area (km^2)	% of world total
Atlantic and Caribbean	**21 600**	**7.6**
Caribbean	20 000	7.0
Atlantic	1600	0.6
Indo-Pacific	**261 200**	**91.9**
Red Sea and Gulf of Aden	17 400	6.1
Arabian Gulf and Arabian Sea	4200	1.5
Indian Ocean	32 000	11.3
Southeast Asia	91 700	32.3
Pacific	115 900	40.8
Eastern Pacific	**1600**	**0.6**
Total	**284 300**	

● Reef growth is influenced by sea temperature.

Reefs in areas where temperatures vary considerably in space and time will often show highly intermittent patterns of growth. Such an area is the Galapagos Islands, where several ocean currents carrying water of different temperatures meet. During warm periods there may be rapid accretion of reefs, but during cooler periods, reefs stop growing and are eroded by a variety of animals, a process known as bioerosion. However in regions such as West Africa other forces are at work, and land-derived siltation also inhibits reef formation. In 10.3 we look more closely at factors affecting growth of corals.

● Fringing reefs grow on shelving coastlines.

Corals and other carbonate-forming organisms form several types of reef, depending on the availability of underlying substratum (existing carbonate reef or igneous rock), long-term changes in sea level, light levels, and wave action. On the shelving shores of most rocky tropical islands, fringing reefs develop (Fig. 10.2). These reefs develop because corals settle and grow on well-illuminated and shallow areas of the rocky substratum. As the corals grow towards the sea surface, they create a shallow platform and, over several thousand years, the platform reaches a height where it is just exposed by low tides. Further upward growth of the reef then stops because the corals cannot tolerate drying and intense wave action. However, the fringing reef continues to grow horizontally, as the edge of the platform provides shallow and well-lit areas for more corals to settle and grow. Horizontal development will continue until the base of the platform is too deep and poorly illuminated to support coral growth.

Patch reefs or bommies are small reefs that grow in shallow lagoonal areas and are often surrounded by sand. Patch reefs may grow upward until they are just below the surface at low tide, but they are often

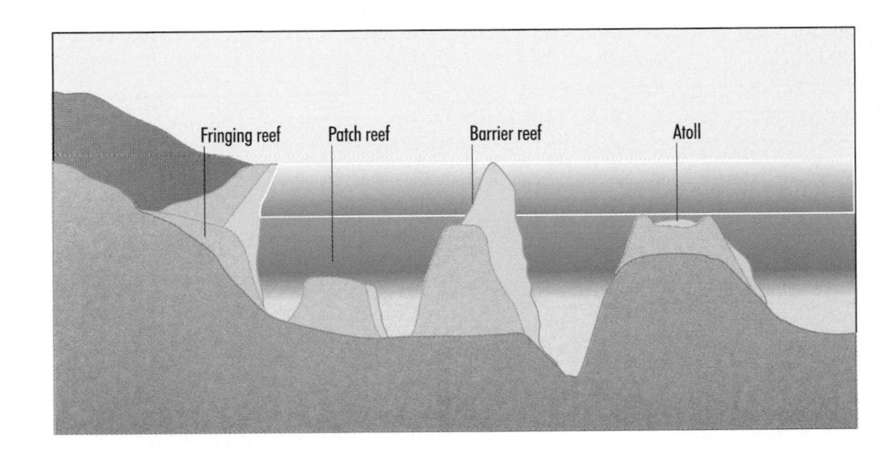

Fig. 10.2 Types of coral reef.

Fig. 10.3 A lagoonal patch reef in Fiji. (Photograph: Simon Jennings.)

deeper (Fig. 10.3). **Barrier reefs** surround many tropical islands and are usually separated from the land by lagoons with patch and fringing reefs. The lagoons are typically 1–10 km wide. Barrier reefs on one side of an island will be exposed to the prevailing oceanic swells, and are visible from the land as a line of breaking surf (Fig. 10.4). Barrier reefs may develop from fringing reefs following periods of sea level rise relative to the land, in which land subsidence may also be important. Barrier reefs grow quickly because the exposure of the corals to prevailing wind and currents can enhance growth (Fig. 10.5). **Atolls** are perhaps the classic example of a coral reef, a ring of reef with low-lying islands surrounding a central lagoon. Atolls can be 10 km or more across, and many are famed for their biological diversity, seabird

● Other types of reef are patch, atoll and barrier reefs.

Fig. 10.4 A lagoon and barrier reef in Vanuatu. Note the line of surf breaking on the barrier reef in the distance. (Photograph: Simon Jennings.)

Fig. 10.5 An outer reef slope on the Great Astrolabe Reef in Fiji. (Photograph: Simon Jennings.)

● The theory of atoll formation was first described by Charles Darwin.

colonies, and military significance (Box 10.1). Charles Darwin was the first scientist to speculate on the formation of atolls, and his theory is widely accepted today (Fig. 10.6). When a volcanic island is formed, it is quickly colonized by corals and a fringing reef forms. Then, as the island subsides or sea level rises, corals grow most rapidly on the outer edges of the former fringing reef and a barrier reef is formed. Eventually, the island may be lost altogether, but the ring-shaped atoll remains.

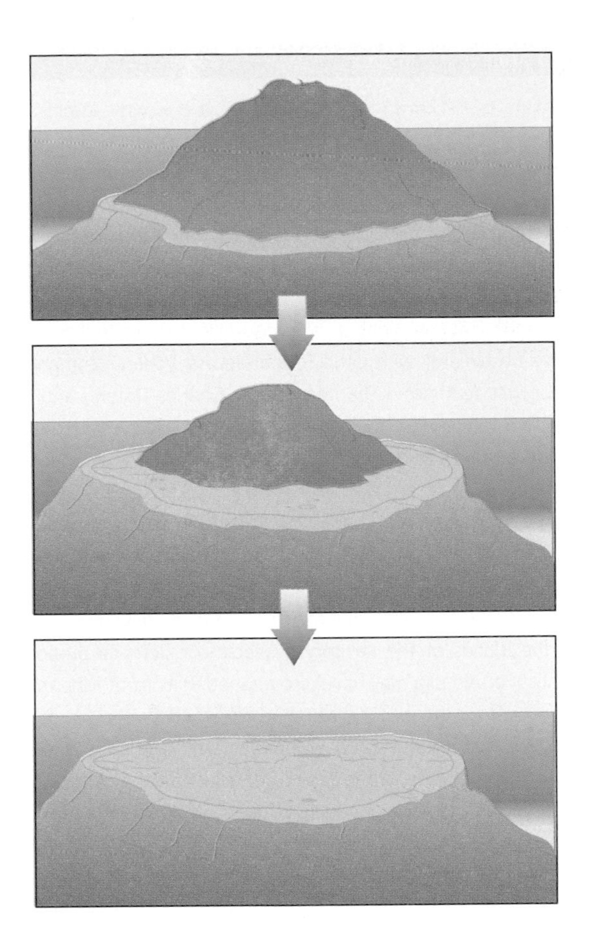

Fig. 10.6 The formation of an atoll. A volcanic island is colonized by corals and a fringing reef develops (upper). As the land subsides (middle), so a barrier reef is formed, which, as the land subsides further or sea level rises (lower) becomes an atoll.

Historical patterns of reef growth are assessed by examining cores cut into reefs. Reef growth is rarely continuous, and periods of fast growth are often interspersed with periods of **bioerosion**. The fastest recorded rates of net reef growth are around 20 m in 1000 years, but 3 m per 1000 years is more typical today. On the Great Barrier Reef and at Enewetak Atoll coral growth involves an average deposition of 4200 tonnes of limestone $km^{-2} y^{-1}$. The linear extension of individual corals can reach $20 cm y^{-1}$, and for short periods reef accumulation can be rapid, but such rates of annual growth are not sustained over geological time. In geological time, reef growth has been primarily affected by sea level fluctuations of up to 140 m, caused by repeated accumulation and melting of glaciers in the northern hemisphere and the associated expansion and contraction of the area of warm tropical seas (Veron 2000). Falls in sea level expose reefs, while increases in sea level promoted reef growth as corals grew quickly to remain well illuminated. However, sea levels have also risen by 10 m or more in 1000 years, which is faster than maximum coral growth rates; in such cases, corals have been left behind in deep water, unable to

● Fluctuations in sea level have a key effect on reef growth.

Box 10.1 Aldabra Atoll

Aldabra Atoll, 34 km by 14 km in size and one of the world's largest coral atolls, is situated in the Indian Ocean, approximately 200 miles north of Madagascar (9°25′S 46°25′E). The scientific importance of Aldabra Atoll was recognized by 1850, since, unlike most other atolls and despite regular visits from fishermen in preceding years, there was no permanent settlement on the island and it had been left relatively undisturbed. By 1850, giant tortoises had virtually disappeared from other Indian Ocean atolls, due to direct exploitation and habitat modification following guano digging and logging. The guano deposits on Aldabra were insufficient to make digging commercially viable and it was a hostile atoll with difficult boat access. When a company proposed to lease the atoll for woodcutting in the late 1800s, Charles Darwin and other eminent scientists argued for protection of the tortoise population. No protective legislation was passed, but private arrangements were reached with the lessor in return for assistance with rent.

The famous expedition diver Jacques Cousteau visited Aldabra Atoll in 1954, and the popular media coverage that followed publicized its biological importance. However, with the 'cold-war' at its height following the Cuban Missile Crisis, and Aldabra forming part of the British Indian Ocean Territory, the British and US governments signed a 1965 treaty to make the islands of the territory available for defence purposes. While the isolation and hostility of Aldabra may have discouraged visits from fishermen, they would have been no deterrent for the military, who had already developed another atoll, Diego Garcia in the Chagos Archipelago south of India as a military base. The British government were clearly unappreciative of the ecological importance of Aldabra. Denis Healey, a senior British politician, had responded to a question in the House of Commons by stating that 'the island of Aldabra is inhabited – like Her Majesty's Opposition Front Bench – by giant turtles [*sic*], frigate birds and boobies; nevertheless, it may well provide useful facilities for aircraft.' The Royal Society, Smithsonian Institution, National Academy of Sciences and other bodies vigorously opposed development. The protest culminated with the British Royal Society mounting an expedition to Aldabra in 1967. They aimed to obtain all possible information on the islands before any development started. Devaluation of the British currency rather than the views of environmentalists eventually rendered the military scheme too expensive and subsequent changes in international relations and defence strategy meant that development never occurred. Partly as a result of the military threat, however, Aldabra is one of the most thoroughly studied oceanic atolls and its ecological significance led to its designation as a World Heritage Site by the IUCN in 1982. It is now managed by the Seychelles Island Foundation (**http://www.sif.sc/aldabra/**).

Source: Stoddart (1984).

keep pace. Maximum rates of reef expansion and growth usually occur in the warmer interglacial period.

Reef growth is also rapid when shallow shelf areas subside, provided that the rate of subsidence does not carry corals out of the brightly illuminated zone.

10.3 Corals and Coral Communities

Corals build vast limestone structures but, in evolutionary terms, the corals are quite simple organisms. During reef formation, their most important features are the capacity to live colonially and to deposit calcium carbonate skeletons. Corals, and hence reef growth, are very sensitive to the physical environment.

● Reef building corals are known as hermatypic.

Hermatypic corals are largely confined to the tropics, while ahermatypic corals are found worldwide. The polyps of hermatypic corals usually contain symbiotic algae or **zooxanthellae**, which photosynthesize and provide energy for the coral polyp (Fig. 10.7). The relationship is symbiotic because the polyps provide protection and nutrients for the zooxanthellae while the zooxanthellae fix carbon and produce energy for the polyp (Table 10.2). A number of stressors, such as elevated temperature, ultraviolet light, and nutrient levels, can cause the symbiosis to break down and ultimately kill the corals.

● Hermatypic corals contain symbiotic algae that fix carbon dioxide and produce energy.

Zooxanthellae are **dinoflagellate symbiotic algae**, mostly belonging to the genus *Symbiodinium*. They contain characteristic dinoflagellate pigments as well as chlorophyll. Zooxanthellae can live independently of the host coral in a free-living motile stage, but in the coral polyps they lose their flagellae and motility. The symbiotic algae can be transmitted during coral reproduction, but corals can also obtain them directly from seawater (Muller-Parker 1997). Successful coral growth requires that the

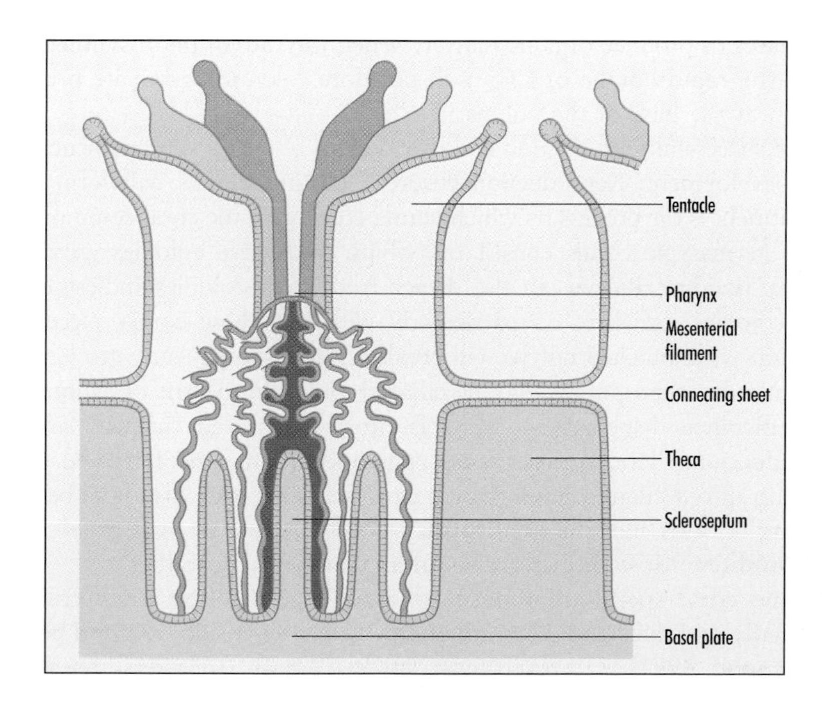

Fig. 10.7 A coral polyp. The zooxanthellae are found in the gastrodermis. After Muller-Parker et al. (1997) and others.

Table 10.2 Benefits of the symbiosis between zooxanthellae and corals. Modified from Muller-Parker and D'Elia (1997).

For the animal
Supply of reduced carbon providing low respiration costs and conservation of metabolic resources
Increased growth and reproduction
Increased calcification rate
Conservation of nutrients
Sequestration of toxic compounds by algae
For the zooxanthellae
Supply of CO_2 and nutrients from host
Maintenance in photic zone
Protection from UV damage by animal tissues
Maintenance of a high population density of a single genotype by host under uniform environmental conditions

corals can gather nutrients from oligotrophic tropical waters. The tentacles of the coral polyps can capture nutrient-rich zooplankton directly while the zooxanthellae provide photosynthetic by-products.

The zooxanthellae are **autotrophic** and require nutrients, light, and carbon dioxide to fix carbon. These symbiotic algae can obtain the nutrients by recycling waste products from their animal host, from the zooplankton caught by their animal host, and through direct uptake from seawater. During photosynthesis, carbon dioxide is removed from the water to produce organic matter. When the rate of photosynthesis is high, the rapid uptake of CO_2 will promote calcium carbonate precipitation at the base of the polyps.

The successful reproduction and recruitment of corals is fundamental to reef development. Reproduction ensures new individuals will form and recruitment is the process by which young corals join the coral community. Most hermatypic corals consist of polyps, and these colonies grow by asexual budding. Polyps can also detach from some colonies and establish new colonies elsewhere or, particularly with branching corals, pieces of skeleton with attached polyps will break off and subsequently grow.

● Corals can reproduce sexually and asexually.

While detached polyps may be ciliated and able to drift or swim, the long distance dispersal of corals is mostly achieved through sexual reproduction. Here, corals release eggs and sperm, and fertilized eggs develop into a ciliated larva, which can drift over long distances before settling. Corals may be **gonochoric**, **simultaneous**, or **sequential hermaphrodites**. Most species are simultaneous hermaphrodites.

Some coral species reproduce by brooding; the eggs are fertilized internally and the ciliated larva (**planula**) develops inside the coral polyp (Richmond 1997). However most species are broadcasters, releasing eggs and sperm into the water column for fertilization (Fig. 10.8). Spawning corals exhibit mass releases of gametes during particular periods of the year, usually related to lunar cycles. On the Great Barrier Reef gamete release by over one hundred species is synchronized, presumably because the corals are responding to common cues. Synchronization no doubt increases the probability of fertilization within species and also has the effect of swamping egg and larvae predators.

Once eggs are fertilized they develop into planulae that are capable of settlement in 1–3 days. These larvae generally only acquire their zooxanthellae after settlement and metamorphosis, whereas the larvae of brooders already contain zooxanthellae from the parent colony. During metamorphosis the larva secretes a calcified basal plate and the form of the polyp develops. Factors such as suspended sediment load, temperature, and salinity will affect rates of coral egg fertilization and larval survival. Interestingly, even distantly related corals seem to require the presence of crustose-coralline algae for successful metamorphosis and settlement. These algae in turn need the grazing of fishes and invertebrates to keep them clear of other organisms, thus within the ecosystem there is an indirect connection between the amount of grazing and coral establishment.

The reproduction and growth of hermatypic corals is strongly influenced by the physical environment. Corals are principally affected by temperature, light, depth, salinity, wave energy, sediment, and pollution.

● The physical environment determines coral distribution.

Fig. 10.8 Coral spawning showing the eggs emerging from the polyps. (Photograph: Greg Bunch/gbundersea)

Light is vital to corals, as it allows the zooxanthellae to photosynthesize. Most reef accretion occurs at depths of 0 to 10 m and even in clear oceanic water hermatypic corals are rare below 30 m. We saw in Chapters 2 and 6 how quickly light attenuates as it enters the water, although light penetration is good in shallow water when the sun is almost vertically overhead and the water is clear. Indeed, working underwater on coral reefs at a depth of 3 to 4 m there is still sufficient ultraviolet light to cause sunburn.

● Corals need light to grow and are rarely found at depths >30 m.

Diving down an outer reef slope it is always surprising how quickly the bright colours of the reef and a superabundance of small colourful fish are replaced by more sombre oceanic blues. As the light intensity fades, so corals become increasingly scarce, and are found only in small patches (Fig. 10.5). The availability of light determines the maximum depth at which corals are found. Coral linear extension rates may drop from a centimetre per year at 2–5 m depth to a few millimetres or less at 10 m. In the presence of bioeroders, it is not surprising that almost all reef accretion takes place in the shallow well-lit zone.

We have already seen that temperature governs the global distribution of reefs, with optimum reef development occurring at 26 to 28 °C. Corals have quite low tolerance to temperature variation, although tolerance is species and site specific. Corals that have evolved in very stable temperature regimes are typically less tolerant to change than those that have evolved in more variable environments. Sheppard et al. (1992) for example, describe reef corals in the western Arabian Gulf that survive maximum temperatures of 38 °C and minimum temperatures of 12 °C. Such temperature changes in the central Indian Ocean would cause extensive coral death and, in many cases, sustained increases in temperature of 2 to 3 °C for a few weeks, as occur during El Niño events (Chapters 6 and 7), may be enough to lead to mass mortality.

● Most corals are sensitive to temperature change

Reefs grow fastest when salinity is 33 to 35, typical of offshore oceans. At lower salinities, in the vicinity of river estuaries, for example, extensive coral reefs do not develop, even if sediment loads are low and light penetration through the water is good.

Wave action generates currents around reefs and enhances the transfer of nutrients and dissolved gases such as oxygen. However, some corals cannot withstand the turbulence induced by waves. For this reason, wave action has an important structuring force on coral communities, especially on outer reef slopes. Many tropical regions are also affected by powerful waves during hurricanes. The history of hurricane damage has affected the distribution of modern reefs, since corals in areas impacted by hurricanes will usually be tolerant to intense disturbance or be short-lived species that can recolonize an area between hurricane events. In the Caribbean for example, the distribution of reef types reflects historical patterns of storm and hurricane disturbance (Hubbard 1997).

Some corals can remove sediment from their surface, but in general sediments cause smothering, abrasion, shading, and recruitment inhibition in reef corals. Pulse sediment loads are often storm related and reef organisms may be only acutely affected by them, but suspended sediments that result from agricultural practices on land, dredging, and coastal development have had a significant impact on reefs in recent times. One of the main effects of increases in suspended sediment load is to shade corals, and they can no longer grow in the reduced light levels.

- The zonation of corals is partly explained by the effects of wave action.

- Suspended sediment can kill corals.

10.4 Coral Reef Productivity and Food Chains

Swimming up to coral reefs from the surrounding waters it soon becomes clear that reefs are highly productive. The productivity is usually expressed per unit of projected surface area of the reef. This is very different from the true surface area, which accounts for the complexity of the reef habitat, and can be 15 times greater than projected area (Hatcher 1988, 1997).

- Coral reefs are very productive, but most of the production is recycled within the reef.

Corals are responsible for a lot of the carbon fixation on coral reefs, but less than half the carbon they fix is available to consumers, because most is respired, recycled or accumulated within the coral colony. In fact, only a few reef species, such as the gastropod *Drupella*, the crown-of-thorns starfish *Acanthaster*, and some parrotfishes (Scaridae) or butterflyfishes (Chaetodontidae), feed on coral polyps directly. Reef algae are extremely productive, and the carbon that they fix plays a much greater role in supporting reef food chains because they are directly grazed by many reef species and because across whole reefs coral typically cover only a small proportion of the total substratum.

- Not much coral production is eaten directly.

Shallow and well-lit coral reefs provide an ideal substratum for algal growth. Most reef areas not covered with corals are colonized by turfs of filamentous and small fleshy algae that are directly grazed by fishes and invertebrates. The algae and photosynthetic bacteria that account for most reef production across whole reefs ultimately rely on the surrounding seawater to supply organic nutrients and disperse the organic compounds they leak or excrete. Most turf algae are constantly grazed by herbivores, keeping them in the most productive phase of growth through the intense feeding and probably local recycling of nutrients via defecation and excretion. Rates of gross primary production by reef algae range from 1 to 40 $g\,C\,m^{-2}\,d^{-1}$, 20 to 90% of what is left after respiration and export being grazed by herbivores, even though some algal species have evolved anti-grazing mechanisms.

- Reef algae can be very productive and support many fish and invertebrate food chains.

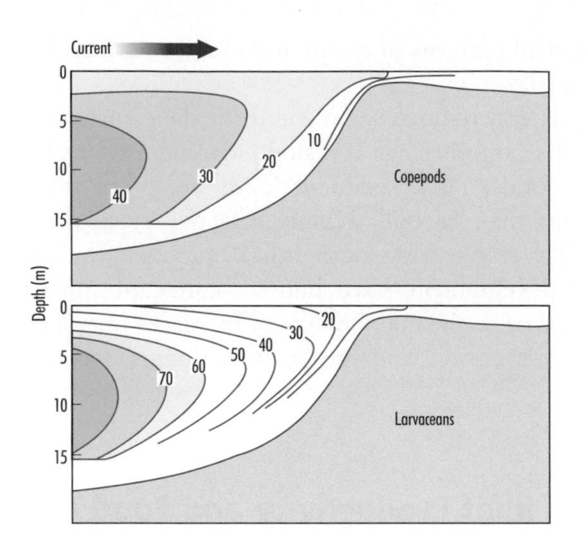

Fig. 10.9 Zooplankton densities at sites on the reef front. Zooplankton abundance falls to almost zero as water crosses the reef because fishes and other planktivores are grazing the zooplankton. After Hamner et al. (1988).

Phytoplankton production in the water surrounding reefs also makes an important contribution to the production of reef animals. The phytoplankton support zooplankton that are grazed by reef animals as currents carry then across the reef. The extent of plankton use by reef fishes was dramatically illustrated by Hamner et al. (1988). As water flowed over the reef slope, virtually all copepods and other zooplankters were picked off by planktivorous fishes (Fig. 10.9). Although production in the waters surrounding reefs is low on a per unit area basis, the continuous flow of water over the reefs means that reef planktivores can eat plankton from a very large area of ocean (Polunin 1996).

● Reef animals also eat plankton that drift over the reef.

A 24-hour cycle is used to measure net community production on reefs because photosynthetic production takes place in daylight hours and respiration occurs at night. Net production over the 24-hour period is known as the **excess production** (E), such that an excess production of zero equates to a **photosynthesis/respiration** ratio (**P/R**) of one. On most reefs, E is very low in relation to measured rates of production. This indicates strong competition among animals for the food produced in the ecosystem. The low E and prevailing oligotrophy of surrounding waters also imply nutrient recycling within the system, but it is unclear how this might occur. For entire reefs, gross community production is typically 2.3 to 6.0 $g\,C\,m^{-2}\,d^{-1}$, while E is -0.01 to 0.29, and P/R is 1.0 to -1.11 (Hatcher 1988).

In spite of their primary productivity approaching that of upwelling systems (Chapter 6), the capacity of reefs to produce fish is probably no greater than that of other continental shelf areas. Thus in the South Pacific, fishery yields of 14–35 $t\,km^{-2}\,y^{-1}$ have been reported from many locations and this compares favourably with 20 to 50 $t\,km^{-2}\,y^{-1}$ from temperate shelves.

10.5 Reef Fauna

While 100 000 or so species are now known from coral reefs, when these communities have been fully studied they may ultimately prove home to over a million species. However we already know that on an area-specific basis, the species richness of animals on coral reefs is exceptionally high. At least 4000 species of marine fish, almost one third of the global total, are found on coral reefs.

- Coral reefs are among the world's most diverse ecosystems.

We also know that the diversity of reef animals is far from uniform, with higher species richness of most groups in the Indo-Pacific rather than the Atlantic, and a centre of diversity in South-East Asia. In the case of some 794 species of Scleractinian reef building corals, for example, the number of recorded species peaks in South-East Asia and falls at higher and lower latitudes and longitudes. Remote reefs in higher latitudes often have the fewest species. The same general patterns of species richness apply to almost all other groups of reef animals, for example between the Indo-West Pacific, which contains South-East Asia, and the East Pacific (Table 10.3).

- Diversity is determined by reef history and the local physical environment.

Local variations in the diversity of reef organisms are attributable to the regional pool of species, the variety of habitats and spatial differences in environmental factors that determine which species are represented in a particular patch of habitat. When a specific site on a reef is followed over time, it will be evident that the mix of species changes. This dynamism is determined by the interaction between processes introducing species new to the site (e.g. larval drift, survival, and settlement) and those leading to local demise (mortality of all kinds). Area is important in this; large areas have more species than small areas, and there are remarkable similarities in their make-up (Bellwood & Hughes 2001).

- On geological time scales there have been mass extinctions of reef animals.

Reef-building corals evolved over 200 million years ago, and their distribution was effectively circumglobal in shallow seas for at least 100 million years. Given the position of the continents at this time (Chapter 1) the only

Table 10.3 The diversity of reef animals expressed as the number of known species. Modified from the compilations of Spalding et al. (2001) and Paulay (1997).

Group	Indo-West Pacific	Eastern Pacific	Western Atlantic
Scleractinian corals	719	34	62
Alcyonarian corals	690	0	6
Cypraeid gastropods (cowries)	178	24	6
Bivalves	2000	564	378
Echinoderms	1200	208	148
Reef fish	3000	300	750

major barrier to coral distribution was the proto-Pacific ocean, extending from Asia to the Americas. Subsequently, the continents moved apart, and created oceans that were largely unconnected, and the coral reefs and associated communities in the Indo-West Pacific, East Pacific and West Atlantic, and East Atlantic evolved in different ways (Veron 2000). Then, around 3 to 4 million years ago, the Isthmus of Panama emerged and severed the connection between the East Pacific and West Atlantic, and these communities also evolved independently. Based on the fossil record, it is clear that the coral reefs we see today are not the result of a long history of evolution in a stable environment. Rather, the diversity we observe is a result of large-scale extinction events and rapid periods of evolution. Some reefs were effectively eliminated by mass extinction every 20 to 30 million years, with a delay of 3 to 10 million years before they appeared again. Other reefs were less affected by these extinctions and evolution continued.

● Modern patterns of large-scale diversity are influenced by the isolation of reef provinces following continental drift.

Events during the last 2 million years appear to have led to the major differences in the diversity of reef animals in the East Pacific, West Atlantic, East Atlantic, and Indo-West Pacific (Paulay 1997). The East Pacific reef fauna proved vulnerable to the variable environmental conditions in this region and there were several large-scale extinctions after the East Pacific and West Atlantic faunas became isolated. This isolation prevented recolonization from the West Atlantic although, despite the distances involved, there was some recolonization from the central Pacific. Today, therefore, more of the East Pacific fauna is shared with the Indo-West Pacific than the Atlantic.

In the West Atlantic there were several mass extinctions, including those associated with oceanographic changes that followed the closure of the Isthmus of Panama. Despite subsequent speciation, diversity in the West Atlantic remains relatively low. Rates of extinction in the Indo-Pacific region were lower than elsewhere, at least in the last 5 million years, and almost all extant coral genera are found in the fossil record. Rates of speciation were also low, but coupled with the low rates of extinction they contributed to the high levels of diversity observed today.

When scientists first began large-scale studies of coral reefs in the 1950s and 1960s, they sent back pictures of colourful flourishing reefs supporting abundant and diverse communities of animals. Early research focused primarily on the systematics and ecology of the corals and other animals, and the first collections of time-series data on coral abundance and diversity were initiated. It became clear that reefs were not the stable communities, which were structured by competition, that many scientists thought they were looking at. Rather, reefs were dynamic ecosystems.

● Coral reefs are dynamic systems.

In contrast, the geology of reefs has tended to emphasize continual change because of the huge time-spans involved, but there is geological evidence also of consistency in community structure over considerable periods of time. Thus, reefs formed successively at the same location by repeated land uplift in northern Papua New Guinea exhibit remarkable within-habitat similarity in the coral species present over a 95 000 year time-span. This is in spite of there being more than seven times as many coral species available in the regional pool of species (Pandolfi 1999).

Thus reefs do change over time, but at the same time demonstrate some characteristics of stability. Many questions surround the nature of these processes structuring this intricate ecosystem, and the extent and nature of its resilience to natural and anthropogenic environmental change.

10.6 Threats to Coral Reefs

In the 1970s many scientists had begun to ask whether contemporary changes in reefs were faster and larger than those seen previously, and whether these changes were the result of human impacts. Decline in coral cover on Caribbean reefs and on the Great Barrier Reef since the 1970s and 1980s respectively is cause for great concern and indicates that reef ecosystems are not sufficiently robust to withstand the major human disturbances (Bellwood et al. 2004).

Most reef research is still relatively young and most 'long-term' time series of ecological data began in the last 35 years. Since many of the first scientists to work on reefs considered that they were rather stable ecosystems, the first observations of disease and damage recorded by reef scientists were considered to be unprecedented and catastrophic. In the longer term, a more pluralistic view of reefs has arisen. It is clear that reef development is a cyclical rather than a continuous process, with episodes of rapid coral growth interspersed by episodes of coral death and erosion.

- Coral reef research is relatively young.

Despite the natural variability and evidence of resilience over large time-spans, it is also clear that human impacts are a threat to coral reef ecosystems. Today, the relative roles of natural and human activities on reefs are a focus of much research, which seeks to understand what drives the balance between coral reef growth and bioerosion, the causes of coral bleaching, and what has been recognized as the worldwide degradation of reef ecosystems. The threats are both direct, from increased sedimentation, destructive fishing practices, ship groundings, or pollution, and indirect, due to climate change and over-fishing (Chapters 12 and 14).

- Humans are having detrimental impacts on coral reefs.

10.6.1 Disease

Corals and other reef organisms are susceptible to diseases caused by pathogens or parasites (Richardson 1998). The likelihood and effects of disease may be aggravated by the direct and indirect impacts of human activities. One of the first recorded coral diseases was the so-called 'black band disease' that affected brain corals in the Caribbean. This is caused by the cyanobacterium *Phormidium corallyticum*, which invades the coral tissue and produces a fine black band that spreads across the surface of the coral. The cyanobacterium and associated microorganisms create anoxia, which kills the living coral tissue. The bacteria utilize the organic compounds released by the dying coral cells. The black band will move a few millimetres every day, leaving a bare coral skeleton, which is usually colonized by filamentous algae. It appears that physically damaged corals are more susceptible to black band disease. Several other diseases have since been recognized in corals, although accurate diagnosis is often difficult.

● Corals can suffer from a number of diseases.

10.6.2 Storms, cyclones, and hurricanes

In shallow water, storm-induced waves can lead to destruction of branching corals and even to destruction of corals on the reef slope due to debris falling from the reef crest. Storms can also cause the death of reef fish when they try to seek shelter in corals, which are then smashed by wave action. Coasts exposed to frequent storms will have entirely different coral communities from those found in sheltered areas and close to the equator, where cyclones and hurricanes are relatively rare. In the cyclone belt, typically 10° to 25° north and south of the equator, many shallow reefs are dominated by fast-growing *Acropora* corals that quickly recolonize reefs following hurricane damage. With hurricanes occurring at intervals of years to decades, slow-growing corals rarely have time to develop. The recovery of reefs following storms may be further delayed by effects of human activities.

● Fast growing corals dominate reefs exposed to cyclone and hurricane activity.

10.6.3 Climate change

It is widely predicted that climate change will lead to rising sea levels over the next century. Future sea-level rise is expected to be 3–10 mm y^{-1} and evidence suggests that healthy corals could accrete at this rate in the past. However, there is concern that other impacts will weaken corals, degrade coral communities, and impair their capacity to accrete during periods of sea-level rise. These other factors include climate-linked increased frequencies and intensities of storms or ENSO events (Chapters 6 and 14; Box 10.2) and also direct human impacts such as nutrient inputs from poor land management. In deeper water corals will not photosynthesize or deposit carbonate as fast, and their ability to keep up with sea-level rise and resist bioerosion will be reduced.

Box 10.2 Natural archives of environmental change in corals and reef ecosystems

Coral reefs have been profoundly impacted by humans but it is not always easy to discern the relative roles of natural and human processes in long-term changes. We expect that the changes wrought by humans cannot decline in the foreseeable future (e.g. McClanahan 2002) and in some respects records of past change can inform us about forms of future change. We are lucky that through coring into reef structures and individual corals (figure below) much information has been derived about rates of change in the surrounding environment and responses to this of the corals themselves and the wider community. Four examples illustrate this remarkable historical data.

Massive corals from the Huon Peninsula of northern Papua New Guinea retain an annual growth pattern and fluorescence of the growth bands varies with rainfall, while the relative natural abundances of the stable isotopes of oxygen in the skeleton are indicative of El Niño-related changes in water temperature since 1880. This in turn has made it possible to infer that El Niño is not just a modern phenomenon, but the magnitude of recent El Niño events tends to be larger than over the last 130 000 years (Tudhope et al. 2001). In western Sumatra, cores from massive corals show that a mass mortality of corals and fishes on reefs coincided with upwelling and a bloom of algae, the latter probably driven by inputs to the sea of iron derived from burning forest during a dry year (Abram et al. 2003).

Coral coring into whole reef structures has shown that at Discovery Bay in Jamaica, the modern changes in the community are unprecedented over the last 1260 years (Wapnick et al. 2004). In Panama, the community of reef corals has in the most recent decades converged with that of corals reefs some 1500 km distant in Belize, which were distinct biologically for more than 2000 years (Aronson et al. 2004).

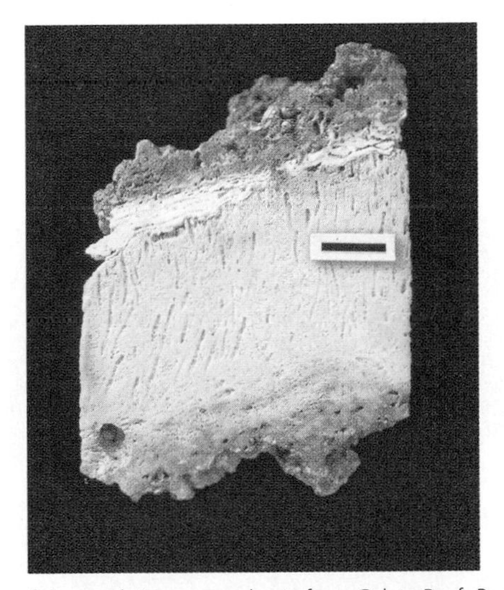

Example of a core of the coral *Acropora palmata* from Galeta Reef, Panama. The excellent preservation, upward-curving growth pattern of the individual corallites, and light-coloured layer of crustose-coralline algae at the top all indicate that this coral has not been overturned and probably was part of a framework formed in place. Submarine microcrystalline calcite coats the coralline crust, a red encrusting foraminifer *Homotrema* tops this and penetrates the unprotected base of the coral. Scale bar 0.5 cm (Macintyre & Glynn 1976). Photograph: Ian Macintyre.

10.6.4 Coral bleaching

Coral **bleaching** is a generalized response of corals to stress (Coles & Brown 2003). Bleaching is so called because the zooxanthellae are ejected from the corals and the skeleton, which is typically white, becomes visible (Fig. 10.10). Bleaching is also associated with decreases in growth and reproductive output, and prolonged bleaching leads to coral death. Following bleaching the coral framework is often taken over by bioeroding animals and reef growth is replaced by bioerosion. Moreover, bleached corals will be colonized by filamentous green algae which although more productive tend to take over the reef from corals and thus reduce reef accretion. Since fish and invertebrates are also commonly dependent on the refuge that coral provides, it is likely that many species will be lost and herbivorous species will become more abundant.

● Coral bleaching has major implications for the future of coral reefs.

Many of the most severe bleaching events have been attributed to high water temperatures, which occur seasonally and may be further enhanced in certain years by regional climatic events. Bleaching can be induced by short term exposure to temperatures that are elevated by 3 to 4 °C and longer term exposure to elevations of 1 to 2 °C, although vulnerability to bleaching depends on the thermal regime to which corals have already adapted.

The 1997–8 ENSO event caused increases in sea temperature and bleaching on most of the world's coral reefs outside the central and western Pacific. Ultimately this led to large-scale coral mortality and losses of 70–80% of coral cover on many reefs. The effects were particularly pronounced in the Indian Ocean, where more than 90% of coral died. The biological

Fig. 10.10 Bleaching of massive corals in the Galapagos Islands. (Photograph: Simon Jennings.)

and economic consequences have been profound on many islands that rely on reefs to act as coastal sea defences, to attract tourist divers and to provide fishing grounds (Jones et al. 2004). While the ENSO is a natural event, the frequency and intensity of ENSOs may be affected by human-induced climate change. The frequency of coral bleaching is expected to increase in the future.

● The frequency of coral bleaching events is expected to increase in the future.

10.6.5 Crown-of-thorns starfish and other coral predators

Crown-of-thorns starfish (*Acanthaster planci*) feed on living coral (Fig. 10.11), and at high abundance they can kill large areas of reef. Crown-of-thorns starfish are free-spawning sexually reproducing species, and a single female may produce 12 to 60 million eggs in a spawning season. The eggs and larvae are planktonic for 9 to 23 days before settlement. Following settlement, the juvenile starfish grow to 1 cm diameter in 4 to 5 months and shortly afterwards, they start to feed on live coral. The juveniles grow fast and reach maturity in 2 years or so, feeding on coral by everting their stomachs over the live coral and secreting an enzyme that breaks down the coral tissue, allowing them to absorb the products. Crown-of-thorns starfish often move across reefs in fronts, large groups of animals that kill the most of the corals in their path.

● Crown-of-thorns starfish (COTS) are a major agent of change on Pacific reefs.

Typically, the abundance of adult crown-of-thorns starfish is 1 to 20 km^{-2}, but during outbreaks they can reach densities of 500 km^{-2}. The first major crown-of-thorns **outbreak** that was witnessed on the Australian Great Barrier Reef led to massive media interest and a race by

Fig. 10.11 A crown-of-thorns starfish. (Photograph: Jeffrey Jeffords/Divegallery.com.)

scientists to understand and control the population explosion. Outbreaks have been reported from many other regions. There are many possible causes of starfish outbreaks and no single theory seems to apply in all circumstances. One popular theory suggests that fishing has reduced the abundance of fish predators on juvenile starfish, and this has allowed the juvenile starfish to proliferate, leading to outbreaks of adults in subsequent years (Moran 1986). Other theories suggest that outbreaks are due to natural phenomena such as exceptional larval recruitment. In reality, outbreaks probably have different causes on different reefs. There is evidence that outbreaks occurred prior to human impacts, but equally, there is evidence for an increased incidence of outbreaks in some heavily disturbed or fished regions. Recovery of reefs may take 20 to 50 years.

- Starfish outbreaks are often linked with disturbance associated with fishing.

10.6.6 Pollution, sediments, and nutrients

The main causes of increased sediment loadings on coral reefs are runoff following deforestation, port development, and dredging. Erosion rates from deforested farmland can be 100 times greater than from natural forests, often reaching 1000 tonnes or more $km^{-2} y^{-1}$. Coastal development also increases sediment loading, as road run-off is often discharged directly to the coast. It is hard to separate the effects of nutrient inputs from those of sedimentation on coral reefs, because they often go hand in hand. However, nutrients derived from industrial discharge, vegetation removal, and lack of sewage treatment are thought widely to have accelerated algal growth on many reefs at the expense of corals. Corals are also directly sensitive to nutrients such as phosphate. Locally, oil pollution from tanker or ship groundings has killed areas of reef and there are major concerns about increased nutrient levels from farmland and sewage inputs.

- Nutrients from agricultural runoff and sewage discharge have accelerated algal growth at the expense of corals.

10.6.7 Fishing

Burgeoning human populations with inherent needs for food and income have driven the development and expansion of fisheries on coral reefs. Fishing leads to direct damage to reef habitat and a reduction in the abundance, biomass, and mean size of species targeted by the fishery, and local extinction of the most vulnerable species. Fishing can also have more profound and complex impacts on the structure and function of reef ecosystems.

Habitat destructive fishing methods may be used on reefs, such as reef drive netting and trapping, blast and chemical fishing. The latter methods are highly unselective and often adopted by fishers desperate to meet immediate requirements for food or income. Blast fishing is practised on many reefs, even though it is often illegal. Commercially produced explosives from mines or armaments are frequently used, but mixes of charcoal and oxidizing agents are a common alternative. Completed bombs are dropped into the water after igniting a short delay fuse. Given that charges are deployed in different locations, detonate at different depths, and contain different types and amounts of explosives, it is hard to quantify their impact. However, repeated explosive fishing reduces actively growing reef to dead coral rubble. Much of the kill is wasted, since fish and invertebrates will be eaten by other fish, invertebrates, and birds before collection begins and fishers only collect a small proportion of the remaining fish. Explosive fishing thus impacts the reef ecosystem at all levels and recovery from damage takes many years.

● Much of the 'kill' associated.

The ecological relationships between sea urchins and fishes, which we discuss in 10.7, are readily affected by fishing (McClanahan et al. 2002). The persistence of herbivorous fishes appears to depend on the presence of sea urchin predators, which maintain sea urchin populations at a level where their low gross production makes them inefficient competitors. A fishing-induced shift towards a herbivore community dominated by one species of urchin, rather than many species of fishes and urchins has been a major factor in the degradation of reef ecosystems in the Caribbean. These impacts are further discussed in Chapter 12.

● Fishing has driven sea urchin population explosions in some parts of the world.

Our examination of separate human impacts on coral reefs is, of course, rather misleading. Even though scientists tend to study single impacts rather than synergistic or cumulative impacts, impacts are rarely separable. Most over-fished reefs, for example, will be found close to highly populated areas, where agricultural development leads to high sediment loadings, discharge of human and animal waste leads to high nutrient loadings and where land is reclaimed for port developments and possibly housing. One of the main criticisms of many human impact studies on reefs is that there is too much emphasis on finding a single cause of change in the reef ecosystem, and not enough emphasis on looking for synergistic and cumulative causes (McClanahan 2002).

● Human impacts need to be studied holistically.

10.7 Reef Growth and Bioerosion

While corals are laying down limestone and building reefs, many other reef processes are eroding the limestone and producing coral rubble and coral sand even on undisturbed reefs (Fig. 10.12). Reefs are presumed to be 'healthy' and showing net growth when corals have the upper hand. However, in many cases erosion is the dominant process and the reef fails to grow. Even on growing reefs, erosion rates are usually high and net accretion only just exceeds net loss. Relatively small shifts in the

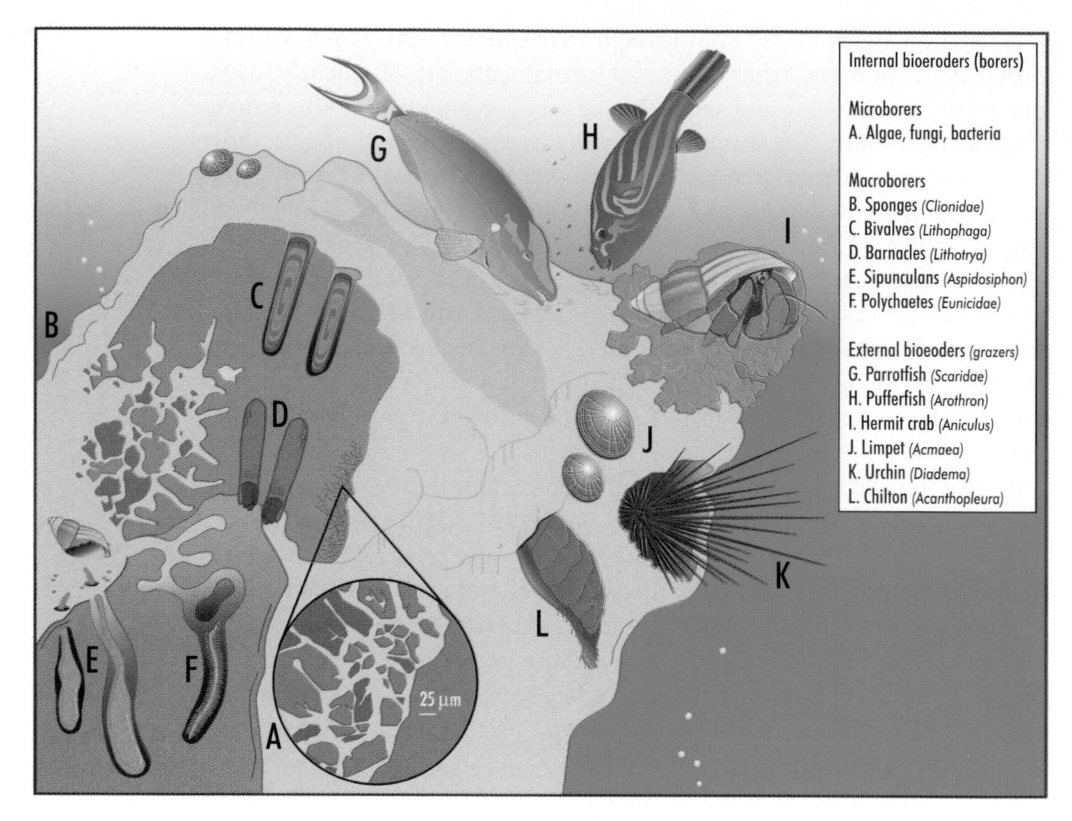

Fig. 10.12 Internal and external bioeroders on a coral reef. After Glynn (1997).

● Reef growth is a balance between accretion and bioerosion.

structure of reef communities can thus shift the balance from net accretion to erosion (Glynn 1997).

Most of the erosion of the reef substratum is by organisms. Many of these bioeroders are not visible on the reef surface, since they bore into coral skeletons. Internal borers include species of sponge, polychaetes, crustaceans, and molluscs. While rates of bioerosion are very variable, and will depend on the solidity of the carbonate matrix, borer abundance, and body size, bioerosion rates of over 2 kg $CaCO_3\,m^{-2}\,y^{-1}$ have been recorded for individual species. The principal external grazers are the larger species of mollusc, echinoderm, and fish. Depending on abundance, individual species can bioerode 3 kg $CaCO_3\,m^{-2}\,y^{-1}$ and more (Glynn 1997).

● Many bioeroders are hidden in the reef.

Very few bioeroders invade coral through living tissue. Rather, they tend to attack dead skeletons. Thus natural and human impacts that lead to the death of coral tissue (10.6) allow bioturbators to colonize. There are positive feedbacks here, of course, since the death of corals allows bioturbators to proliferate and the proliferation of bioturbators further weakens a reef and makes it more vulnerable to storm damage. Storm damage will result in further coral death, allowing more bioturbators to colonize.

Some reefs have entered long-term cycles of net bioerosion. In the Galapagos Islands, for example, where conditions are not ideal for coral growth and where many corals died after the 1982–3 ENSO, most reefs are slowly crumbling to rubble with little sign of recovery. Even when some recovery was observed, this was further stalled by more coral bleaching and death. In Galapagos, the main bioeroder is a sea urchin *Eucidaris*, which erodes carbonate as it grazes and reached very high densities (up to $30\,m^{-2}$) following the 1982–3 ENSO.

● Sea urchins are important bioeroders of coral reefs and are key 'ecosystem engineering' biota.

The relative dominance of herbivorous fishes and invertebrates has profound influences on rates of reef accretion and bioerosion. The grazing activities of herbivorous fishes clear space for coral settlement and enhance the survival and growth of young coral colonies. Conversely, urchin grazing leads to bioerosion, which may exceed the rate of carbonate accretion. The significance of urchin grazing was illustrated following a major die-off of urchins in the Caribbean, when carbonate began to accrete in some (less intensively fished) areas previously subject to bioerosion. If urchin populations are established, they can reduce algal biomass so markedly that they out-compete herbivorous fishes. While grazing, the area-specific rates of bioerosion due to urchins are higher than those of carbonate eroding fishes such as parrotfishes, and urchins have been implicated in limiting reef growth in many locations.

10.8 Dynamics of Reef Animals

Coral reefs provide superb natural laboratories for the study of ecology, behaviour and population dynamics because the inhabitants are usually so much more visible than in many other marine ecosystems (Fig. 10.13). Thus it is perhaps not surprising that scientists have used reefs to test many hypotheses about the processes that structure animal communities.

● Reefs provide ideal natural laboratories for the study of behaviour and ecology.

One nice example of the ways in which coral reefs have been used as laboratories is for the study of the factors structuring reef fish communities. In the early years of reef science, when many scientists were still coming to terms with the species richness and range of adaptations observed in reef fishes, they were quick to suggest that the high diversity of reef fishes must be a consequence of communities that were structured by very strong competition for resources. This was assumed to result in population sizes close to carrying capacity and limited by the availability of shelter or food. Moreover, the supply of larvae was assumed to be almost limitless in comparison with the capacity of established populations to accept new individuals. Careful experimentation and study of the dynamics of reef fish populations soon falsified this hypothesis, and it was clear that population sizes were often determined and limited by the rates at which larval fish recruited from the plankton to the adult

(a)

(b)

Fig. 10.13 The clear warm water overlying coral reefs means that they are ideal places for the study of marine ecology. Many fish and corals are highly visible (a) and thus divers can easily conduct underwater studies of marine life (b). (Photographs: Simon Jennings)

populations; the '**recruitment limitation**' hypothesis. This also implied that one species would not in time out-compete another, because the local abundance of juveniles was unlikely to be related to the number of spawning adults at the same location. In these terms, population sizes and thus whole communities would be less orderly, being determined by chance events in the plankton.

However, no one factor can adequately account for the complex structure of reef fish communities. For one thing, there may be significant recruitment to local populations derived from local adult spawning (Jones et al. 1999). There is also evidence that communities are not necessarily chaotically structured; in some respects patterns are repeated in space and in time. Recruitment limitation is likely to be observed when the

Fig. 10.14 Striped cleaner wrasse (*Labroides dimidiatus*) attending a monocle bream (*Scolopsis* sp.). These wrasse use their thick lips and front canines found at the front of both jaws to remove ectoparasites. (Photograph: E. Schlögl.)

potential of the community to absorb settling individuals is high. In this case, the subsequent abundance of a cohort will be a function of the abundance at settlement and density independent mortality. Conversely, when the capacity of the community to absorb settling individuals is lower, density dependent mortality may be detected and result in modified relative abundance (Doherty 2002). In overall terms the structure and dynamics of reef fish communities are determined by many of the same factors that determine the abundance of much less visible communities in deeper, muddier and superficially less attractive ecosystems.

- No one factor can adequately account for the complex structure of reef fish communities.

Of note amidst the great wealth of coral reef species is the large number of very close associations such as symbioses between species. We have already considered the coral-zooxanthellae relationship that is a basis for reef construction, but there are many more. These include fish that live with burrowing shrimps and in sea anemones, and delicately coloured crabs that live within coral heads. There are shrimps that live deep inside the shells of hermit crabs, and small fish such as the 'cleaner wrasse' that remove **ectoparasites** from fish that could easily eat the cleaner. Such close associations help to explain how so many species can coexist; they also highlight how in many cases the loss of one species may have knock-on effects on others (Fig. 10.14).

10.9 Reefs and Human Society

Reefs are more than a source of fascination and a scientist's and tourist's playground. For millions of people, often in the poorest countries of the world, reefs provide their food and income.

- Reefs are a source of food and income for millions of people.

● Subsistence fishing is the main source of dietary protein in countries where GDP is low.

On many tropical coasts, subsistence fishing is the main source of dietary protein and fish protein accounts for a greater proportion of total dietary protein in countries where per capita gross domestic product (GDP) is low (Kent 1998). In the fishing villages of the Pacific Islands, fish consumption of 200 to 300 g per person d^{-1} is common and is far greater than the consumption of other protein. For the South Pacific Islands as a whole, with a population of around 6.5 million, per capita annual fish production is approximately 17 kg. In the heavily populated coastal regions of Southeast Asia and Central America, people treat fishing as a fallback source of food and income when no alternatives are available.

Many reef fish caught for subsistence are not traded and are not recorded in conventional fishery statistics. Thus their value to society may be overlooked by policy-makers. For the Pacific Islands, Dalzell et al. (1996) calculated the replacement value of subsistence fisheries production as $US180 million, substantially more than the $US82 million of commercial fisheries production.

● Reef tourism has brought significant economic benefits to many countries.

Coral reefs are also valuable because they attract tourists, especially divers and snorkellers. Over ten million people visit the Great Barrier Reef in Australia each year, providing annual tourist revenue of around $US700 million and making it one of the most commercially significant ecological sites on the planet. Over 25% of all tourists visiting the Seychelles Islands pay to enter one of the Marine National Parks, and the income from this park has exceeded the budget of the government's Division of Environment. Whole towns, such as Sharm El Sheikh in Egypt are supported by revenue from tourist divers in the Red Sea, to the extent that a ship grounding in the nearby Strait of Tiran resulted in compensation payments for reef damage of $US1765 m^{-2}. Valuation of coral reefs as a source of tourist revenue is a complex process, depending on expected rates of recovery and loss of value to tourist divers (Spurgeon 1992).

● Reefs act as coastal defences and hence provide an important 'service' for human society (see also Preface).

Reefs also provide important **ecosystem services** in terms of coastal defence. Shallow reefs cause waves to break and dissipate wave energy. Loss of these defences as a result of coral mining, coral bleaching, or sedimentation may leave costs vulnerable to wave erosion and storms. In the Indian Ocean Maldive Islands, the mining of corals as building materials has removed sections of reef and left the coast exposed to wave action. The cost of replacing these reefs with concrete blocks that dissipate wave energy would be $US2 million km^{-1}. The government recognized that it would be ecologically and economically efficient to use the concrete for building and to leave the coral reef where it was! Rates of coral mining have been cut from a peak of 1 million tonnes $year^{-1}$ in the early 1990s to a few thousand tonnes at designated sites today.

CHAPTER SUMMARY

- Coral reefs are some of the most diverse and productive communities in the marine environment. They are created from tiny coral polyps but are easily visible from space. With the exception of deep water corals, most corals grow at temperatures of 18–36 °C and most are found 30 ° North or South of the Equator
- Coral reefs are rarely stable communities living in benign environments. Rather, they are continuously evolving and subject to physical disturbance on time scales from seconds to centuries.
- Contemporary distributions of reef corals and associated animals reflect 200 million years of evolution and the separation of oceans due to continental drift.
- Rates of reef growth only just exceed rates of bioerosion, while fixation of carbon only just meets the community respiratory demand for fixed carbon. Small shifts in reef community structure due to human and environmental impacts can lead to net bioerosion and loss of net productivity.
- Coral reefs constitute natural laboratories for the study of ecology. Thus reef science has made important contributions to the wider understanding of marine ecology.
- Reefs provide millions of people, often in the world's poorest countries, with a vital source of food, income and other environmental services.
- Coral reefs are threatened by human impacts, including fishing, pollution, sedimentation and climate change.

FURTHER READING

The edited book by Birkeland (1997) *Life and Death on Coral Reefs* is an excellent summary of the history of reefs, reef growth, and the impacts of natural and human activities. Spalding et al. (2001) provides detailed maps of global reef distribution, protected areas, and human threats to reefs. This is an essential reference text for anyone with an interest in coral reefs. Sale's (2002) edited book on reef fishes gives a good description of the ecology of fishes on coral reefs and shows how reefs provide a perfect natural laboratory for the study of fish ecology.

- Birkeland C. 1997. *Life and Death on Coral Reefs*. Chapman and Hall, New York.
- Sale P. F. (ed.) 2002. *Coral Reef Fishes: Dynamics and Diversity in a Complex Ecosystem*. Academic Press, San Diego.
- Spalding M. D., Ravilious C., Green E. P. 2001. *World Atlas of Coral Reefs*. University of California Press, Berkeley.

Chapter 11
Polar Regions

CHAPTER SUMMARY

The polar oceans and seas have been a lure for the explorer and scientist ever since the first whalers and sealers identified them as rich hunting grounds. Organisms living here are adapted to low temperatures, long periods in the year of poor light, and an ecosystem dominated by the seasonal formation, consolidation, and subsequent melt of a layer of frozen seawater that can be more than 10 m thick. The effects of global climate change are likely to have profound influences on these regions, but our understanding of their ecology, and therefore the potential threats of climate change, is still rudimentary due largely to the hostile working conditions. Like other extreme environments, the metabolic and physiological adaptations that enable life to go on are a potential source for novel biotechnological applications, and polar organisms, in particular microorganisms, are also potential proxies for life on extraterrestrial systems.

11.1 Introduction

Cold, hostile, barren white wastelands would be a fairly typical impression of the frozen pack ice that covers polar oceans and seas. This image is reinforced during the long polar winter when, if the sun does rise above the horizon, the fleeting light is no more than dull dreary half-light. Even in summer, snow blizzards can descend with no warning, forming conditions of such poor visibility that it is impossible to see just a few centimetres in front of you. It is of course not all gloom and despondency and even in the bleak winter ethereal displays of aurora borealis (in the North) or aurora australis (in the South), light up the sky in electromagnetic pyrotechnic displays that enhance the feeling of being in a part of a world governed by physical forces quite disconnected from the rest of the Earth.

At its maximum extent area pack ice (Fig. 11.1) can cover an area comprising 13% of the Earth's surface making it a biome that compares with those of the tundra and deserts. Outside of the polar circles, sea ice is also an ephemeral feature of the Baltic, Caspian, and Okhotsk Seas. Despite

Fig. 11.1 The seasonal dynamics of polar seas and oceans are dominated by the annual formation and consolidation of millions of square kilometres of frozen seawater that drive ocean circulation, climate, and the ecology of the regions (Photograph: David Thomas).

this, the study of these regions still lags behind what is known from temperate and tropical seas. Relatively few oceanographic research expeditions venture into polar waters each year, and there are limited numbers of coastal research stations, so that despite over 200 years of scientific endeavour in the Arctic and Antarctic, polar marine research is still a rather young research area: Vast expanses of the polar oceans and coastlines remain unvisited, and we know very little indeed about the polar systems in winter, simply because even now we cannot conduct large-scale research programmes during winter months.

● Pack ice can cover areas as large as the tundra and deserts, although because of hostile working conditions we know relatively little about these frozen realms.

Sea-going research is always expensive, but specially strengthened ice-breaking ships that can sail into pack ice (Fig. 11.2), and whose powerful engines consume vast quantities of extra fuel to break through even modest ice fields, are even more expensive than the norm. There are permanently manned coastal stations dotted around the Antarctic continent, as well as stations that are just manned for the summer. Likewise on Arctic coasts there are several research centres that conduct scientific research throughout the year, although for logistical reasons most of the activity takes place during summer months. However, these stations are few, and the hostile conditions and immense effort in simply keeping the stations running mean that research opportunities are limited during winter months. There have been attempts to study pack ice processes by establishing floating ice camps. These are either set up by freezing a ship into the ice and building a camp around the ship, or deploying the equipment and camp personnel by air. In 1992 a floating base on the sea ice was established for the joint Russian-American expedition, Ice Station Weddell, in the Weddell Sea. The ice camp started in the February drifting

● Ice camps and drift stations are good opportunities for collecting temporal data sets in pack ice regions.

Fig. 11.2 Modern icebreaking research ships, such as the German *RV Polarstern*, enable scientists to study deep into pack ice, throughout the year. They become floating laboratories and are key to the rapid progress made in our understanding of these hostile regions of the past 50 years. (Photograph: David Thomas).

● Conditions at the poles are not uniformly harsh. In winter it is difficult to imagine any life existing at the poles. In summer life flourishes during the long days, and temperatures can be above freezing.

northwards with the pack ice until finally breaking up in the June. A similar concept was used for the year long Arctic SHEBA project, October 1997 to October 1998, where 35 scientists, technicians, and a ship's crew collected valuable time series data on the physics, chemistry, and biology in and under the pack ice. Studies such as these have highlighted the need for multidisciplinary research teams combining forces to maximize the understanding of the key processes taking place.

There is a longer history of drift stations in the Arctic than in the Antarctic, and probably the most famous ice drift in ships were the pioneering drifts by Nansen on the *Fram* from 1900 to 1906, and Amundsen from 1918 to 1925 in the *Maud*. Since these early ship drift experiments, there has been a large number of ice camps in various parts of the Arctic. The longest series of camps were those by the Russian scientists who started annual pack ice camps in 1937 that continued until 1991. The stations were generally constructed on multiyear ice floes about 2 km in diameter and 3 to 5 m thick. The series was started again in 2003 with station NP-32, which ended in March 2004 when the researchers had to be rescued as their ice floe began to break up. These drift stations have collected some of the most comprehensive data sets about weather patterns, ocean processes, and of course the sea ice on which they are floating.

Polar regions are places of stark contrast. It is not without reason that the polar oceans are considered to be some of the most hostile places on Earth for life to survive, but surprisingly they support considerable life. Contrary to the picture described above, in the summer the polar experience is dominated by long days, where at the height of summer the sun does not sink beneath the horizon. Although seldom hot, temperatures rise to such an extent that it is possible to work in lightweight clothing, and one of the greatest problems is ensuring that the correct precautions are taken to avoid severe sunburn (Box 11.6).

The distinctive seasons at the poles (Chapter 2) are of course the reason for the obvious migrations and behaviour of the larger biota: polar bears, arctic foxes, walrus, seals, whales, penguins, and a plethora of other seabird species. Intriguing as the behaviours and ecologies of these larger animals are, the seasonal freezing over of polar oceans, the consolidation of an ice layer on the surface of the water, and its subsequent melt have far-reaching consequences on the whole of the ecosystem, in particular the seasonal dynamics of the plankton and the seasonality of coastal benthic communities. In this chapter various aspects of polar ecosystems will be discussed within the framework of seasonal changes in the extent of sea ice.

Pack ice not only influences the polar regions themselves, but also helps drive global ocean thermohaline circulation patterns through the formation of deep water masses (Box 11.1), as well as influencing global weather patterns. Over the past decade there has been increasing interest

Box 11.1 Sea ice and ocean circulation

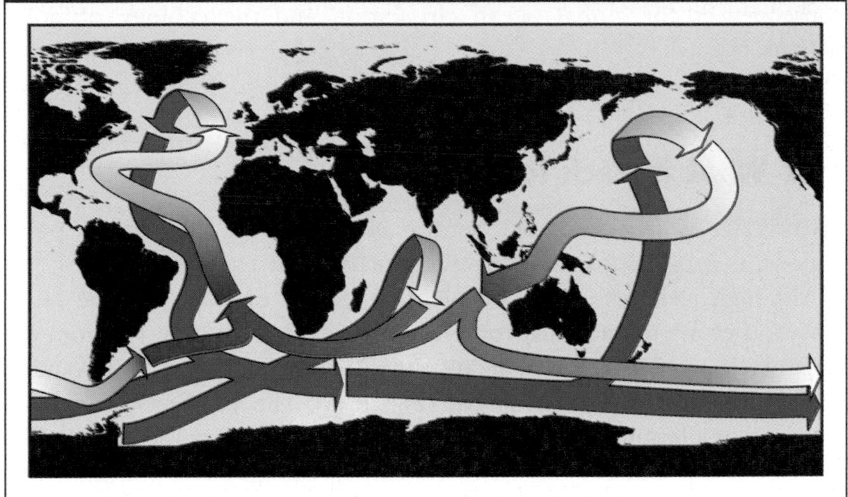

The sinking of cold saline water in the Arctic and Antarctic Weddell and Ross Seas drives large-scale ocean circulation in the so-called Global Thermohaline Conveyor Belt.

Despite their very different characterstics and the vast expanses covered by the oceans of the world, they are all interconnected by a large-scale movement of water that is referred to as *The Global Thermohaline Conveyor Belt* (Broacker 1997, Broecker et al. 1999, Clark et al. 2002).

The basis of themohaline circulation is that a kilogram of water that sinks from the surface into a deeper part of the ocean, displaces a kilogram of water from the deeper waters. As seawater freezes in the Arctic and Antarctic and ice sheets consolidate, cold, highly saline brines are expelled from the growing ice sheet (see 11.2) increasing the density of the water and making it sink.

In the conveyor belt circulation, warm surface and intermediate waters (0–1000 m) are transported towards the northern North Atlantic, where they are cooled and sink to form North Atlantic Deep Water that then flows southwards. In southern latitudes rapid freezing of seawater during ice formation also produces cold high-density water that sinks down the continental slope of Antarctica to form Antarctic Bottom Water. These deep-water masses move into the South Indian and Pacific Oceans where they rise towards the surface. The return leg of the conveyor belt begins with surface waters from the north-eastern Pacific Ocean flowing into the Indian Ocean and then into the Atlantic.

It is not just the temperature and salinity of the deep-water formation in the polar regions that is crucial to the ocean circulation. These water masses are oxygen rich, and so are fundamental for transporting oxygen to the ocean depths where respiration by deep-sea organisms consumes oxygen. The transport of dissolved organic matter and inorganic nutrients are also fundamentally governed by this transport, increasingly nutrients being remineralized (see Chapters 2 & 3) during the transfer of the deep-water masses. Therefore water rising at the end of the conveyor belt in the north-eastern pacific has higher nutrient loading, and lower oxygen concentrations than North Atlantic waters at the beginning of the conveyor belt (Sarmiento et al. 2004).

● The freezing of seawater to form the pack ice of the Arctic and Antarctic is fundamental for the deep water circulation of water masses.

in the effects of global climate warming on sea-ice dynamics, and the consequences for global ocean circulation and the ecology of polar regions.

11.2 What is Pack Ice?

● Icebergs are not derived from seawater, but are broken-off pieces of freshwater glaciers.

Because ice is the dominant feature governing the ecology of polar regions, it is prudent to understand exactly what pack ice is.

Although icebergs are a conspicuous feature of polar marine landscapes, they are not actually formed in the sea itself, but derive from the large glaciers and ice sheets that cover the Antarctic continent or Ellesmere Island and Greenland in the Arctic (Fig. 11.3). The ice sheets grow from snow accumulation and spread out, eventually reaching the coasts. The ice sheets, less dense than seawater, can spread to cover thousands of square kilometres of the ocean. Tides, currents, wind, and wave action cause stresses in these huge masses of ice that result in them breaking up, or **calving**, into icebergs that vary in size from a few square metres to many hundreds of square kilometres. Icebergs in shallow coastal regions can ground, causing severe disruption to benthic communities (see also Chapter 1), whereas others are carried on ocean currents where they are slowly broken into smaller components and of course melted. Due to the physical pressures and strains that form in these huge blocks of ice, they can sometimes explode in spectacular fashion.

Pack ice is instead formed from the freezing of seawater in autumn when the surface of the water is cooled to below $-1.8\,^{\circ}\mathrm{C}$ the freezing point of seawater (Eicken 1992, 2003). The resulting ice crystals rise to

Fig. 11.3 Sea ice (foreground) is frozen seawater, and not the same as icebergs, which are formed when the freshwater ice sheets flowing off the Antarctic continent (cliffs in background) break up (Photograph: David Thomas).

the surface where they aggregate into dense slicks of **grease ice**. It is at this early stage that the ice is '**inoculated**' with organisms because plankton stick to, or are '**scavenged**' by the ice crystals as they rise through the water. After a few hours of further freezing the ice crystals accumulate to form loosely aggregated discs, **ice pancakes**, 5 to 10 cm in diameter. These grow larger becoming 20 to 50 cm thick '**super pancakes**', several metres across. Wind and wave action raft the pancakes together, and often several end up lying on top of one another. They freeze together and after one or two days a closed ice cover has formed, strong enough to support working on the ice with heavy sledges, generators, and ice coring equipment. Further freezing and the subsequent thickening of the ice take place by **congelation** ice growth where water molecules freeze onto the bottom of the ice sheet. However the pack ice zone is still dominated by water currents, wind, and wave action, and **ice floes** are continually being moved apart so that areas of open water appear between the floes, or rafted under great pressure so that huge ridges over ten metres thick comprised of ice boulders span large distances.

● The pack ice undergoes a characteristic formation process including grease and pancakes.

Unlike freshwater ice, frozen seawater forms a semi-solid matrix, permeated by a network of channels and pores (Fig. 11.4). Salt does not enter the ice crystal structure and so, as the ice forms, salts and other dissolved constituents of seawater are expelled and collect as a highly concentrated **brine solution** within the labyrinth of brine channels (Fig. 11.7) and pores in the ice matrix (Weissenberger et al. 1992, Krembs et al. 2000). The morphology of these channels and pores, the total volume of the ice occupied by them, and the salinity of the brines contained within them is

Fig. 11.4 Pancake ice forms when sea ice crystals coagulate and grow together in turbulent water conditions (Photograph: David Thomas).

governed by temperature and the age of the ice: In colder ice the volume occupied by the channel system is less, and the salinity of the remaining brine is higher than in warmer ice. As the ice grows and ages there is a continuous loss of brine due to brine expulsion and gravity drainage, resulting in a gradual reduction of bulk ice salinity (Box 11.1).

● When seawater freezes it is not solid but rather forms a structure more akin to a sponge filled with a highly concentrated brine.

11.3 Arctic vs Antarctic Pack Ice

There are profound differences between the Arctic and Southern Oceans with major implications for the organisms living in the two regions (Spindler 1994, Comiso 2003b). The total area covered by the Southern Ocean (between 50 to 60 and 70°S) is about 36 million km^2, much larger than the 15 million km^2 of the Arctic Ocean (between 70 to 80°N). At its maximum extent sea ice in austral winter forms a girdle covering 20 million km^2 around the Antarctic continent. This is an area greater than that covered by North America. By summer this has melted to a mere 4 million km^2. In contrast the Arctic Ocean is almost totally covered by about 14 million km^2 of frozen seawater in winter, although by summer only about 7 million km^2 melts (Fig. 11.5).

● The largely land-locked Arctic Ocean has considerably more ice, which lasts several years compared to the larger Southern Ocean, which has predominantly ice that lasts just one year.

The main reason for these differences is that the Arctic basin is a Mediterranean-type sea almost enclosed entirely by land. There are only two broad openings, at the Bering and Fram Straits, where exchange with other Oceans are possible. In contrast the Southern Ocean is a circumpolar girdle that is open to, and exchange is possible with, the Atlantic, Indian, and Pacific Oceans. Therefore a much higher percentage of Arctic sea ice lasts for longer than one year (multiyear ice) even surviving up to 10 years or more, whereas most of the ice in the Antarctic lasts less than a year. Consequently the Arctic ice has the greater mean ice thickness of around 3.5 m compared to 1.5 m for Antarctic sea ice. Rafting and deformation processes can result in significantly thicker floes being formed in localized areas (Haas, 2004).

An important feature of the Arctic Ocean is that about one third of the ocean is taken up by shelf seas with depths of 100 m or less. In fact the Arctic Ocean only has a mean depth of 1800 m, the smallest of any ocean. In the Antarctic the continental shelves are very narrow, and so the Antarctic pack ice zone is largely over deep oceanic basins between 4000 and 6500 m deep.

● There are no river inputs into the Southern Ocean, in stark contrast to the Arctic basin into which flows considerable freshwater run-off from large river systems.

The Southern Ocean is largely cut off from terrestrial influence, except for limited aerial deposition, whereas the Arctic basin is characterized by a high input of fresh water from many large river systems such as the MacKenzie, Ob, Lena, and Yenisey. These river systems naturally input high amounts of suspended solids, dissolved organic matter, and inorganic nutrients (Dittmar & Kattner, 2003). Naturally these rivers also discharge

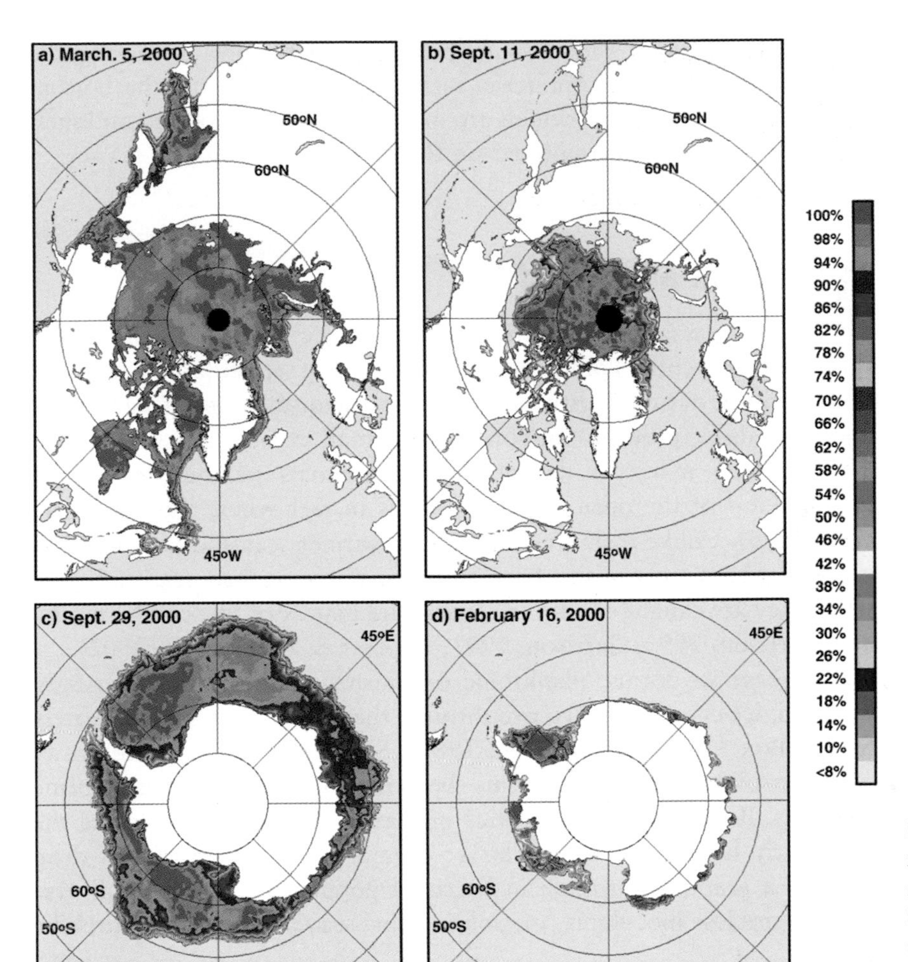

Fig. 11.5 The maximum and minimum extent of sea ice cover in the Arctic Ocean (top) and Southern Ocean (bottom) (Images: Josefino Comiso, NASA/GSFC).

pollutants such as PCBs, heavy metals, and biological contaminants, which will be absent from the Antarctic. As a result of such large amounts of freshwater input, much of the surface waters of the Kara, Laptev, East Siberian, and Beaufort Seas have low salinities. Arctic ice formed in coastal areas can be heavily laden with sediments, which are then transported large distances in the moving ice fields carried by the transpolar current. Even **allochthonous** material such as tree trunks, and soil turfs become encased in Arctic ice floes and are eventually released many thousands of kilometres from the place they were initially caught up in the ice.

Even during the melt seasons there are significant differences in the ways in which melting takes place. As summer approaches more solar radiation is absorbed by the ocean, and the surface layer of the ocean warms up, increasing the rate of melting of the ice. However, in the Arctic the sea ice begins to melt at the surface and extensive **melt pools** form decreasing the **albedo** (reflectance of light) from the ice surface,

● Surface melt ponds are a conspicuous feature of the surface of Arctic ice in summer. These are only rarely found in Antarctic ice.

allowing more solar radiation to be absorbed thereby increasing the melting process. In the Antarctic, melt occurs mainly from the bottom and sides of the ice floes, which are in contact with the ocean, and melt pools are less often encountered on the surfaces of Antarctic sea ice.

11.4 Life in a Block of Ice

● Often sea ice is turned brown due to a rich microbial flora and fauna that grows within the semi-solid ice matrix. See **www.awi-bremerhaven.de/eistour.**

Underneath a pristine white snow cover, the ice is often a light brown through to rich coffee colour, caused by a thriving **sea-ice biota** composed of a multitude of tiny (mostly microscopic) organisms comprising viruses, bacteria, algae, protists, flatworms, and small crustaceans (Fig. 11.6). They include many of the plants and animals of the phyto- and zooplankton of the open ocean, which is in fact where they originate (Chapter 6). Unlike their counterparts in warmer seas and oceans, these polar planktonic organisms have a unique phase in their seasonal cycle when they are caught up into the semi-solid matrix of the ice (Palmisano and Garrison, 1993; Garrison, 1991, Thomas & Dieckmann, 2002).

As grease ice forms, planktonic organisms, stick to, or are caught between, ice crystals as they rise through the water when surface waters freeze (Box 11.2, Garrison et al. 1989). Subsequently as the ice grows and consolidates, the organisms become trapped within the brine channels. Bacteria are conspicuous in sea-ice assemblages, entering the ice mostly on the surfaces of larger organisms. As temperatures drop there is a gradual transition in bacterial populations from the diverse mixed species inoculums originating in seawater to **psychrophilic**

Fig. 11.6 The growth of photosynthetic diatoms can reach such proportions that the ice floes are frequently coloured brown (Photograph: David Thomas).

Box 11.2 Where do ice organisms come from?

For several of the sea-ice fauna the mechanisms by which they are incorporated into sea ice is unclear, since many do not have an obvious planktonic life. The most obvious vector is that residual multi-year ice contains populations of organisms that act as innocula for newly formed ice. Whereas this is fine in the Arctic, in the Antarctic only a very small percentage of the sea ice lasts for more than one season, so this seems unlikely.

Nematodes are apparently not present in Antarctic sea-ice (only one record by Blome & Riemann 1999), a stark contrast to Arctic sea ice, where free-living species, especially belonging to the superfamily Monhysteroida, are found in abundance (Riemann & Sime-Ngando 1997). No rotifers have been found in Antarctic sea ice either, even though these too are also common in Arctic sea-ice samples. The reasons for these Arctic/ Antarctic differences are unclear, and it is possible that they are simply sampling artefacts: in time, more comprehensive sampling may produce more complete faunal records for Antarctic sea ice. Foraminiferans, which are very abundant in Antarctic sea ice, had for many years remained unknown from Arctic sea ice but since the late 1990s have now been found there, albeit only in a few samples. Alternatively, it may be that suitable vectors for colonization of sea ice are not present in the Antarctic.

In coastal regions with shallow water depths it is not difficult to imagine colonization of the sea ice from the benthos by larval stages, even in species with poor swimming capabilities. Another commonly cited vector in shallow water is lifting of organisms from the benthos attached to anchor ice (see 11.4.1), platelets of ice formed in shallow coastal waters. However, most of the pack ice, especially in the Antarctic overlies water several thousand metres deep and here mechanisms of colonization by non-planktonic organisms remain enigmatic.

Antarctic turbellarians spawn in sea ice in austral summer. Eggs, juveniles, and adults will be released into the water column upon ice melt. Although sea-ice turbellarian species can swim, none have been reported in the plankton and it is presumed that they sink to the sea floor. It has been suggested that sea-ice turbellarians may have an adhesive disk by which they attach to crustaceans before being released from the ice. Swimming crustaceans, including amphipods that migrate from the sea floor to the ice peripheries or the common ice copepods may act as vectors to transfer the flat worms to different ice floes (Janssen & Gradinger 1999).

Many ciliate species have been described from sea ice, with no equivalent planktonic form being described (Petz et al. 1995). It is speculated that for many of these species colonization takes place via resting spores, although there is no direct evidence for this. There is a similar conundrum for the Arctic nematodes, and even whale baleen plates that contain thriving populations of nematodes have been cited as possible vectors for bringing nematodes into close contact with sea-ice floes.

species (minimum growth temperature ≤ 0 °C, optimal growth temperature <15 °C, and maximum growth temperature <20 °C). The growth of psychrophilic heterotrophs (Deming 2002) is fuelled by large pools of dissolved organic matter (DOM) consisting largely of carbohydrates produced by the death and **lysis** (breaking apart) of sea-ice organisms, as well as by the exudation of organic polymers by algae and bacteria (Fig. 11.7). Much of this organic matter is produced as

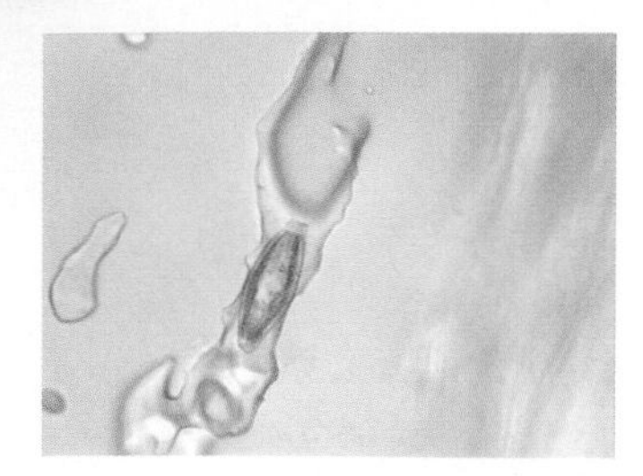

Fig. 11.7 A sea-ice diatom in a sea-ice brine channel (Photo: Christopher Krembs).

extracellular polymeric substances (EPS, Krembs et al. 2002), in such quantities that in dense sea-ice microbial assemblages there is evidence that the organisms are living in a situation more akin to biofilms (see Chapters 3 and 7).

The most conspicuous organisms in the ice are pennate diatoms (unicellular photosynthetic microalgae), which reach such concentrations that their photosynthetic pigments discolour the ice brown (Box 11.4). Diatom standing stocks up to 1000 μg chlorophyll l^{-1} have been measured in sea ice, which compare with typical values of 0 to 5 μg chlorophyll l^{-1} for surface waters in the Southern Ocean (Legendre et al. 1992, Lizotte 2001). Primary production in Antarctic sea ice (63 to 70 Tg C y^{-1}) is estimated to contribute 5% to the total annual primary production (1300 Tg C y^{-1}) in the sea-ice influenced zone of the Southern Ocean. This apparently minor contribution is in fact highly significant, in that the sea-ice organisms provide a concentrated food source that extends the short period of primary production in the water column (Box 11.3).

● Up to 5% of the total Southern ocean primary production takes place within the sea ice. However, this forms a highly concentrated food source for grazing zooplankton in winter when the water column is devoid of food.

Heterotrophic protists and metazoans, such as rotifers, nematodes, turbellarians, foraminferans, and small copepods. can accumulate in sea ice at concentrations ranging from hundreds to thousands of individuals per litre; several orders of magnitude higher those in the water (Schnack-Schiel et al. 2001, Michel et al. 2002). The smaller protozoans and metazoans graze the dense growths of bacteria, flagellates, and diatoms, while the narrow brine channels exclude larger species from exploiting these food sources (Krembs et al. 2000). Small channels with diameters <200 μm have been shown to be effective refuge from grazing organisms. Only rotifers and turbellarians can traverse '**channels**' significantly smaller than their body diameter. Turbellarians change their body dimensions in response to salinity changes, shrinking (because of osmotic water loss) with the increasing salinities found in the smaller channels. Rotifers can penetrate channels just 57% the width of their body diameter, whereas most other organisms simply congregate in the narrowest of the tubes into which they can physically fit according to body size.

● The smallest brine channels within the ice provide microscopic refuges from grazers for diatoms.

Within the ice there is therefore a complex microbial food web, based on a concentrated source of primary production within the ice. Although ice is an effective barrier to light, the light conditions within the ice are certainly more favourable to algal growth in the top 2 m of the ocean's surface than when the algae are mixed through the upper mixed water layers in open waters. Sea-ice algae also have well developed **photo-acclimation** and adaptation (see chapter 2), resulting in them having efficient light harvesting mechanisms for growing in low light (Kirst & Wiencke 1995). As long as re-supply of inorganic nutrients is possible, either by exchange or **remineralization processes** (Chapters 2 & 3) within the ice itself, dense algal biomass is ensured (Fig. 11.8).

● Sea-ice organisms are adapted to low light, low temperatures and high salinities, all prerequisites for life within a block of ice.

Fig. 11.8 Food web based around the Antarctic pack ice. (a) pennate diatoms; (b) autotophic flagellates; (c) foraminifers; (d) ciliates; (e) turbellarians; (f) copepod nauplii; (g) harpacticoid copepods; (h) calanoid copepods; (i) euphausiid larvae; (k) Antarctic krill, *Euphausia superba*; (l) young notothenoid fish such as *Pagothenia borchgrevinki*. Note not to scale. (Image: Sigrid Schiel.)

Even despite the physiological and biochemical prerequisites for survival and growth within the ice matrix there have been surprisingly few obligate/endemic microalgal or protozoan species associated with sea ice. A particularly exciting discovery was the new dinoflagellate species, *Polarella glacialis* isolated from sea-ice brine in McMurdo Sound. The cysts of *P. glacialis* are remarkably similar to fossil Suessiaceae cysts dating back to the Triassic and Jurassic. To date, *P. glacialis* is the only extant member of the Suessiaceae (Montressor et al. 1999).

● Surprisingly not many endemic organisms have been isolated from Antarctic or Arctic sea ice.

Naturally this growth has implications for pelagic organisms that remain free of the ice matrix. Amphipods, copepods, krill, and ice fish are examples of types of organism that graze on the rich food source associated with the ice (Schnack-Schiel 2003). This food source is collected on the peripheries of ice floes or in cracks and crevices of rafted ice, or ice that has recently been broken up or begun to melt. For many of these organisms their seasonal dynamics and life histories are tightly coupled to a source of food in the ice, and even the interannual distributions of krill, the pivotal organism in the Southern Ocean food web, is strongly correlated with pack ice cover because of food contained within it (Box 11.3).

11.4.1 Anchor ice

Anchor ice is plate-like ice that forms underwater attached to an object that is not frozen itself. It forms at depths between 0 and 30 m, and the sheets of ice fasten to submerged objects such as rocks, gravel, and

● Krill population dynamics are reviewed by Brierley and Thomas (2002).

Box 11.3 Southern Ocean krill

Euphausia superba stocks in the Southern Ocean are estimated to exceed 1.5 billion tonnes (cf., the total mass of people on the Earth approximates to 0.5 billion tonnes) and are the primary food for squid, penguins, seals, and baleen whales. In fact they are often referred to as being central to the Southern Ocean food web (see reviews in Nicol & de la Mare 1993, Brierley & Thomas 2002). Krill feed voraciously on phytoplankton and also feed carnivorously on copepods. Krill populations can form dense swarms, which, if they come to the surface, can turn the water a spectacular blood red.

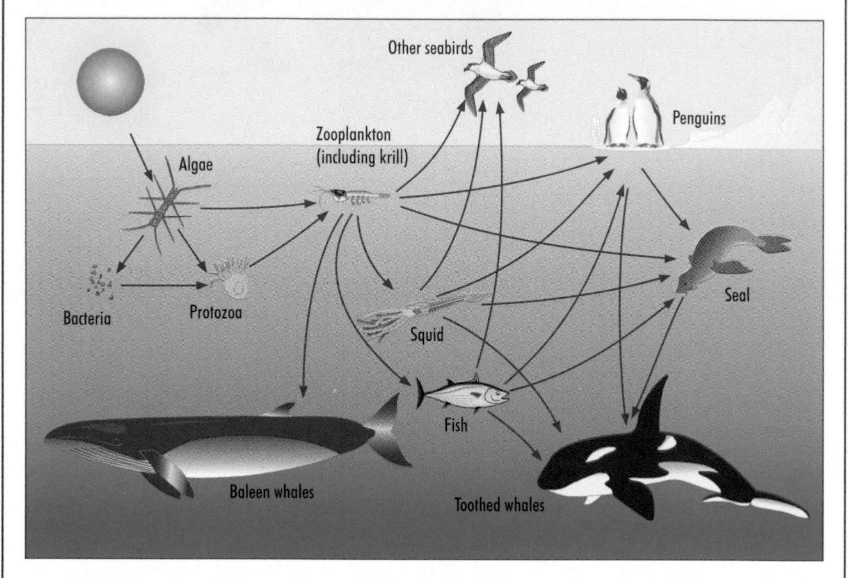

Simplified drawing of Antarctic pelagic food web showing the central role of krill as a 'keysone species' (After Priddle & Medlin 1990)

Adult krill can survive extended periods of starvation, even shrinking over winter periods. However, juvenile krill require a constant supply of food. The dense growth of sea-ice organisms often on the periphery of ice floes provides such a source of food throughout the winter when food in the water column is not present. When the ice melts the organisms released from the ice can initiate ice edge algal blooms that are rich grazing grounds. Even if blooms are not initiated, the large amounts of particulate material released from the ice upon melting are valuable food sources.

It is now clear that the distribution patterns of krill are closely linked to sea-ice conditions. In years when sea-ice cover is prolonged there is significantly higher krill recruitment, and in some regions of the Antarctic krill abundance can be predicted on the basis of cyclical variations in sea-ice extent (Atkinson et al. 2004).

The salp (*Salpa thompsonii*) is thought to reach high densities in years following reduced ice extent. Salps live for less than one year, and feed by filter feeding phytoplankton. They do not feed on ice organisms. In the absence of krill, the salps are able to exploit the spring phytoplankton bloom and undergo explosive population growth. In good sea-ice years, the krill have the upper hand over the salps because the sea ice has provided good feeding grounds over the winter resulting in good gonad development

continues

BOX 11.3 continued

and possibly allowing multiple spawning to take place. In these years the krill exploit the phytoplankton bloom resulting in poor food stocks for the salp populations.

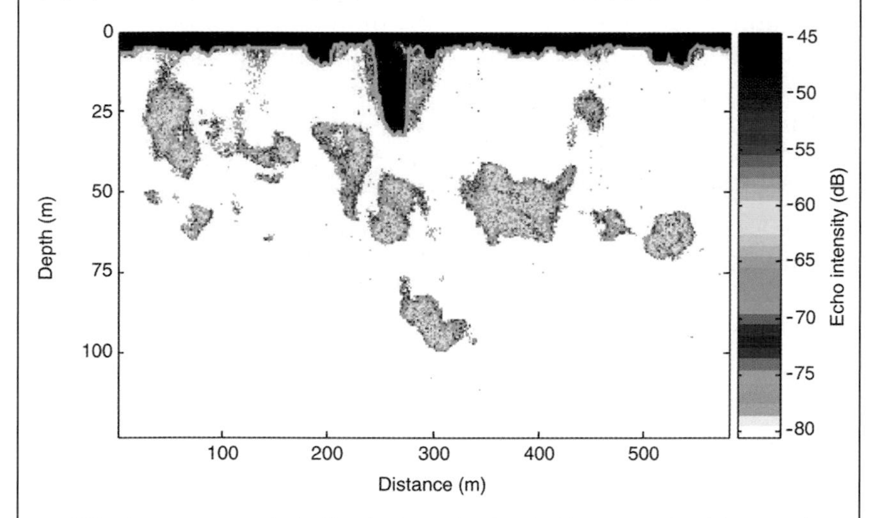

600 m long transect under ice floes by Autosub euipped with an upward and downward looking echo-sounder. The huge swarms of krill are swimming mainly at a water depth of 50 m. (Image: Andrew Brierley, from Brierley et al. 2002).

One of the problems with the studies about the distribution of krill under sea ice is the difficulty in actually measuring the distribution of the krill. Recent developments in autonomous underwater vehicles (AUVs) mean that it is now possible to deploy them reliably under pack ice to conduct extensive echo-sounder-surveys covering areas of many hundreds of square kilometres. In a study in the Weddell Sea in January 2001 the density of krill was three times higher under the ice compared to open water, with the vast majority of the krill within a band 1 to 13 km south of the ice edge (Brierley et al. 2002).

animals. It is a major physical disturbance in shallow benthic polar regions. Anchor ice can entrap large benthic organisms including echinoderms, sponges, fish, as well as macroalgae, rocks, and sediments (Dayton et al. 1969, Reimnitz et al. 1987). The ice does not necessarily kill the organisms, and can even act as a refuge from predators by forming cracks and crevices in which benthic organisms, including fish can hide.

Anchor ice is buoyant when detached from its nucleating site and can then rise from the seafloor and if present in enough of a mass can carry trapped organisms with it to the surface. Masses of macroalgae and animals up to 25 kg in weight can be detached by anchor ice. These elevated masses often get caught on the underside an overlying sea-ice floe. Some organisms get incorporated into the sea ice, or may fall back to the benthos a considerable distance from where they were lifted from the bottom.

● Anchor ice rising from the benthos to the surface can lift masses of macroalgae and animals up to 25 kg in weight.

11.5 Sea-Ice Edges

When large volumes of sea ice melt the fresh water released lowers the salinity waters at the ocean surface can stabilize the upper water column, effectively setting up a frontal system, which is associated with algal blooms (e.g. Strass & Nöthig 1996; also see Chapter 2) and subsequent higher rates of grazing. In the Southern Ocean, the spring phytoplankton bloom tracks polewards, beginning in October, in the wake of the melting ice edge. Rates of sea-ice decay are equivalent to movements of the sea-ice edge of up to 1.6 km h^{-1}. With the exception of tides over mudflats, the passage from day to night or the spread of forest fires, it is difficult to conceive of a faster changing biological environment (Squire et al. 1995). The **marginal ice zone** (MIZ) extends up to 100 to 200 km wide. the extent of which is determined by the penetration of waves and swell that break the pack into numerous smaller floes. Floe size increases rapidly with distance from the sea-ice edge (Fig. 11.9). Although more pronounced in the Antarctic, MIZs are also important features in the relatively limited zones where Arctic sea ice meets warmer waters.

● Melting sea ice releases vast volumes into the surrounding sea, often forming a stabilized low salinity surface water layer in which algal blooms take place.

Although the stabilized water column has often been observed at melting ice edges, it is not always the case. When there are strong prevailing winds and pronounced wave activity in the MIZ the released fresh water will be effectively mixed with surface waters (Bathmann et al. 1997, Murphy et al. 1998).

The blooms of algae take place because there is a reduced likelihood of algae being mixed downwards into light-limited depths. The inoculum is often species being released from the ice matrix. It is not just

Fig. 11.9 The concentration of ice floes in the marginal ice zone (MIZ) decreases the further towards the open ocean. The surface waters of MIZs can be highly productive, if stabilized by melt water from melting sea ice. (Photograph: David Thomas.)

an inoculum of algae of course, and inorganic dissolved nutrients, DOM, bacteria, metazoans, and protozoans released from the ice all contribute to a highly dynamic biological system. In turn the blooms are a rich feeding ground for zooplankton and in turn larger predators. Swarms of krill feeding in 'frenzies' have been recorded at ice edges, and in turn the MIZ is often a very intense feeding ground for seabirds, whales, and seals. Birds and mammals also utilize the ice edges in their migrations, exploiting the rich food sources as they migrate.

- Ice-edge algal blooms in turn attract zooplankton grazers, often in a 'frenzy' of feeding activity. In turn birds and mammals are attracted to ice edges for the rich sources of food.

Not all of the material released from the ice is incorporated into ice edge blooms (Riebesell et al. 1991). There have been studies to show that much of the biology actually aggregates into large accumulations and sinks quickly to depth as marine snow (Chapter 3). The feeding of zooplankton on the ice organisms can also effectively package a high percentage of the ice biota into rapidly sinking faecal pellets (see Fig. 11.2). Mean settling velocities for faecal pellets of between 60 and 200 m d^{-1} are common although values up to 1500 m d^{-1} are reported (Leventer 1998). There are numerous reports of faecal pellets from copepods and protozoans containing unbroken ice diatom frustules, often of monospecific origin reaching the sediments. Krill faecal pellets contain mostly broken/digested diatom frustules and are easily broken. Therefore their efficiency as a major flux-mediator to great depths is questionable, despite them having potentially high settling velocities. However, a sediment trap under a krill swarm recorded a flux of 660 mg C m^{-2} d^{-1}, which is the greatest flux recorded for faecal matter of herbivorous plankton (Cadée 1992).

- Aggregating ice organisms and faecal pellet production results in a large downward flux of biogenic material to the underlying sediments.

The major characteristic of this flux from the ice is the highly seasonal nature of short pulses of organic material that result in adaptive feeding strategies for benthic organisms, which may receive only occasional pulses of material. This effect is compounded by the short intense periods of primary production that are characteristic of open waters in polar regions (Chapter 2), quite different from the fluxes of material in tropical or temperate regions.

In both the Arctic and Antarctic summer, light is not the limiting factor to phytoplankton growth in the open ice-free waters when day lengths reach up to 24 hours. In winter light is severely limiting. Low temperatures are not considered to be limiting to phytoplankton production in either of the polar regions. Despite such similarities the productivity of the open Arctic and Antarctic waters are quite different. Over much of the open Arctic Ocean, the water is strongly stratified with a distinct **pycnocline** at between 20 and 50 m (Chapters 2 and 6). Inorganic nutrients in the surface waters are quickly exhausted following phytoplankton growth. Nutrients are only replaced by regeneration processes in the surface waters, or when the pycnocline is disrupted and nutrient-rich water from below introduced into the surface waters. In coastal waters or regions of nutrient upwelling, such as along the western Bering Strait, primary production yields are similar to temperate regions.

- The Southern Ocean has high concentrations of inorganic nutrients at all times of the year. In contrast the Arctic surface waters become nutrient limited.

Box 11.4 Washing Powders to 'Life in Space'

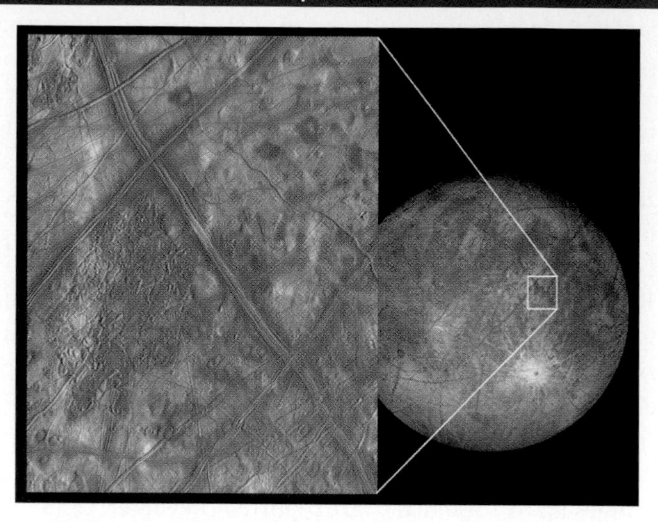

The brown coloration of the ice on the surface of the moon of Jupiter, Europa, has raised the tantalizing link with ice systems on Earth. It is unlikely that the coloration is due to diatoms, although organisms that live in sea ice are useful proxies in the growing field of astrobiology. (Image: Courtesy of NASA/JPL-Caltech).

In recent years, the study of sea-ice organisms has intensified, fuelled in part by the realization that the physiological and biochemical acclimations that micro-organisms thriving in the ice have to undergo may have considerable potential for biotechnological applications. These include the production of polyunsaturated fatty acids, which in ice organisms keep membranes fluid at low temperatures, for aquaculture, livestock, and human food. Research is also concentrating on the use of cold-adapted enzymes for a host of industries, ranging from cleaning agents to food processing (Cavicchioli et al. 2002).

Some of the keenest interest in sea-ice organism biology comes from the new breed of astrobiologists who are enthusiastically scrutinizing the ice-covered seas of Jupiter's moons Europa and Ganymede, and the surface of Mars, as well as speculating about life in the ice-covered neoproterozoic 'snowball earth'. Despite the tantalizing lure to compare the brown coloration observed on Europa's surface to diatom-coloured sea ice, the extraterrestrial ice systems, tens to hundreds of kilometres thick, are substantially different to the 1–10 m thick ice we know from Earth's polar oceans. If life forms do exist, or have existed on these bodies, it seems that they will be quite unlike those that dominate the sea ice found on Earth today (Cavicchioli 2002, Chyba & Phillips 2002, Marion et al. 2003).

In contrast, the Southern Ocean is an example of a high-nutrient, low chlorophyll region. Inorganic nutrients rarely become limited in open waters, but at the same time phytoplankton growth is moderate. Open waters rarely become stratified, except for the stratification at ice edges discussed above. Recent large-scale iron fertilization experiments indicate that for many regions of the Southern Ocean a paucity of iron is the major factor limiting the potential for phytoplankton growth (discussed in more detail in Chapter 2).

11.6 Polar Benthos

Blocks of ice, especially on a shore with a large tidal range, can be very effective at removing attached intertidal organisms (Fig. 11.10). It is therefore a general feature of polar intertidal shores that very little life is supported from autumn through to late spring when ice thaws again. Tide pools, even large ones, are no refuge, since these just form solid blocks of ice. In some sites, it is not uncommon for macroalgae and seagrasses to become encased in the sea ice together with fauna that was not able to migrate to deeper waters (Fig. 11.10). When ice melts in the late spring, melt ponds, consisting of fresh water form, and big sheets of melting ice can result in lowering of the salinities of the surface waters of coastal waters for significant periods of time. Fast-growing, short-lived opportunistic macroalgae such as *Enteromorpha* dominate the resulting floras of the ice-scoured parts of shores.

● Tide or wind-driven blocks of sea ice are particularly effective in clearing intertidal areas of all sessile organisms.

The effect of ice scour and the clearing of benthic habitats by seasonal drifting ice can reach down to depths of 10 m or more in extreme cases, and ice disturbance is possibly the major structuring element of polar nearshore biological communities (Gutt 2001; and Chapter 1). Therefore sublittoral communities can be highly disturbed, at times showing characteristics of a highly ephemeral flora and fauna (Box 11.5).

● The keels of icebergs can cause dramatic destruction of benthic assemblages growing at depths between 10 and 500 metres.

In particular in shallow coastal regions in both the Arctic and Antarctic the grounding of icebergs causes considerable damage to benthic communities (Gutt 2000, Gerdes et al. 2003, Gutt & Piepenburg 2003). However, the keels of large icebergs can be very deep and damage from large icebergs scraping across the benthos has been recorded by remotely operated vehicles carrying camera and video equipment. Sessile

(a)

(b)

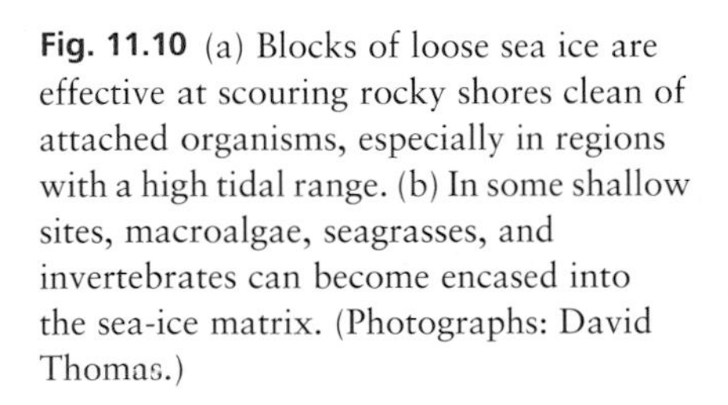

Fig. 11.10 (a) Blocks of loose sea ice are effective at scouring rocky shores clean of attached organisms, especially in regions with a high tidal range. (b) In some shallow sites, macroalgae, seagrasses, and invertebrates can become encased into the sea-ice matrix. (Photographs: David Thomas.)

● Kirst and Wiencke (1995) review many of the responses of polar macroalgae to low light and low temperatures.

Box 11.5 Macroalgae in Polar Regions

Some macroalgal species, including the red *Iridaea cordata*, occur at latitudes of 77 °S, where they experience up to 10 months of ice cover. Polar macroalgae are generally highly adapted to low light conditions often growing under sea ice for much of the year, and only experiencing short windows of light in the summer (reviewed by Kirst & Wiencke 1995). Many species are severely photoinhibited (see Chapter 2), even at low incident irradiances.

Several species of macroalgae grow in seasonally ice-covered waters, are very efficient primary producers at low light, and have seasonal growth strategies to best utilize the short windows of light in summer months (Photograph: David Thomas).

Several species, including the brown *Laminaria solidungula* from the Arctic and the red *Palmaria decipiens* from the Antarctic, actually begin to grow during periods of darkness at the end of the winter, when still covered by the ice. These species have not found a way of growing without light. Instead they begin to grow by using the starches and other metabolites that they stored in the previous year's growth period. The new tissues produced are ready to begin photosynthesis as soon as light becomes available when the ice breaks up. This kick-start maximizes the growth period during the short summer months. The development of new blades in the dark is probably controlled by circannual rhythms governing seasonal growth patterns. This hypothesis is supported by indications of free-running growth rhythms when these seaweeds are grown under constant conditions.

organisms are eradicated and pioneer species begin to grow in high abundances on the disturbed substratum. In some areas major iceberg scour events have been estimated to take place over periods of every 50 to 200 years. The consequence of this periodicity in disturbance, combined with the slow growth of many species (particularly in Antarctica), means that areas disturbed in this manner are likely to be characterized by a continuous natural fluctuation between extreme disturbance and recovery. Communities can be held at early successional stages, or even completely destroyed by scouring, and these effects occur from the intertidal to depths down to 500 m in Antarctica. The wide scales of disturbance intensity are thought to contribute to the overall high levels of Antarctic benthic biological diversity. Iceberg scour effects are less prevalent in the Arctic where there are far fewer icebergs compared to the Antarctic.

11.7 Polar Bentho-Pelagic Coupling

● Suspension feeders make up the vast majority of shelf and deep sea benthos of polar regions.

The largest proportion of the Antarctic shelf benthos fauna is made up of sessile suspension feeders such as sponges, cnidarians, bryozoans, ascidians, and echinoderms (Fig. 11.11). Suspension feeders can feed on a whole range of particulate organic matter, extending from bacteria, pico- and nanoplankton through to zooplankton several centimetres in length, a particular species having adaptations for a particular size class of food (reviewed by Arntz et al. 1994). Any extensive assemblage comprised of different types of suspension feeder will be a three-dimensional community with a range of feeding strategies and types that have the ability to exploit this huge spectrum of food classes.

Such suspension feeders survive in an environment of high primary production in spring and summer, but almost nothing in winter, and so the scarcity of primary production and diverse abundant benthic assemblages of high biomass are somewhat anomalous. It was long thought that these organisms simply feed on the short pulses of phytodetritus, and then survived long periods of feeding inactivity. However, it now

Fig. 11.11 In many regions surrounding the Antarctic, the benthos is covered by diverse assemblages of suspension feeders. The bentho-pelagic coupling between seasonal primary production in the water column and transport of this food source to the benthos is key to the survival of such assemblages. (Photograph: Julian Gutt, Alfred Wegener Institute.)

Fig. 11.12 Faecal pellets produced by zooplankton and protozooplankton are important in transporting organic matter from surface waters to the benthos. The faecal pellet shown was collected at 400 m, and is full of sea-ice diatoms consumed at the surface. Interestingly many of the diatoms seem to be intact. (Image: David Thomas.)

seems that this is not as straightforward. In particular, **resuspension** of sediments and **lateral advection** of organic rich material, are important for the transport of biogenic material from shallow sites to deeper basins. Resuspended and advected material is thought to be the main food source during winter months when the influx of photodetritus and faecal pellets from waters above is negligible (Klages et al. 2001).

● Although there are only short pulses of food descending from upper water layers, resuspension and lateral advection of sediments are important processes for feeding sessile organisms for the rest of the year.

There is evidently no uniform pattern in the degree of coupling of Antarctic invertebrate benthic animal reproduction and the extreme seasonality of primary production and concentrated food source. However, there are many examples where reproduction takes place uncoupled from the summer supply of food. One explanation is that the organisms utilize the high influx of food to best build up gonadal tissues to the maximum.

11.8 Endemism in Polar Benthos

The benthic floras and faunas of the Arctic and Antarctic are quite different. In general the numbers of species in the major groupings of macrobenthic organisms are between 1.5 and 6 times greater in the Antarctic than the Arctic. In fact the Antarctic species diversity is akin to the same levels of diversity recorded for tropical regions (Brey & Gerdes 1997; Chapter 1). The differences are also compounded by the fact that there is a much higher degree of **endemism** among the Antarctic flora and fauna compared to that of the Arctic (Briggs 2003). The degree of endemism varies greatly between different groupings of organism: up to 70% of the fish genera and 95% of fish species found in the Antarctic are endemic, whereas only 5% of polychaete genera and 57% of the polychaete species in Antarctic waters are endemic. Only about 5% of the macroalgae found in Arctic waters are endemic, and in the Antarctic 30% of the macroalgal species found are endemic. Interestingly crabs, flatfishes, and balanomorph barnacles, common in the Arctic, are missing from the Antarctic fauna (Eastman 2000). Of course these numbers are only approximate and change considerably as more taxonomic research takes place, and of course by the introduction of powerful molecular techniques to unravel phylogenetic conundrums.

● There are a much higher number of endemic genera and species in the Antarctic compared with the Arctic fauna and flora.

The explanations for the differences in degrees of endemism between the Arctic and Antarctic are mainly due to the greater age and longer isolation of the Antarctic (Crame 1999): Arctic benthic floras and faunas arise from the last 2 million years. The Antarctic was effectively cut off from the world ocean 25 million years ago by the formation of the circum-Antarctic current. The Arctic benthos is more akin to cold-temperate Atlantic or Pacific forms, it being more accessible to invasion from these regions. In the Southern Ocean, the polar front and lack of shallow sea linkages effectively isolate the system from recruitment from elsewhere.

● The Antarctic was cut off from other waters about 25 million years ago, whereas the Arctic is far younger, having been established about 2 million years ago.

11.9 Gigantism in Polar Waters

One of the characteristics of many benthic organism groups (particularly molluscs and crustaceans) in polar regions is the fact that they reach much greater sizes than their counterparts in warmer waters. Of course small species are still present, and not all polar invertebrates are large.

Antarctic sea spiders up to 400 mm across are a hundred times the size of the common European sea spider. Isopods, such as *Glyptonotus antarcticus*, found throughout Antarctica, the Antarctic peninsula, and sub-Antarctic Islands from the intertidal to 790 m depth, are up to 20 cm in length and weigh 70 g. In comparison, isopods in other parts of the world may reach a maximum size of just several centimeters. Other 'giants' include 2 to 4 m tall sponges and ribbon worms over 3 m long. It is thought that this '**gigantism**' is brought about by a combination of factors (Fig. 11.13; see also Chapter 8).

Low water temperatures certainly slow metabolic rates to the extent that growth rates are slow enough to enable them to live longer. Respiration rates are barely measurable in many of the benthic organisms by standard laboratory techniques. These result in some polar organisms having lifespans that are considerably longer than allied species from warmer waters (Brey 1998).

It appears that the primary cause leading to variations between polar organisms and those from warmer climes is due to differences in the

Fig. 11.13 Giant sponges are a conspicuous component of Antarctic benthic communities. These grow at extremely slow rates, and are thought to be the longest living organisms on earth, although their colonial nature means that not all cells are ancient. (Photograph: Julian Gutt, Alfred Wegener Institute.)

● 'Gigantic' invertebrate species are thought to result from slow growth rates in low temperature waters and the increased dissolved oxygen concentrations in polar water.

dissolved oxygen content of the water (Chapelle & Peck 1999, 2004). The physiological limit on the size of a particular organism is the amount of oxygen it can get into its blood, and in oxygen-rich waters this is largely dependent on the efficiency and length of the circulation systems supplying oxygen to the tissues. It is also true that in colder waters the oxygen demands of tissues are less, contributing to the possibility of large size.

Gigantism in polar waters is not restricted to organisms with highly developed circulation systems for transfer of oxygen. Ctenophores, anemones, sponges, copepods, and pteropods all have representative species that are much larger than allied species from temperate and tropical waters. These larger organisms have more tissue volume to which they must supply oxygen, but relatively less surface area with which to sequester the oxygen. Therefore the higher oxygen contents of the low temperature polar waters are ideally suited to support these larger species.

11.10 Birds and Mammals

The birds and mammals that inhabit of polar regions, are of course the most conspicuous organisms of the pack ice, and attract an inordinate amount of attention of the public and policy-makers interested in the tourism, conservation and exploitation of the regions. Historically it was these organisms that stimulated much of the early enthusiasm for the exploration of the Arctic and Antarctic.

Evocative as these animals are, an interesting comparison is that the total biomass of bacteria (each weighing less than 1 picogram) in the Southern Ocean is around 3×10^7 tonnes, whereas that of whales (each approximately 100 tonnes) is only around 8×10^6 tonnes. Even a single krill superswarm extending over 450 km^2 is estimated to have a biomass of 2.1×10^6 tonnes (Fogg, 1998).

● Sea ice is an important platform for raising young and hauling out for many seal species.

Sea ice is a vital requirement as a platform for reproduction for many mammal and bird species (reviewed by Ainley et al. 2003). Antarctic seals such as the leopard (*Hydruga leptonyx*), crabeater (*Lobodon carcinophagus*), Ross (*Ommatophoca rossi*), and Weddell (*Leptonychotes weddellii*) seals, and emperor penguins (*Aptenodytes forsteri*) in the Antarctic give birth to their young on the ice in late winter/early spring. This ensures that young become independent by late summer when food is most available. Sea ice is also an important platform for birds to renew their plumage. Penguins and flighted birds such as snow and Antarctic petrels all use icebergs and/or ice floes as platforms to avoid predators while replacing their insulating feathers that are vital for surviving the cold, as well as flight feathers (Fig. 11.14).

In the Antarctic many of the penguin species and seals such as the crabeater rely on the vast stocks of krill as a food source. Crabeaters

(a)

(b)

(c)

Fig. 11.14 (a) Adult Weddell seals (*Leptonychotes weddelii*) breed further south than any mammal on earth. They spend much of the winter in the water, keeping breathing holes open in the overlying ice by grinding the ice with their teeth. (b) Crabeater seals (*Lobodon carcinophagus*) are thought to be the most numerous seal species on earth (estimates up to 30 million). They feed on krill using interlocking teeth to 'strain' krill from the water. (c) Penguins are a conspicuous feature of Antarctic pack ice, such as these Adélie penguins (*Pygoscelis adeliae*), the most numerous of the Antarctic penguins. They eat small fish and krill. (Photographs: David Thomas.)

feed almost exclusively on krill, and therefore the links between sea-ice distribution and krill discussed previously have important implications on the foraging ecology of the seals. Leopard seals feed on krill, but also feed on penguins and seal pups. The Ross seal preys on squid, whereas the Weddell and elephant seals feed on fish, which in turn feed on the krill, such as the Antarctic toothfish (*Dissosthichus mawsoni*) and Antarctic silverfish (*Pleuragramma antarcticum*).

● The rich sources of food associated with ice edges and under ice floes provided major feeding grounds for birds and mammals alike.

In the Arctic the main food item for many of the birds and mammals is the Arctic cod (*Boreogadus saida*), a fish that feeds on crustaceans, often associated with sea-ice floes. However, pelagic amphipods, especially of the genus *Parathemisto* also form an important part of the diets of seals such as the ringed (*Phoca hispida*), harp (*P. groenlandica*), hooded (*Cystophora* cristata), and bearded (*Erignathus barbatus*) seals. These amphipods feed on phytoplankton, including the dense accumulations of ice algae on the peripheries of ice floes.

Walruses (*Odobenus rosmarus*) do not feed on ice-associated organisms, however they also use sea ice for hauling out, often using their distinctive tusks as ice axes. Their prey is mainly made up from molluscs such as bivalves (*Mya truncata*) and whelks (*Buccinum* spp.) and so they are restricted to hauling out on ice covering shallow (<70 m) coastal regions.

Probably the most evocative of all the mammals associated with sea ice is the polar bear (*Ursus maritimus*), which are confined to seasonally ice-covered areas of the Arctic. The ice is a vital platform for them to forage on their prey, which comprises seals, walruses, and whales (Stirling 1999). Although polar bears do spend time on land, it is a true marine mammal depending on the sea and pack ice for its existence and it is the largest of the non-aquatic carnivores. Many polar bears prefer to remain within the pack ice all year long (if possible) since the ice floes provide a platform from which the bears can catch seals, which are their main prey.

● The southern limits of polar bears are basically governed by the southernmost extent of sea ice.

Polar bears are found throughout the ice-covered waters of the Arctic Ocean. The greatest majority of bears roam near pack ice that is thinner or breaks open on a regular basis. Generally bears avoid heavily ridged, rough sea ice and thick multiyear ice, mostly because the densities of seals are low in these ice types. The southern limits of polar bears are basically governed by the southernmost extent of sea ice, and a few have been reported as far north as close to the north pole, although generally it is thought that very few stray further than 80°N since the ice is generally thick multiyear ice floes.

In both the Arctic and Antarctic baleen whales are able to forage under ice only as far as their breath-holding allows. These are open water species and generally follow the seasonal retreat of ice edges, and subsequently leaving the polar regions before the seas freeze over in autumn. Therefore although blue (*Balaenoptera musculus*), fin (*B. physalus*), and right (*B. glacialis*) whales all migrate to the Southern Ocean to feed on zooplankton such as krill, minke whales (*B. acutorostrata*) are the only cetaceans to occur deep in the Antarctic pack ice. Some populations of minke whales even live in the pack all year round. These 'small' whales that feed on krill use a hardened rostrum on top of their heads to break the ice to enable them to breath.

● Few whales are capable of living deep within the pack ice, and most species that frequent polar waters feed in open waters or in the relatively diffuse ice-edge regions.

In the Arctic, beluga (*Delphinapterus leucas*), bowheads (*Balaena mysticetus*), and narwhals (*Monodon monocerus*) also break sea ice with their backs to form breathing holes, and their lack of dorsal fin is considered an adaptation to having to break ice with their backs. However, in general, cetaceans in the pack ice of both hemispheres tend to be mostly found in leads and open water areas towards the periphery of pack ice. Bowheads and minke whales are the only whales to be found truly within the Arctic pack ice.

● Open water is vital for breathing mammals, and seal species living deep within the pack ice expend much energy in keeping breathing holes open.

The toothed sperm (*Physeter macrocephalus*) and killer whales (*Orcinus orca*) are found in sea-ice regions. The former largely feed on squid, consuming up to 3% of their body weight per day. Killer whales predate on seal, birds, or cetaceans in the pack ice, but they also feed on fish including the large Antarctic toothfish. Certainly it seems that some whales including bowheads, narwhals, minke, and belugas may shelter in the pack ice in an effort to avoid groups of killer whales.

For all the mammals living in pack ice breathing holes or areas of open water are vital for survival. This is the reason why most whales are restricted to the diffuse marginal ice zones, or areas of open water within the pack. Antarctic Weddell seals maintain their breathing holes by gnawing away the ice at the periphery of the hole with their teeth. Older Weddell seals have worn teeth as a result of blow-hole excavation, and starvation following broken teeth is evidently a common cause of mortality. Arctic ringed seals use heavy claws on their foreflippers to maintain breathing holes in ice up to 2 m thick.

There are several flighted birds that are highly adapted to an almost exclusive life in the pack ice: Ross's and ivory gulls (*Pagophila eburnean* and *Rhodostethia rosea* respectively) in the Arctic and the snow petrel (*Pagodroma nivea*) in the Antarctic. These are all white in colour, an adaptation allowing them to approach their prey more closely as well as avoiding predators, as long as their background is white.

Box 11.6 Ozone holes and the poles

Seasonal depletion of ozone in the troposphere over both the Arctic and Antarctica and increased ultra violet radiation reaching the ocean's surface is a well-reported phenomenon (Uchino et al. 1999). The consequences of this increase have been the source of much speculation and include the possibility of harmful effects on organisms living in surface waters of the polar oceans (reviewed by Brierley & Thomas 2002).

In terms of the ecology of sea ice and ice-associated waters, the ecosystem components that may suffer the largest influence of an increase in UV radiation are the blooms that occur in the stabilized waters of the marginal ice zones during spring and summer. The melt-water-stabilized shallow mixed depths may be 20 m or less, and UVb can penetrate to depths in excess of 50 m. Sea ice itself, especially if covered by snow is a particularly good barrier to UV light, and so sea-ice organisms and those living under ice cover are unlikely to be severely damaged by UV effects.

In Antarctic waters links between UV radiation increases and reduced viability of natural bacterial assemblages have been shown, although bacterioplankton is often not very sensitive to increased UV radiation due in part to rapid repair of UV damage (Winter et al. 2001). Relatively swift population changes in bacterial species composition and acclimation to high UV radiation may also have contributed to low UV induced inhibition of bacterioplankton. After strong UV exposure there can actually be increased bacterial production resulting from increased phytoplankton mortality and inactivation of viruses

continues

● Seasonal ozone holes result in increased harmful doses of ultraviolet radiation reaching the surface of polar oceans.

BOX 11.6 continued

that reduce marine bacteria. There are reports of increased concentrations of bacterial grazers, ciliates, heterotrophic nanoflagellates and choanoflagellates following high UV radiation (Davidson & van der Heijden 2000).

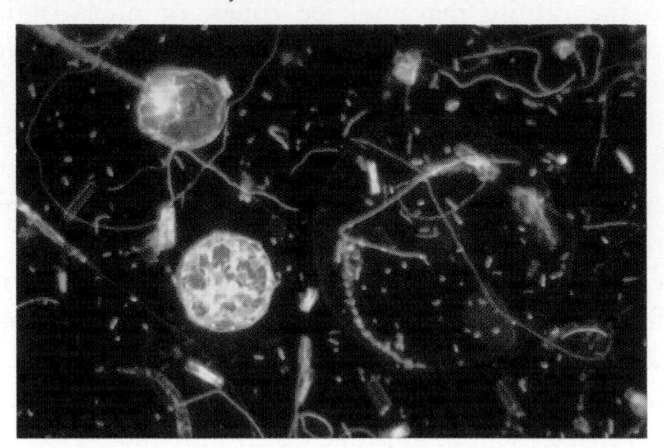

Image of a stained concentrated sample of sea-ice bacteria viewed under the microscope. The fluorescent stain makes it easy to count the numerous bacteria. The large structures are sea-ice diatoms. (Photograph: Sönnke Grossmann)

It has been proposed that one of the environmental factors influencing krill abundance in the Antarctic is increased UV damage due to ozone depletion following evidence that *Euphausia superba* may be more susceptible to damage than other Antarctic zooplankton (Jarman et al. 1999). In the pelagic, krill swarms are most often found well below the sea surface during daylight and are unlikely to be exposed to high UV levels. Interestingly, it has recently been shown that vertical migration by zooplankton is influenced by UV: deeper migrations occur in the presence of high levels of UVb. Therefore the depth and duration of vertical migrations may be influenced by ozone-related increases in UV, in turn impacting predators that feed on these species.

● UV radiation may induce chances in the vertical migration patterns of zooplankton.

11.11 Climate Change and Pack-Ice Regions

● Despite over two decades of satellite data on sea-ice distribution, we are only just beginning to discern long time trends that may be related to global climate change.

● The latest evidence indicates that Arctic ice is getting thinner in places, whereas in the Antarctic there are distinct shifts in the distribution of sea ice from year to year, even though the total sea-ice extent is not decreasing.

There are now over two decades of satellite data that can be used to study the long-term changing characteristics of sea ice in the Arctic and Antarctic (Fig. 11.15). However, it must be stressed that these data sets are at the borderline for making long-term predictions about ice and temperature changes at the poles. The most up-to-date analyses show that there has been a reduction in the average thickness of sea ice and changes in the overall characteristics of the Arctic ice cover since the 1980s that correspond with warming of surface temperatures measured by satellite borne thermal infrared sensors (Comiso 2002, 2003a). In fact recent estimates are that by the 2050s there will be almost no summer sea ice in the Arctic Ocean (Fig. 11.15).

The trends in the Antarctic are not so clear-cut, and there are some years that have significantly more ice, and others with significantly less

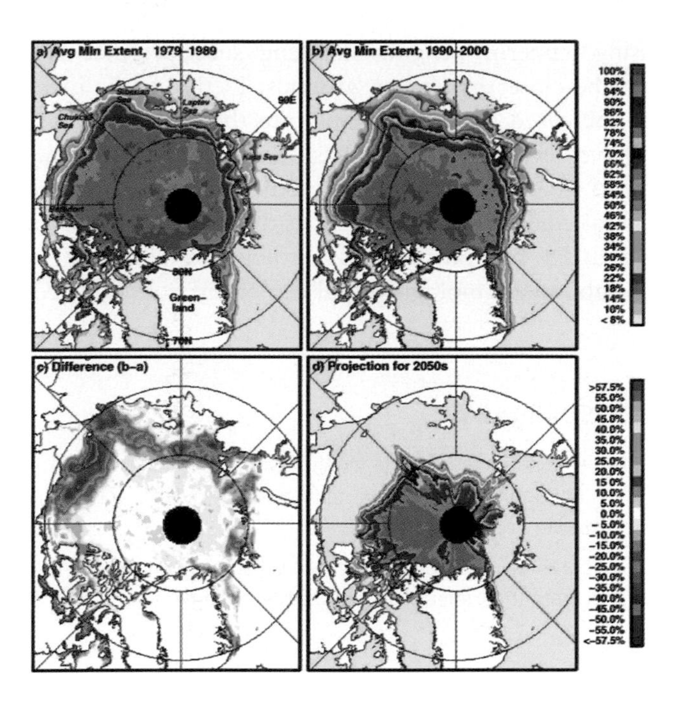

Fig. 11.15 The minimum summer ice extent in the Arctic is reducing. (a) The average summer ice extent from 1979 to 1989. (b) The average summer ice extent from 1980 to 2000. (c) The difference between the 1979–1989 and 1980–2000 averages. (d) Predicted summer minimum ice extent for the 2050s (Images: Josefino Comiso NASA/GSFC, from Comiso 2002).

ice. In fact there is evidence of a slight decadal increase in ice cover in the Antarctic (Parkinson 2002). However, there have been significant regional shifts in the distribution of sea ice, with significant decreases in the perennial ice cover of the Amundsen and Bellingshausen Seas being matched by increases of a similar magnitude in the Ross Sea.

One of the most intriguing findings in recent years has been the identification of the **Antarctic Circumpolar Wave** (ACW) that propagates eastward around the periphery of the Southern Ocean sea-ice zone. This results in increases followed by decreases in ice extent with a periodicity of 3 to 4 years (White & Peterson 1996, White & Annis 2004). Given the substantial interannual variation in such systems, it makes prediction of the consequences of global climate change a difficult task to accomplish.

It is very difficult to guess the effects that global climate change may make to the highly seasonal polar seas. Naturally, there are some organisms, such as the krill, whose recruitment is directly linked to sea-ice extent, and it is likely that krill spawning and recruitment will be significantly reduced. However, the effects may not be all bad. Thinner sea-ice floes may result in increased primary production within the ice and greater inocula of seeding organisms into the water on ice melt. Naturally increased food sources in the ice will possibly enhance zooplankton stocks in under ice waters. It is also possible that extended periods of primary production may result in the summer open waters, especially in the Southern Ocean where inorganic nutrients are available in excess. Decreases in ice extent have even been proposed to be beneficial for increasing emperor penguin hatching success, since the breeding colonies will be closer to feeding grounds. In the Arctic increased river run-off and increased nutrient loading into the coastal waters where ice forms will have very significant effects on the biology of these waters.

Not only do we not really know what the effects of climate change are on the physical environment that defines the polar regions, but with our present knowledge base about polar ecology we can only make crude speculations about the biological and ecological consequences.

CHAPTER SUMMARY

- Polar regions are characterised by highly seasonal changes in day length and temperatures. Although extreme low temperatures prevail, this is not the main limiting factor to life in polar regions.
- The annual formation, consolidation and subsequent melt of sea ice is the fundamental process that determines the unique ecology of the Arctic and Antarctic oceans.
- A significant primary production takes place in the ice itself, providing a vital source of food at times of the year when food in the water column is sparse.
- The stabilisation of the water column by melting ice can induce ice edge algal blooms that induce high grazing activity.
- Benthic assemblages in Polar regions are dominated by suspension feeders, which not only utilise the highly seasonal pulses of food from waters above, but also from resuspended and advected sediments.
- Sea ice, anchor ice and the scraping by icebergs have considerable destructive effects, down to below 300 m, on polar benthic assemblages.
- High dissolved oxygen concentrations, low temperatures and longevity in polar waters enhances the phenomenon of 'gigantism'.
- Many bird and mammal migrants frequented Polar waters in summer months. However, only few obligate species live throughout the year deep in the pack ice.
- Seasonally high doses of UV radiation has mixed effects on Polar ecosystems, although damaging effects may be far less than often perceived.

FURTHER READING

For general texts on the biology of polar regions, consult Thomas (2004), McGonigal and Woodworth (2001), Fogg (1998), and Woodin and Marquiss (1997). For more information on the biology and physics of sea ice, consult Brierley and Thomas (2002), Knox (1994), Leppäranta (2001), Thomas and Dieckmann (2003), and Wadhams (2000). Arntz et al. (1994) reviews the zoobenthic processes. Mastro and Wu (2004) present an inspiring view of Antartic benthos, ice and mammals. A good introduction to remote sensing of sea ice is given by Parkinson (1997). Smetacek & Nicol (2005) describe vividly the potential effects of global climate change on plankton through to whales living in Polar Oceans.

- Arntz, W. E., Brey, T. & Gallardo, V. A. 1994. Antarctic zoobenthos. *Oceanography and Marine Biology, An annual review:* 32: 241–304.
- Brierley, A. S. & Thomas, D. N. 2002. Ecology of Southern Ocean pack ice. *Advances in Marine Biology:* 43: 171–276.
- Fogg, G. E. 1998. *The Biology of Polar Regions*. Oxford University Press, Oxford.
- Knox, G. A. 1994. *The Biology of the Southern Ocean*. Cambridge University Press, Cambridge.
- Leppäranta, M. 2001. *Physics of Ice Covered Seas*. Helsinki University Press, Helsinki.
- Mastro, J & Wu, N. 2004. Under Antarctic Ice. University of California Press.
- McGonigal, D. & Woodworth, L. 2001. *The Complete Story Antarctica*. Frances Lincoln, London.
- Parkinson, C. 1997. *Earth From Above*. University Science Books, Sausalito.
- Smetacek, V. and Nicol S. 2005. Polar ocean ecosystems in a changing world. *Nature*: 437: 362–368.
- Thomas, D. N. 2004. *Frozen Oceans – The floating World of Pack Ice*. Natural History Museum, London.
- Thomas, D. N. & Dieckmann, G. S. (eds) 2003. *Sea Ice – An Introduction to its Physics, Chemistry, Biology and Geology*. Blackwell Publishing, Oxford.
- Wadhams, P. 2000. *Ice in the Ocean*. Gordon and Breach Science, Reading.
- Woodin, S. J. & Marquiss, M. 1997. *Ecology of Arctic Environments*. Blackwell Science, Oxford.

Part 3

Impacts

Chapter 12

Fisheries

CHAPTER SUMMARY

Fishing ought to be the perfect example of an industry based on a renewable resource, since fishers take harvests they need not sow. Fisheries are a key contributor to global food security and economic activity, but fisheries are often subsidized, wasteful, cause excessive environmental damage, and ignite conflicts between otherwise friendly nations. Scientific understanding of fish populations and their management is good, but much of this understanding is never translated into effective management and relatively few fisheries realize their potential benefits to fishers and society. In past decades, the main objective of fishery management was to maximize catches without compromising future catches. For biological, social, and economic reasons, this objective was rarely met. Some fisheries collapsed and others threatened species and habitats of conservation concern that were not their intended targets. Today, fishery managers still face great challenges as they start to take account of the impacts of fishing on the ecosystem and to integrate conservation concerns into management plans.

12.1 Introduction

Fishing provides food, income, and employment for millions of people, but fishing is also one of the most widespread human activities in the marine environment and has unsustainable environmental costs that threaten rare species and marine ecosystems. Fisheries, of course, do not only target fish. The term is loosely used to refer to humans capturing many wild marine species, ranging from algae to invertebrates, fish and whales.

● Fisheries are a vital source of food, income, and jobs for millions of people – especially in the developing world.

In this chapter we describe the scale and impact of fisheries, the biological processes that support fish production, how sustainable levels of catch can be predicted, and how fisheries and marine ecosystems are best conserved. Our emphasis is on fisheries that target fish and invertebrates, as these groups account for the majority of global landings.

12.2 Global Fisheries

Global fish landings rose steadily from 1950 to 1990, but have since hovered around 80 million tonnes (Fig. 12.1). The cessation of the long-term increase in global fishery landings reflects the over-fishing of many species, and landings are likely to remain stable or decline in future. Fisheries contribute significantly to the global supply of protein, particularly in the world's poorer countries, and sustainable management of fisheries brings benefits to society and environment alike.

- Fisheries yield around 80 million tonnes of fish each year with a first sale value of $US50 billion.

Global fish catches are dominated by remarkably few species. Five species of fish, out of more than 15 000 found in the sea, usually account for more than 15% of annual landings. These species are the Peruvian anchovy (*Engraulis ringens*), the Alaskan pollock (*Theragra chalcogramma*), the Chilean jack mackerel (*Trachurus murhyi*), the Atlantic herring (*Clupea harengus*), and the chub mackerel (*Scomber japonicus*). All these fishes are found in upwellings or on continental shelves, areas where levels of primary production are very high (Chapters 2, 6 and 7). Indeed, the majority of global landings come from productive shelf ecosystems and upwellings, and not from the open ocean where production is generally low (Fig. 12.2).

- Global landings statistics are updated annually at **www.fao.org** and spatial distribution of global landings can be plotted interactively at **www.seaaroundus.org**.

- For further information on all the fished species listed in this chapter, see **www.fishbase.org**.

12.2.1 A brief history of fisheries exploitation

The sea has been fished since ancient times, and there are many references to fisheries in Greek, Egyptian, and Roman texts. In many small island nations, fish have been a vital source of protein, and Polynesian and Melanesian people migrated across the Pacific thanks to the ubiquity of fishes and coconuts. As seafaring and navigational skills improved, so vast offshore fish resources were discovered. By the early 1500s, fishers from France and Portugal were regularly crossing the

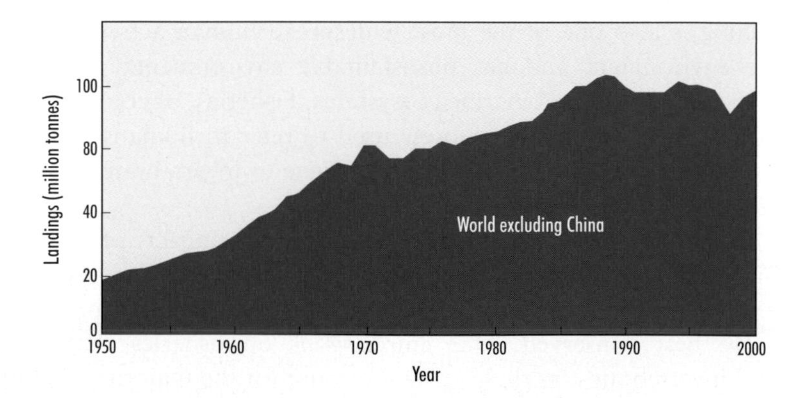

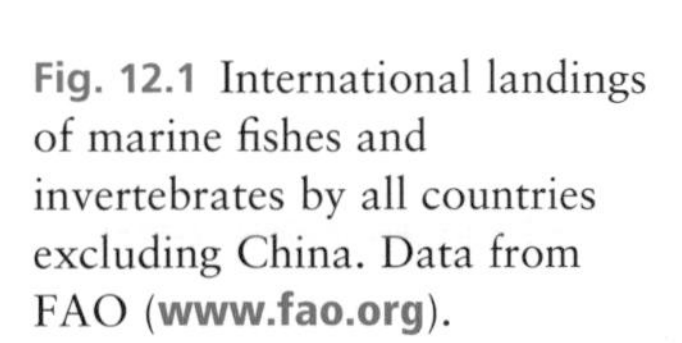

Fig. 12.1 International landings of marine fishes and invertebrates by all countries excluding China. Data from FAO (**www.fao.org**).

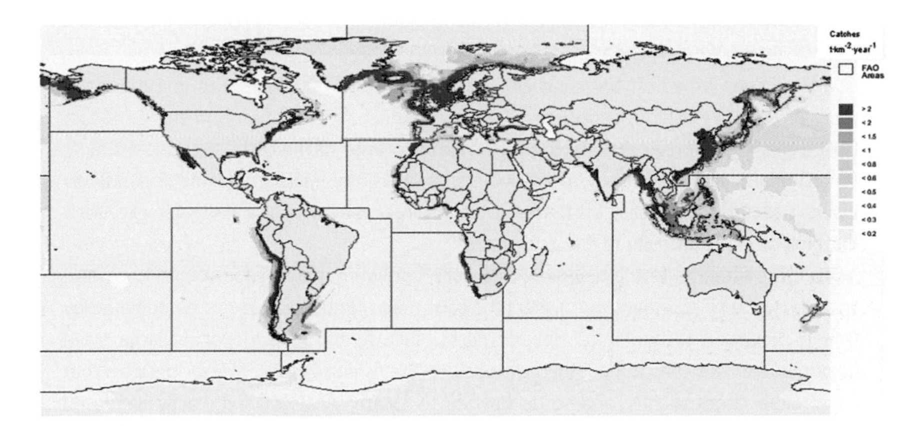

Fig. 12.2 Spatial distribution of marine fish landings in 2001. From FAO data processed by **www.seaaroundus.org**.

Box 12.1 Catches and landings

Fish and other animals that fishers bring ashore are often called **catches** or **landings.** Although these terms are often used interchangeably, this is not correct. Catches describe the fish caught by the fishers but, as well as the fishes they are trying to catch, catches contain a range of other species, some valuable and some not. These are known as **by-catches.** By-catches have caused social and political controversy because they can include species such as dolphins, seabirds or turtles.

Valuable by-catch species will be kept, but the remainder of the by-catch is usually discarded at sea because it contains animals smaller than legal size limits or not suitable for selling and eating. The discarded animals are known as **discards**. Since many discards will have died or been wounded during capture, they do not survive when returned to the sea. **Landings** describe fish that are brought ashore, and are thus a measure of catches minus discards. Since discards account for around 27% of the *catches* in global fisheries, landings data, such as those published by the FAO, greatly underestimate the numbers of fish and other animals killed by fishing.

Source: Alverson et al. (1994)

Atlantic Ocean to fish for cod (*Gadus morhua*) off the coast of Canada (Box 12.2). Many wars were fought over fish, and Kurlansky's fascinating book *Cod: the biography of a fish that changed the world* (1997) tells the story of the burgeoning cod fishery and the international struggle to control it.

● Fish have been caught from the sea by humans since prehistoric times, and many wars have been fought over fish.

While fish were depleted by pre-industrial fisheries, fishing really began to affect fish stocks and the marine ecosystem on a global scale following the 'industrial revolution'. With industrialization, the power and range of fishing vessels increased rapidly, as did the capacity to

Box 12.2 Collapse of a fishery: the Canadian cod

Cod have been fished off the coast of Newfoundland for 500 years, but in 1992, after years of intensive fishing, cod were fished to commercial extinction and the Canadian Government was forced to close the fishery. The collapse of the cod is a classic example of too many fishers chasing too few fish, and as the fishery became increasingly mechanized and technically efficient, so the fishing capacity of the fleets far exceeded the biological productivity of the cod stock.

From the fifteenth to the nineteenth century, cod were mainly caught with hook and line, and were so abundant that fishers thought their numbers could not be depleted by fishing. However, technological developments, such as the introduction of large trawl nets, powered winches to haul the nets back to the fishing vessels, freezer trawlers that could store large catches, and sonar devices to locate shoals of cod, vastly increased fishing capacity and the cod were threatened. Annual cod landings peaked at around 810 000 tonnes in 1968, when many large European freezer trawlers regularly travelled across the Atlantic to fish. By 1977, when Canada took control of the fishing grounds within 200 miles of her coast, and effectively closed the European fishery, landings had already fallen to 20% of their maximum.

Even with the cod stock under national control, social and political pressures still made it very difficult to reduce fishing effort, and the demise of the cod continued. The abundance of adult Canadian cod fell from an estimated 1.6 million tonnes in 1962 to 22 000 tonnes in 1992, when the fishery was finally closed. The closure was a disaster for the Canadian fishing industry and many fishers were left struggling to survive on welfare payments, while service and processing industries, that once thrived in many coastal towns, had to shut or relocate. Ten years later, the cod stock has shown little sign of recovery and many boats now fish for other species such as shrimp.

Source: Hutchings & Myers (1994)

catch and store fish. Methods for the large-scale canning and freezing of fish products were developed, and the new intensive farming industry needed large volumes of cheap fish-meal as animal feed. Countries such as the former USSR, Japan, Spain, Korea, and Poland had massive government **subsidized** fleets that ranged the world's oceans in search of fish. Until the 1970s, there was little control of these fleets, since most of the world's oceans were not under national control. However, in the early and mid 1970s, the increasing competition for fish led many countries to claim national jurisdiction up to 200 miles from their coast. Since 90% of world catches were taken within 200 miles of the coast, these claims brought 90% of fisheries under national control. The demise of the subsidized high seas fleets followed, but most of the fisheries under national control were still poorly managed and collapsed in the subsequent decades (Cushing 1988, Burke 1994).

● With industrialization, the power and range of fishing vessels increased rapidly.

● Subsidies led to massive over-capacity in global fishing fleets.

12.2.2 Over-fishing and the need for management

There are many detailed reasons why individual fisheries are over-fished but, in basic terms, over-fishing is a consequence of too many fishers and vessels chasing too few fish. Why does this happen? Studies of many fisheries around the world have shown that when a potential fishery is discovered, new fishers start to fish as fast as they can. Indeed, on an individual basis, the faster fishers fish, the more money they will make. Other fishers who find out about the new fishery will also join the race to fish. Soon, there will be too many fishers chasing too few fish, the stock will be depleted and catch rates and profits will fall because the rates of fishing exceed the biological capacity for replenishment.

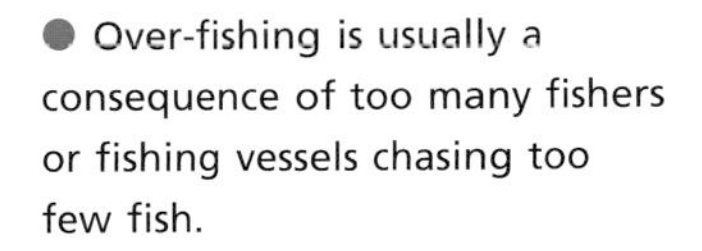
● Over-fishing is usually a consequence of too many fishers or fishing vessels chasing too few fish.

The range of phases that a fishery can go through can be described as fishery development, full exploitation, over exploitation, collapse, and recovery. During each of these phases, there are clear trends in the abundance of fish, fleet size, catch, and profit (Fig. 12.3). Good fishery management should ensure that catch rates do not exceed potential rates of replenishment and that the sustainability of the ecosystem is not compromised.

● Good fishery management should ensure that catch rates never exceed the rates of biological replenishment of fish stocks.

Fishery managers would also like to prevent overexploitation and ensure that catches and profits are sustainable in the long term. However, these goals are hard to achieve in practice. Thus 47% of the world's main stocks or species groups are fully exploited, while 18% are overexploited and 10% are severely depleted or recovering from depletion. Only 25% of stocks are under- or moderately exploited (FAO 2002).

Ensuring that catches and profits are sustainable are not the only goals of fishery management. Today, managers are increasingly seeking

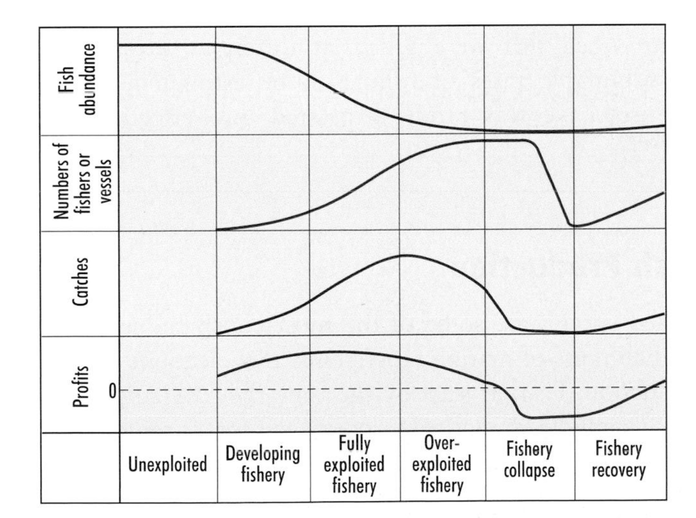

Fig. 12.3 Patterns of exploitation in an unregulated fishery. Ideally, fisheries are regulated to hold them in the developing or fully exploited phase. After Hilborn & Walters (1992).

to implement **ecosystem-based management**, where management decisions take account of conservation and ecosystem concerns, such as the wider effects of fishing on biodiversity, rare species, birds, and mammals. Later in this chapter we will look in detail at the effects of fishing on the environment and what can be done to mitigate these effects.

● Despite the potential benefits of fisheries to society, the effects of fishing on ecosystems and rare species are a growing concern.

12.2.3 Fishery science

Fishery science was first treated as a scientific discipline in the late 1850s, when the Norwegian Government hired scientists to find out why catches of cod fluctuated from year to year. Cod had been caught in northern Norway, and then preserved by salting and drying for many centuries. The money made from exporting dried cod was used to import other foods and to pay traders and bank loans in southern Norway. However, cod were not always a reliable source of income and, when the cod fishery failed, the people of northern Norway would start to default on their bank loans. For the Norwegian Government, fluctuations in the cod fishery caused economic and political hardship, and so the Government hired scientists to explain why the catches fluctuated and what might be done about it.

● Fishery science began in Norway during the nineteenth century, because the Government wanted to know why cod catches fluctuated.

By 1900, most of the developed economies were employing fishery scientists. These scientists described the biology and migrations of fishes and had shown that abundance fluctuated from year to year because the numbers of young fish entering the fishery were very variable. Moreover, it was clear that the abundance of a year class of young fish was established in the first few months of life, and depended on conditions in the ocean environment. By the 1960s, fisheries science was an advanced field and scientists had developed models that predicted fishery landings and profits when fish were killed at different rates. These models allowed sustainable rates of fishing to be estimated, but the advice that scientists gave was rarely translated into effective management (Smith 1994).

● By the 1960s, fishery scientists could predict potential fish catches and profits when fish were killed at different rates.

12.3 Fish Production

While fished species are some of the most conspicuous and intensively studied inhabitants of marine ecosystems, they account for a fraction of total production. Fish, at least by the time they are sufficiently large to be caught, usually feed relatively close to the top of long food chains. As we saw in Chapter 6, these food chains are quite inefficient, and only a fraction of prey production is converted to predator production at each step in the chain. This means that the production of fished species per unit of primary production falls as the length of food chains increases, and that higher fisheries yields can be taken from species lower in the food chains, like plankton feeding sardine or anchovy, rather than from their predators such as hake. The effects of inefficiency in the food chains are such that, while global marine primary production is thought to range from 30 to 60 $\times$ 10^9 tonnes y^{-1}, fish production is unlikely to exceed 2 $\times$ 10^8 tonnes.

● Fish account for a fraction of total production in the sea because the food chains leading from primary production to fish production are quite inefficient.

Globally, phytoplankton account for almost 90% of primary production and thus support most fish production. Ecosystems with higher primary production invariably support higher fish production and so fish production is highest on the continental shelves and in upwellings (Fig. 12.2). More than 20% of primary production is probably required to sustain fisheries on the continental shelves and in upwellings, and thus fishing must have a major impact on patterns of energy flow in marine ecosystems (Pauly & Christensen 1995).

● Ecosystems where primary production is higher will support higher fish production.

12.4 Fished Species and Their Fisheries

Knowledge of the life histories and distribution of fished species is key to understanding how they are affected by fishing and the environment. Thus large species with slow growth and late maturity tolerate much lower rates of fishing than smaller and faster growing species, and the abundance of species with short life cycles and a small number of age classes in the population will fluctuate more from year to year.

● Large and slow growing species are the most vulnerable to fishing.

Most fish and shellfish species form a number of stocks across their geographical range. These stocks have characteristic life histories and migrations, and will therefore respond differently to fishing. Stocks of the same species can interbreed, but mixing between them is low, and so they respond more or less independently to fishing. This means that stocks are treated as management units in most fisheries.

● Fish stocks are treated as the primary management units in most fisheries. They have characteristic life histories and respond more or less independently to fishing.

Fish stocks do not respect national or administrative boundaries, and they migrate on many scales, from local to oceanic (Chapter 7). Migrations allow stocks to remain in the best possible conditions for feeding, growth, or reproduction. Since migrating fish often cross the boundaries of nations, international management areas and marine reserves, catches of fish in one area can affect catches of fish in another. This can lead to international competition for 'shared stocks' and, for this reason, many fisheries are assessed and managed by groups of nations. For example, the Inter-American Tropical Tuna Commission (IATTC) was founded in 1949 to manage tuna fisheries in the tropical

● Because stocks may migrate between the national waters of many countries, many fisheries are jointly managed by several countries.

Pacific Ocean and includes member countries from North, Central and South America, France, Japan, and Vanuatu.

Most fished species have pelagic larvae and will grow in mass by 5 to 6 orders of magnitude in the course of their 2 to 20 year life cycles (Fig. 12.4). Most fished species are also very **fecund**, but the majority of eggs and larvae will perish, from starvation or predation, in the first few weeks of life. Thus, although many fished species have

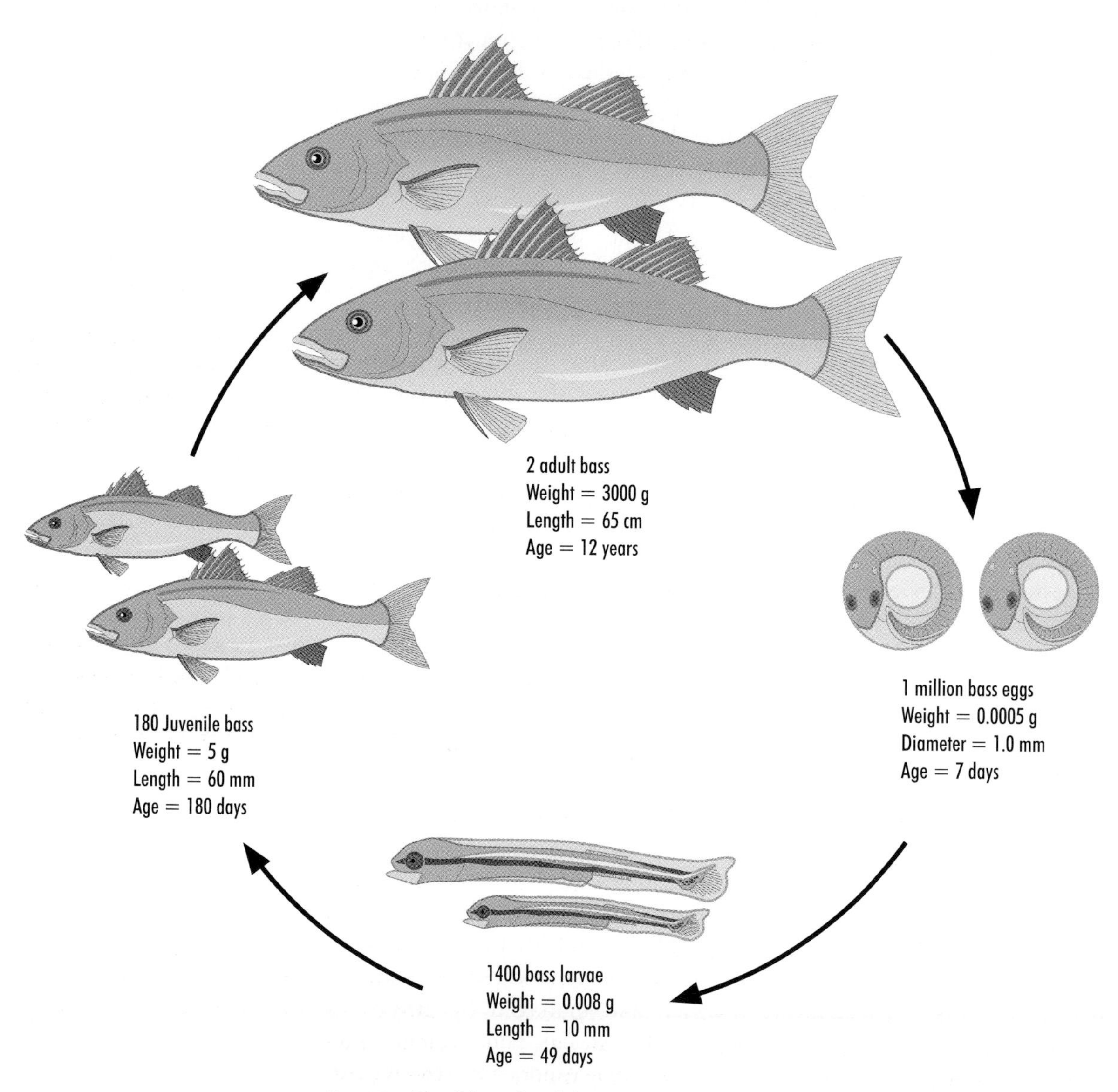

Fig. 12.4 The life cycle of the sea bass (*Dicentrarchus labrax*).

fecundities of 10^3 to 10^6, the mean number of adults produced per adult per year usually falls in the range 1 to 7. The extent of larval mortality is shown, for example, by the West African rock lobster (*Jasus islandii*). For every 1 million eggs laid, only one, on average, survives to adulthood.

12.5 Fish Population Biology

Fish abundance, like that of other species, fluctuates in space and time. Abundance fluctuations are driven by physical and biological processes that affect the production and survival of eggs and larvae, and by growth and mortality during the juvenile and adult phases. When scientists discovered that they could age fish by counting growth rings on ear bones (otoliths) or scales, it became clear that only a few age classes accounted for most of the biomass in fish stocks, even when many year classes were present. The fish in these abundant age classes often dominate catches for several years.

● Fish are aged from bands in their scales and otoliths.

Recruitment is the term used to describe the number of fish that reach a specified stage in the life cycle, usually the age of joining the fishery. Large variations in egg and larval survival mean that recruitment is very variable from year to year, and often varies by a factor of 20 or more.

● Recruitment in fish stocks is highly variable. This is the main reason why the abundance of fish stocks fluctuates from year to year.

On average, more adult fishes would be expected to produce more recruits. If there were no limit to this relationship, populations would grow indefinitely. In fact, compensation occurs, and the number of recruits produced per spawner is higher at intermediate than large population size (Fig. 12.5). At large population size, density dependent effects, such as spawners consuming their own larvae (common in plankton feeding fishes like anchovy) or reduced reproductive output due to competition for food or space, lead to reduced average recruitment. Relationships between spawner and recruit abundance (often known as stock-recruitment relationships) determine how fish stocks respond to fishing; because they describe the capacity of the stock to maintain abundance in the face of fishing pressure.

● Stock-recruitment relationships, which describe the average numbers of recruits produced by a given number of spawners, are needed to assess the capacity of a fish stock to tolerate fishing.

Various mathematical models are used to describe spawner-recruit relationships. These all feature high recruitment at low population size and lower recruitment at large population size. Variance around the average model is as important as the model itself. Such variance is driven by physical and biological processes that are largely unpredictable, such as the stability of the water column and the timing of phytoplankton or zooplankton production peaks in relation to the timing of spawning and

● View stock-recruitment relationships for many of the world's major fish stocks at **www.dal.ca/myers**.

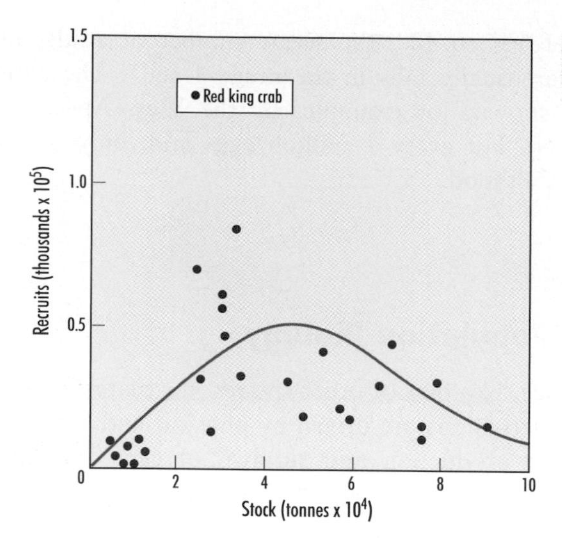

Fig. 12.5 The spawner-recruit relationship for Alaskan red king crab. Note that more recruits per spawner are produced at low abundance and the considerable variance around the relationship that is largely due to the effects of the environment on egg and larval survival.

larval development (Chapter 6). The impossibility of predicting annual levels of recruitment creates uncertainty for the fisher, the fisheries scientist, and anyone else who depends on a fishery.

12.6 Fishing Methods

Humans use a staggering variety of innovative and not so innovative methods to catch fish. These range from simple baited hooks on a handline to dynamite bombs that are thrown into the sea to stun fish (Fig. 12.6). When fish are very valuable, complex and expensive methods are worth using. Thus off Cape Cod in the USA, large bluefin tuna (*Thunnus maccoyii*) that can sell for $US50 000 each are tracked by spotter planes and shot from fast boats with electrified harpoons (see also Chapter 13).

● Humans will do almost anything to catch fish.

Fishing gears are conveniently categorized into two groups, active and passive. Active gears are usually towed across the seabed or used to encircle fish. These include the familiar trawl and beam trawl nets that are towed behind fishing vessels on most of the world's continental shelves, and the giant encircling seine nets that are used to catch oceanic fishes such as anchovies and tunas (Fig. 12.7). Active gears also include dredges that are towed across the seabed to catch shellfish, and gears such as spears, harpoons, and explosives.

● The main fishing gears can be described as active or passive.

Passive gears are put in the water and fish are caught when they move into the gear. Examples of passive gears are nets that are set on the seabed or left to float in the open ocean, and entangle fish by the gills (Fig. 12.8). Other passive gears in widespread use are the baited pots

Fig. 12.6 Examples of fisheries: (a) tuna line fishery, (b) tuna purse seine fishery, (c) squid jig fishery and (d) dynamite fishing on a coral reef. (Copyrights: M. Marzot (FAO Photograph) (a), *Fishing News International* (b, c) and Thomas Heeger (d).)

used to catch crabs, lobsters, or prawns, and various types of fish traps (Sainsbury 1996).

Fishers would like their gears to catch target species with great efficiency, and to catch little or no by-catch. However, gear design relies on many compromises and factors such as price and efficiency are taken into account. As a result, many gears catch target species but also impact the environment and non-target species. Thus trawl nets crush rare animals on the seafloor, and may catch non-target sea turtles or skates, and gill nets can catch seabirds and marine mammals. Only a few gears efficiently catch target species and few others. These include the large purse seine nets used to surround shoals of anchovy, herring, or sardine.

● Fishing gears are very selective, but never so selective that they catch the target species and nothing else.

Not surprisingly, fishers are highly innovative people and will always strive to increase the efficiency of their gears. As legislation is introduced, either to reduce catching efficiency or to reduce the environmental impacts of fisheries, so fishers usually alter their behaviour and gear to maintain their income.

● Most fishers are highly innovative people and will always strive to improve the efficiency of their gears.

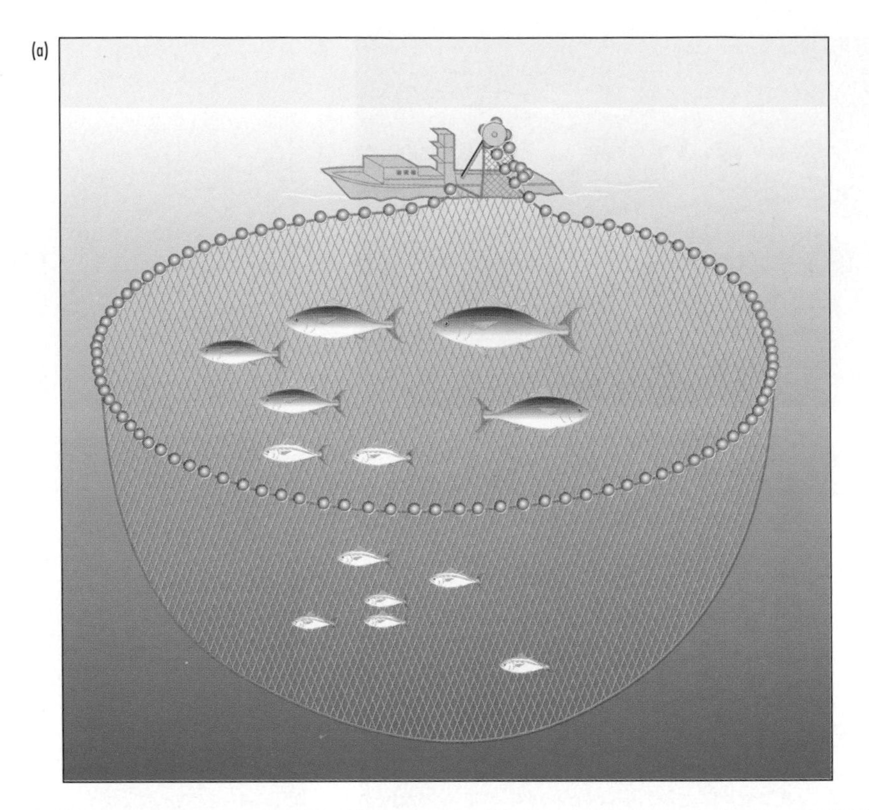

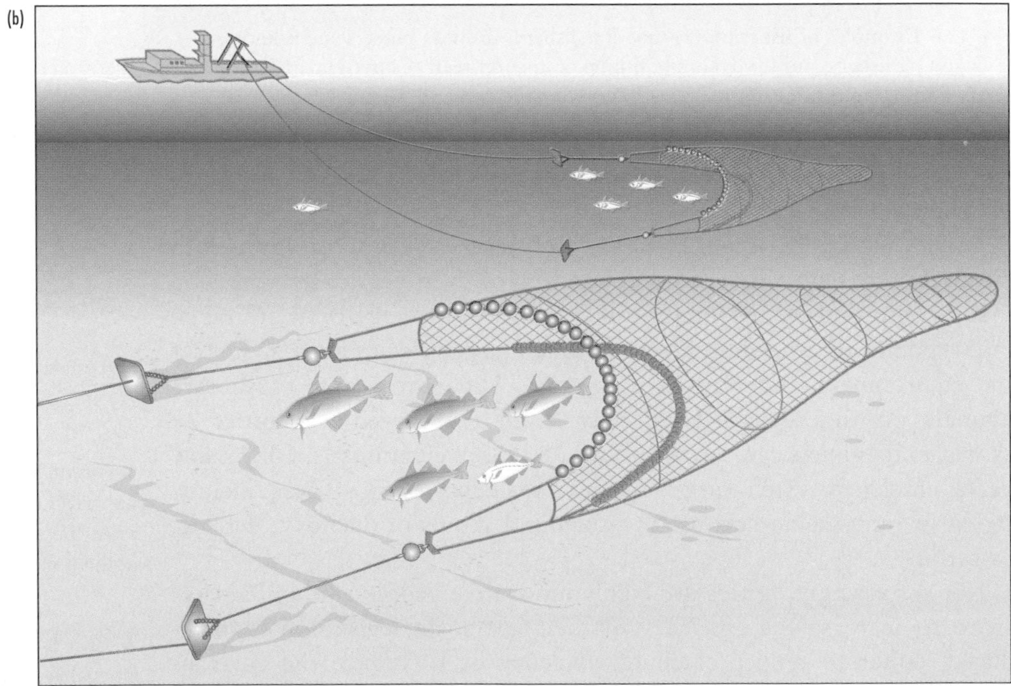

Fig. 12.7 Examples of active fishing gears: (a) purse seine and (b) bottom trawl net.

Fig. 12.8 Examples of passive fishing gears: (a) gill net and (b) long-line.

12.7 Fish Stock Assessment

A good understanding of the population biology of fished species is needed to predict the levels of catch that can be taken from fisheries, but without compromising catches in future years. For fishing to be sustainable, the fishing mortality that reduces population biomass must be balanced by the levels of recruitment and growth that maintain it (Fig. 12.9). Methods of fisheries stock assessment seek to estimate the levels of fishing mortality that will safely be balanced by recruitment and growth. If these levels of mortality are exceeded, then the size of the stock and the capacity of the stock to maintain fishing mortality will fall. Excessive fishing mortality is the main cause of stock collapse.

● Fish stock assessments are used to predict the catches that can be taken from fisheries without compromising catches in future years.

Many methods are used to assess sustainable levels of fishing. Potential yields of fisheries are initially calculated on a per-recruit basis because, as we saw in 12.5, recruitment is unpredictable from year to year. However, once the abundance of recruits in a year class has been measured, the yields per recruit can be converted to a total yield that can be taken from the fishery.

Yield-per-recruit models are used to determine the rates of fishing that give the best trade-off between the size of fish caught and the numbers in the stock. Because fish will die from both natural and fishing mortality, the numbers of fish in a year class declines over time. However, individuals will also be growing over time. As a result, there will be an optimal age for catching fish, the age when their biomass is highest (Fig. 12.10).

● Yield-per-recruit models predict when the potential yields per recruit are highest, but they ignore the potential effects of fishing on recruitment.

Since yield-per-recruit models only predict the age when population biomass is highest, they ignore the potential effects of fishing on

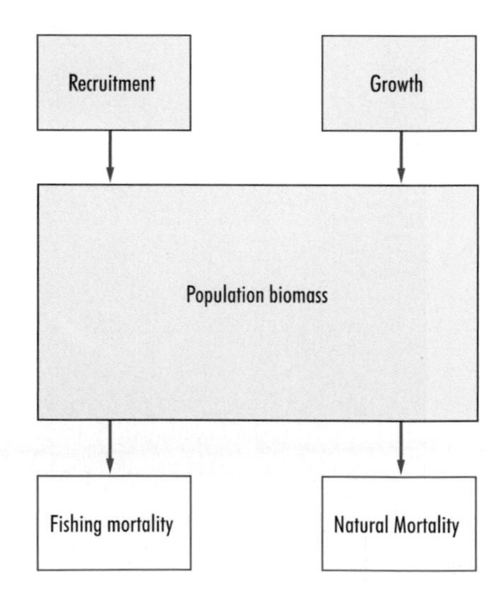

Fig. 12.9 The population biomass of fish is increased by the processes of recruitment and growth and is decreased by natural and fishing mortality.

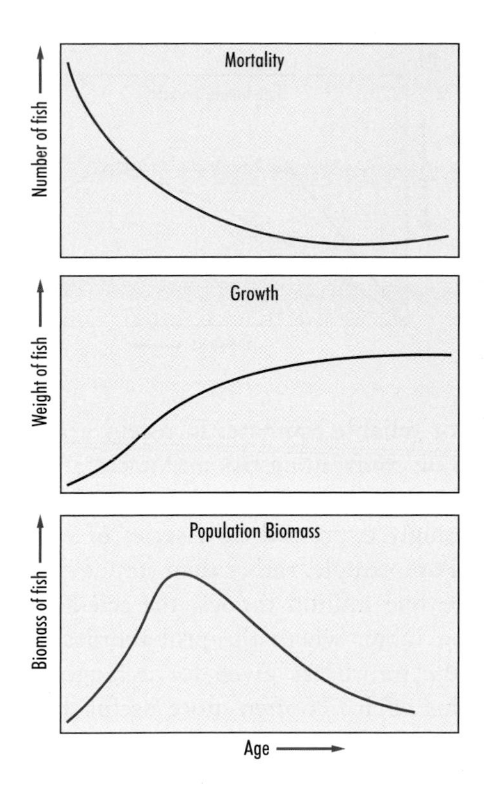

Fig. 12.10 As fish get older their numbers decline due to mortality. However, their individual body size will increase due to growth. As a result, population biomass (size × numbers) is greatest at an intermediate age.

recruitment. The age at maximum population biomass could well be less than the age at maturity. Thus intensive fishing could impair recruitment. Indeed, we have already seen how recruitment falls when spawning stock biomass is low in our analysis of stock-recruitment relationships.

Replacement lines can be used to test whether fishing will lead to recruitment failure and stock collapse. A replacement line shows the replacement level where average recruitment rates are high enough to balance spawning stock size. Using the replacement line, and a stock-recruitment relationship such as introduced in 12.5, Figure 12.11 shows two replacement lines, corresponding to low (a) and high (b) rates of fishing. We predict the fate of the population from the starting stock size by using the stock-recruitment relationship to predict recruitment in the next generation and taking this as the new stock size '*S*'. In our first example, the replacement line crosses the spawner-recruit curve and the population will reach a theoretical equilibrium. In the latter example, the replacement line is steeper because fishing mortality is higher, so the stock cannot replace itself and tends towards collapse.

● Replacement lines can be used to assess the risk of recruitment over-fishing.

The diagram in Figure 12.11 represents average and idealized conditions. In reality, there are uncertainties in the data used to conduct population analyses and a best single estimate of the potential yield from

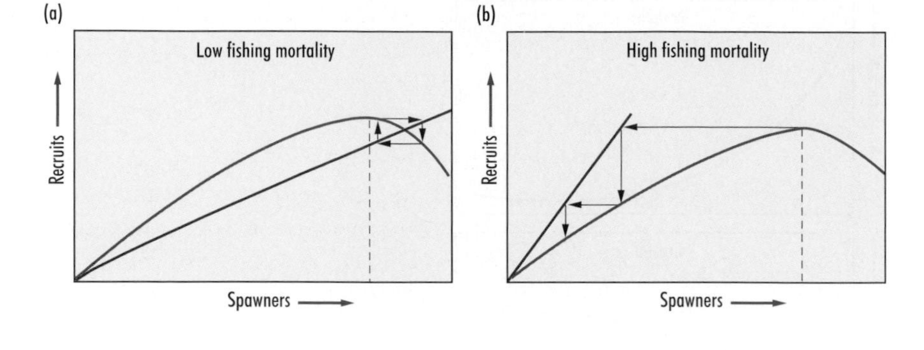

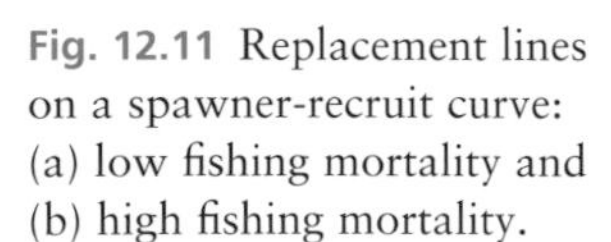
Fig. 12.11 Replacement lines on a spawner-recruit curve: (a) low fishing mortality and (b) high fishing mortality.

a fishery is not necessarily a good or reliable estimate. In recent years, there has been much more emphasis on confronting risk and uncertainty in fisheries science, and the single best estimates that were once used to advise fishery managers are increasingly expressed as a series of estimates and associated probabilities. For example, rather than stating that the total allowable catch should be one million tonnes, the scientists would provide advice as a **decision table**, where the probabilities of different stock sizes occurring in the future are given for a range of capture options (Hilborn 1996). This advice is often more useful than advice based on single best estimates as it allows managers to assess the potential consequences of their decisions.

● Modern fisheries assessment methods have confronted the problems of risk and uncertainty.

Reference points are increasingly used to guide management (Caddy 1998). Reference points may provide targets for stock biomass or fishing mortality that are sustainable or avoid serious and irreversible harm to a fish stock. An example of a reference point is B_{pa}, the minimum stock biomass that is consistent with the precautionary approach. This means that if the biomass is above B_{pa}, then there is a specified probability that biomass is above B_{lim}, the limit reference point identifying the biomass that is needed to ensure stock persistence.

● Biological and precautionary reference points are increasingly used to guide management.

12.8 The Management Process

There are many possible goals of fishery management, from maximizing catches, employment, income supply, or profit, to ensuring that conflicts are avoided and rare species are not threatened. Methods of analysis, such as those already described for optimizing yields from single species stocks, have been used to assess how other management objectives can be met (Hilborn & Walters 1992). Many fisheries are still managed to meet social and economic objectives, although there is increasing emphasis on fishery conservation and protection of marine ecosystems.

Once management objectives are identified, science provides the basis for setting a management strategy and advising on a management action. Figure 12.12 gives examples of management objectives, strategies, and actions. We have already seen how science can be used to advise on management strategy and action.

● Effective fisheries management requires that clear management objectives, strategies and actions.

Fishery management actions usually consist of catch controls, effort controls, or technical measures. Catch controls are implemented as total allowable catches or individual catch quotas and effort controls are usually implemented by limited licensing schemes for fishers or vessels, effort quotas, or vessel and gear restrictions. Technical measures include controls on the size of mesh in nets in order to let young fish escape, and closed seasons or area closures in fisheries. All these management actions have advantages and disadvantages, and may be difficult or expensive to enforce. Most fisheries are managed using a combination of all three approaches.

● Most management action is taken using catch controls, effort controls, or technical measures.

In modern fishery management, there is increasing emphasis on a so-called precautionary approach, as managers try to learn from lessons of the past and to apply this wisdom in the future. The precautionary approach is based on the idea that management actions taken today should not compromise the needs of future generations, avoid changes that are not reversible, and identify unwanted outcomes of management actions in advance, and the measures needed to correct them. The precautionary approach applies both to the management of the target stock, and to managing the impacts of fishing on the ecosystem (FAO 1995).

From ecological, economic, and social perspectives, fisheries management has often been unsuccessful. An FAO analysis suggested that the main factors leading to unsustainable fishing included inappropriate incentives, high demand for limited resources, poverty and lack of alternatives, to fishing and lack of governance. These factors are often paraphrased as 'too many fishers chasing too few fish'. The FAO analysis also showed that scientific advice on the status of fish stocks and

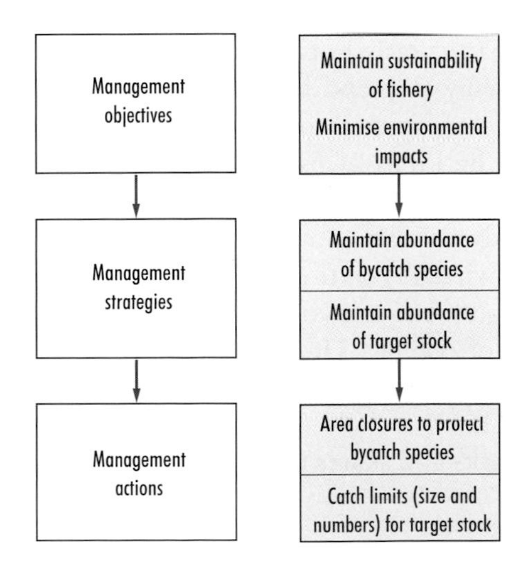

Fig. 12.12 From a management objective to a management strategy and action.

the effects of fishing made only a small contribution to a complex management and decision-making process, and often carried little weight in relation to immediate social and economic considerations.

12.9 Environmental Impacts of Fishing

While fishing provides food, income, and employment, the ever-increasing intensity and diversity of fishing operations has environmental costs that threaten marine ecosystems. Here, we see how non-target fishes, birds, mammals, and turtles are affected by fishing and the impacts of fishing on habitats such as seamounts, kelp forests, and coral reefs. These impacts are a growing concern for fishery managers and conservationists.

● Fisheries can have significant environmental costs.

12.9.1 Fish communities

Species vary in their **vulnerability** to fishing. Apart from the simple fact that a large hook or net with a large mesh will only catch large individuals, vulnerability depends on behaviour and life history. Life history determines the levels of fishing mortality that a population can tolerate, and species with slow growth, late maturity, and large body size are very vulnerable to fishing. In many cases, species with large body size are also the favoured targets of fishers.

● Large slow-growing species are most vulnerable to fishing.

Fishing has consistently reduced the abundance of large fishes in the world's oceans. Thus communities once dominated by large fishes are now dominated by small fishes and invertebrates (Fig. 12.13). Since larger fishes, on average, feed at higher trophic levels, fishing has led to a reduction in the **mean trophic level** of fish landings and fish communities, an effect that has been dubbed 'fishing down the food web' (Pauly et al. 1998).

Many fisheries are '**mixed**' fisheries and catch many species of target and non-target fishes. This applies particularly to trawl net fisheries that catch bottom-dwelling fishes and many trap and line fisheries on coral reefs. The differential vulnerability of species to fishing means that some will be dramatically depleted by levels of fishing that others can withstand. The North Sea, for example, has been heavily trawled for over 100 years, and species with large body sizes and slow growth, such as the largest skates and rays, have been virtually **extirpated** in that period. The biomass of the largest fishes is now one to two orders of magnitude less than expected in the absence of fishing. Throughout the world's seas, there are other examples of large slow-growing species becoming locally and regionally extinct. These include groupers and large parrotfishes on tropical reefs and sharks and skates in the Atlantic (Dulvy et al. 2003).

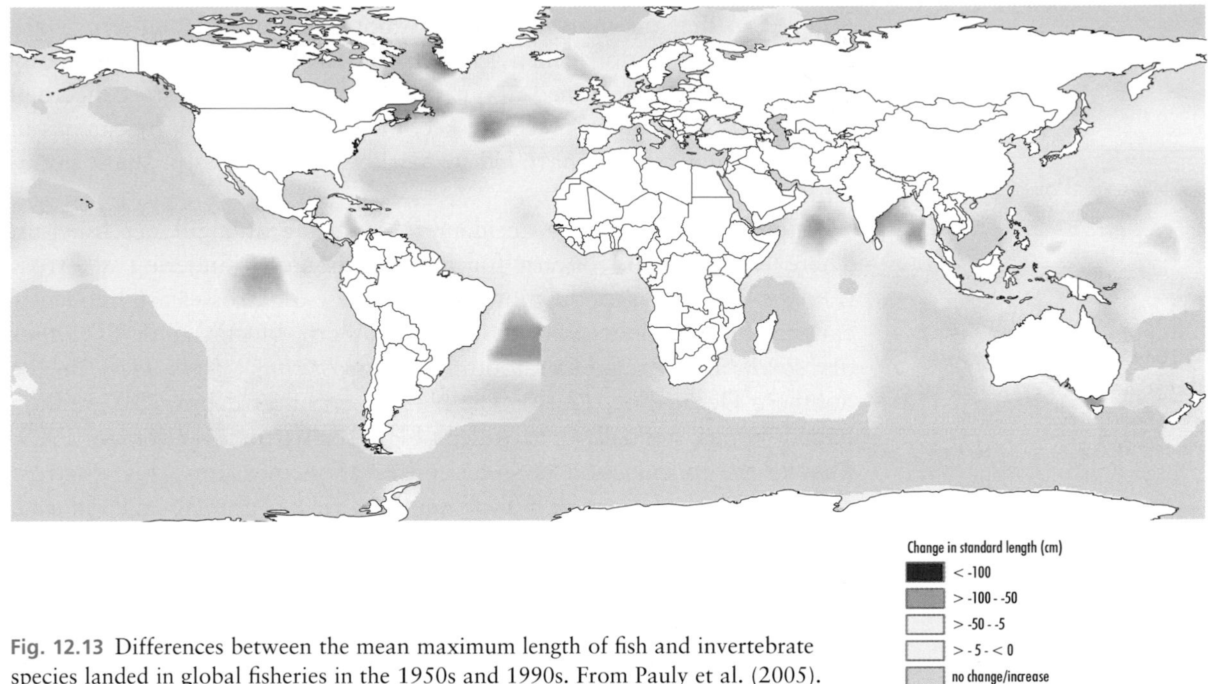

Fig. 12.13 Differences between the mean maximum length of fish and invertebrate species landed in global fisheries in the 1950s and 1990s. From Pauly et al. (2005).

12.9.2 Birds and marine mammals

Fishing is most vehemently criticized when birds, marine mammals, and turtles are taken as by-catch. These impacts have consistently attracted more public attention than the impacts of fisheries on fish, and have forced fishery closures and significant changes in fishing methods. Mass dolphin mortality in the eastern tropical Pacific tuna purse seine fishery during the 1960s and early 1970s attracted considerable public attention and provoked widespread criticism of the environmental impacts of fishing. Hundreds of thousands of dolphins were killed annually because purse seiners set their nets around pods of dolphin to catch the tuna that swam with them. When the nets were hauled, the dolphins were trapped and usually died. By 1980, the population of eastern spinner dolphins (*Stenella longirostris*) was reduced to one fifth of its original abundance. Intense public pressure and consumer boycotts of tuna products led to new fishing practices to release trapped dolphins alive and the use of observers to monitor dolphin mortalities on all purse seiners.

● Bird and marine mammal by catches have become a dominant management issue in some fisheries.

Today, annual dolphin mortality is less than 5% of that in the 1970s, and consumers increasingly purchase pole and line caught 'dolphin friendly' tuna. Levels of marine mammal by-catch in most fisheries are now set so that the mortality rates do not threaten the long-term viability of the population (Allen 1985, Hall 1996). However, while many consumers were concerned about dolphins, far fewer were concerned about sharks, and it has recently become clear that shark by-catches in tuna long-line fisheries have led to worldwide collapses in shark populations (Baum et al. 2003).

● Some albatross species are threatened with extinction by long-line fisheries.

Seabirds are caught by accident in long-line and gill net fisheries. There is particular concern for the viability of wandering albatross (*Diomedea exulans*) populations because these albatrosses are frequently caught on long-lines used to catch southern bluefin tuna (*Thunnus maccoyii*) and Patagonian toothfish (*Dissostichus eleginoides*) in the Southern Ocean (Fig. 12.14). The bluefin tuna fleet deploys 107 million baited hooks annually and killed 44 000 albatrosses prior to 1989. Despite the introduction of some conservation measures, this albatross could decline to extinction if hook numbers are not dramatically reduced.

Seabird catches on long-lines are an inconvenience for the fishers as well. They lose bait and have to reset the lines. Conservation measures introduced include bird scarers and devices that set the lines below the diving depth of the birds. However, some bird populations have already been so depleted by fishing that any further by-catches are likely to threaten their viability (Weimerskirch et al. 1997).

12.9.3 Trawling impacts on the seabed

When trawls are dragged over the seabed, they are bound to kill animals and damage habitats. The significance of trawling impacts depends on

(a)

(b)

Fig. 12.14 (a) A drowned black browed albatross caught up in the meshes of a trawler fishing in the Southern Seas. (b) These shags became entangled in a ghost-fishing bottom-set gill net while trying to extract the gadoid fish (middle). (Photographs: (a) Richard Woodcock (b) Blaise Bullimore.)

the habitats where they are fished and levels of natural disturbance on the fishing grounds. On sandy seabeds in shallow, tide-swept, and wave-impacted areas, most of the animals living on and in the seabed are very well adapted to high rates of mortality and disturbance and the effects of trawling are relatively minor (Chapters 7 and 14). Conversely, in deep areas where wave and tidal action are low and the seabed habitat is dominated by habitat forming (biogenic) species, trawling can have major and long-term impacts that reduce both biomass and diversity (Jennings & Kaiser 1998, Kaiser & De Groot 2000). Typical examples of vulnerable habitats are maerl beds, coral reefs and the biogenic habitats on seamounts. The impacts of trawling on seamounts are perhaps the most dramatic described to date (Box 12.3).

● The impacts of trawling depend on the balance of fishing and natural disturbance.

Simple empirical and modelling studies have shown that many seabed species will be extirpated if trawling frequencies exceed 2 to 3 times per year. However, large-scale studies of the distribution of fauna on many major fishing grounds show that many vulnerable species persist. The reason for this is that trawling disturbance is actually very patchy (Fig. 12.15). While many shelf seas are swept, on average, at least twice each year, fishers tend to return to favoured fishing grounds and trawl tows that they know to be free from obstructions. As a result, some areas may be trawled 10 to 15 times each year while adjacent areas are virtually unfished. Provided that spatial patterns of effort are consistent from year to year, animals can recolonize fished areas from those that remain unfished. Management measures that increase the long-term homogeneity of effort, such as rotating closed areas, may increase the cumulative effects of trawling.

● Trawling effort is very patchy and this can allow vulnerable species to persist in heavily trawled areas.

In addition to the changes to ecosystem processes caused by bottom fishing disturbance, alteration of the seabed may also change its functional importance for associated target species. In other words, not only does fishing deplete fish stocks, but it may also adversely affect their habitat. This realization has sparked a flurry of activity to identify what has become termed 'essential fish habitat'. Amendments to the *US Magnuson-Stevens Act* require fisheries managers to consider the wider ecological impacts of fishing gears upon fish habitat in addition to issues such as fishing effort control and total allowable catches (Auster & Langton 1999). Essential fish habitat (EFH) is a description open to interpretation, but it is clearly defined within legislation to mean 'those waters and substratum necessary to fish for spawning, breeding, feeding or growth to maturity'. This legislation is an important step towards an ecosystem based fisheries management (12.10) as it recognizes the potential negative effects of the direct and indirect effects of fishing on fish populations. The introduction of this legislation caused a hiatus among managers, who were faced with the task of

Box 12.3 Trawling impacts on seamounts

Technological development of fishing gears and fishing boats is making more and more habitat accessible to trawling. Seamounts, once difficult to locate and too rocky to trawl, have been heavily fished in recent decades, as fishing vessels can now 'mine' them for stocks of long-lived and slow-growing species such as orange roughy (*Holplostethus atlanticus*). Electronic positioning systems on the vessels and nets can be used to position the net precisely and to trawl on the relatively tiny seamounts that rise from depths of several kilometres to within 500 to 2000 m of the surface.

Seamounts are unique deep-sea environments, and, because they rise from great depths, their topography enhances local water currents. These currents carry nutrients and prey over the seamounts, and rich complex communities of suspension feeders such as corals develop, in contrast to the low biomass communities of deposit feeders on the surrounding seafloor. Many benthic species living in association with seamounts are new to science, long-lived, and vulnerable to trawling.

When seamounts are fished for the first time the nets can take by-catches of several tonnes of coral! In fact, as the first orange roughy fisheries developed off New Zealand, there was interest in starting a precious coral industry based on black coral by-catch! The damage to coral is the first and most dramatic effect of trawling on seamounts and has immediate knock-on consequences for fish and invertebrates that use coral habitat.

When trawled and untrawled seamounts are compared, it is obvious that the untrawled mounts have a diverse invertebrate fauna of high biomass, dominated by suspension feeders including hard and soft corals, hydroids, sponges, and brittlestars. Of the invertebrate species studied on untrawled seamounts south of Tasmania, 24–43% were new to science and 16–33% were probably restricted to the seamount environment. The benthic biomass on untrawled seamounts was 106% greater, and the number of species per sample 46% greater, than on trawled seamounts. The catastrophic effects of trawling on these habitats provided good justification for establishing a 'Marine Protected Area' around twelve seamounts.

Source: Koslow et al. (2001)

identifying EFH for a multitude of species. This revealed large voids in our knowledge about the basic biology of many of the species currently under management.

Some elements of EFH such as spawning and nursery areas are relatively well known for the majority of commercially exploited species, and in many cases these are already afforded some protection by permanent or seasonal area closures. Nevertheless, for some particularly vulnerable species such as rays, we know where breeding aggregations occur, but the habitat into which the eggs are laid remains open to speculation. However, of equal relevance are the habitat quality issues that affect the acquisition of food and the avoidance of predators. The identification of those habitats that may have an important or 'essential'

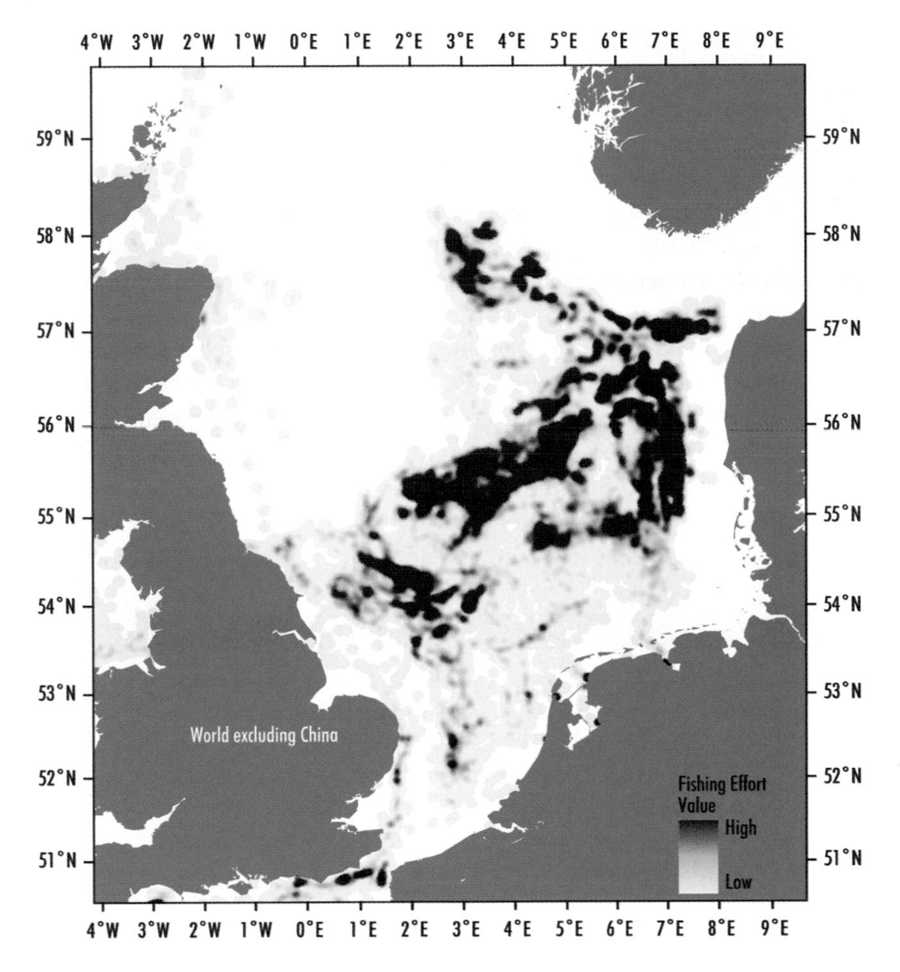

Fig. 12.15 The patchy spatial distribution of beam trawling effort in the North Sea in 2002. Each point is the record of a vessel location from a satellite vessel monitoring system. Vessels travelling at speeds between 5 and 8 knots are assumed to be fishing. (Image: Craig Mills.)

functional role for individual species is likely to be complex for those species that utilize different habitats at particular stages of their life history. For example, many fish species start life close inshore, moving further offshore as they increase in body size (Rijnsdorp & Leeuwen 1996). It might be relatively straightforward to identify EFH for fish strongly associated with seabed structures such as reefs or biogenic habitats (seagrasses, mangroves, oyster beds), however many fish in temperate systems exhibit highly flexible lifestyles and utilize a wide range of different habitat types. Nevertheless, for such species, it is possible to identify areas that consistently attract the greatest proportion of the population, and presumably these areas fulfil some functional role.

Previous studies of the relationship between fish assemblages and their environment have focused on relationship between fish distribution and environmental variables such as salinity, depth, and substratum type (e.g. Smale et al. 1993). In some cases these variables are good correlates of some fish assemblages. They do not necessarily define the essential features of a specific habitat; rather they constitute a component of that habitat that may act as a surrogate for some other more important habitat feature. Habitat complexity and structure appear to be important physical features for some fish species (Auster et al. 1997). Many studies have demonstrated the relationship between flatfish species and the sediment particle composition of the seabed (Gibson & Robb 1992). For example, small plaice are better able to bury themselves in sediments that have a particular grain-size composition and choose particular sedimentary habitats accordingly. Hence, a specific particle size-composition may be an essential habitat requirement for flatfish, whereas the presence of large sessile epifauna or rocky substrata might be considered non-essential. In contrast, structural complexity greatly increases the survival of juvenile roundfishes as it provides refuge from predation (Tupper & Boutillier 1995) (Box 12.4).

Just because we find fish associated with a particular habitat does not necessarily indicate that it has an essential functional role. For example,

Box 12.4 Essential fish habitat

Biogenic (living) and physical features of seabed habitats can have a critical function for certain commercially important species. Mud habitats appear to have few large-scale topographic features, but large cerianthid anemones provide shelter for animals such as spider crab (*Lithodes* sp.) (a). Depressions in the seabed formed by other organisms such as sea scallops or the troughs of sand waves provide cover for silver hake (*Merluccius biliniaris*) from where they ambush prey (b). Attached fauna such as sponges provide complex three-dimensional structure in which fish such as sculpin (*Myoxocephalus* sp.) can shelter from predators and ambush their prey (c). (Photographs: © Peter Auster.)

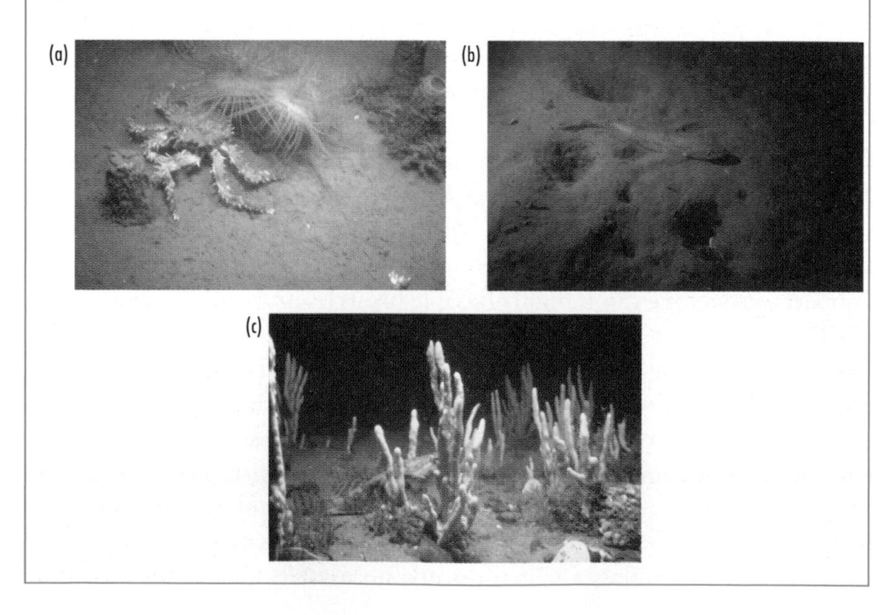

there is great interest in the relatively recent discovery of deep-water coral reefs on the continental shelf edge of the North Atlantic and elsewhere. These reefs provide prominent habitat features in a relatively uniform seabed landscape and act as foci for fish such as redfish (*Sebastes* spp.) for which there is a commercial fishery. Conservation groups argued for the protection of the reefs on the basis that they provided EFH for the redfish (Malakoff 2004). However redfish are found in a wide variety of habitats across a large geographic area, and despite a concerted research effort evidence of a critical link between redfish ecology and deep-water corals remains uncertain (Fossa et al. 2002, Kaiser 2004).

12.9.4 Effects of fishing on coral reefs

In Chapter 10, we saw that coral reefs are among the most delicate and highly structured marine habitats. The topographic complexity of reefs provides refuge and feeding opportunities for many fish and invertebrate species. Areas of actively growing reefs that have not been damaged by human activities can sustain fishery yields of up to 5 tonnes km^2 $year^{-1}$ (Fig. 12.16). These fisheries are usually prosecuted by line and spear fishers, who cause minimal damage to coral and catch a range of species from many trophic levels. However, growing human populations have driven the development and expansion of fisheries on tropical coasts, and fishers are increasingly turning to habitat destructive fishing methods to maintain yields. For many fishers, who have few, if any, alternative sources of food and income, there is no choice but to do this.

● The lack of alternate sources of food and income has driven many fishers on tropical coasts to adopt habitat destructive fishing methods.

Fig. 12.16 Selling fishes caught in a coral reef fishery in Fiji. Photograph: S. Jennings.

Muro-ami drive netting, bottom set gill nets and heavy traps have all caused habitat destruction on reefs. Muro-ami fishing is widely practised in south-east Asia and involves fishers driving fish towards a net with weighted scare-lines that are dropped repeatedly onto the coral. However, fishing with explosives has even greater effects on reefs. Explosive, blast or dynamite fishing is practised in many regions, and although it is often illegal, this is unlikely to deter fishers desperate for food and income. Repeated blast fishing, where explosive charges are detonated over and on the reef, soon reduces large areas of actively growing reef to rubble. Needless to say, fisheries of this type are not sustainable, killing many juvenile and non-target species and destroying their habitat.

● Reef fisheries have indirect effects on the reef food webs that help corals to keep growing.

Fisheries also have indirect effects on coral reefs. Sea urchins and fishes, such as the colourful parrot and surgeonfishes, are the dominant herbivores on reefs. The persistence of herbivorous fish on reefs depends on the presence of fish that eat sea urchins. These urchin predators maintain sea urchin populations at a size where their low gross production makes them poor competitors with herbivorous fish. If populations of urchin-eating fish are reduced by fishing, then urchin populations expand and the urchins graze reef algae to such low levels that herbivorous fish can no longer compete. The dominance of urchin grazers on heavily fished reefs has knock-on consequences for corals. Thus, herbivorous fish clear space for coral settlement, and enhance the growth and survival of young coral colonies, while urchins erode the substrate as they graze and prevent coral recruitment and growth. On heavily fished reefs where urchins dominate the grazer community, bioerosion leads to loss of structural complexity, reduced fish biomass and disruption of the ecological processes responsible for fish production (McClanahan 1992) (see also Chapter 7).

12.9.5 Effects of fishing on kelp forests

As our last example of the indirect effects of fishing we look at another situation where urchins play a keystone role. Kelp forests are found in many cool, shallow, and nutrient-rich coastal waters (Chapter 7), and support diverse fish and invertebrate fisheries. Sea urchins graze on kelp, and when sea urchins become abundant, either as a result of high levels of recruitment or the indirect effects of fishing, they can graze kelp until the kelp forest ecosystem shifts to an alternative state, known as 'urchin barren ground'!

● Fisheries also have indirect effects on kelp forest ecosystems.

The extent of *Macrocystis* kelp forests off southern California has shrunk dramatically since the mid-twentieth century. This is partly the result of changing oceanographic conditions and pollution, but is also due to the depletion of urchin predators. Both the spiny lobster

(*Panulirus interruptus*) and the sheepshead (*Semicossyphus pulcher*) were once sufficiently abundant to help limit urchin populations, but this is no longer the case. Ironically, a large fishery that targets the red sea urchin (*Strongylocentrotus franciscanus*) has now developed, depleting urchin populations and allowing the re-establishment of kelp in some areas (Tegner & Dayton 2000).

● Compare this example of a fishing effect with that of hunting of sea otters in Chapter 7.

12.10 Ecosystem-based Fishery Management

Fisheries can provide society with huge benefits in terms of income, food, and employment, but instead, fisheries are often unsustainable and an environmental threat. Globally, fishing effort needs to be cut to realize the real benefits of fisheries. Some estimates suggest that cuts of 40 to 50% are needed.

The ecosystem approach to fisheries management (EAFM) is a new way of looking at fisheries management, and treats fishery management as just one part of environmental management, potentially placing the environmental standards expected from fisheries on a par with those applying to other human uses of the marine environment such as aggregate extraction or oil and gas exploration. This is a major change, when fishing is often cited as one of the main human threats to marine ecosystems. The EAFM is part of the ecosystem approach and is consistent with the concept of sustainable development, which requires that the needs of future generations are not compromised by the actions of people today.

The broad purpose of the EAFM is to plan, develop, and manage fisheries to meet the multiple needs and desires of societies, but without jeopardizing the options for future generations to benefit from the full range of goods and services (including, of course, non-fisheries benefits) provided by marine ecosystems. The success of the EAFM will depend on whether high-level governmental commitments to an EAFM can be turned into specific, tractable, and effective management actions that will lead to better fish stock management and remedy the unwanted impacts of fishing on non-target species, habitats, and ecological interactions. Moreover, the EAFM will not remove the high short-term costs of moving towards sustainability (principally the costs of capacity reduction and providing alternative employment) and ways of meeting and mitigating these costs still need to be explored.

● The ecosystem effects of fishing are increasingly considered as important by fisheries managers, who are starting to adopt an ecosystem-based approach to fisheries management.

Existing moves towards the implementation of the EAFM have been characterized by management action to mitigate the environmental impacts of fishing on a case-by-case basis. Thus turtle excluder devices (TEDs) have been fitted to trawls to stop turtles being killed as by-catches, or fisheries for small pelagic 'forage' fishes have been closed in the vicinity of seabird breeding colonies. These management actions are likely to be the first steps in a long process of integrating environmental concerns into fisheries management, and many commentators now see the EAFM as an evolutionary rather than revolutionary process. In the longer term it is likely that indicators of the impacts of fishing on habitat and non-target species will be used to set targets for ecosystem-based fishery management and to assess the success of management action. The choice of indicators is likely reflect the growing priorities of species and habitat conservation as justifiable goals of management, and it is likely that increasingly large areas of the sea will be closed to fishing in order to meet conservation objectives (Chapter 15).

● At present, most ecosystem-based management means dealing with the adverse impacts of fisheries on a case by case basis.

12.11 The Future of Fisheries

Many fisheries are currently fished beyond sustainable limits and expected reductions in fishing effort will provide long-term benefits, by creating more productive and more profitable fisheries and reducing environmental impacts. Capacity for improved fisheries management will be greater in the developed world where there will be some commitment to meeting the short-term costs of effort reduction. The prospects for sustainability in poorer nations are not so good, unless the international community makes the financial commitment to provide alternative livelihoods for fishers.

● CHAPTER SUMMARY

- Fisheries are a vital source of food, income and employment, especially in the developing world. Around 80 million tonnes of fish are landed each year and 20 species of fish account for over 40% of these landings.
- Fishing gears can be described as active or passive. Active gears are towed towards the fish while passive gears are not. Fishing gears catch a range of species, including bycatches, that will be discarded by the fishers if they are not worth eating or selling.
- Stocks fluctuate in abundance as a result of natural and fishing effects. Variations in egg and larval survival have a major impact on the size of year classes recruiting to the fishery.
- Fish stocks are usually assessed to predict the size of catches that can be taken without compromising future catches. Fishery management actions consist of catch controls, effort controls and technical measures.
- Fisheries have environmental costs which can threaten rare species and the sustainability of the fished resource. There have been increasing attempts to mitigate these costs by adopting the precautionary approach and ecosystem-based fishery management.

- Large reductions in global fishing capacity are needed to get the greatest social and economic benefits from fisheries and to ensure that biodiversity conservation and sustainability goals are met.

FURTHER READING

A wide-ranging introduction to marine fisheries ecology is provided by Jennings, Kaiser and Reynolds (2001). This book describes fisheries exploitation, conservation and management in tropical, temperate and polar environments and focuses on issues of contemporary concern such as marine reserves, the effects of fishing on coral reefs and the incorporation of the precautionary principle into management advice.

- Jennings S., Kaiser M. J. & Reynolds J. D. 2001. *Marine Fisheries Ecology*. Blackwell Science, Oxford.

Chapter 13
Aquaculture

CHAPTER SUMMARY

Aquaculture in marine systems can be traced back to well before Roman times. Even then it was clear that the culture of marine species was only feasible at great expense. While freshwater products continue to contribute most in terms of landings and value, marine products are becoming increasingly important and technological advances make it more viable to produce fish and other biota at reasonable cost. Aquaculture creates its own environmental problems and has lead to disease outbreaks, over-harvesting of forage fishes to generate fish-meal, genetic dilution of wild stocks from farm escapees, and has caused ecological problems in areas where local carrying capacity has been exceeded. A better understanding of these issues will help to ensure that aquaculture is undertaken in an environmentally sustainable manner.

13.1 Introduction

Marine aquaculture is a rapidly expanding global industry and is probably one of the best examples of our ability to take biological knowledge and apply it to improve the biological and economic yield from the marine environment. As current yields from capture fisheries taken around the world have either reached a plateau or are diminishing, there is a growing need to consider aquaculture as an alternative source of obtaining protein from the sea. As many wild capture fisheries have collapsed and local economies have suffered the consequences of the ensuing economic hardship, many have pointed to aquaculture as the solution to the impending shortfall in food production from the sea. Some even speculate that aquaculture could remove the need to fish wild stocks altogether. As our understanding of the life-history requirements of an increasing range of species expands, so previously 'difficult' species have become more feasible subjects for cultivation. In addition to these advances, increasing automation, improved feed technology and high technology waste treatment systems have contributed to a decline in the costs of production of many species. As a result, fish-farmed cod (*Gadus*

● As current yields from capture fisheries taken around the world have either reached a plateau or are diminishing, there is a growing necessity to consider alternative means of obtaining protein from the sea.

morhua) landed on the shelves of major retail outlets for the first time in 2001.

Although aquaculture can be traced back several millennia, it is still an industry very much in its infancy and consequently is associated with its own social and environmental problems. Indeed some sectors, such as shrimp cultivation, have witnessed their own '**gold rush**' as a result of high market values of shrimp. As with any new industry, the primary concern of participants has been to invest in technology and research designed to overcome the problems of production and stock control such that the enterprise becomes economically viable. In contrast, less attention has been given to the environmental consequences of cultivation practices. Yet it is ironic that in many parts of the world the environmental impact of aquaculture practices has become one of the impediments to the further expansion of the industry. This chapter aims to introduce the historical development of aquaculture, the variety of techniques used to rear a range of different organisms, the technology employed to increase productivity, and the environmental and biological consequences of aquaculture in marine ecosystems.

● The environmental impacts of aquaculture have become one of the impediments to its further expansion.

13.2 Aquaculture Past and Present

Fishes, molluscs, crustaceans, and plants have been cultivated by humans for several thousand years. Archaeological evidence is strongly suggestive of the culture of carps (Cyprinidae) by early civilizations as widely distributed as China, Japan, and Italy. Indeed, carp are recurrent motifs in art from these regions. Carps were no doubt cultivated for their food value but also for their aesthetic appeal, which gave rise to varieties such as koi carp much valued by aquarists in present day cultures. Carp are an ideal species for cultivation due to their tolerance of high-stocking densities and low dissolved oxygen levels, their omnivorous diet and high growth rates. However, marine species present a much greater technological challenge because of their requirement for high water quality and in many cases their carnivorous diets, a fact recognized long ago by the Romans (Box 13.1).

● Fish have been cultured for many thousands of years. However, the culture of marine species is far more demanding than that of freshwater species.

Box 13.1 Aquaculture in Roman times

There are two kinds of fish-ponds, the fresh and the salt. The one is open to common folk, and not unprofitable, where the Nymphs furnish the water for our domestic fish; the ponds of the nobility, however, filled with sea-water, for which only Neptune can furnish the fish as well as the water, appeal to the eye more than the purse, and exhaust the pouch of the owner rather than fill it.

From *Rerum Rusticarum* by Marcus Terentius Varro (127–126 BC), translated by Hooper & Ash (1934).

● In past and present times aquaculture has provided an easily harvested and reliable source of food.

In past and present times, aquaculture has provided a reliable and easily harvested source of protein that is independent of natural fluctuations in capture fisheries. The term '**aquaculture**' has only been in use for about 40 years, which coincides with its increasing **industrialization** most notably in the salmonid sector of the industry. The term 'culture' implies human intervention that enhances production of the cultivated species. This may be very simplistic at one level (e.g. the introduction of nutrients and organic material from animal wastes as in Chinese **polyculture** systems) ranging through to highly sophisticated computer-controlled culture in intensive water recirculation systems. In the early 1970s the production of freshwater and marine fish and shellfish was the equivalent of only 12% of the annual world fish landings for human consumption (*c*.6 million tonnes). At that time, this source of protein provided 4% of the world's supply of animal protein excluding the contribution of milk. In Asian countries, aquaculture makes a far greater contribution to the overall supply of animal protein due to the prevalence of freshwater aquaculture in this area (Allsopp 1997). Since the early 1970s, the production of finfish has increased tenfold while the contribution of capture fisheries to world protein supply has only doubled (Fig. 13.1). As a result, in 1998, aquaculture contributed 26% to the global production of fishes worth $US47.1 billion.

● Since the early 1970s, the production of finfish has increased 10 fold while the contribution of capture fisheries to world protein supply has only doubled.

While marine capture fisheries are governed by environmental factors that affect recruitment and hence are unpredictable and vulnerable to over-exploitation, aquaculture is largely dependent upon inputs of feed and upon good management of the cultivation system to avoid disease and environmental degradation. These criteria for successful aquaculture in some way appear to be more attainable than a more rational use of wild fisheries. Despite this, many aquaculture practices are themselves dependent upon wild capture fisheries to provide protein from lower down the food chain (e.g. smaller fishes such as sandeels, *Ammodytes* spp.) to make fish-meal. However this irony has been brought sharply into focus by some that have made a case that aquaculture is partly the cause for the demise of certain world capture fisheries (Naylor et al. 2000). For this and other reasons, plant based fish-feeds are seen as the most attractive solution to disengage aquaculture from its dependence upon wild fisheries that provide an unpredictable product of variable quality. With the increasing move towards genetically engineered crops, it should be possible to manufacture feeds that contain the particular protein and lipid requirements of marine piscivorous species.

● Aquaculture is often dependent upon feeds produced from fish at low trophic levels such as sandeels.

In terrestrial systems, the domestication and farming of animals has centred on cattle, pigs, sheep, and goats and chickens, turkeys, geese, and duck. The restricted range of farmed species means that technology is easily transferable between species in terrestrial systems. This restricted range of species contrasts sharply with the hundreds of fish, mollusc,

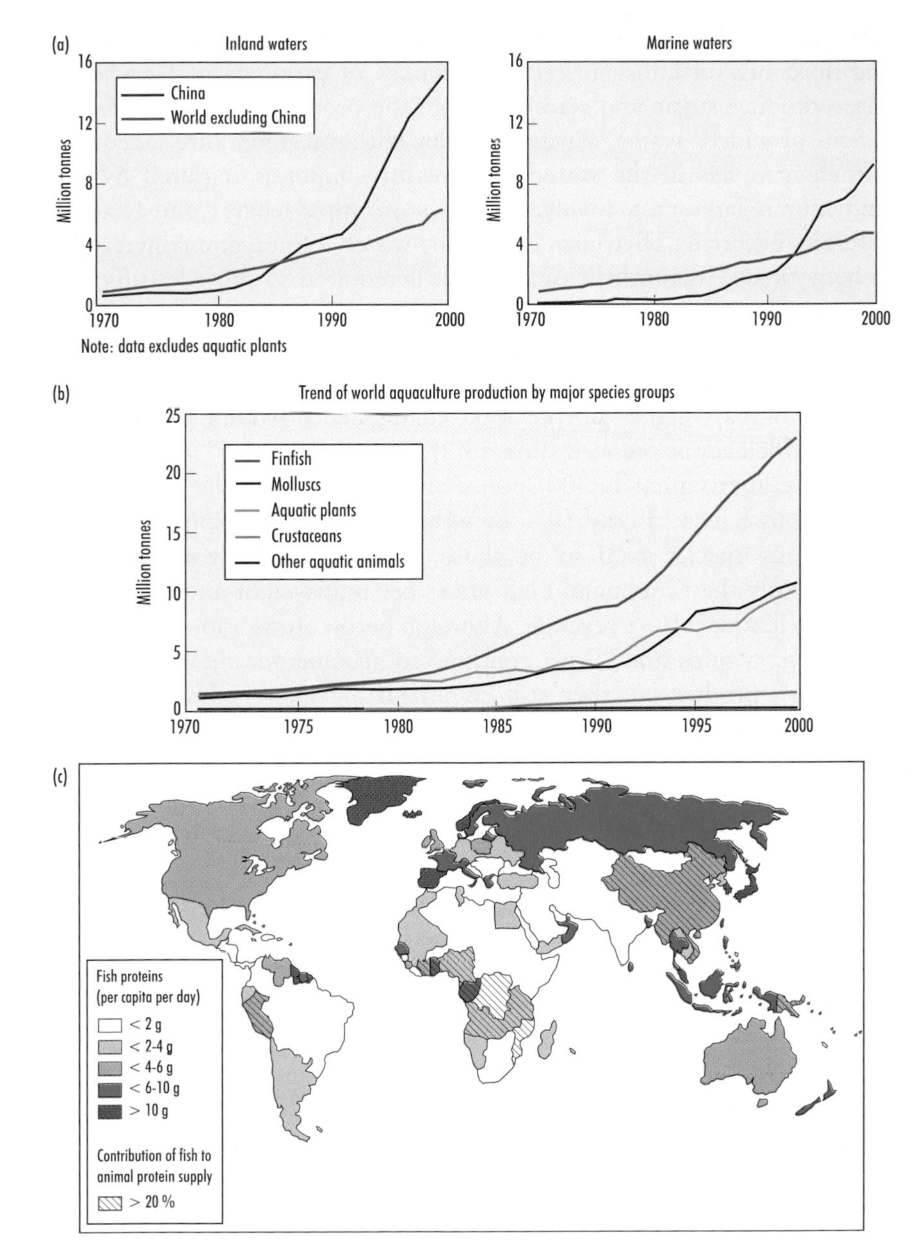

Fig. 13.1 (a) Aquaculture production in marine and inland waters (millions tonnes) showing the contribution of China and the rest of the world. (b) Trends in aquaculture production since 1970 by major species groups. (c) The contribution of fish to animal protein supply in terms of g per capita per day for different areas of the world (Source: FAO 2002).

crustacean, and latterly echinoderm species that have become the subject of widespread cultivation. The diversity of farmed marine species makes their cultivation an exciting though onerous area of research as the individual requirements of such a wide range of taxa require specific research effort.

● A much greater range of animal species is cultivated in marine systems compared with land-based systems.

Algae and bivalve molluscs are among the most attractive candidates for cultivation as these are the most efficient in converting energy into biomass that is available for consumption, and they are ranked second and third behind finfish in terms of tonnes of production (Fig. 13.1). Algae produce sugar and starch through the process of photosynthesis, hence, provided water clarity is sufficient and there are adequate nutrient supplies in the water column, little input is required by the cultivator (Chapter 2). Bivalve molluscs are filter feeders and feed on naturally occurring phytoplankton in the water column and convert this to body tissue. Again, little or no input is required to provide sufficient food for the cultivated species although care is needed to ensure that bivalves are not overstocked to the extent that the carrying capacity of the local environment is exceeded. In contrast, most fish and crustaceans feed on animals higher up the food chain; hence overall energy conversion efficiency is reduced (Box 13.2).

● Carrying capacity is the amount of biomass that can be supported in terms of oxygen and food requirements by a particular volume of water.

To date approximately 300 species are cultivated. Of these, 20% are carnivorous but yield only 10% by weight of all production. However, carnivorous species tend to be those most prized by gourmets and consequently they command high prices per unit weight and contribute 40% of all aquaculture revenue. Although herbivorous and omnivorous fishes (e.g. tilapias and carps) continue to account for the majority of global fish production, they still command a relatively low price per unit weight. Nevertheless, they are the best subjects to fulfil basic world food requirements as they do not require sophisticated cultivation systems and are inexpensive to rear. They also tend to be species that are highly tolerant of high stocking densities and low dissolved oxygen levels.

● Herbivorous and omnivorous freshwater fish continue to dominate world fish aquaculture production.

To this day, fish form the largest component of world aquaculture production both in terms of tonnage and financial value (Fig. 13.1). For fish production, it is apparent that freshwater and diadromous (e.g. salmonids) fish products remain the dominant sector of the market, while in the marine and brackish water sectors, molluscs, algae, and crustaceans dominate (Fig. 13.2). However, freshwater fish products are in general relatively low value per unit of production compared with marine fish products that are worth at least 3 times as much as

Box 13.2 **Conversion efficiency**

Conversion efficiency refers to the passage of energy through food-chains and webs. As one organism consumes another, energy is used to create body tissue, for metabolism and reproduction, which results in net energy loss. As more and more links are added to a food-chain, so the process of the conversion of energy (kJ) into body mass becomes less and less efficient. Hence the conversion of light into the tissue of plants is the most efficient process, while the food chain that links micro-algae to large carnivorous fishes such as sharks is one of the most inefficient.

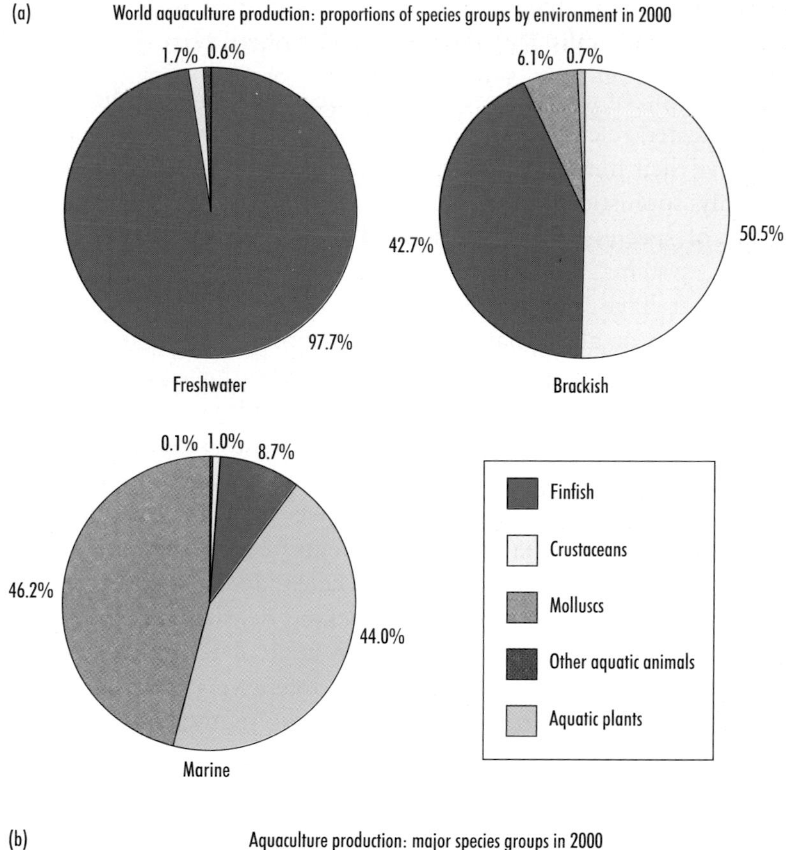

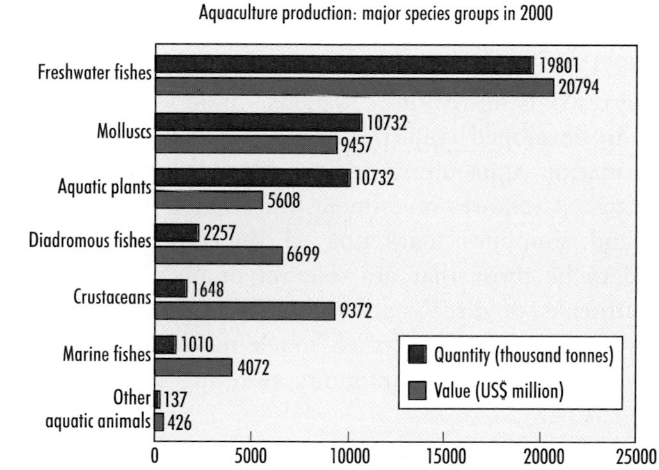

Fig. 13.2 (a) Global aquaculture production in terms of proportion of species groups subdivided among different aquatic environments for major species groups. (b) Aquaculture production in the year 2000 broken down by major species groups in terms of tonnage (1000s tonnes) and value (US$ millions) (FAO 2002).

freshwater equivalents (Fig. 13.2), although, marine products are more risk prone and hence costly to produce, as the marine sector of aquaculture remains a new and emergent industry (FAO 2002).

13.3 How Do We Produce Food from the Sea?

● Extensive cultivation systems rely on primary production to fuel the growth of cultivated species.

The production of marine species spans the spectrum from the relatively unsophisticated release of species into naturally enclosed systems where they derive their nutrition from natural sources (extensive cultivation) to the highly sophisticated and controlled cultivation and high density stocking of species in totally enclosed intensive recirculation systems. Extensive systems are typically based upon brackish water coastal lagoons that have high levels of primary production in the form of phytoplankton or algae that stimulate the production of high numbers of food organisms required by fish or prawns that are released into these systems. Another example of extensive cultivation would be the deposition (known as relaying) of the seed of bivalve molluscs harvested from the wild in areas with enhanced productivity and easier access for harvesting. The productivity of extensive systems can be further enhanced by the additional input of nutrients or fertilizers (e.g. the input of animal faeces). Production may also be enhanced with some human manipulation of feeding regimes or stocking density and species composition. As soon as human intervention is involved in the cultivation process it is deemed to be semi-intensive. Intensive systems are typified by high stocking densities and usually require the use of artificial feeds that are dosed with vitamins and antibiotics to ameliorate the effects of disease and stress that are symptomatic of high stocking density. These systems are also subject to close environmental control (often using computerized systems) and require constant monitoring (Allsopp 1997).

● Intensive systems require sophisticated monitoring and environmental control.

As in terrestrial agriculture systems, monoculture is typically undertaken in developed countries and generates the greatest income within the marine aquaculture sector. **Monoculture** is attractive to farmers because it requires investment in a limited range of equipment and food and simplifies marketing of the product. Monocultured species tend to be those that are tolerant of high stocking densities, however outbreaks of disease are the greatest threat to monoculture systems as the organisms are often much more vulnerable to infection due to their suppressed immunity resulting from elevated stress levels.

● Monoculture yields the highest financial income but is prone to outbreaks of disease.

Polyculture tends to be practised in countries with less well developed economies and is often associated with subsistence farming or fishing. Some of the best examples of polyculture come from freshwater systems in which carp, tilapia, or crustaceans are cultivated. In this case the aquatic organisms are cultivated in combination with plant or animal husbandry. For example, tilapias are farmed from rice-paddies, or

● Polyculture is usually practised with species that feed lower down the food-chain.

domestic animals housed in enclosures suspended over cultivated ponds enrich the water with their faeces. Those species best suited to this form of culture tend to be either herbivorous or consume prey low down the food chain (i.e. zooplankton or arthropod grazers).

13.4 What is Cultivated and Where?

Historically, the majority of aquaculture has developed in freshwater environments, which probably reflects the limited nature of the natural resource and also the perception that the cultivated species are unable to escape an enclosed body of water (Figs. 13.1 and 13.2). This contrasts with the reluctance of mariculturists to explore the possibility of ranching due to the inability to contain released individuals. Freshwater aquaculture continues to dominate world production of aquatic animals with 18.1 million tonnes produced in 1998 (FAO 2002). In addition to this a further 1.9 and 10.8 million tonnes were produced in brackish and marine waters respectively. In addition to animal production, a further 8.6 million tonnes of aquatic plants were produced in 1998 of which 99.8% were cultivated in marine waters (FAO 2002).

● Go to **http://www.fao.org** for the latest facts and figures for world aquaculture production.

Production is dominated by countries in Asia. China is by far the world's leader in terms of overall production with 20.8 million tonnes produced in 1998 with a value of $US21.7 billion. The relatively low value per unit production is due to the dominance of low-value carps in the Chinese aquaculture sector. This contrasts sharply with Japan's production of 0.77 million tonnes, which was valued at $US3.06 billion, comprising of high-value species such as scallops, oysters, and amberjacks. Within the last decade the low-income food deficit countries have maintained an upward trend in production due to their active promotion of aquaculture as a means of subsistence. However, in contrast to areas such as Asia and South America, Africa has been slow to develop its potential for aquaculture despite considerable investment. Production in Africa has risen from 37 000 tonnes in 1984 to 189 000 tonnes in 1998, however this is comprised mainly of low-value carps and tilapias.

● While Asia and Central America have expanded marine aquaculture rapidly, Africa is yet to realize its full potential.

Although nearly half of all aquaculture production is composed of finfish, more than 90% of this fish production is derived from freshwater systems. The remainder of the production is divided mainly between plants and molluscs. Crustaceans contribute approximately 5% to world production in terms of weight but their economic value is considerably more than this (Fig. 13.2). The cultivation of salmonids is so successful that production has overtaken the amount of fish generated by salmonid wild-capture fisheries (Fig. 13.3).

Fig. 13.3 Fish that make ideal subjects for aquaculture are tolerant of high density stocking levels, but high stocking densities inevitably leads to physiological and behavioural problems. These sole *Solea solea* are prone to caudal fin damage as they instinctively try to bury themselves into the base of the rearing tank. Technological advances in feed and rearing systems mean that many fish species are now much less expensive to produce than wild caught fish.

However, as we will see later, the proliferation in salmonid production has led to a plethora of environmental problems. Cultivated marine fish species are exclusively carnivorous species that command a high market value. Bass (*Dicentrachus labrax*), sea bream (Sparidae), and turbot (*Scophthalmus maximus*) are now commonly cultivated throughout southern Europe, where warm waters promote high growth rates. Humans go to real extremes to cultivate some of the most prized fish species (Box 13.3).

● Carnivorous species are favoured for marine aquaculture and they command higher prices than herbivorous species.

Brown and red algae are the main plants that are cultivated. Red algae are somewhat more valuable than brown algae as they are produced for food consumption whereas brown algae (mostly kelp species) supply the alginate industry. Production in 1997 reached 4.17 million tonnes and was worth $US2.70 billion. Nearly all cultivated mollusc species are bivalve molluscs of which mussels contribute most in terms of production, followed by oysters, clams, and scallops (Table 13.1). Although in many cases bivalves command a relatively low price per unit product they require no investment in feeds or antibiotics. The cultivation of crustaceans is largely based on penaeid shrimp (89% by weight of freshwater and marine global production) which are currently valued at approximately $US5 billion per year at first sale. Although a high value species, shrimp are costly to produce and their cultivation has been associated with environmental degradation of coastal margins.

● Alginates are derivatives of marine algae that are used in everyday products ranging from shaving foam to food thickening agents

Box 13.3 Cultivation of southern bluefin tuna

Harvesting southern bluefin tuna from holding pens is a delicate operation to ensure that the fish are maintained in pristine condition. Any damage to the fish will ultimately lower their value at market. As soon as the fish are harvested they are returned for immediate processing and shipping to markets in Japan. Photograph reproduced with the permission of the Tuna Boat Owners of South Australia/Australian Fishing Management Authority.

Undertaking research on methods to improve the growth of southern bluefin tuna is an expensive undertaking when each fish is valued in US$1000s. Here, scientists measure the dimensions of a fish. (Photograph: Brett Glencross.)

● Southern bluefin tuna are captured hundreds of kilometres out to sea and then transferred to holding pens close to the coast where they are conditioned prior to processing.

13.5 Food Requirements and Constraints

In marine plants, good growth will be achieved with sufficient light, oxygen and carbon dioxide, and nutrients. Over-stocking marine plants will lead to competition for all of these resources and will yield lower production per kilogram of plant mass. In terrestrial systems, domesticated herbivores convert vegetable matter into highly desirable protein products. This contrasts sharply with aquatic systems in which only bivalves and gastropods convert phytoplankton, particulate and particular organic matter and algae into high quality protein. In contrast,

● Individual bivalve species dominate the production of marine products by aquaculture.

Table 13.1 World aquaculture production in 1999 of the six most important marine species in terms of production (tonnes). The value (US$ 000s) of each of these species is also shown. The relative value per unit product is shown as the ratio of the value to total production. Statistics reproduced from the FAO (2002).

Species	Common name	Production	Value	Ratio
Crassostrea gigas	Pacific oyster	3 600 459	3 312 713	0.92
Ruditapes philippinarum	Manila clam	1 820 413	2 194 522	1.21
Patinopecten yessoensis	Scallop	928 724	1 252 448	1.35
Penaeus monodon	Prawn	575 842	3 651 783	6.34
Mytilus edulis	Mussel	498 461	272 419	1.83
Chanos chanos	Milkfish	381 930	599 957	0.63

herbivorous fish are low-value (but high production) species. Cultivation of high-value carnivorous fish, crustacean, and mollusc species depends upon a dependable and consistent high-quality protein diet. Artificial feeds are the staple diet of most cultivated carnivorous species, but these still depend on fish-meal derived from low-value by-catch species for their protein component. Half of the production costs for salmonids and Asian shrimp production is spent on artificial feed (New et al. 1993). Approximately 65% of the cost of artificial feed in the salmonid industry is attributed to fish-sourced products. However, the idea of catching fish to feed to fish (or terrestrial animals) is increasingly controversial (Naylor et al. 2000) as well as energetically inefficient (Pitcher & Hart 1982; Box 13.2). As a result there is great interest in the development of artificial feeds based on plant-derived protein.

● The cultivation of carnivorous species depends upon a consistent supply of high-quality protein diets.

Artificial diets based on the protein derived from soybean, lupin, and canola (a type of rapeseed) appear to yield acceptable growth rates and are half as costly to produce. Plant-based feeds offer benefits in addition to cost savings. Plant crops are more reliable than the wildly fluctuating stocks of small pelagic fishes used for fish-meal, the nutrient composition is more easily controlled and the shelf life of plant-based feeds is superior. With the current trend for retailers to offer sustainable products with ingredients that can be traced to source, plant-based feeds offer a much simpler solution in terms of product tracking. Another consideration is the potential to introduce pathogens through imports of feed. Frozen pilchards imported from South America and then fed to farmed southern bluefin tuna kept in suspended fish cages (Box 13.3) introduced a virus which then infected South Australian stocks of pilchard and caused high mortalities (Thorpe et al. 1997).

● Plant-based feeds are seen as the future for more sustainable farming of marine carnivores.

13.6 The Role of Biotechnology

The production per unit volume of water of marine aquaculture systems has increased with biotechnological advances. Often the species that have the highest unit value are the most demanding in terms of husbandry or dietary requirements. For example, halibut (*Hippoglossus* spp.) require very cold water at elevated pressure for their eggs to hatch, after which the newly emerged fish larvae are particularly problematic to wean onto artificial feeds. In another example, the formulation of an appropriate pellet feed is currently a major problem for the cultivation of southern bluefin tuna in South Australian fish farms, partly due to the expense of having a laboratory subject with a price tag of thousands of US dollars per individual (Box 13.3).

- Technology has enabled us to overcome the particular developmental needs of species such as halibut.

Biotechnological advances have enabled the **cryopreservation** of viable gametes, which has eliminated the breeding constraints associated with species that have seasonal reproductive cycles and ensures a reliable supply of larvae year round or when conditions (biological or economic) are most favourable. This is particularly important for markets in developed countries where retailers expect a predictable source of products throughout the year. Controlled sex differentiation has enabled the production of sexually sterile fish. This can be achieved by a number of techniques, including the induction of **triploidy** in females, that yield **monosex** stocks of fish. Sexually sterile fish put more energy into tissue growth and yield a greater proportion of consumable tissues at market size. Genetic selection has been used to reduce head and fin size. This helps to further reduce waste in the final product. Artificial diets now include growth stimulants that improve feed conversion rations. They are often dosed with antibiotics to reduce the risk of infection in intensive finfish and crustacean aquaculture systems. Finally, in common with terrestrial farmed systems, marine organisms are now the subject of trials using recombinant DNA methodologies to produce fish with altered body shape and size characteristics.

- Cryopreservation enables gametes to be stored until they are required.

- For more information on genetically modified fish go to **http://www.ems.org/salmon/genetically_engineered.html.**

13.7 Negative Effects of Biotechnology

Biotechnology has certainly helped to improve our ability to cultivate marine organisms more efficiently and to provide better quality products. However, there is considerable concern regarding the deleterious effects of escapees of normally farmed fish and latterly **genetically modified organisms**. The genetic diversity of wild stocks is threatened by escapes and accidental releases of hatchery-reared individuals that can competitively exclude wild fish. A reduction in genetic diversity has the potential to lower resistance of wild stocks to environmental change and pathogens. This issue has been particularly prominent in the salmon industry of North America. Large-scale hatchery release programmes have been undertaken to counter the effects of habitat loss and over-fishing. However, hatchery-reared fish compete with the existing wild fish for habitat and food, prompting Hilborn (1992) to argue that large-scale hatchery programmes for salmonids in the Pacific Northwest posed the greatest single threat to the long-term maintenance of salmonid stocks (but see Hilborn 1999 for a recently revised opinion).

● Large-scale releases of hatchery-reared fish into the wild may pose one of the greatest threats to the long-term maintenance of salmonid stocks.

Artificial feeds are a convenient means for administering medication to captive organisms. Although the use of antibiotics reduces infection rates in intensive cultivation systems, their excessive use can lead to a buildup of drug resistance in the pathogen. As uneaten feed accumulates on the seabed beneath fish cages, the encapsulated antibiotics can affect microbial communities in the immediate vicinity, leading to a reduction in their diversity. Additional concerns associated with the use of antimicrobials in fish feeds include: the spread of drug resistant plasmids to human pathogens; transfer of resistant pathogens from fish farming to humans; and presence of antimicrobials in wild fish.

● Encapsulated feeds can lead to the release of large amounts of antibiotics into the marine environment.

13.8 Cultivation Systems

Terrestrial farmers need to contain and protect their livestock and crops using enclosures and barriers. Equally, mariculturists need to ensure that their livestock is secure. Mobile biota (e.g. fishes, snails, shrimp) have different containment requirements to those that are sessile (e.g. bivalves, algae). Semi-enclosed coastal water bodies such as lagoons and saline ponds provide naturally enclosed or semi-enclosed culture systems. However, these coastal resources are limited, and often they are associated with a unique fauna that may be of conservation interest such that they clash with the needs of the aquaculturist. Shrimp cultivation in Central America and Asia has led to the creation of artificial coastal ponds in areas where there formerly existed mangrove forests. These ponds need to be in close proximity to seawater so that the pond water can be exchanged on a frequent basis. However, the coastal margin is a finite resource and the demand for space has resulted in such a high density of shrimp ponds that the effluents from one pond have contaminated the intake water of the adjacent pond. Not surprisingly, this has resulted in a severe reduction in coastal water quality and serious

● Cultivated species need to be contained within some form of enclosure either artificial or of natural origin.

● The coastal margin is a finite resource that is easily over-exploited.

outbreaks of disease that have caused entire shrimp harvest failures. We will return to this subject later in the chapter.

The use of natural ponds and lagoons is usually associated with extensive culture or semi-intensive culture. Often, wild juveniles may be captured and released into the ponds to on-grow by feeding on the natural food within the pond. Alternatively hatchery-reared juveniles are released into the ponds in conjunction with supplemental feeding or enhancement of pond productivity using natural or artificial fertilizers. The stocking density of natural and semi-natural pond systems has to be monitored carefully to avoid exceeding the carrying capacity and oxygen supply within the system. In some cases, additional aeration of the pond water is undertaken using paddle aerators or by pumping air directly through air-blocks located in the base of the pond.

● The excessive development of coastal aquaculture has led to a reduction in coastal water quality in some regions.

13.9 Cultivation of Fish in Cages

Ponds and lagoons do not permit farmers to have close control on their stock; they are less easily maintained and disease problems are not so easy to treat. Suspended cages provide an excellent means of retaining the cultivated species in open well-aerated seawater while liberating farmers from the constraint of limited coastal pond resources. Suspended fish cages are used in all aquatic systems. Essentially, they consist of a large net bag with buoys at the top, which is anchored to the seabed using anchors and chains (Fig. 13.4). Nevertheless, suspended cages are themselves limited to coastal areas that are sheltered from adverse weather conditions and are within a short distance of the shore for easy of maintenance. For this reason, the greatest concentration of caged fish farming tends to occur in coastal fjordic systems (sea lochs, rias, inlets) that provide natural shelter from wind and wave action. Such areas are only found in a limited number of locations in a particular country, and many countries lack such an environment. The most prolific producers of suspended-cage farmed fish include Norway, Scotland, Canada, New Zealand, Chile, and Greece. All of these countries have complex coastlines that are either riddled with inlets or fjords or have a high number of offshore islands that provide natural shelter. Once again the limited nature of the coastal resource has led (in the past) to over-expansion of this sector of the industry, with high densities of fish farms developing within these semi-enclosed water-bodies.

● Suspended-cage culture is usually confined to coastal areas that are sheltered from adverse weather conditions.

This has led to problems with carrying capacity in systems in which the exchange of water is reduced in comparison with open coastal systems (Gowen & Bradbury 1987, Black & Truscott 1994). Single, isolated fish farms will increase the demand on oxygen supplies in the surrounding water; they will reduce current flow beneath the fish cages

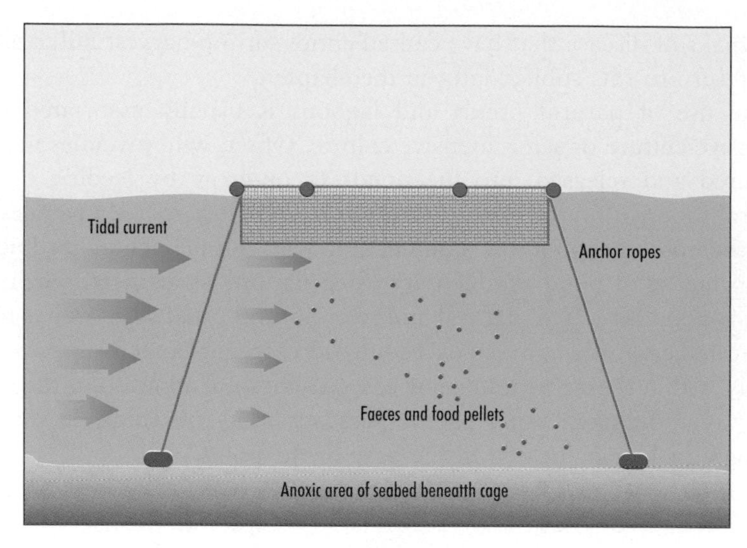

Fig. 13.4 A typical suspended fish cage showing the buoys on the surface of the sea, the anchor ropes and the zone beneath the cage that will be affected by falling uneaten food pellets and fish faeces. A build up of organic matter beneath the cage can lead to anoxia and an impoverished benthic fauna. In extreme cases of organic pollution only bacterial mats persist beneath fish cages. The presence of the cage and the ropes that anchor it to the seabed will increase friction and hence the turbulence of the flow of water around the cage. This will tend to cause a zone of lower velocity water beneath the cage that will further concentrate faeces and uneaten pellets as they fall through the water column. A proportion of the solid particles will be carried down stream by the prevailing tidal currents leading to zones of decreasing organic enrichment around the cage.

● In cases of severe organic enrichment, benthic communities that are found beneath fish cages can be replaced by mats of bacteria.

and increase the supply of organic matter to the seabed. Typically this increases the physiological stress on the benthic community beneath the cage and leads to a reduction in species diversity or in extreme cases the elimination of all macrobenthos, which are then replaced by bacterial mats of *Beggiatoa* spp. (Pearson & Rosenberg 1987; Chapter 14).

When farms occur in isolation it is relatively simple to alleviate the environmental problems by moving the location of the farm to allow the affected seabed to recover over a number of years (Fig. 13.5). However, some enclosed water bodies have been overpopulated with fish farms that have magnified the adverse effects of cage farming and eliminated the possibility for simple management solutions to the associated problems. Typically, these systems have been associated with greatly reduced dissolved oxygen concentrations in the water column, which has increased stress levels for the fish and increased their vulnerability to disease outbreaks. The environmental changes in the water column have been linked with outbreaks of toxic algal blooms and mass fish mortalities (Smayda 1990).

● When fish farms occur in isolation they have relatively little effect on the wider environment.

Fig. 13.5 Isolated fish farms are less likely to cause severe environmental effects as large concentrations of fish farms in enclosed water bodies. (Copyright: *US National Oceanic and Atmospheric Administration.*)

13.10 Cage Cultivation: a Lousy System?

Intensive cultivation of any species is likely to encourage the rapid spread of disease or parasites. It is not surprising then that sea lice infestations are common in most coastal areas in which marine salmonid aquaculture occurs. Sea lice (*Lepeophtheirus salmonis*) are isopod parasites that feed on the blood of marine fishes. In salmonids, a parasite load of only 10 sea lice can lead to mortality of the fish as they migrate to sea. As cage-farmed fish are held in high density, the density of sea lice within the vicinity of the cages is also high. Recently, concern has been raised that wild migrating salmonids (salmon and sea trout) are more susceptible to infestation as they pass close by salmon farms that are found on their migration route. This potential for elevated infestation rates has been linked by some to the decline in wild salmonid populations, however the evidence to support this is currently lacking (Northcote & Walker 1996).

● Intensive cultivation practices encourage the rapid spread of disease and parasites.

The losses of fish stock to sea lice infestation led to the development of chemical treatments that would kill the sea lice. However, these were limited to two main treatments, dichlorvos (an organophosphorus derivative now banned for use in the fish aquaculture industry) and hydrogen peroxide. The frequent and widespread use of these compounds led to reduced efficacy caused by resistance that developed in the sea lice. In more recent times, there has been a tendency to look to biological control methods. Intensive research efforts have been

● Fish such as cleaner-wrasse could be a natural method of controlling sea lice, but they are not presently the most effective solution.

● See the website **http://www.ecoserve.ie/projects/sealice/** for more detailed information about research on sea lice.

made to examine the utility of so-called 'cleaner-wrasse' the goldsinny (*Ctenolabrus rupestris*). However, it is not practical to contain the small-bodied wrasse within fish cages. Under normal circumstances, sea lice remain attached to a fish until it dies, hence any pathogen-infected blood taken up by the sea louse would remain in close proximity to the infected fish. However, consumption of the sea lice by wrasse means that they could act as a vector for any fish pathogens ingested by the sea lice. As a result there is concern among commercial farmers that the wrasse could spread disease and parasites from one farm to another (Costello et al. 1994, Gibson & Sommerville 1996).

13.11 Breaking Away from the Coastal Margin

With the ever-increasing demands on the coastal margin the prospects for the expansion of suspended-cage cultivation is finite. This has led to research into alternative technologies for intensive marine fish cultivation. Two developments are most noteworthy; the development of offshore automated fish cages and the development of recirculation systems. Automated fish cage systems are a recent development in the last decade (Fig. 13.6). The essential features of these systems are that the fish are fed automatically, they can be lowered to the seabed such that they are beyond the effects of wave action and cause less of a navigation hazard to shipping, and they can be resurfaced for repair, maintenance, and harvesting (Dahle 1995) (Fig. 13.6).

● To find out more about offshore fish farm cultivation visit **http://www.masgc.org/oac/**.

Recirculation systems present the most sophisticated level of current aquaculture rearing techniques. Enclosed tank and raceway systems have been used to cultivate marine fish for a number of decades, but in the past they have been restricted to coastal locations by the necessity to

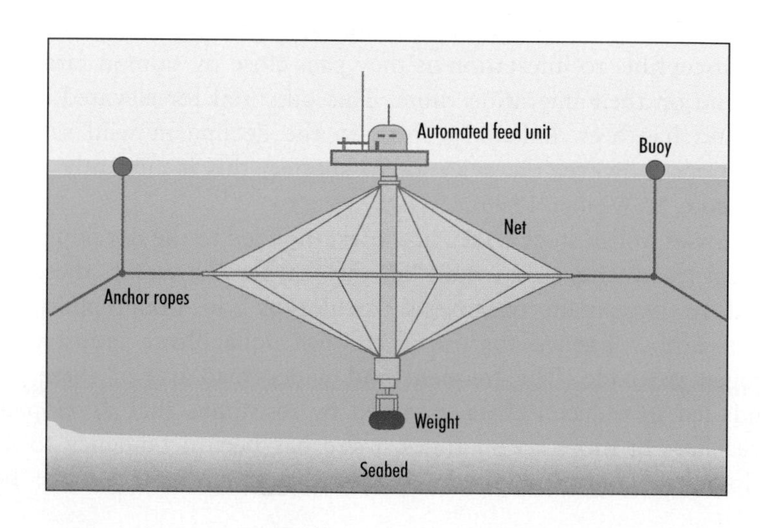

Fig. 13.6 A schematic of an offshore fish cage as currently in use in the North Atlantic and the Gulf of Mexico. These 15 m deep nets are entirely automated and sunk to the seabed to avoid the effects of extreme weather. Apart from occasional servicing and reloading of the feed hopper, they are relatively low in terms of maintenance needs.

pump a sufficient through-put of seawater. This incurred costs both from the pumping of water and in some cases the need to heat it when warm-water species were cultivated in temperate areas. With the development of recirculation systems, the cultivation of marine species is no longer tied to locations adjacent to the coast (Fig. 13.7). In addition, non-indigenous species can be cultivated as there is no risk of accidental introductions into the local environment. The highly controlled environment of recirculation systems increases productivity per unit area by reducing stress induced by disease and water contamination. Recirculation systems are divorced from the environmental problems associated with cage and pond cultivation as the waste products can be collected and disposed of in an appropriate manner and even used as fertilizer. There is even the potential for diversification into areas such as hydroponics in which high-value plant crops could be grown in conjunction with semi-recirculation systems such that they remove excess nitrogen and phosphorus products and yield a vegetable crop as a bonus (Fig. 13.8).

● Recirculation systems mean that marine species can be cultivated almost anywhere.

● Biological filtration systems can also incorporate hydroponic cultivation.

(a)

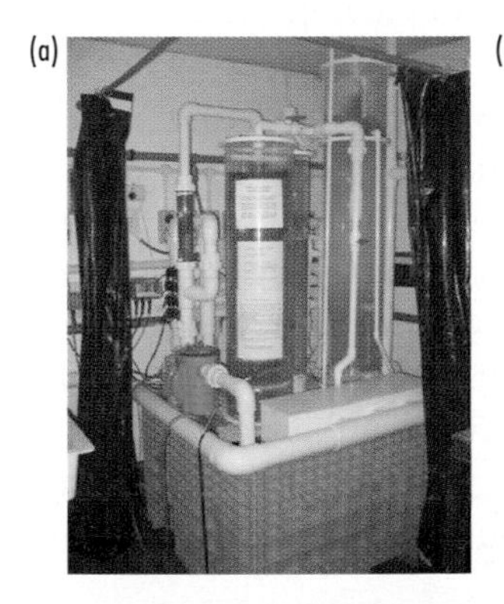

(b)

Fig. 13.7 Recirculation systems (a) can be constructed on land far from the sea as they require minimal inputs of fresh seawater and can be maintained using artificially made seawater if necessary. Land based systems have the advantage that they can be enclosed which gives greater potential for environmental control (b). (Photographs: Michel Kaiser.)

Fig. 13.8 A fine crop of salad greens growing on the waste products of aquaculture. Hydroponics has the advantage of using natural plant growth to remove excess nitrogen and phosphorous from cultivation systems. (Photograph: Kevin Fitzsimmons, University of Arizona).

13.12 Shrimp Cultivation: the Gold Rush

Shrimp are a high value crop that is prized throughout the world. The high by-catches (16 kg of by-catch for every 1 kg of landed shrimp) associated with shrimp fisheries (Chapter 12) mean that their cultivation is an attractive way of overcoming this adverse side-effect of the wild capture fishery. Once the technological difficulties in rearing shrimp had been overcome, shrimp culture increased rapidly in Asia and Central America. Consequently, shrimp farming has become a multi-billion US$ business worldwide. However, uncontrolled and unplanned expansion in the early days of the industry has resulted in widespread crop failures due to disease and poor water quality (Fig. 13.9). These chronic problems have arrested the further expansion of shrimp aquaculture. Disease is mostly responsible for recent production declines. Nowadays, shrimp production is seriously affected by rapidly spreading viral epidemics across the Asian region. Of the 20 known types of virus that affect shrimp, the **white-spot virus** in Asia and **Taura Syndrome virus** in the Americas are the most threatening.

● Shrimp fisheries yield high value catches but generate large by-catches.

In 1987, Taiwan was among the world's leading shrimp producers with 115 000 tonnes of *Penaeus monodon* produced by small family-run intensive farms. However, a combination of environmental and disease factors crippled the industry with a two-thirds decrease in production by 1988. Despite attempts to revive the industry by cultivating alternative *Penaeus* species, production was down to 25 000 tonnes by 1994 and many farmers turned their attention to marine fish (Rosenberry 1994). The main culprit was the excessive number of shrimp farms that developed in the coastal margin. Farms were packed together so tightly that the effluent from one farm polluted the intake water of the adjacent

● The rapid and uncontrolled expansion of shrimp farming has resulted in severe disease problems.

Fig. 13.9 Deforestation along the coastline of Java has occurred to give the land over to the development of shrimp cultivation ponds. This story is repeated around the world and has often led to severe environmental, water quality, and disease problems. (Photograph: L. LeVay.)

farm. As this process was repeated many times along the coastline, so the quality of the coastal water deteriorated and elevated the physiological stress for the cultivated shrimp. This in turn probably made the shrimp more vulnerable to disease, which then spread uncontrolled from one farm to the next. This collapse in the Taiwanese shrimp industry left a gap in the world market that was quickly filled by production from China, which rose at a rate of 80% per year and reached 199 000 tonnes in 1988. Again, a similar story unfolded and resulted in the spread of white-spot virus from China to other Asian countries causing mass mortality in a wide range of shrimp species (Nakano et al. 1994).

● White-spot virus spread from China to the other Asian countries causing mass mortalities of shrimp.

Similar problems have occurred in the shrimp cultivation industry in the western hemisphere. In Ecuador, extended drought elevated the salinity and nutrient levels in the Guayas River estuary, which supported a large number of shrimp farms. *Vibrio* spp. thrive in this type of environment and infected the shrimp held in the farms bordering the estuary. This infection causes shrimp to swim at the surface of the ponds where they fall prey to seagulls. The name of this syndrome is *Gaviota*, which is the Spanish word for seagull. These outbreaks of disease were treated with antibiotics or by encouraging the growth of harmless bacteria that out-competed the harmful *Vibrio* spp. but had mixed results. Fortunately, El Niño rains lowered the salinity in the affected estuaries and diluted nutrient concentrations and eliminated the disease in 1993 (Chamberlain 1997).

● Climatic conditions can change such that they favour the outbreak of certain diseases.

13.13 Shrimp Farming and Mangroves

As we have seen in Chapter 9, mangroves act as nursery areas for estuarine fishes and invertebrates, and provide a feeding and breeding ground for birds and mammals. They also prevent coastal erosion by storm and wave action and prevent soil erosion. In the early years of the shrimp farming industry, areas occupied by mangroves were considered ideal for development for shrimp farming as they were considered to have little economic value. As they were the natural habitat for many of the cultivated shrimp species they seemed to be the best location for shrimp farms (Fegan 1996). However, it turns out that mangroves have acid sulphate soils with an acidity of pH 3 to 4 when dried out, which is very poor for shrimp cultivation. In a global context, the development of shrimp farms has accounted for only 5% of mangrove destruction to date, but on a localized scale the impact of shrimp farming may be far more severe. For example, the construction of ponds used for aquaculture has destroyed 20% of the mangrove forests in areas of Ecuador (Phillips et al. 1993).

● Although mangrove areas are natural habitats for many shrimp species, areas cleared of mangrove forest are far from ideal for their cultivation.

Fig. 13.10 (a) Shrimp pond constructed in a deforested mangrove swamp. (b) The new approach to shrimp cultivation that incorporates the development of mangrove forest in conjunction with shrimp ponds. (Photograph: L. LeVay.)

While most shrimp-producing countries now recognize the ecological importance of mangroves, the high revenues earned from shrimp cultivation are likely to maintain the pressure for development of some areas of mangroves. Recent attempts to model the social and biological benefits of mangrove conservation indicated that only 12% of the mangrove outside Thailand's mangrove conservation areas should be given over to shrimp farming and that 61% of these mangroves should be maintained in their natural state (Pongthanapanich 1996). Nowadays, the importance of mangrove forest for the sustainable use of coastal areas is widely acknowledged. Hence new shrimp farm developments incorporate mangrove plantations that naturally remove some of the excess nutrients and organic matter produced by the cultivation process (Fig. 13.10).

● Socio-economic and biological modelling indicates that only a small proportion of mangrove forests should be given over to shrimp cultivation.

13.14 Cultivation of Molluscs

Bivalve aquaculture is confined to intertidal and shallow subtidal areas. One of the biggest attractions of bivalve species is that they require no inputs of food by the farmer, as they feed on phytoplankton. In addition, bivalves generally require relatively little husbandry and hence are suitable for cultivation in areas of the world where the use of high technology is unfeasible at present. Despite the obvious attractions of bivalves as a potential crop, as in other sectors of the aquaculture industry, environmental problems have occurred due to over-exploitation of limited coastal resources (e.g. space). Various environmental effects may result from different stages of the bivalve cultivation process: seed collection, seed nursery and on-growing, and harvesting (Table 13.2).

● Bivalves require relatively little husbandry and convert primary production (phytoplankton) into protein.

The cultivation of bivalves can help to redress the adverse effects of other aquaculture activities. For example, the effluents from fish farms

Table 13.2 A summary of the environmental effects associated with different stages of bivalve cultivation, adapted from Kaiser et al. (1998).

Seed collection
Ecological effects of harvesting wild seed (e.g. reduction of food for birds, physical impacts on the seabed by fishing gear, trampling across shores).
Alteration of habitats by use of spat settlement materials such as crushed shells, gravel or man-made products (known as cultch).
Effluents from seed hatcheries may contain antibiotics and chemical pollutants.
Introductions of alien species with imported brood-stocks.
The on-growing phase
Over-stocking leading to problems with carrying capacity of enclosed systems such as bays.
High density of intertidal structures leading to poor water circulation and build up of organic matter and reduced species diversity in local benthic community.
Spraying of chemicals on intertidal areas to remove 'pest species' such as burrowing shrimps.
Alteration of habitat by addition of gravel and other material to provide firm substratum on which to grow bivalves.
Exclusion of birds, fish and crustaceans from feeding areas due to use of protective netting.
Harvesting
Disturbance to wading birds.
Physical disturbance of substratum by mechanical harvesting devices.
By-catch mortality of invertebrate species associated with bivalve culture sites.

are nutrient enriched, and can lead to the development of toxic algal blooms. The filter-feeding activities of bivalves remove algae and particulate matter from the water column and hence reduce the likelihood of bloom formation (Shpigel et al. 1993). Mussels are particularly effective organisms that remove excess particulate-bound nutrients from eutrophic systems (Haamer 1996), and have been used for the restoration of enclosed water masses that are heavily polluted (Russell et al. 1983). While these mussels are unfit for human consumption they represent a cost-effective and self-perpetuating means of maintaining water quality.

● Bivalves can improve water quality by removing toxic algae and pollutants in the water column.

Suspended rope cultivation of bivalves generates some similar environmental problems to suspended fish cage cultivation. The location of mussel farms is critical if these environmental effects are to be minimized. Mussels require a good exchange of water to replenish phytoplankton biomass to achieve adequate growth rates for commercial purposes. Flow is severely reduced toward the centre of a set of mussel ropes. The reduced flow enhances sedimentation rates, hence faeces and pseudofaeces tend to accumulate beneath the mussel ropes leading to organic enrichment of the sediment. The degree to which this occurs is strongly related to the flow regime at the site of cultivation. Hartstein and Rowden (2004) studied three contrasting sites of suspended mussel cultivation in New Zealand. They found that the site with the highest mean flow regime had the lowest percentage organic content and that organic content was always higher directly within the mussel farm area compared to the surrounding sediments (Box 13.4).

While there is now wide concern regarding the introduction of alien or exotic species into different geographical areas of the world through ships' ballast water, humans have been deliberately moving bivalve species around the world for many years. For example, Asian Manila clams (*Tapes philippinarum*) and Pacific oysters (*Crassostrea gigas*) have been grown in Europe for at least the last thirty years. The importation of adult and bivalve seed has been responsible for the introduction of a wide range of marine organisms that are alien to countries where formally they did not occur (Reise et al. 1999). These organisms can have severe ecological consequences and financial implications for the aquaculture industry when they include competitors of bivalves, such as the slipper limpet (*Crepidula fornicata*), bivalve predators, such as the American whelk tingle (*Urosalpinx cinerea*) and diseases, such as *Bonamia*, which infects the blood cells of flat oysters *Ostrea* spp. (see also Chapter 14).

● Bivalve cultivation has been linked to the introduction of a large number of alien species in certain countries.

13.15 Ranching at Sea

Technological advances in rearing techniques have considerably increased the range, quality, and quantity of fish that can be cultivated and has provided a new impetus for stock enhancement programmes worldwide. Reseeding and stock enhancement programmes are more likely to be financially viable if the species in question have a high financial or resource value. Examples include red sea bream (*Pagrus major*) and Japanese flounder (*Paralichthys olivaceus*) in Japan, and Atlantic cod in Norway (Watanabe & Nomura 1990, Svånstad & Kristiansen 1990). The rearing environment does not expose hatchery fish to the natural variability of temperature and food availability that they encounter in the wild. They are also protected from predators. As a result, hatchery-reared fish of an equivalent age to wild fish are inferior in terms of their behaviour and physiology. Recent research indicates that their chances of survival are greatly increased if they are conditioned to natural temperature regimes and food in advance of release. For example, the technique required to eat an artificial food pellet will differ greatly from a natural prey item that will exhibit cryptic behaviour and escape responses (Brown & Laland 2001). Hatchery-reared fish that have never experienced a predator threat can be trained to avoid predators more effectively. Brown and Laland (2001) describe approaches for using wild 'trainer' fish that demonstrate the appropriate behavioural responses to predator stimuli in view of hatchery individuals.

● Fish reared for mass release into the wild need to be 'taught' how to behave in response to predators and prey.

Box 13.4 Environmental problems associated with suspended rope cultivation of mussels

Suspended rope cultivation of mussels is practised extensively in New Zealand. As in other localities, the environmental effects of cultivation are closely linked to physical parameters. In low energy environments such as Catherine Cove, organic material accumulates beneath the cultivation site partly due to the reduction in water flow among the mussel ropes. Clear changes in the benthic community occur beneath the mussel ropes (inside) compared with surrounding areas (outside). At this site, the opportunistic dorvallid polychaete *Schistomeringoes loveni* is numerically dominant beneath the mussel ropes, compared to the species-rich surrounding areas, which are dominated by more bioturbating fauna such as *Amphiura* spp. At the higher energy Blowhole Point, the organic content of the sediment is far lower than at Catherine Cove, nevertheless significant subtle differences are still apparent in benthic community structure (adapted from data in Hartstein & Rowden 2004).

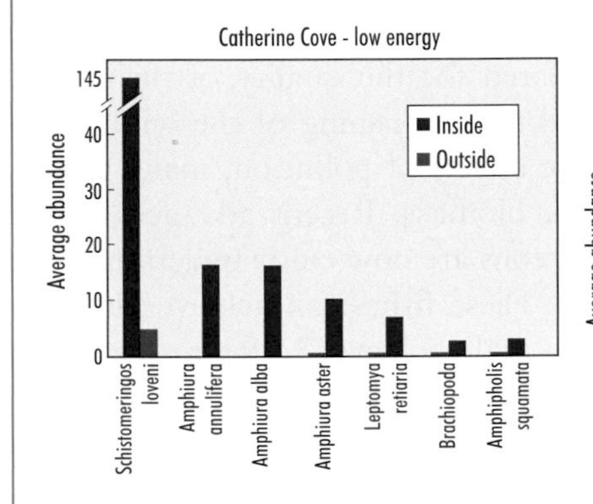

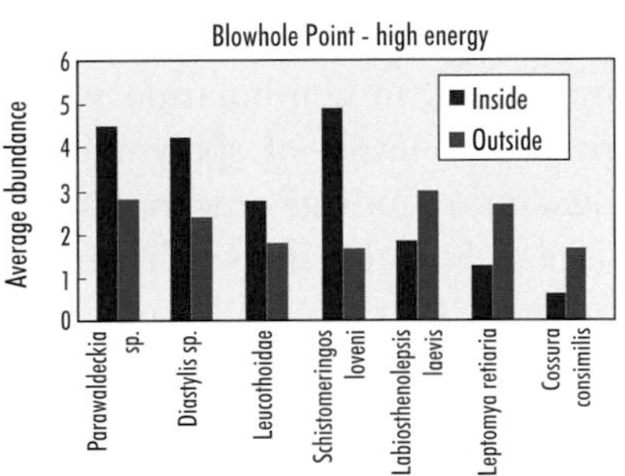

Table 13.4 The environmental conditions at Catherine Cove and Blowhole Point upstream and downstream of the cultivation sites.

	Current (cm s^{-1})		Percentage organic content	
Catherine Cove (low energy)	3.41	Upstream	10.6	Inside
	3.16	Downstream	5.4	Outside
Blowhole Point (high enrgy)	10.2	Upstream	3.2	Inside
	9.7	Downstream	2.9	Outside

Releasing many millions of small fish or shellfish into the sea is, to an extent, an act of faith. How do we know that the released individuals make any contribution to overall catches? For sedentary taxa such as bivalves this is not so much of a problem, but for mobile taxa such as fish it is necessary to tag them in some way so that future returns can be ascertained. This is usually achieved by chemically tagging the fish such that a chemical signature is incorporated in the bony body parts such as the otoliths. In Hawaii, hatchery-reared striped mullet (*Mugil cephalus*) eventually accounted for 75% of the sampled fish population in nursery areas. Thorough pilot studies that identified the optimum release-size of fish, the best sites and season for release, contributed to the success of this programme. Bivalves, such as scallops, need to be seeded at a density that is sufficient to permit effective fertilization at spawning. Not only must the density at which the animals are released be considered, but also local hydrography, which affects larval dispersal.

● Successful mass release programmes require detailed knowledge about what constitutes a favourable environment for a particular species.

13.16 A Conservation Role for Aquaculture?

Sturgeon (Acipenseridae) were originally harvested for their eggs or caviar. Today they are also appreciated for the quality of their flesh. However, the demand for caviar led to over-fishing of the wild populations and, in combination with the effects of pollution, many species are at low levels of spawning stock biomass. Recent advances in the aquaculture industry mean that sturgeons are now cultivated around the Caspian Sea and in North America. These fishes can achieve phenomenal growth rates and some species will achieve 3.8 kg in their first two years with a conversion ratio of 1.5 : 1 or better. As improvements in aquaculture techniques continue to develop there is great potential to alleviate pressures of over-fishing on the wild stocks of sturgeon, although these continue to be threatened by changes in climate and environmental quality (Elahimanesh 2001).

● Aquaculture could alleviate fishing pressure on wild stocks of sturgeons

Syngnathids (sea horses and pipefishes) are threatened by over-exploitation due to the demands of the Chinese medicine trade (Fig. 13.11). These fishes have unique life-history strategies that involve male incubation of the eggs in a brood pouch. They have a low reproductive output and in some species they pair-bond for life. Hence these fishes are highly vulnerable to sources of additional mortality. As a result, syngnathid aquaculture is seen as a potentially lucrative commercial venture, which may aid the conservation of wild populations. Cultured syngnathids could help to meet any future increases in global demand and possibly alleviate the effects of over-fishing. The extent to which syngnathid culturing reduces the exploitation of wild syngnathid populations will depend, in part, upon its effects on subsistence fishing communities that currently depend upon syngnathids for their livelihood. Syngnathid fishers are commonly so poor that they cannot stop catching syngnathids unless they earn money in other ways. One outcome of syngnathid culture in countries that do not traditionally exploit them might be reduced prices for syngnathids in source countries. This could force fishers to catch more syngnathids in order to meet their basic needs or deflect them from

(a)

(b)

Fig. 13.11 Sea horses are harvested and then sold dry for the traditional medicine trade in China (a). Conservation of seahorses requires a combination of education regarding their vulnerability to exploitation, and the use of aquaculture to replace a wild-capture industry (Photograph: Patrice Ceisel/Shedd Aquarium and Project Seahorse). Such approaches have been used successfully to supply the aquarium trade with reef fishes such as clownfish (b).

one diminished resource to another, creating new conservation problems. Aquaculture is likely to have the greatest conservation value when syngnathid fishers become syngnathid farmers, thereby directly reducing pressure on wild syngnathid populations. Once such a shift in perspective is achieved, the value of live syngnathids in the wild as a supply of brood-stock will exceed their value as a dead product.

● To find out more about the biology and cultivation of syngnathids go to **http://www.projectseahorse.org**.

● CHAPTER SUMMARY

- Global aquaculture activities continue to increase in terms of their overall contribution to fish-based sources of protein. Nevertheless, marine species are primarily carnivorous and are unlikely to achieve levels of production seen in species that feed at lower trophic levels (e.g. carps and tilapia).
- The cultivation of carnivorous species requires the manufacture of feeds from other sources of protein. While much effort has been centred on replacing fish-based protein with plant-based protein, the former remain the primary component of fish feeds and create their own environmental problems associated with industrial fisheries.
- The cultivation of marine algae and bivalve molluscs are among the most environmentally sustainable sectors of the cultivation industry as neither require inputs of feeds. However bivalve cultivation can exceed the carrying capacity of relatively enclosed marine systems (e.g. lagoons).

- Most cultivation sites are located in coastal areas where aquaculturists compete for space with other stakeholders. Constraints on carrying capacity and space have necessitated the development of automated offshore fish cages that remain on the seabed with minimal maintenance.
- Developments in modern recirculation systems mean that land-based cultivation of marine species may become economically viable on a large scale within the next decade.
- Aquaculture of genetically modified species has the potential to dilute the genetic integrity of wild populations when cultivated fish escape holding nets. Sea lice infestations associated with salmonid cultivation sites have been blamed for declines in wild fish that are infested as they pass cultivation sites.
- The movement of cultivated species around the globe has led to introductions of 'exotic' pest species, diseases and parasites into new environments.

FURTHER READING

For a comprehensive overview of the environmental effects of aquaculture see Black (2001). Naylor et al. (2000) highlight the additional impacts of harvesting fish to grow fish. Hilborn (1992, 1999) provides an insight into the debate surrounding the use of hatcheries to supplement declining stocks of salmonids.

- Black, K. D. 2001. *Environmental Impacts of Aquaculture*. Sheffield Academic Press, Sheffield.
- Hilborn, R. 1992. Hatcheries and the future of salmon in the Northwest. *Fisheries* 17: 5–8.
- Hilborn, R. 1999. Confession of a reformed hatchery basher. *Fisheries* 24: 30–31.
- Naylor, R. L., Goldburg, R., Primavera, J., Kautsky, N., Beveridge, M. C. M., Clay, J., Folke, C., Lubchenco, J., Mooney, H. A. and Troell, M. 2000. Effect of aquaculture on world fish supply. *Nature* 405: 1017–24.

Chapter 14

Disturbance, Pollution, and Climate Change

CHAPTER SUMMARY

Human activities around the globe have been linked to a diverse range of ecological changes in marine ecosystems. Many of these activities involve exploitation of biological and mineral resources, while other forms of interference are linked to industrial and agricultural discharges into river basins. Often, the changes in marine systems that result from these activities are catastrophic, while others are subtler and only become apparent after many years. In order to assess the relative ecological importance of human interference, it is necessary to understand how marine communities respond to ecological disturbance and environmental change. It is also imperative to appreciate the temporal and spatial scale at which these processes are relevant. Ascertaining whether ecological changes have occurred in response to disturbance or environmental changes requires a rigorous approach to experimental design and monitoring and a careful consideration of the issue of statistical power to detect change.

14.1 Introduction

Human activities are having an ever-increasing influence on global marine ecosystems. While the debate continues about the causes and significance of present global warming, few would now deny that the world's climate is changing more dramatically than at any time in the last century, and scientists are beginning to debate the proximity of the next ice age (King 2004, Weaver & Hillaire-Marcel 2004). Although the subtle effects of changes in weather patterns might not cause immediate alarm, significant events such as the break-up of a large area of the Antarctic ice shelf in 1995 made global media headlines (Vaughan & Doake 1996). With the exception of desert and polar regions, coastal areas are densely populated, particularly at the mouths of estuaries and

● Global climate is currently changing faster than at any time in the last 100 years and has already caused dramatic changes in polar ice cover.

sheltered bays that provide safe anchorages and convenient access inland. In contrast to the vast tracts of open oceanic waters, the coastal margins of the world's land masses are the focus of intensive human activities.

Human activities modify the marine environment both through the removal of biomass and habitats and via the addition of contaminants and physical structures. Marine biological resources are heavily exploited for consumption and economic gain, and habitats often altered incidentally. Rivers convey terrestrially derived material loaded with sewage, agricultural, and industrial pollutants onto the continental shelf. Understanding the ecological responses to these human activities requires an appreciation of both watershed and marine environmental processes. Marine traffic, oil and gas extraction, and dredging are all concentrated in shelf areas. Further offshore in the mid-ocean, direct human influences are limited to oceanic crossing by marine vessels and fishing activities. The decreasing influence of human activity with distance from the coast is related to the physical limitations imposed by the environment (wave height and depth) and the logistics of getting there. Nevertheless, these activities are expanding further offshore as coastal resources are depleted and technological developments enable exploitation of more extreme environments.

● Understanding the ecological consequences of human activities in the coastal zone requires knowledge of land–ocean interactions that include watershed and marine environmental processes.

● Extreme environments offshore in deeper water are made more accessible with technological advances that reduce costs, increase profitability, and increase safety.

In this chapter, the role of human activities in causing ecological disturbance is considered, and their significance is gauged against the scale and frequency of natural sources of disturbance. To understand the ecological importance of human activities in the marine environment, we need to be able to detect changes in measured ecological characteristics using appropriate observational or experimental techniques. An understanding of the scale at which ecological processes operate should underpin the ultimate selection of metrics chosen for study (e.g. diversity, abundance, biomass). We discuss the relevance of a selection of measures and the importance of appropriate experimental design. The wide range of human impacts on the marine environment dictates that we are selective in our coverage. Issues surrounding aquaculture and fisheries were addressed in previous chapters (Chapter 12 and 13).

14.2 Ecological Role of Disturbance

Earlier chapters have discussed the spatial and temporal scale at which various small and large-scale processes operate in the marine ecosystem (Chapter 1). These dynamic processes affect ecological processes and the structure of communities and habitats, such that they are in a continuous process of change. These natural fluctuations form the backdrop against which the relative importance of human activities should be assessed. Almost any human intervention in the marine environment,

● Most human activities in the marine environment cause some form of ecological disturbance. These occur against a background of natural disturbances that occur at a variety of spatial and temporal scales.

whether positive (e.g. habitat restoration) or negative (e.g. dumping of waste), leads to some measure of ecological disturbance. Pickett and White (1985) defined disturbance as 'any discrete event in time that disrupts ecosystem, community, or population structure and changes resources, substratum availability, or the physical environment'.

14.2.1 Sources of disturbance

Disturbances act at different scales and frequencies. Changes in sea level, ocean temperature, and water circulation modify habitats and their associated fauna over large areas and usually over long timescales (>20 years). Natural phenomena such as cyclones and hurricanes have regional impacts on a seasonal basis and affect a wide range of marine habitats to different extents (Hall 1994). Periodic outbreaks of ecosystem engineering organisms such as starfishes and sea urchins can lead to periods of prolonged habitat modification (Chapter 7). Rising ocean temperatures may facilitate the proliferation of non-indigenous species, which can alter the existing community assemblage through the process of competition for space and other resources. Human agents of change vary from the direct effects of an oil spill, habitat damage by bottom fishing on the seabed, eutrophication of sea basins, discharge of toxic substances, to trampling across the seashore. Whatever the source of disturbance, it is a fundamental process that contributes to the maintenance of diversity in all ecosystems.

● Disturbance is one of the most important ecological processes in the maintenance of diversity by opening up resources for colonization by opportunistic species.

14.2.2 Scale of disturbance

There is an intimate link between the spatial and temporal scale at which environmental change occurs. At the smallest scales (μm) chemical reactions between sediment particles, flocculation of organic matter in the water column, viral and bacterial processes occur on a very short timescale. At the next scale up, minute-by-minute disturbances occur as the result of the feeding activities of macrofauna in sedimentary habitats and as a result of their movements upon or through the sediment, creating burrows or furrows in their wake. At slightly larger scales up to one metre, larger-scale bioturbating processes become important (e.g. faecal mound formation, burrow chamber excavation) and the feeding activities of megafauna such as birds, fishes, and crabs become important agents of disturbance that occur every tidal cycle. Thus the seabed is a mosaic of patches in different stages of alteration and recolonization (Grassle & Saunders 1973). Each modification on its own is relatively small compared to the total habitat resource, however the summed total of these effects is such that they have a significant role in determining the characteristics of that habitat (Fig. 14.1).

● Small-scale physical disturbances may seem relatively insignificant when considered on their own, but their additive effects may influence wider-scale community structure.

Fig. 14.1 The physical disturbance created by individual fauna such as estuarine fiddler crabs, may be small in scale (cm), but the additive effects of the disturbance created by the entire population within an estuary can lead to the complete reworking of the surface of these sediments in just one low tide (Photograph: M.J. Kaiser).

The additive effects of small-scale natural disturbances are well illustrated by the effects of the burrowing and feeding activities of the soldier crab (*Mictyris platycheles*) in south-eastern Tasmania. As soldier crabs forage across intertidal mudflats they create intensely disturbed areas of sediment in discrete patches interspersed with undisturbed areas. These small-scale disturbances have the most profound effect on the meiofaunal assemblage, such that nematodes, species richness, species diversity, and evenness were significantly reduced in disturbed as opposed to undisturbed areas, although total abundance was unaffected. Changes in community structure were subtle and resulted from an overall change in the balance of relative abundances of many species, rather than from changes in a few dominant species (Warwick et al. 1990). Hall et al. (1993) undertook a similar study in a Scottish sea loch. They hypothesized that the pit-digging activities of a large predatory crab (*Cancer pagurus*) would generate significant community changes in the soft-sediment assemblage. Yet despite a carefully designed experiment no ecological effects were apparent. They concluded that larger-scale natural disturbance processes such as wave disturbance, masked any subtle effects that pit digging by crabs might generate.

● Even the additive effects of multiple small-scale disturbances may be masked by much larger-scale environmental disturbances such as storm-induced wave perturbation.

14.2.3 Recovery rate

The scale of physical disturbances created by micro-, meio-, and macrofauna are relatively small and the recovery time of the habitat in response to those disturbances is rapid, occurring within seconds to weeks (Table 14.1). Recovery times in response to disturbance increase with scale. At smaller scales (<1 m) recolonization occurs through active movement and passive transport of adults into the disturbed area. At larger scales of disturbance typical patterns of recolonization occur. For example, physical disturbance of the seabed by towed bottom

Table 14.1 The relationship between the spatial scale of a disturbance and the time taken for recovery to occur. Recovery is defined as the point at which the biological community within the disturbed area is no longer significantly different from that in other similar and proximate areas that were not the subject of the disturbance (control areas).

Spatial scale	Temporal scale (frequency)	Processes
μm	milliseconds–seconds	chemical reactions, virus and bacterial driven processes
mm–cm	seconds–minutes	meiofaunal processes, macrofaunal sediment reworking, predation, herbivory, faecal production
0.1 m–1 m	minutes–days	bioturbation, diatom mat formation, megafaunal disturbances (e. g. feeding pit excarvation), precipitation of minerals (calcium carbonate mounds)
1 m–100 m	days–months	Recolonisation and redistribution of small microbiota, biomass and population fluctuations, sediments resuspension and settlement, bedload transport tidal scour and currents
100–10 000 m	months–years	Hurricane and strong events, iceberg scouring, submarine landslide events, siesmic activity, recolonisation of large macrobiota
>10 000 m	years–decades	volcanic and seismic activity, anoxic events, submarine landslides, global warming, El Niño events, coral bleaching, global warming, recolonization of large slow-growing macrobiota

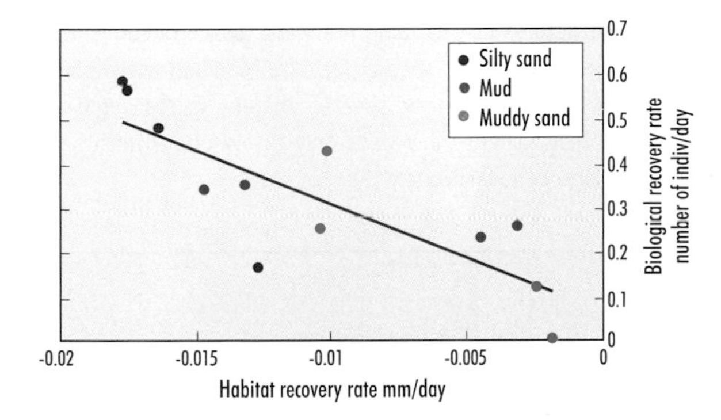

Fig. 14.2 The relationship between the rate at which sedimentary habitats recover (quantified as how quickly standard pits dug into the sediment filled in (decrease in depth of pit cm d^{-1})) and the rate of biological community recovery (measured in terms of the rate at which the difference between control and dug plots decreased in terms of the total number of individual organisms inhabiting the sediment). Muddy sand treatments filled in more slowly than any other treatment and consequently recolonization was much slower.

fishing gear disrupts the surface of the seabed and kills, damages, or digs out many of the infauna and epifauna in the path of the trawl. The immediate post-trawling effects are typified by a reduction in resident species number and abundance (Kaiser et al. 2002). However this is

● Recolonization after disturbance events follows a typical pattern, with short-term immigration of active and scavenging fauna followed by longer-term (months to years) recolonization through larval recruitment. The scavenging fauna disperse within a few days after the biota killed by the disturbance have been consumed.

always accompanied by a short-term (2 to 3 days) influx of scavenging species, which are attracted to the carrion generated in the disturbed area (Ramsay et al. 1997). Thereafter, the recolonization rate is linked to the supply of larvae to the disturbed area and habitat stability. If disturbance involves disruption of the habitat, a process of habitat restoration is required before recolonization of the associated biota occurs in full (Fig. 14.2). For those habitats formed by living biota with slow growth rates and irregular recruitment of larvae recovery times are measured in years or even decades (Collie et al. 2000).

14.2.3 Intermediate disturbance hypothesis

The responses of species diversity and assemblage characteristics to disturbance can be described by an important ecological paradigm – the

Box 14.1 Intermediate disturbance hypothesis

Although not the first, Connell's (1978) application of the intermediate disturbance hypothesis (IDH) is perhaps the most quoted. He used the IDH to explain changes in species diversity under different scenarios of disturbance in two habitats of high diversity, tropical rainforests and coral reefs (Fig. 14.1). At low levels of disturbance, succession processes eventually lead to a climax community dominated by relatively few species of large biomass (3). When this state is disturbed, e.g. by a tree fall in a rain forest, or the damage of coral reef structures by a passing hurricane, space becomes available which permits colonization by opportunistic species (2). The modified assemblage now has a combination of climax and opportunistic species, thereby increasing diversity. As the severity or frequency of disturbance increases only a few opportunistic species persist resulting in an assemblage of low diversity (1).

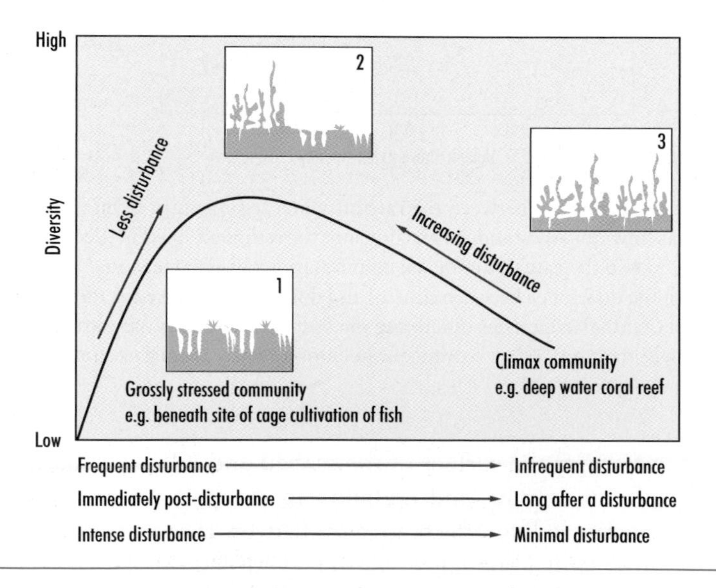

(a)

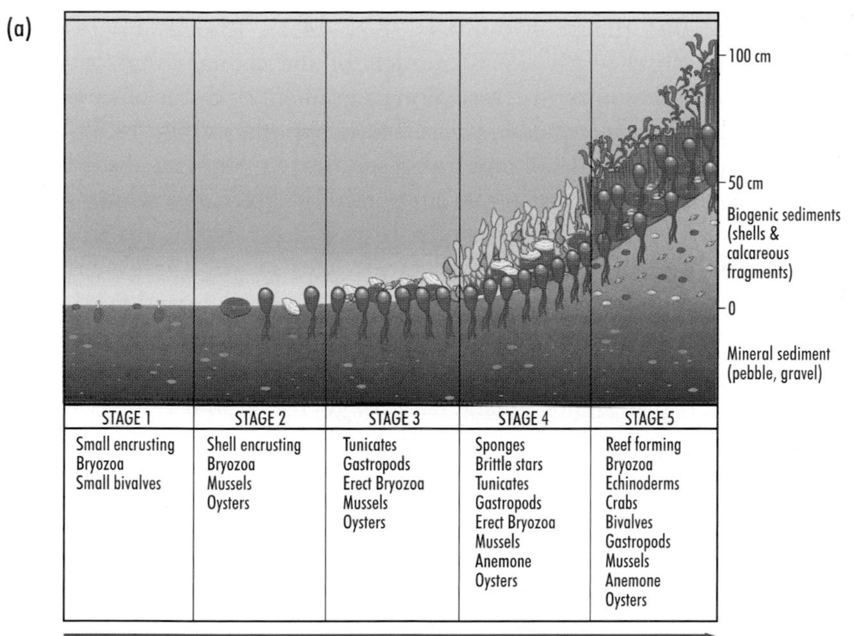

(b)

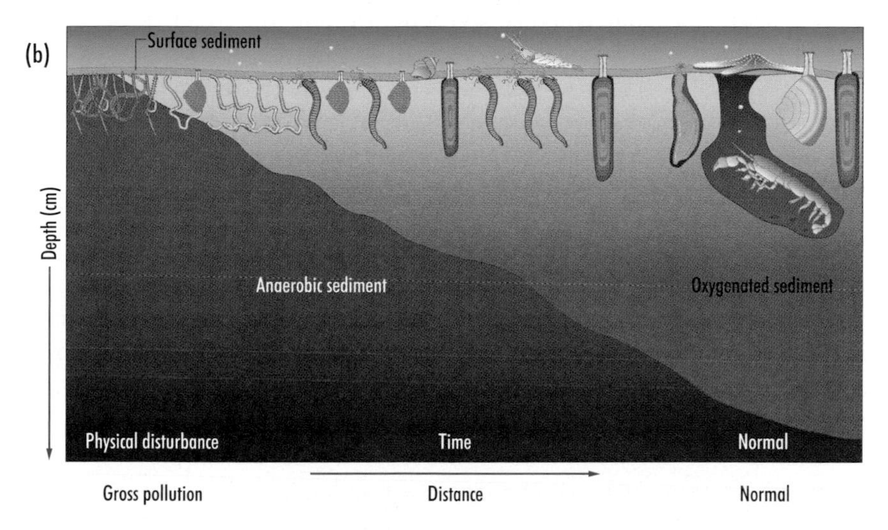

Fig. 14.3. The Pearson and Rosenberg (1978) model of the response of seabed communities to physical disturbance and a gradient of gross pollution (a + b). Towards the left of the diagrams the benthic communities experience the greatest intensity of stress and the community is characterized by low diversity and typically is dominated by only oligochaete and polychaete worms and mats of bacteria. (b) Only the top few mm of the sediment is oxygenated. As stress decreases to the right of the diagram, so the body size and longevity of the fauna increases (typified by large burrowing bivalves and echinoderms). The depth of oxygenated sediments gradually increases with decreasing stress levels and increasing bioturbatory activity. The same gradient occurs in response to organic enrichment usually along a gradient of distance from the source of input. The same model can be applied to the response of epifaunal communities (a) to dredging as in this example of macrofaunal succession on the seafloor of the Foveaux Strait, New Zealand (Cranfield et al. 2004). (Figure (a) reproduced from Journal of Sea Research, Vol. 52, 2004 p. 109–126 with permission from Elsevier.)

intermediate disturbance hypothesis (Box 14.1). Rhoads (1974) and Pearson and Rosenberg's (1978) models of the changes that occur in macrofaunal community structure along a gradient of disturbance follow the principles of the intermediate disturbance hypothesis (Fig. 14.3). Their models of macrofaunal community responses to physical disturbance and organic enrichment are amply supported by direct observations. The Pearson and Rosenberg (1978) model is well illustrated by the situation that occurs beneath suspended fish farm cages in low energy environments such as fjords. The fallout of uneaten food pellets and faeces that descend to the seabed results in an organically enriched environment, which in extreme cases is suitable only for the growth of bacteria or a few species of small opportunistic taxa (oligochaetes and *Capitella* spp.) that are highly tolerant of low oxygen conditions (Chapter 13). Such a community has very low diversity. At increasing distances from the area beneath the cage the loading of organic enrichment decreases, dissolved oxygen levels increase, and environmental stress is reduced, such that the opportunistic species are gradually replaced by larger-body-sized biota such as echinoderms (sea urchins and brittlestars) and bivalves, many of which perform important ecological functions such as mixing sediments that enhance oxygenation and microbial production (Fig. 14.3).

● The intermediate disturbance hypothesis describes the response of community diversity across gradients of disturbance. Highest diversity occurs at intermediate levels of disturbance that result in assemblages with both opportunistic and climax species.

14.3 Measuring the Effects of Human Activities

In order to study the effects of human interference on the ecology of marine communities and their components, we need to be able quantify appropriate measures or metrics that respond in some way to the human activity of interest. These metrics must be amenable to quantification in a reliable and consistent manner. Processes that respond over decadal timescales require long-term strategies of sampling if these are to be quantified. The metrics to be measured may be univariate responses of single factors such as change in abundance or species richness; distributional techniques that measure the distribution of individuals or biomass among an assemblage of organisms; or community responses that measure the responses of more than one species, known as multivariate responses.

14.3.1 Univariate measures

When investigating the response of a single species to an environmental gradient or disturbance treatment in an experiment, it is typical to measure changes in abundance, biomass. or some other physiological, morphological, or behavioural response. However, measuring the response of a single species takes no account of the effect that changes in the abundance or body-size class distribution of that species might have on community structure. This is particularly important when the trophic status of taxa changes through their life history. This is typified by the relationship between cod (*Gadus morhua*) and whiting (*Merlangius merlangus*). Small cod are consumed by adult whiting, which in turn are eaten by adult cod. Single species that are the focus of long-term studies tend to be those that are easily identified and counted *in situ* (e.g. seabirds, pinnipeds, limpets, barnacles), or show strong morphometric changes in response to a stress agent (e.g. development of male characteristics in female whelks). Often, such taxa are considered to be indicators of environmental change that may (or may not) be reflected in other parts of the ecosystem (Fig. 14.4).

● Measures of changes in body-size class describe the relative abundance of small and large individuals in a population. This is particularly important when trophic status changes with increasing body size e.g. the transition from juveniles considered as prey, to adults that become predators.

In most cases, the effects of human activities are not species-specific and have variable effects on different components of an assemblage of organisms. For example, release of a contaminant from a point source of discharge (e.g. chemical discharges from an outfall pipe) is likely to have greater cumulative effects on a range of sedentary fauna and less severe effects on mobile fauna that are transient within the area affected by the discharge. In the case of direct physical disturbance (e.g. by bottom

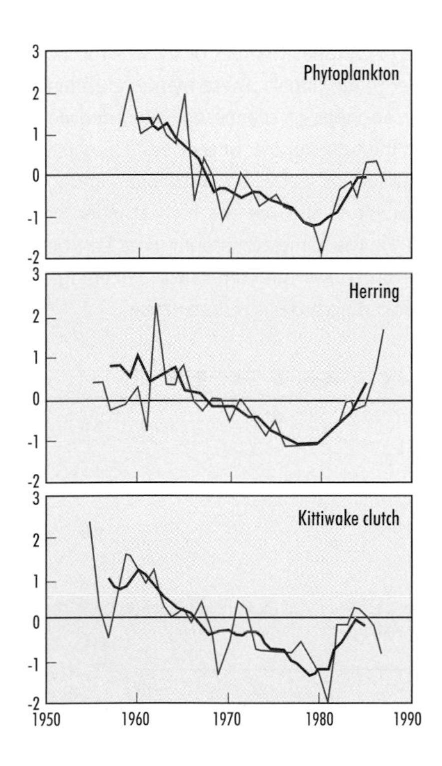

Fig. 14.4 Parallel long-term trends in phytoplankton biomass, herring abundance, and kittiwake clutch size driven by patterns in weather in the North Sea. The *y* axis is a relative scale showing the degree of change around a mean value (0 = no change from mean). (From Aebischer et al. 1990.)

Box 14.2 Univariate and distributional community characteristics

		Sample		
		A	B	C
Abra alba		150	2	1
Acanthocardia echinata		5	1	3
Acteon tornatilis		1	1	1
Chamelea gallina		1	3	6
Corbula gibba		1	1	4
Donax vittatus		55	5	2
Dosinia sp.		2	1	1
Ensis ensis		1	15	2
Fabulina fabula		2	350	1
Total number of species	S	**9**	**9**	**9**
Total number of individuals	N	**218**	**379**	**21**
Species richness (Margalef)	$d = (S - 1)/\text{Log}\,N$	**1.49**	**1.35**	**2.63**
Pielou's evenness J	$J' = H'/\text{Log}\,S$	**0.4**	**0.18**	**0.9**

This example shows three samples of a benthic assemblage with only the abundance of the bivalve taxa recorded. The distribution of individuals within each of samples A, B, and C is very different. In samples A and B, *Abra alba* and *Fabulina fabula* are numerically dominant, while in sample C the individuals are more evenly distributed among the species. Below the community data are shown two univariate metrics of these samples, the total number of species, and the total number of individuals. These metrics are then used to calculate a diversity index (Margalef) or an index of evenness (Pielou) derived from the Shannon Wiener diversity index (H′) and the total number of species (*S*). Sample C has the highest species richness as the same number of species are distributed among a smaller number of individuals. Accordingly, sample C also has the highest index of evenness (measured on a scale between 0 and 1). The same community data can be displayed graphically in a *k*-dominance plot, which shows the cumulative percentage dominance of each species in the assemblage ranked according to dominance.

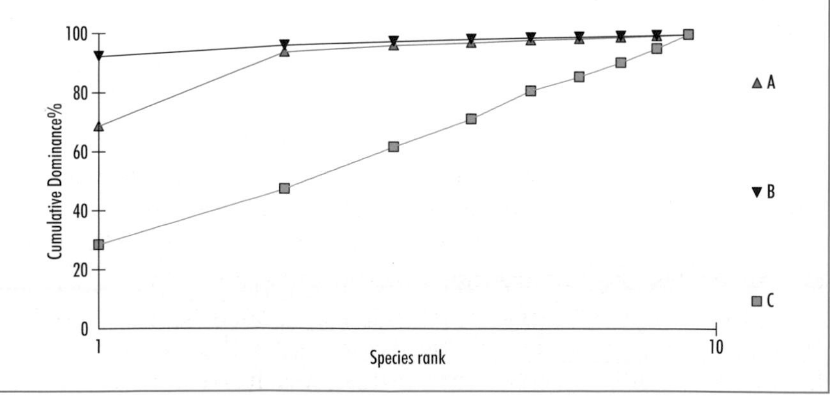

trawling or aggregate dredging) certain fauna may be more vulnerable than other constituents of the biological assemblage. The resultant community changes are often subtle and might not be revealed by the examination of the response of individual species. In these cases it is more usual to take multi-species samples that are representative of the biological assemblage. Within each sample, counts of multiple species can be collapsed into a single coefficient such as a diversity index. For example, some diversity indices give a measure of the extent to which a relatively small number of species account for a large proportion of the total number or biomass of individual organisms within a sample (Box 14.2, see also Chapter 1).

● Diversity indices summarize counts of multiple species within a sample in a single coefficient. Different indices are sensitive to either changes in rare or dominant species within the assemblage.

While diversity indices are a useful tool for detecting major community changes that occur as a result of gross disturbance or along steep environmental gradients, they can be insensitive if the changes are manifested as species replacements with similar levels of abundance. In addition, if diversity indices are to be used effectively it is essential that the scale at which community diversity metrics are measured is appropriate for the scale at which an agent of change may operate. A good example of this scale effect is the response of diversity metrics to bottom fishing disturbance. Fishing gears disturb large areas of the seabed. Even so, highly abundant small-body-sized animals (e.g. polychaete worms) are relatively resilient to physical disturbance, while less common larger fauna such as burrowing sea urchins and bivalves are highly vulnerable to disturbance. Typically, marine ecologists collect many small samples of the seabed to try and improve the chance of detecting responses to environmental change (14.3.5). However Kaiser (2003) found that it was not possible to detect the effects of fishing on community diversity when small-scale ($0.1\,m^2$) samples were collected. Only when larger-scale samples were considered, which sampled more effectively the less common and more vulnerable species, did the effects of fishing on diversity emerge (Fig. 14.5). This example underlines how sampling at the wrong scale can lead to erroneous conclusions about changes in community diversity!

● While collecting many small samples may increase statistical power, fewer larger samples reveal changes in rare species more effectively.

The use of diversity indices can fail to detect large changes or differences in the composition of biological communities. For example, two communities may have the same diversity index even though their species compositions are entirely different. Alternatively, the abundance of one species might decrease while another increases by an equivalent amount in response to environmental stress. These limitations have been overcome to an extent by the introduction of new measures such as taxonomic diversity and taxonomic distinctness (Somerfield et al. 1997) (Chapter 1). These indices are more sensitive to environmental and other disturbance gradients than traditional measures (Fig. 14.6). However, in order to calculate taxonomic distinctness it is necessary to have

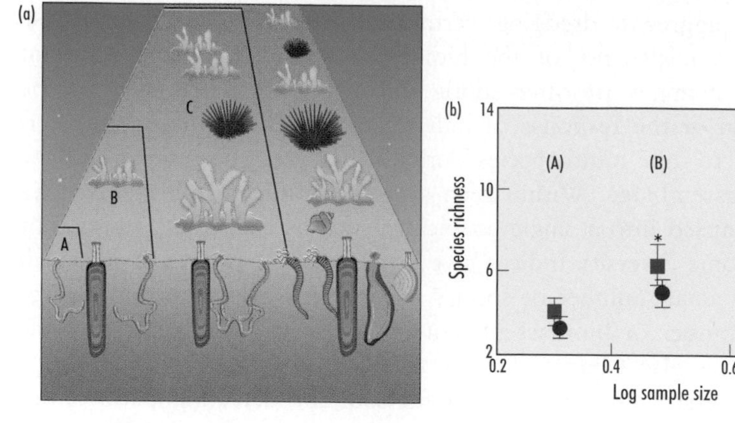

Fig. 14.5 The size of sample units collected (note: do not confuse with the number of samples collected) is critically important if rare species are those most vulnerable to a particular form of disturbance such as bottom fishing (a). If the size of samples collected is too small (A) then species richness will not be sampled with sufficient power to detect the effects of the disturbance. This is illustrated in an example of a study of the effects of sample size on the ability to detect the effects of fishing on benthic diversity (b). As the size of the samples collected increases (A, B and C) the effects of fishing on species richness (means ± 2 S.E.) suddenly become apparent as indicated by the significant differences (denoted by *) between fished and unfished areas.

● New diversity indices such as taxonomic distinctness are sensitive to changes in abundance or biomass of particularly species and have overcome some of the insensitivity of other indices.

a detailed knowledge of the taxonomy of most of the components of the assemblage, a feature that is lacking or variable for many areas of the world.

14.3.2 Distributional measures

These techniques summarize a set of species counts from a single sample as a curve or histogram. For example, *k*-dominance curves rank species in decreasing order of abundance, the steeper the initial part of the curve the greater the contribution of one or a few species to the overall abundance (Box 14.2). Counts of each taxon are converted to percentage abundance relative to the total number of individuals in the sample. The cumulative percentages are then plotted against species rank (Lambshead et al. 1983). A similar plot can be generated for biomass and superimposed on the abundance plot to give ABC (abundance-biomass-comparison) curves as defined by Warwick (1986). In a relatively pristine or low energy environment, a community typically will be dominated by large-bodied individuals (e.g. sponges, soft corals, large fishes) and the biomass curve would lie above the abundance curve (relatively few individuals contribute to a relatively high proportion of the biomass, giving a steep biomass curve). At increasing levels of environmental or physical stress the large-body-sized biota are replaced by high abundances of very

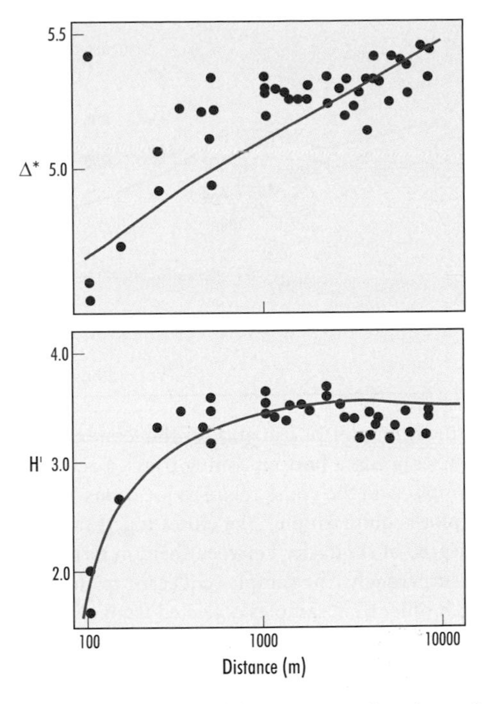

Fig. 14.6 The example shown is calculated from species abundance data for an oilfield in the North Sea. Various diversity indices were calculated and plotted against distance from the centre of the oilfield to a distance of 10 km. Taxonomic distinctness (Δ*) was found to be the most sensitive univariate measure of community structure, increasing linearly as samples were collected further away from the centre of hydrocarbon contamination while the Shannon-Wiener diversity index (H′) was insensitive beyond 100 m from the centre of the oilfield (Somerfield et al. 1997).

small biota (many individuals contribute a relatively small proportion of the total biomass but a relatively high proportion of total abundance, giving a steep abundance curve) and the biomass and abundance curves eventually cross over such that the abundance curve lies above the biomass curve.

● Distributional techniques such as ABC curves provide a useful insight into changes in biomass distribution relative to the abundance of biota in response to disturbance.

14.3.3 Multivariate techniques

These techniques compare two or more samples based on the extent to which these samples contain the same or different species at varying levels of abundance. Multivariate analytical techniques incorporate both species identity and abundance or biomass. A comparison between pairs of samples will yield a similarity coefficient that indicates the degree (percentage) of similarity between each pair of samples. When there are more than two pairs of samples, pair-wise comparisons are made between each possible combination of samples. The resulting similarity coefficients are used as the basis to classify or cluster mutually similar

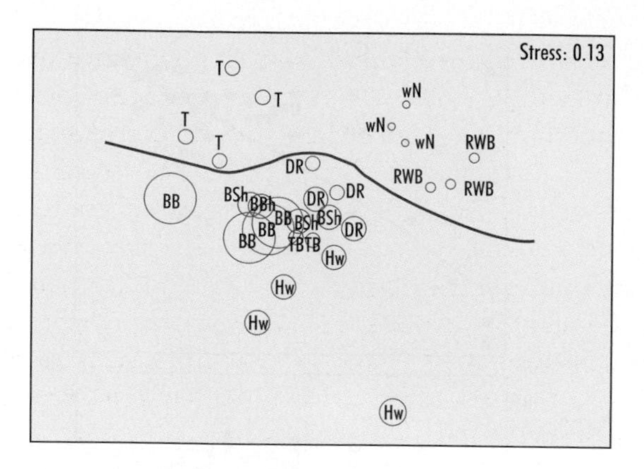

Fig. 14.7 An MDS ordination plot of fish and benthic communities sampled from sandbanks in the Irish Sea using a bottom fishing trawl. Each point represents one standardized trawl sample, and the codes relate to locations around the coast of Wales. Interpretation of the plot is quite simple. The closer together the samples appear in the plot the greater the degree of similarity between them in terms of species composition and abundance. Not surprisingly, the samples collected from the same location tend to be more similar to each other than samples collected from elsewhere. In addition, a key environmental parameter (carbonate content of the sediment) can be overlaid on the plot. This shows that locations shown above the solid line of the plot have a relatively low carbonate content (small bubbles), while those below the line have a relatively high carbonate content (large bubbles). (From Kaiser et al. 2004).

● Multivariate techniques such as multidimensional scaling ordination (MDS) can be used to visualize the literal similarity between multiple samples in relation to experimental manipulations or environmental gradients. The closer together two sample points appear in the ordination plot, the more similar their assemblage composition; the further apart they appear in the plot, the more dissimilar their constituents.

groups of samples. This can be taken on a further stage to yield an ordination plot (Fig. 14.7), which maps the samples in either two or three dimensions such that the distances between pairs of samples reflect their relative dissimilarity of species composition (Clarke & Warwick 1994). In a similar manner, the same samples can be analysed in terms of their environmental characteristics (depth, salinity, bottom current speed, etc.) and another ordination plot generated. The similarity matrices on which the species and environmental data plots are generated can then be compared to investigate the extent to which environmental similarities among samples reflect patterns in community data (Fig. 14.7). These analyses are complex but have been made accessible to non-specialists through the development of software packages such as PRIMER (Plymouth Routines in Multivariate Ecological Research).

14.3.4 Detecting change

As marine ecologists, we are interested in detecting ecological changes in response to human activities in the marine environment. This might be the change in biomass of plankton or fish abundance over several decades or might be the change in community characteristics in areas where aggregate extraction occurs. To stand any chance of success we need to ask the correct question at the outset. This is a problem that besets many of the long-term data sets that continue to be maintained today. While they are an invaluable source of data, many were never designed to answer some of today's ecological challenges. Long-term data sets often track environmental factors such as seawater temperature, which is reflected in the response of ecological metrics such as phytoplankton biomass (Fig. 14.4). However, many other variables are also correlated to these fluctuations and may be either synchronous or out of phase. Long-term ecological changes are strongly driven by environmental fluctuations and it is difficult to disentangle the relative importance of the influence of multiple environmental factors. For example, to ascertain the non-lethal effects of chemical contaminants on fish larval survival, it would be desirable to conduct a series of controlled laboratory experiments that manipulated only this variable while environmental conditions were held constant. Adding additional variables into the experimental design is possible but much more demanding on resources. Similarly, the ecological consequences of activities such as dumping of waste, channel dredging, aggregate extraction, and bottom fishing require an experimental approach to determine the degree of change attributed to these activities and the rate at which restoration of community attributes occurs (if at all).

● Long-term data sets are an invaluable tool to elucidate long-term trends and responses in biological metrics. However, many of these data sets are unable to answer conclusively some of today's pressing environmental questions due to their original design and sampling limitations.

Long-term data sets of fish abundance were set up specifically to measure fluctuations in spawning stock biomass (Chapter 12) and provide key guidance when setting conservation priorities for fish populations. However, as fish become less abundant with over-exploitation, it becomes more difficult to sample the population with accuracy (i.e. there simply are not enough fish to sample, or they are too elusive). Knowledge of the statistical power of surveys to detect trends is essential, since the consequences of not detecting a real trend can be profound (Box 14.3). In other words, a population may be heading for extinction but we might not detect this critical decline because the survey design does not have sufficient statistical power. This situation is particularly acute for species that are highly vulnerable to fishing effects or that are uncommon within an assemblage under normal circumstances. Maxwell and Jennings (unpublished data), examined these effects for North Sea fish survey data and found that the power of the survey to detect abundance decreases on timescales of <10 years was low. Hence, there is a real danger that some species could become extinct many years before we become aware of their demise.

● Vulnerable species could become extinct many years before we realize their demise if survey design is inadequate to detect declines in their populations.

Box 14.3 The importance of statistical power

Statistical power is not about flexing your analytical prowess. Every experimental study has some level of error whether from deficiencies in sampling or due to natural variability. Statistical tests are associated with a probability value, which is often set at 0.05. What this means is that there is a 1 in 20 chance that the result obtained occurred by chance alone. Hence, there is always a danger that our apparently significant effect is in fact false and occurred by pure chance. This is known as a **Type I** error. This is particularly likely to occur when undertaking multiple pair-wise tests, the more tests that we perform the greater the chance of getting an erroneous result by pure chance (e.g. testing for differences in each environmental parameter measured at each of two sites). The chance of making a Type I error is reduced by correcting or lowering the level at which significance is deemed to have been achieved (e.g. from 0.05 to 0.01).

When significant differences occur between two treatments, but we fail to find this difference, a **Type II** error has occurred. Non-significant outcomes are equally as important and just as informative as significant effects. Consider the example of the study of the effects of a shellfish fishery on benthic communities, the outcome of which may decide whether fisheries managers decide to close, or permit harvesting, and possibly whether the fishermen continue to make a living or not. When no effect of the fishery is reported it is important to examine whether the statistical power of the experiment was sufficient to detect the effects of interest (see Hall et al. 1993 for an elegant example). This underlines the value of **preliminary sampling** that can be used to calculate the amount of replication required to detect a certain level of effect (e.g. a 10% change in abundance of a particular species or community metric) before the main experimental manipulation is performed. When reading scientific papers that report non-significant results, always examine the error bars surrounding the mean or median values that are reported. If these are large compared to the mean or median (i.e. they have a high **error : mean ratio**) then treat their declared 'non-significance' with caution.

14.3.5 The experimental approach

The ability to design and interpret appropriately designed experiments is one of the fundamental tools of the ecologist. There are a number of crucial steps that must be considered to avoid common pitfalls made by both students and professionals alike. First, what is the question you are trying to answer? Let us say we are interested in knowing the effects of bivalve shellfish harvesting on the associated benthic fauna in a sandflat habitat. So the question is 'What changes does shellfish harvesting cause in the associated benthic community?', or to put it as a null hypothesis 'Shellfish harvesting causes no change in associated benthic community structure'. Thus we have identified that we need to manipulate a benthic assemblage containing an appropriate target species (e.g. clams, cockles, mussels) by subjecting it to a typical harvesting methodology (e.g. hand raking, suction dredging). This is the experimental **treatment**. The manipulated treatment will need to be compared with a **control** that is not subjected to the treatment.

● Good experimental design is dependent upon asking the correct question at the outset.

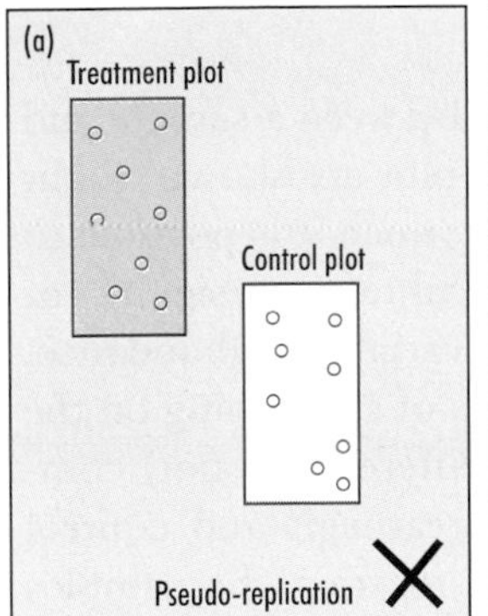

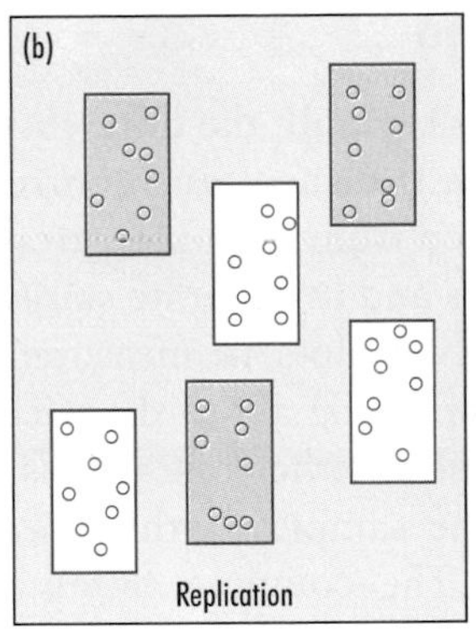

Fig. 14.8 (a) The wrong way to do it! One treatment plot and one control plot each sampled eight times. The eight samples are not replicates of the treatment, they are simple multiple samples of a plot. Multiple samples are taken to ensure that a representative selection of the biological assemblage is sampled at the correct spatial scale. (b) The right way to do it! Multiple replicated treatment and control plots. Each plot is a replicate in an investigation to determine, for example, the effects of shellfish harvesting on benthic biota (c).

An appropriate site needs to be selected, ideally this should be relatively uniform in nature, with no obvious additional sources of variation (e.g. riverine discharge at one end) or these will need to be taken into account in the eventual experimental design. The chosen site will also have a representative fauna, clearly there is no point investigating the effects of shellfish harvesting at a site with no shellfish. The treatment (the act of harvesting) needs to replicate as closely as possible the intensity and scale of activity that is normally conducted (Fig. 14.8), otherwise there is a danger that the results will be open to the accusation that they are not truly representative of the real activities they purport to represent. Other factors to consider might be the seasonality, intensity, and frequency of the normal harvesting activities.

- If an experiment is designed to answer questions about the effects of a particular human activity, it is important that the scale and techniques used to create the treatment effect are representative of the activity of interest.

Having identified a likely site for the experiment it is important to ensure that there are no strong biological gradients by taking preliminary samples in a random grid pattern across the proposed study site. A strong biological gradient (i.e. much higher densities of shellfish at one end of the site) will complicate the process of ascertaining the effects of the shellfish harvesting treatment but can be accounted for by using a Latin squares approach that replicates the control and treatment plots across the environmental gradient. However, time and budget constraints do not always permit the luxury of preliminary sampling, hence it is important to build additional treatment and control replicates into the design to ensure that any anomalous replicates do not jeopardize the validity of the experiment. It is much better to sample an excessive number of replicates, as it may not be possible to return and collect additional samples at a later stage once the experimental manipulation has been completed.

- Preliminary sampling often identifies environmental or biological gradients or anomalies that may complicate later analysis, and enables the calculation of the sampling effort necessary to detect a given level of change in a variable.

14.3.6 Pitfalls to avoid

This brings us onto a classic pitfall: the difference between a sample and a replicate. In Fig. 14.8a a treatment and control plot are shown. Eight samples have been collected from each plot, from which it is possible to calculate means and errors and to generate statistical tests for significant differences between the two plots in measured variables (abundance, biomass). But this tells us nothing about the effects of shellfishing on the benthos. It can only tell us whether one plot is different (or not) compared with the other. The samples within the treatment and control plots are *not* replicates. The common pitfall is to treat the samples within the plots as if they were replicates of the treatment and control. This is known as pseudoreplication. A properly replicated design is shown in Fig. 14.8b. Clearly more sampling effort is used, but the treatment and control plots are replicated (2 treatments × 3 replicate plots × 8 samples). So what can one do if time and budget dictate that this is too many samples? It is important to preserve or increase replication at the level that addresses best the initial question. In this example, if we had to reduce our sampling effort by half, it would be better to take only 4 samples from each of 3 control and 3 treatment plots rather than taking 6 samples from each of 2 control and 2 treatment plots. The greater the level of replication, the greater the statistical power to detect differences among treatments (Box 14.3).

What can we do if it is not feasible to replicate the treatment? Abandon the experiment is one alternative, but not very helpful if a scientific study is required for policy advice. Typically such an example might occur if the cost of creating the treatment is prohibitive, as in the study of the ecological effects of aggregate extraction undertaken by Kenny and Rees (1996). Using ships to generate experiments can be very expensive, typically $US25 000 per day and clearly places a constraint on what can be achieved on a fixed budget. Kenny and Rees (1996) overcame this problem by using a Before After Control Impact (BACI) design in which both a designated control and treatment plot are studied prior to any experimental disturbance (Underwood 1991). Once the similarity of these plots was confirmed the experiment proceeded. The design is still pseudoreplicated (there is only one treatment and one control plot) but we have more confidence that the difference detected immediately after aggregate extraction is indeed attributed to the effects associated with dredging (as opposed to some other random factor). However, an examination of this study reveals the weakness of such designs as the authors were unable to account for changes recorded some 24 months later and were only able to speculate that these were site-specific changes caused by local physical forcing due to wave action (Kenny & Rees 1996).

● In some cases, a Before, After, Control, Impact experimental design is the only viable option, but such a design has far weaker powers of inference than a fully replicated experiment with equal numbers of control and treatment replicates.

BACI designs are appropriate for the examination of short-term responses to human interference, but are highly unsatisfactory for detection of long-term effects. This basic design could be improved by introducing multiple replicate control plots that can be studied for the effects of random factors that might also lead to changes in community structure. For example, if four control plots were studied, three might exhibit similar changes over time while one of these might experience a chance settlement of larvae of a particular species that did not occur at the other sites. This design tells us that we can be more confident that the difference between the treatment and control sites that we studied are real effects and not attributed to a chance occurrence.

14.4 Agents of Change

14.4.1 Riverine input and land use

The composition and quantity of riverine discharge into the shelf seas will affect habitat composition and biology in the immediate area of discharge. Globally, this is an important process given that 70% of the sediment input to the sea is from riverine sources, although this varies from region to region (Milliman 1991 and see Chapter 7). Dam construction, fluvial management regimes, and the effects of changing patterns in precipitation linked to global climate change will all affect the rate and quantity of riverine inputs in the marine environment. The amount of sediment discharged is related to drainage basin size; the surface area of a small drainage basin provides lower storage capacity for sediment than a larger basin. Thus smaller rivers discharge a proportionately greater load of sediment per unit of water discharged than much larger systems.

● Some 70% of sediment input into the sea is derived from rivers, hence changes in land-use and rainfall are likely to have a significant impact on marine communities in coastal regions.

Some of the world's largest rivers are located in South America where they are known to play an important role in structuring the topography and sedimentary conditions of the adjacent seabed (Amazon and Papuan continental shelves). Close to the river mouth the water column is heavily laden with sediment and results in a distinct layering of sediment reflecting discrete periods of deposition of riverine sediment and erosion of the seabed. Consequently the fauna is highly impoverished in this environment and tends to be dominated by bacteria (Rhoads et al. 1985). Field studies demonstrate that the body size, abundance, and depth distribution of benthic fauna are at their lowest across the continental shelf coincident with peaks of riverine discharge and maximum trade-wind stress. During periods of low riverine discharge and minimal wind stress, macrofauna abundance was highest and bacterial biomass increased by a factor of two (Aller & Stupakoff 1996).

The quality and quantity of sediment transported by river systems is affected by the land-use practices that occur in the river catchment.

Deforestation and other agricultural practices have led to substantial soil erosion such that it is estimated to be 3.7 times greater than 2500 years ago, prior to the period when forests were cleared for agricultural purposes. Widespread cultivation of land in northern China is considered to have caused the tenfold increase in sediment discharge from the Yellow River. The increasing demand for greater areas of agricultural land to support subsistence farming in developing countries seems likely to increase the sediment discharge from regionally important rivers (Hall 2002).

● Deforestation for agricultural purposes has been associated with a 3.7 times increase in soil erosion.

14.4.2 Eutrophication

High nutrient loads in river discharge are associated with phytoplankton blooms that generate mainly diatomaceous phytodetritus. This material descends to the seabed where it fuels high rates of microbial respiration and depletion of oxygen (anoxia) at the sediment–water interface (Fig. 14.9). Such events tend to coincide with periods of calm weather when inshore waters undergo stratification. The Baltic Sea, the Adriatic, and the Gulf of Mexico are well-documented examples of locations where such

Fig. 14.9 Although this looks like the aftermath of an oil pollution incident washing ashore, it is in fact the product of a phytoplankton bloom die-off off the coast of Cape Cod (NE United States), lovingly known as 'munge' by local people (the word 'munge' or 'mung' means to mash into a mess). The genus responsible for the bloom is probably *Phaeocystis*. (Photograph: D. A. Kaiser.)

events occur on a regular basis (Turner & Rabalais 1994; Chapter 7). Systems that are particularly susceptible to eutrophication would include those with restricted tidal inundation, poor water exchange or those susceptible to stratification. Typically such systems include fjordic areas, enclosed bays, and areas with relatively deep water.

● High nutrient inputs into enclosed bays and other areas prone to water column stratification have been associated with eutrophication of coastal waters.

While periodic anoxic events occur naturally, their extent and severity appears to be related to human influences. Nitrogen loading doubled in the River Mississippi between the 1900s and 1980s, and it is thought that the increasing scale of anoxic events in the Gulf of Mexico is linked to eutrophication (Turner & Rabalais 1994; Box 14.4). Evidence for this hypothesis arises from studies of stable isotope signatures and organic tracers in sediment cores, which indicate that nutrient levels started to increase in the 1950s and finally levelled off in the 1980s, coinciding with a threefold increase in the use of chemical fertilizers by the agriculture industry in the latter half of the twentieth century (Eadie et al. 1994). The sources of nitrogen that flow into the Mississippi would appear to be diffuse and can originate as much as 1000 km upstream, thus it is almost impossible to pinpoint a single major source of contamination. Arresting the activities of an entire industry will take considerable scientific proof that requires the elimination of other possible causative agents (e.g. climate change). For example, hydrocarbon fuel consumption appears to be the primary source of atmospheric inputs of nitrogen, which have increased by 50 to 200% over the last 50 years (Paerl 1995).

● Sources of nitrogen input into river systems are diffuse and may occur many hundreds of kilometres from the sea, but evidence suggests that agricultural run-off is the primary culprit.

Box 14.4 Anoxic events in the Gulf of Mexico

Weather conditions, large-scale physical oceanographic processes, biological productivity, the amount of nutrient discharge, and other factors affect oxygen depletion in coastal waters. The largest zone of oxygen-depleted waters in the entire western Atlantic Ocean, occurs in the northern Gulf of Mexico on the Louisiana continental shelf adjacent to the outflows of the Mississippi and Atchafalaya Rivers. In the last decade the extent of bottom water hypoxia (16 000 to 18 000 km^2) has been greater than twice the surface area of the Chesapeake Bay, rivalling extensive hypoxic/anoxic regions of the Baltic and Black Seas. Depending upon environmental conditions, hypoxia occurs from late February to early October, but is most widespread, persistent and severe in June, July, and August. See also Justić et al. 2005. Figure reproduced with permission of Nacy Rabalais.

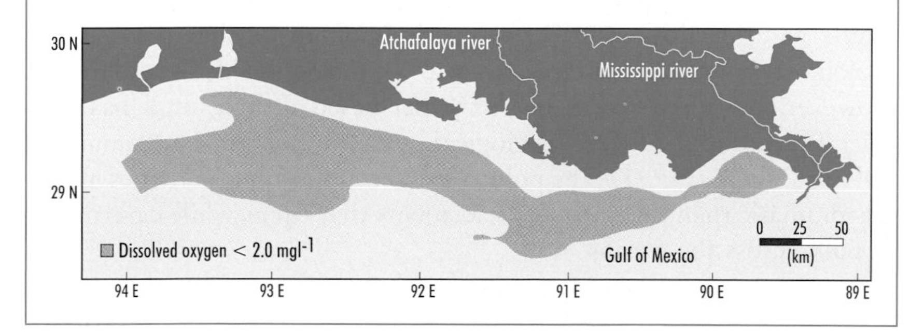

14.4.3 Water use

Clearly the inputs derived from riverine discharge have an important influence on shelf sea ecosystem processes. Consequently, any activity that alters riverine discharge is likely to have ecological consequences for near-shore ecosystems. Major dam schemes are responsible for some striking changes in marine ecosystems. For example, in the Adriatic Sea, the output of the River Po and adjacent river systems has been lowered by 12% in the recent years. The reduction in nutrient inputs has been associated with a decrease in primary production in the local shelf water mass (Alongi 1998). The structure of many river delta regions is maintained by the supply of suspended riverine sediment. Accretion of the Nile delta was reversed after the construction of the barrage on the River Nile in 1868, and further exacerbated with the construction of the Aswan dam. The reduction in supply of riverine sediment to the coastline led to coastal erosion rates of between 5 and 240 m per year. The associated reduction in nutrients transported out to sea was linked with a decline in landings from fisheries for both pelagic and demersal fisheries (Box 14.5). In later years, fishery production increased as the reductions in nutrients supplied through the Nile discharge were replaced through coastal urbanization and the concomitant increase in discharge of sewage into the River Nile (Nixon 2004).

● Reductions in riverine discharge as a result of water abstraction and dam construction affect coastal erosion, primary production, and commercial fisheries.

14.4.4 Hydrocarbon exploitation

The extraction of oil and gas from the sea inevitably releases hydrocarbons into the marine environment as a result of drilling activities, and in the past due to the use of contaminated drilling mud (Olsgard & Gray 1995). The ecological effects of the addition of fine sediment to the seabed beneath drilling platforms and the contaminants within drilling mud are associated with a localized reduction in species diversity. These effects are ameliorated with increasing distance from the drilling platform. In the case of gas platforms in the Gulf of Mexico, community changes are confined to within 100 m of the drilling platform (Montagna & Harper 1996). However, in the North Sea, the cumulative effects of using drilling mud over periods of up to nine years, led to ecological changes at distances of up to 6 km from the drilling platform. However, the more recent use of water-based drilling mud has considerably reduced adverse ecological effects on seabed communities (Olsgard & Gray 1995). The pollutant effects of drilling activities are far less dramatic than the impacts of accidents that occur while carrying oil in bulk across the oceans.

● Understanding the negative ecological effects of the contaminants in drilling mud has prompted the oil and gas industry to seek more environmentally friendly alternatives.

Box 14.5 Effects of damming the River Nile

Catches of commercially important fish species (a) fell dramatically after closure of the Aswan High Dam in 1964. Both demersal (jacks, mullets, bass, and redfish) and pelagic (herring, sardine, and anchovy) fish landings remained low until 1980 when they rose steadily. Shrimp landings showed similar trends but have never recovered to levels experienced prior to closure of the Aswan high dam. The declines in fisheries are in part thought to relate to the fall in phosphorus and nitrogen released into the Mediterranean Sea from the River Nile. (b) However, as populations along the river expanded it seems highly probable that inputs from sewage have offset the loss of nutrient input as a result of closure of the Aswan High Dam. Reproduced with permission from American Scientist 92: 156–165.

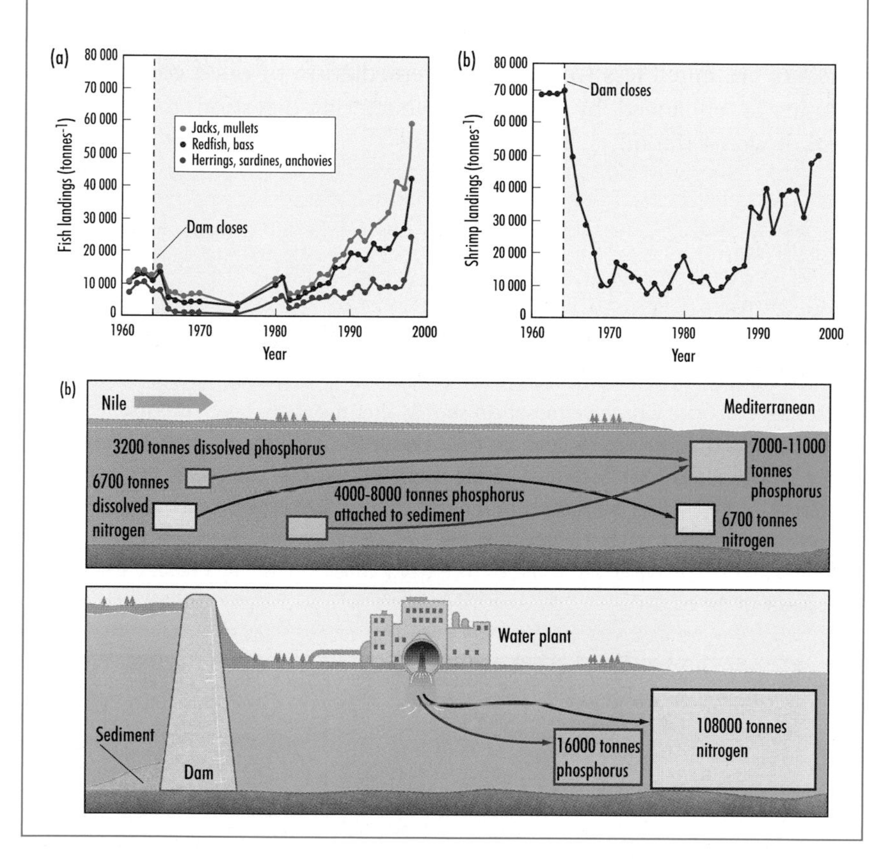

High profile examples of oil spillage incidents include the grounding of the MV *Exxon Valdez* off Canada and the MV *Braer* off the Shetland Islands or the destruction of oil pipelines and refineries during the Gulf War. The grounding of the MV *Braer* differed from most oil spill incidents in that it coincided with extremely severe storm force winds that were maintained for a number of weeks. The resulting sea conditions meant that the oil was highly dispersed by natural wave action (Box 14.6). Technology for dealing with these incidents and our understanding of the ecology of the biological recovery processes that occur after such incidents is much improved. Indeed, it is often ecologically better to leave natural ecological processes to degrade oil spills rather than to intervene with the use of **chemical dispersants**. The use of dispersants is often driven more by aesthetic demands to preserve the appeal of tourist beaches rather than on any scientific grounds. In the early 1970s the oil dispersants used to clean beaches after the MV *Torrey Canyon* oil spill were more harmful to organisms than the oil itself (Southward & Southward 1978). Nowadays, dispersants are much less toxic and bioremediation of oiled contaminated habitats is enhanced by the use of bacterial digestion to speed the breakdown of the oil.

● The use of dispersants on oil spills can have more severe ecological consequences than if the oil were permitted to biodegrade. However, social and economic issues often dictate that unsightly oil needs to be removed from tourist hotspots, and thereby override ecological considerations.

Box 14.6 Impact of oil on the marine environment

Crude oils and petroleum products are complex substances, typically most of the volatile fractions evaporate into the atmosphere shortly after they have been released into the environment. The heavier fractions are then exposed to a variety of physical, chemical, and biological processes according to the prevailing weather conditions. For example aerosols and emulsions are formed as a result of wave action, while exposure to ultraviolet light leads to photo-oxidation at the surface of the sea. Persistent rough seas contributed to emulsification of the oil in the MV *Braer* incident and helped the rapid dispersal of the oil.

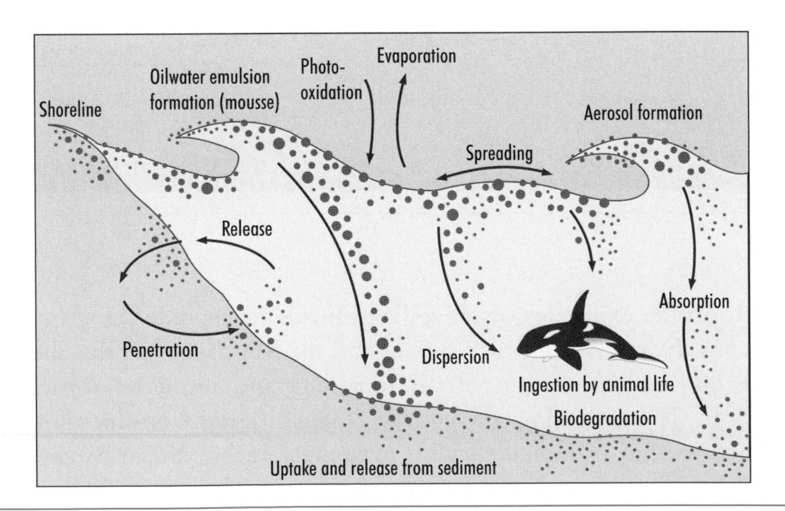

14.4.4 Mining geological resources

Advances in marine technology have made the mineral resources of the seabed accessible in a wide range of different marine environments. These range from relatively shallow coastal waters where sand and **aggregate** (gravel) are extracted, to deep-sea mining of **manganese nodules** on the abyssal plain, to **diamond dredging** off the coast of Namibia. In each of these examples, sediments are removed directly from the seabed (usually via a dredge head) and the desired products extracted on board the vessel at the surface. Not surprisingly, these activities represent an intense localized disturbance to the seabed and completely alter the biological and geological composition of the habitat.

● Most aggregate extraction is permitted under licenses granted for defined areas of the seabed. Improvements in satellite navigation systems have enabled more accurate compliance with these licenses by aggregate companies, but also enable stricter enforcement through the use of 'black box' recorders that log vessel movements.

The demand for raw materials to supply housing and construction projects is unlikely to decline in the foreseeable future, and Charlier and Charlier (1992) predicted that production would reach 200 million tonnes for the United Kingdom alone in the next century. Extraction of marine aggregate in the United Kingdom has increased from 7 million tonnes per annum in 1969 to a relatively constant figure of approximately 22 million tonnes per annum between 1989 and 2003. Dredging of aggregates and the disturbance of the seabed generated by towed bottom fishing gears has some similarities in terms of their ecological effects on benthic biota and habitats. However, while fishing activities are widespread, aggregate dredging is confined to restricted areas that are strictly controlled through licensing schemes. However, dredging is a much more intensive activity, and removes the habitat to depths of up to 1 m. Hence, recovery of the habitat is a key process in determining the rate of recovery of the biota associated with that habitat (Dernie et al. 2003, Fig. 14.2).

● Aggregate extraction causes intense, but localized, disturbance of the seabed that can lead to long-term habitat alteration lasting in excess of several years.

Kenny and Rees (1996) used a commercial gravel dredger to remove 50 000 tonnes of aggregate off the east coast of England. Two years later, the sediment composition in the area from which the aggregate had been removed was still dominated by finer particles than in a comparable control area, and was much more prone to resuspension as a result of wave action. This is perhaps not surprising given that habitat recovery is likely to occur as a result of the active transport of gravel material into the site, a process that requires highly energetic processes (violent storm activity and strong seabed currents). Although species diversity was restored after two years, the biomass of many of the longer-lived species remained greatly reduced (e.g. the horse mussel, *Modiolus modiolus*). The natural disturbance regime at individual sites will determine the rate of habitat and biological recovery that is likely to occur as a result of dredge disturbance (Hall 1994).

Fears of shortages of metal resources in the early 1970s sparked interest in the idea of mining manganese nodules in the deep sea. Manganese nodules contain metals such as iron, manganese, copper, nickel, cobalt, zinc, and silver. Dredging for manganese nodules involves the use of remotely operated seabed vehicles that crawl along the seabed scooping up surface deposits, and thereby create a furrow in the seabed surface. Experimental studies have demonstrated (not surprisingly) that the dredging process reduces both habitat and species diversity within the dredge track, and produces a sediment plume. Sedimentation at these depths occurs at the rate of a few millimetres every 1000 years. As a result, dredge tracks will persist for many years creating long-term changes in deep seabed topography, while filter-feeding organisms may be adversely affected by the unnaturally large pulse of sediment that settles out of the plume (Theil & Schriever 1990).

● Early excitement regarding the feasibility of deep-sea mineral extraction subsided with a reduction in the stock market value of commodities such as copper and cobalt.

Mining minerals at extreme depths is an expensive business and only likely to become economically viable when the combined nickel, copper, and cobalt content of the manganese nodule exceeds 2.5% at an abundance of $>10\,kg\,m^{-2}$. Only a few regions in the Pacific Ocean and from the Central Indian Basin in the Indian Ocean meet these criteria. Economic-grade manganese nodules are generally found in the middle of the ocean in water depths exceeding 4500 m. Manganese nodules would have to be lifted at the rate of 3 million tonnes per year for 20 years at an individual mine site covering an area $>6000\,km^2$ to make such a proposition economically feasible. The lower costs associated with the terrestrial mining of these metals mean that mining in the deep sea is likely to remain on hold for the foreseeable future (Glasby 2000).

14.4.5 Contaminants

In addition to concern regarding the effects of elevated nutrient inputs into coastal waters, there is considerable concern over the long-term and often subtle effects of persistent contaminants in the marine environment. Many of these contaminants, such as radionuclides and organic substances, are derived from industrial and heat generation sources. Polychlorinated biphenyls (PCBs) are particularly persistent and accumulate through the food chain with negative effects expressed in top predators. For example, PCB contamination of prey has been linked to hatching failure for a number of avian predators, while contaminants such as mercury occur in higher concentration in the feathers of seabirds sampled in the latter half of the twentieth century compared with those sampled from museum collections (Arcos et al. 2002). While the effects of such contaminants on apex predators have been well documented, the effects on organisms at lower trophic levels is less well studied. Although invertebrates such as amphipods are used in lethal toxicity tests, the incidence of contaminant mortality associated with bioengineering organisms (Chapter 7) that affect habitat structure remains unknown. The effects of contaminants may be subtle in that they do not necessarily cause direct mortality but may have negative population effects via their influence on recruitment processes and larval viability. For example, Atlantic croaker (*Micropogonias undulatus*) larvae exposed to the chemical Aroclor 1254 have significantly reduced growth rates and are less capable of evading predator stimuli (McCarthy et al. 2003).

● Contaminants can have subtle sub-lethal effects that affect the performance and subsequent survival rate of organisms. Larval stages are particularly vulnerable to these effects that are easily overlooked.

Abnormalities in the growth and reproduction of a number of different molluscs became apparent in the mid 1980s. Symptoms included reduced growth, shell deformation, and declining population abundance in specific localities. In particular, dogwhelk (*Nucella lapillus*) populations had become locally extinct in the United Kingdom and elsewhere, usually in localities close to harbours and ports. The most striking symptom associated with these declines was the occurrence of imposex. Imposex is manifested in female gastropods by the development of a *vas deferens* and in some cases a penis that lowers reproductive success. Not all molluscs exhibited the same symptoms; populations of *Littorina littorea* had lower egg production, while oysters developed thickened and deformed shells (Matthiessen et al. 1995). The causative agent was identified as tributyl tin (TBT) based anti-fouling paint that was used on the hulls of boats. Legislation has since banned the use of this paint on certain categories of vessels and has been associated with a recovery in mollusc populations in countries where its use is restricted.

● See Evans et al. 2000 for the controversy surrounding potential replacements for TBT.

14.5 Climate Change

So far we have examined the effects of human activities that have relatively restricted spatial effects on communities or particular groups of animals. However the potentially significant consequences of global climate change for marine systems and activities affect all regions of the planet at both small and large scales (Hall 2002). Those changes likely to occur in response to global warming include sea level rise, water-column warming, precipitation, wind speed and patterns, water column circulation, and the frequency and intensity of storms. We give only a few examples of environmental changes that might affect marine ecosystems.

14.5.1 Temperature effects

Global warming is likely to affect biological processes and biodiversity in the oceans. Rising temperatures may result in a reduction in the equator to pole temperature gradient. The frequency of El Niño-like conditions in the tropical Pacific will probably increase, with the eastern tropical Pacific warming to a greater extent than the western tropical Pacific (IPCC 2000). In conjunction with hydrographic conditions, temperature plays a significant role in determining the biogeographic distribution of species and we can undoubtedly expect to see changes in species distributions with climate change (Chapter 1). For example, a northward extension of warmer conditions may enable the merger of the currently geographically disjunct distributions of warm temperate species such as the commercially important clam *Mercenaria mercenaria* and the eastern oyster (*Crassostrea virginica*). The present northern limit of these species is Cape Cod on the eastern seaboard of the United States, although isolated populations are found further north in the warmer waters of the Gulf of St Lawrence. These temperature changes are also likely to extend the spread of some non-native species that have been introduced through ship's ballast water or other practices such as aquaculture. Many introduced species are considered to be pests as they predate commercially valuable shellfish, for example the green crab (*Carcinus maenas*) that has become established on the western coast of the United States and in Australia.

● Increasing sea temperatures will extend the poleward distribution of low latitude species and may enhance the geographic spread of non-native invasive species.

We have already seen in Chapters 4, 6, and 7 that semi-enclosed water bodies are prone to anoxic events as a result of stratification of the water column and oxygen depletion at the seabed. The frequency of such events may increase as warmer temperatures lower the oxygen carrying capacity of water while the biological oxygen demand of the biota will increase. Lower oxygen carrying capacity, coupled with eutrophication will increase the likelihood of such events, which may have more frequent adverse effects on coastal fisheries. Temperature effects on fisheries are of particular concern since water temperature is a strong predictor of fish distributions, the timing of spawning, growth rate, and survival of larvae (Heath 1992; Lehody et al. 1997). In the coastal waters of the USA, for example, a reduction in the spatial extent of oxygen rich cool water may cause concomitant changes in the distribution of striped bass (*Morone saxatilis*) (Coutant 1990). Further increases in global sea temperature are likely to cause new combinations of species mixes both in fish and benthic assemblages.

Coral bleaching is an increasing global phenomenon (Harvell et al. 1999) that occurs when the symbiotic zooxanthellae are expelled, algal pigmentation is lost, or both scenarios occur, giving the coral tissues a 'bleached' appearance (Chapter 10). These events have been linked with exceptionally high seawater temperatures, elevated doses of ultraviolet radiation, infection by pathogens, and the alteration of local salinity regimes. The degradation of coral reefs has important economic consequences for local communities that often depend on reef related eco-tourism and associated fisheries as an important source of income.

● Coral bleaching has severe implications for artisanal fisheries and eco-tourism.

14.5.2 Change in rainfall

The effects of climate change on rainfall may be important given the influence of river and land run-off in estuarine and coastal systems

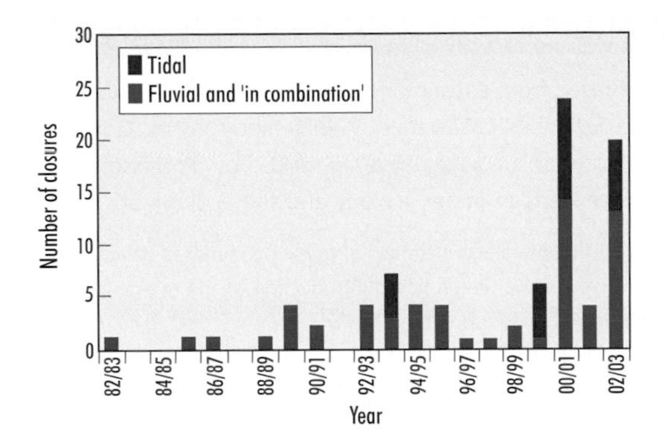

Fig. 14.10 The increasing frequency of closures of the River Thames barrier in recent years is one indication of the increasing effect of elevated rainfall and greater incidence of tidal surges and sea level change (King 2004).

(Chapters 4 and 7). The effects of increasing rainfall can also combine with tidal surges resulting in a greater intensity of coastal and flood plain inundation (Fig. 14.10). Rainfall is a prominent seasonal feature in areas of the tropics especially where monsoon rains occur. Global predictions of changes in rainfall suggest little change in southeast Asia and South America, where riverine influences are especially important, but potential small increases (5 to 20%) for the Indian subcontinent, East Asia, Canada, and Northern Europe and potentially large increases (>20%) for Northern Asia and the Sahara (IPCC 2000). Even if these predictions are accurate, it is difficult to ascertain their consequences given the changing use of water abstraction with industrial growth in many of the countries affected. Changes in rainfall will also have an effect on buoyancy driven water currents that occur in the vicinity of estuaries (Chapter 7). Increasing the rate and volume of water discharged from rivers may alter the position of density gradient fronts, and may drive the ensuing coastal sediment plume further along the coastline (see ROFIs, Chapter 7). Seabed communities, fisheries, and marine mammals associated with these systems would change their distribution accordingly.

14.5.3 Water circulation

Intensification of alongshore wind stress on the ocean surface may already have increased coastal upwelling intensity (Bakun 1990). As many of the most productive marine regions in the world occur in regions of intense upwelling, the potential for climate change to affect fisheries seems highly likely (Bakun 1990). Recent anomalous changes in the upwelling of cool nutrient-rich water off the coast of the western United States led to an extensive anoxic event resulting from unusually high productivity. Extensive areas of the seabed from the continental shelf edge to within 700 m of the surf zone were severely oxygen depleted, resulting in mass kills of commercially important fish and

Box 14.7 Non-native oysters in the Chesapeake Bay

This text is paraphrased from a testimony given to the 108th US Congress (First Session), Subcommittee on Fisheries Conservation, Wildlife and Oceans, Committee on resources, United States House of Representative in 2003 by Professors J.L. Anderson and R. Whitlatch, both experts in oyster ecology and the ecology of Chesapeake Bay.

Good morning Mr Chairman and members of the Subcommittee. Thank you for this opportunity to speak to you about the proposed introduction of the non-native oyster *Crassostrea ariakensis*.

The [native] oyster stock in the Chesapeake Bay has declined dramatically. Harvest is now about one percent of what it was at the end of the 19th century. Fishing pressure and habitat degradation resulting from agricultural, industrial and residential pollution, deforestation, and oyster reef destruction have contributed to the decline. In recent de cades, however, the diseases MSX and Dermo have been identified as the core reasons for further decline. It should be noted that MSX is caused by a parasite that was introduced to the East Coast from Asia. Fisheries management efforts and various restoration programs have not been successful in restoring the oyster stock to date. The loss of the oyster has been devastating to the oyster industry and its dependent communities. Those that remain in the Chesapeake oyster-processing sector now rely on oysters that are brought in from the Gulf of Mexico region and other areas for their economic survival. Furthermore, the loss of oysters has contributed to declines in water quality and clarity.

The introduction of the non-native Suminoe oyster, or *Crassostrea ariakensis*, from Asia has been proposed as a solution to these difficult problems. Indications are that it may grow well in the Chesapeake Bay and it is known to be resistant to MSX and Dermo. Despite the positive results of introductions of some oyster species, some extremely negative consequences have been observed as well. A major risk of introducing a non-native oyster comes from pathogens, such as MSX, or the introduction of other animals or plants that may be attached to oysters. While in Australia and New Zealand, introduced non-native oysters have displaced native oysters.

Aquaculture of sterile non-native oysters, represents an appropriate interim step that possesses least risk (in terms of the available options) to the Chesapeake Bay and its dependent communities. However, limits and controls on aquaculture practices must be implemented to minimize the risk of introducing pathogens or reproductive non-native oysters during this transitional phase. This approach may provide limited benefit to parts of the oyster industry and it provides decision makers with the added information required to make future decisions. Moreover, this option allows more time for innovative, science-based efforts to restore native oyster populations. On the other hand, the option of not allowing any introduction, fails to address fishing industry concerns and will not result in improved understanding of the ramifications of non-native introductions. It may also increase the risk of rogue or uncontrolled introductions. The option of the direct introduction of reproductive non-native oysters, is not advised given the limited knowledge base on *C. ariakensis* and the potential for irreversible consequences of introducing a reproductive non-native oyster into the Chesapeake Bay. It is unlikely that there exists any 'quick fix' to the Chesapeake oyster situation.

● Changes in wind stress may intensify coastal upwelling and have been linked already to anomalously high production and mass kills of commercially important fisheries.

shellfish species (Grantham et al. 2004). Changes in wind patterns may alter wind-induced currents that are important for larval transport, which means that larvae may no longer be conveyed to habitats suitable for further development (Heath 1992).

14.6 Interaction of Multiple Factors

We have provided a sample of the influence of human activities on the marine environment. We have dealt with each of these factors separately for simplicity, but multiple factors may operate at the same time or in sequence. While one factor on its own may not have serious consequences for an organism or community, the synergistic action of several factors may prove catastrophic. For example, organisms under one form of environmental stress (e.g. rising mean sea temperatures) may be less able to cope with the physiological demands of adapting to increasing frequencies of freshwater discharge that might occur with changing rainfall patterns, and their resilience may be further weakened by exposure to industrial contaminants. The effects of multiple and sequential human interference are well illustrated by a testimony given to the United States Congress to encourage it to consider the consequences of introducing a non-native oyster into Chesapeake Bay. Declines in native oyster stocks in Chesapeake Bay are well documented and have been linked to reduced water quality and over-fishing, but attempts to reinvigorate the oyster stocks have repeatedly failed. One solution appears to be to introduce a non-native species that may be more tolerant of present day environmental conditions and parasitic fauna that were accidentally introduced into Chesapeake Bay as a result of past aquaculture introductions or discharge of ship's ballast water. The testimony given in Box 14.7 highlights the complex considerations that face managers of natural resources that have to weigh the ecological risks and potential benefits of such introductions.

CHAPTER SUMMARY

- Natural agents of disturbance to marine communities and systems occur across a full range of spatial and temporal scales.
- The response of community metrics (diversity) to disturbance, can be predicted from ecological paradigms such as the intermediate disturbance hypothesis.
- The relative impact of human activities on marine ecosystems needs to be assessed against a background of natural environmental fluctuations that occur at a variety of temporal scales. Our ability to detect the effects of human intervention critically depends on an appropriate analytical or experimental approach.
- Coastal shelf environments are subjected to the most intensive human activities that include fishing, aquaculture, mineral and hydrocarbon extraction, shipping activities, tourism and discharges of effluents and pollutants.
- Eutrophication that has resulted from elevated inputs of organic matter and nutrients from agricultural run-off have caused wide-spread bloom of toxic microalgae and anoxic events that have resulted in mass mortalities of marine biota.

- Persistent contaminants such as PCBs have subtle effects at the population level by affecting survivorship of larvae and juveniles. The full impact of these contaminants is a subject of speculation.
- The congested inhabitation of coastal margins means that much of the world's population are vulnerable to changes in sea level linked to global climate warming.
- Increases in global rainfall will elevate freshwater and sediment discharge into coastal waters thereby affecting density-driven currents in ROFIs and will change the distribution of biological communities.

FURTHER READING

Clarke & Warwick (1994) provide a readable and sympathetic (to the non-mathematical) guide to the detection of ecological changes in marine (and other) communities. Evans et al. (2000) provides a synthesis of the debate surrounding replacement of tributyl tin with suitable less environmentally damaging alternatives. Hall (2002) provides a review of agents of change in coastal systems in the present and also examines possible future impacts.

- Clarke, K. & Warwick, R. 1994. *Change in Marine Communities: an Approach to Statistical Analysis and Interpretation*. Natural Environmental Research Council, Plymouth Marine Laboratory, Plymouth.
- Evans, S. M., Birchenough, A. C. & Brancato, M.S. 2000. The TBT ban: out of the frying pan into the fire? *Marine Pollution Bulletin* 40: 204–11.
- Hall, S. J. 2002. The continental shelf benthic ecosystem: current status, agents for change and future prospects. *Environmental Conservation* 29: 350–74.

Chapter 15

Conservation

CHAPTER SUMMARY

In 1768, only 27 years after they had been discovered, the giant Steller sea cows (*Hydrodamalis gigas*), which grazed on algae around Bering Island, were driven to extinction by hunters. North Atlantic right whales (*Eubalaena glacialis*) were reduced to near extinction by commercial whaling in the eighteenth and nineteenth centuries, and only 300 remain today. Large fishes of all species are increasingly rare following decades of over-exploitation, and marine habitats are destroyed by reclamation, pollution, and development. Should we, and do we, care? As a reader of this book you probably do, but many others in the world will be more concerned with the source of their next meal. In dealing with issues of conservation and sustainable development, the marine ecologist enters a wide arena, where practice and policy are also swayed by moral, cultural, political, and economic values. However, this is also a challenging and rewarding arena, where science plays a key role in assessing the sustainability of human impacts, prioritizing conservation projects and underpinning the development of effective conservation policy.

15.1 Introduction

In preceding chapters, we have seen how pollution, fisheries, and aquaculture, together with climate change, have impacted and are continuing to impact the marine environment. These are just a subset of possible human impacts (Table 15.1). In many cases, human impacts are unsustainable, threatening species, habitats and ecosystems, and the environmental goods and services they provide. Applied marine ecology is the source of scientific evidence to identify when and where conservation action is needed, both by assessing the sustainability of impacts and developing methods to mitigate unsustainable impacts.

Table 15.1 Principal human impacts on the marine environment.

Principal human impacts
Aggregate extraction and mining
Aquaculture
Dredging
Engineering and construction
Fisheries
Land-based impacts
Military activities
Oil and gas
Reclamation
Recreation
Renewable energy
Shipping

● Applied marine ecology underpins conservation advice and action.

At least in political terms, conservation contributes to sustainable development. Sustainable development requires that the needs of future generations are not compromised by the actions of people today. Sustainable development has many facets, and includes social objectives such as human equity. Conservationists highlight that recognition of the natural limits on human population and economic growth must go hand in hand with attempts to meet social objectives. Increasingly, national and international agreements express a move towards sustainable development, and these underpin much of the conservation and environmental protection legislation we see today. The contemporary treatment of conservation contrasts with the preservationist approach, which focuses on the protection of species and habitats without reference to natural change and human requirements (Agardy et al. 2003).

● Conservation contributes to sustainable development.

In this chapter, marine conservation is considered as a contribution to sustainable management of the seas. Consistent with sustainable development, sustainable management requires human intervention to maintain or create an environment that does not compromise the wellbeing of future generations. The extent to which individuals, towns, regions, nations, and the international community regard long-term human wellbeing as dependent on the presence of a clean, productive, and biodiverse marine environment will determine the strength of ethical and economic support for marine conservation. Conservation potentially involves a whole range of actions. It may include outright protection of populations or areas from human disturbance; but it may also involve zoning and other types of regulation that help to keep human use within sustainable limits.

● Sustainable management requires human intervention to maintain or create an environment that does not compromise the wellbeing of future generations.

In this chapter, we consider the ethical and economic foundations of marine conservation, together with how conservation issues can be identified and prioritized. We discuss the role of poverty in driving the use of the marine environment and highlight how marine conservation has to find a way to accommodate the short-term needs, aspirations, and expectations of humans. We therefore look in some detail at the economics of marine conservation, the strengths and weaknesses of the policies that promote it and the legislation introduced to back it up. We also describe how conservation policy is implemented and give some examples of the successes and failures of conservation initiatives.

15.2 Why Conserve?

Ecology is at the scientific heart of conservation, since population growth is limited by the availability of resources, and species interact with each other and their abiotic environment. Thus the natural environment limits the scope for development and exploitation, and sets an ever-changing baseline (e.g. the effects of climate change). Apart from the fact that some conservation objectives such as protection of particular species and maintenance of water quality are now incorporated into law, conservation is justifiable on ethical and economic grounds. However, perceptions of conservation differ and justification is never a strictly quantitative exercise; one person's sustainable harvesting of renewable fish populations will be another's devastation of a fragile marine environment.

● Population growth is ultimately resource limited.

Conservation of biological diversity has been a major focus of recent conservation efforts. As mentioned in Chapter 1, biodiversity has been defined as 'the variability among living organisms from all sources including, *inter alia*, terrestrial, marine, and other aquatic ecosystems and the ecological complexes of which they are part; this includes diversity within species, among species and of ecosystems' (Convention on Biodiversity; Box 15.1). Some of the key justifications for biodiversity conservation are (1) that humans have moral and **ethical responsibilities** to care for life on earth, (2) that living organisms **enrich** our lives, (3) that '**ecosystem services**' are provided by many species, and (4) that living organisms allow ecosystems to adapt to change and are a source of materials that benefit humans (Kunin & Lawton 1996). At the personal level, belief in the relevance of various justifications will depend on **social circumstances**. Thus moral and ethical responsibilities to care for life on earth may not hold much sway if you are starving.

● Justifications for conservation range from ethical and ecological to economic.

A common justification for the sustainable use of the marine environment is that marine species provide important benefits for humans (Norse 1993). Thus, fisheries and aquaculture are a vital source of food, income, and employment, particularly in the world's poorer countries (Chapters 12 and 13); and marine species are the source of many useful compounds such as the algal polysaccharides, carrageenans, and agars that are used in food manufacture. The sea is also an important source of non-biological resources such as sand and gravel used in building. Such **direct economic benefits** from the sea are probably the easiest to quantify as they have real economic value, but are not necessarily the most important.

● Sustainable use of the marine environment provides long-term benefit for humans, including food and income, and these are relatively easy to quantify.

Benefit may also derive from goods and services that have no **'true' market value**, such as the storm protection afforded by coral reefs, control of carbon dioxide in the atmosphere, and waste and nutrient removal or recycling (Table 15.2). In some attempts to value the marine environment, these services are the most valuable properties of the seas and oceans. However, their value is rarely accounted for when assessing the sustainability of human impacts. We look at issues associated with costing such services in 15.4. The marine environment is widely used for

Box 15.1 International agreements that support marine conservation

At the World Summit on Sustainable Development (WSSD) in Johannesburg (2002), the follow-on meeting to the United Nations Conference on Environment and Development (UNCED) held in Rio de Janeiro, Brazil in 1992, signatories agreed on a plan of implementation for sustainable development. The plan gives high priority to integration of the three components of sustainable development, namely economic development, social development, and environmental protection. It also recognizes the particular circumstances of the major regions of the world, including the plight of small island states, for which coastal zone management and sea level rise are important issues.

Recognizing that marine ecosystems are critical for global food security and sustaining economic prosperity (fisheries and shipping), the WSSD plan of implementation included strong commitments to improved conservation of the marine environment, including application of an Ecosystem Approach by 2010 (Box 15.3), implementation of the FAO code of conduct for responsible fisheries (fisheries that do not overexploit fish populations or harm marine wildlife and habitats), maintenance of the productivity and biodiversity of important and vulnerable marine and coastal areas, establishment of marine protected areas, development of national and international programmes for halting the loss of marine biodiversity, and control of pollution and the spread of alien species. These can be seen as a wish list rather than a series of actions that will be universally implemented on the time scale suggested. However, raising these issues at international summits highlights their importance and underpins improvements in environmental protection at international, national, regional, and local levels.

There can be rapid uptake of high-level commitments in other fora. For example, following WSSD, the 2002 Ministerial Declaration of the Fifth International Conference on the Protection of the North Sea recognized the need to manage all human activities using an Ecosystem Approach that conserves biological diversity and ensures sustainable development. Subsequently, at national level, the United Kingdom Government published 'Safeguarding our Seas', a strategy for the conservation and sustainable development of the marine environment. This included specific commitments to marine conservation, such as the adoption of an ecosystem approach, that were consistent with the requirements of WSSD.

Source: **http://www.johannesburgsummit.org/**

● Other benefits derived from marine systems, which are difficult to cost, are vital to the functioning of the biosphere.

recreation, and coral reef related tourism can be one of the most valuable industries in otherwise poor countries (Chapter 10). People using the sea for recreation generally choose a clean, healthy, and productive environment in preference to one that is polluted and animal poor.

15.3 What to Conserve

At a societal level, the general greening of government policy is intended to encourage nations, companies, and individuals to conduct their activities in a sustainable way. Many of these actions and the ways in

Table 15.2 Examples of ecosystem services from natural marine ecosystems (Costanza et al. 1997).

Ecosystem service	Examples
Gas regulation	Oceans balancing CO_2 content of atmosphere, thus regulating atmospheric temperature
Water regulation	Oceans as ultimate source of water as a basis for agriculture, industry and transport
Nutrient cycling	Gaseous fixation and organic matter decomposition as sources of nitrogen for primary production in all ecosystems
Waste treatment	Breakdown of sewage by micro-organisms in continental shelf waters
Provision of natural refugia	Habitats such as seagrass and saltmarsh as nursery grounds for fisheries
Food production	Coastal waters as generators of fishery products
Provision of raw materials	Marine sediments and rocks as sources of aggregate for building
Sourcing of genetic materials	Provide natural compounds useful in medicine
Provision of recreation	Offer sport fishing habitat, opportunities for ecotourism
Provision of cultural assets	Offer resources for aesthetic, educational, and scientific purposes

which society addresses and fails to address issues of sustainability are covered in 15.5. However, when human impacts are not managed in a sustainable way at source, conservation efforts often focus on halting unsustainable rates of habitat or species loss. Unfortunately, the magnitude of existing human impacts on the marine environment is such that a great deal of effort is focused on fighting rearguard actions rather than planning for sustainability before the impact occurs. It is hoped that the **greening of government policy** will help to encourage proactive rather than reactive conservation. Proactive conservation, for example, might require that measures to mitigate unsustainable impacts are incorporated into plans for development or exploitation before those plans are sanctioned (15.6.1).

Prioritization is needed to identify the conservation actions that make the biggest contribution to sustainability. At the global level, prioritization involves international decisions about the investment in combating threats such as climate change. At regional and local scales, prioritization is needed to decide how scarce resources should be allocated to proactive and reactive conservation, and the specific issues to address. Prioritization involves many difficult and emotive decisions. Thus the last few surviving individuals of a species usually attract great public and media interest and provide impetus for conservation action, but may stretch resources that could be used for large-scale habitat conservation or maintaining the abundance of common species

● Conservation actions must be prioritized, but decisions can be difficult and emotive.

● Societal values play an important role in setting targets for conservation.

that play important roles in the ecosystem and provide direct benefits to humans.

In public perception, marine conservation is often epitomized by concern for marine mammals, seabirds, and sea turtles, even though less conspicuous species and marine habitats are threatened by human activities. In many ways this is also reflected in policy. In Europe, for example, regulations are intended to maintain populations of small cetaceans such as dolphins and porpoises, which are killed as by-catch in fisheries, at 80% of their **theoretical carrying capacity**, while no conservation measures are in place for fish species, such as the common skate and angel shark, that are also taken as by-catch and close to extinction in some areas. However, conservation will never be an entirely logical process when so many value judgements are involved, and since attempts to conserve whales and dolphins often gain strong public support, efforts to conserve them may be more successful. Protecting such **charismatic species** may also have positive and negative implications for habitats and other ecosystem components on which they depend. An example of the former might be the removal of fishing from a particular locality to avoid cetacean by-catch, which may have concomitant beneficial effects for associated species (e.g. seabirds or fish). Negative effects can result when the protected species becomes a 'pest', as exemplified by the conflict between sport fishermen and fish-eating cormorants

● Societal support may often make the difference between the success and failure of conservation action.

Methods for conservation prioritization of individual species are often relatively quantitative. Thus the World Conservation Union (IUCN) produces a '**Red List**' of threatened species, which is intended to be an easily and widely understood system for classifying species at high risk of extinction (Box 15.2). This system is used to classify all the world's animal species for which data are available, and many fishes and marine mammals have been listed.

● Species are prioritized for conservation by the IUCN, **http://www.iucnredlist.org**.

Habitat conservation is now playing a greater role in the marine environment, with widespread calls for greater use of marine reserves. Thus far it has not been possible to create very large marine reserves, so animals such as whales are still managed on a population-by-population basis. Habitats and regions can be ranked for conservation priority according to factors including their biological diversity within particular biogeographical regions (Chapters 1, 6, and 7) and roles in sustaining particular species or groups of species. The process of selecting sites for conservation will typically involve the measurement or description of attributes of a site or series of sites, an evaluation of these measures against a set of criteria and a method of combining the results to enable ranking of sites (Bibby 1998). Measures typically include species diversity and rarity, size of area, representativeness, naturalness (naturally a very hard thing to define), cultural criteria, and vulnerability.

Box 15.2 The IUCN Red List

The IUCN Red List recognizes seven categories of species extinction, threat and endangerment where there are sufficient data (see Figure). The categorization is based on factors such as reduction in population size over a specified time period, geographic range size, and absolute populations size. A species may be listed as critically endangered, for example, because it has decreased in abundance by >90% over the last 10 years or 3 generations. Marine mammals and turtles have been listed for some time, and other marine species are increasingly appearing on the list. Certain sharks, rays, and chimaeras are likely to be added. There has been some debate about whether managed commercial fish stocks should be included in the list. Some fish stock managers argue that the reference points used for management are conservative, and if population biomass were kept at or above those targets then there would be little risk of extinction. Others argue that the observed rates of decline in many exploited fish populations meet listing criteria and that the population biology of fish makes them no less prone to extinction than many bird and mammal species (Hutchings 2001). Regardless of the pros and cons of these arguments, the Red List has a role in highlighting species that are severely depleted or threatened with extinction, and should help to the promote conservation action that is sought by fishery managers and conservationists alike.

Source: **http://www.iucnredlist.org**

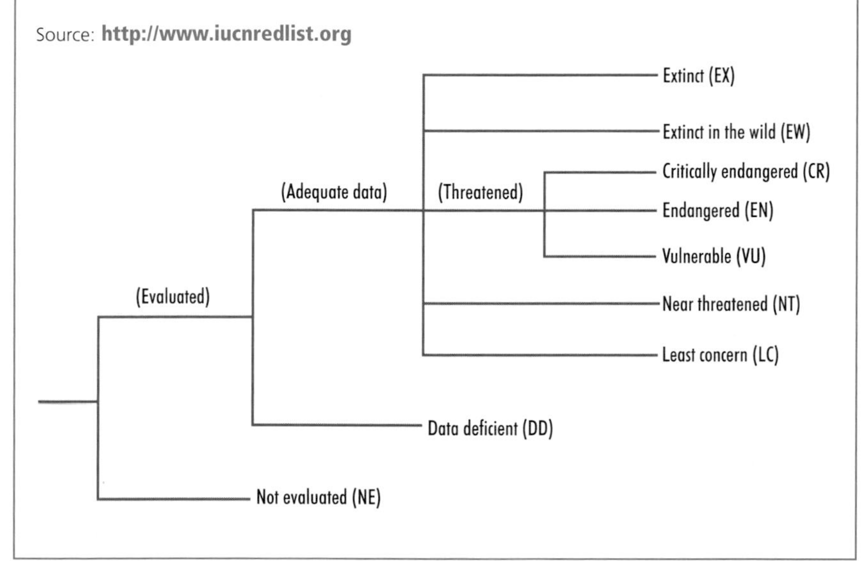

Combining all scores is a problem as some are strictly quantitative (area, diversity) while others are qualitative (naturalness). One approach is to rank sites based on quantitative criteria and then to deal with practical considerations at the final stage.

● Prioritization of habitats and regions can be a quantitative and qualitative process.

Networks of marine protected areas may be proposed to help conserve species and habitats. The idea here is that a set of reserves is selected to meet specific objectives, such as representation of each of the community types known in a region at least once. Algorithms can search a species or attribute by site database to find a minimum set of areas that meets a given objective. If sites are selected sequentially to define a minimum set to meet a stated objective, then the selection of any new site depends on the properties of those already selected. Thus the new set complements the series.

Rigorous quantitative procedures for prioritization are often seen to be essential, and yet the approaches to prioritization we have described can be seen as an intellectual rather than a real-world exercise. Indeed, many sites are protected simply because they were in areas where enforcement was possible, where they brought immediate economic benefits to a local community, or where there was regional or national support for protection, or perhaps more importantly minimal opposition to their creation.

● Sites are also protected just because it is possible to protect them.

It is also important to recognize what site-based conservation can and cannot achieve. The openness of marine systems means that polluting impacts, for example, have to be tackled by reducing polluting inputs, not by setting up reserves. Similar arguments can be proposed for stopping fishing and habitat damage too, because reserves displace activities while outright control of the activities is needed to stop them. It would be inexcusable to lose sight of the activity and thus fail to address the problem at source. This is why conservation of the marine environment involves other actions such as regulation of fisheries and pollution (see Chapters 12 and 14) and integrating the management of multiple uses (15.5). Such linkage between multiple issues, together with the fact that each issue may be addressed by different government departments and nations, has highlighted the need for more joined up conservation objectives and methods of planning. One framework for achieving this is dubbed the 'ecosystem approach' (Box 15.3).

● Habitat and species conservation alone will not ensure sustainable development, impacts have to be controlled as well.

15.4 Economics of Conservation

Economic development has environmental costs, which are rarely paid for by the businesses, governments, and individuals that profit from the development. The assimilation of sewage by ecosystems and side-effects of species declines wrought by fishing are cases in point, the polluter and fisher respectively are rarely asked to bear the costs of their actions.

● The environmental costs of development are rarely paid by the developer.

Failure to take account of the environmental cost of human activities has meant that development is driven by false economic incentives and disincentives, and is unlikely to be sustainable. Moreover, because many human activities in the marine environment, especially fisheries, are subsidized or unprofitable, there are very high short-term economic costs associated with moving towards sustainability. The

Box 15.3 The ecosystem approach

The recognition that sectoral issues that affect the marine environment were often managed in different ways, by entirely different bodies and without sufficient regard for their cumulative or synergistic impacts has led to the recognition that a more 'joined up' approach was needed to ensure environmental protection. The ecosystem approach is intended to provide this. It has variously been defined as 'the integrated management of human activities, based on knowledge of ecosystem dynamics, to achieve sustainable use of ecosystem goods and services, and maintenance of ecosystem integrity' or in less scientific terms as an approach that 'puts emphasis on a management regime that maintains the health of the ecosystem alongside appropriate human use of the marine environment, for the benefit of current and future generations'.

As with the aspirations of international meetings, these definitions are 'high level' and do not help to set operational management objectives. The science support and advisory process must therefore seek to guide the choice of objectives and methods of achieving them. This work is currently in the early stages, but is likely to revolve around the development of a suite of indicators of human activity and ecosystem impacts that can be used to measure the success of management and the response of the ecosystem. With any such approach, the main challenges are separating the ecological impacts of the human activity from the impact of environmental and climatic effects and developing an approach that is sufficiently powerful to detect impacts on the time scales that matter to managers (the lifetime of Governments and often shorter).

The Commission set up by the 1982 Convention on the Conservation of Antarctic Marine Living Resources (CCAMLR) has taken an ecosystem approach to tackling anthropogenic (mainly fishery-related) mortality of marine animals. The problems include loss of albatrosses and petrels to fishery long-lines, entanglement of marine mammals in marine debris, and impacts of fishing on the seabed. In the first case the Commission acted by controlling the timing of fishing and manner of fish offal disposal, as well as requiring scientific observers on all vessels fishing outside national waters in CCAMLR areas. The management of the Great Barrier Reef Marine Park (Box 15.4) also provides an example of an ecosystem approach.

For further details see the CCAMLR and GBRMPA websites: **www.ccamlr.org** and **www.gbrmpa.gov.au**

high short-term costs of changing human behaviour are one of the greatest impediments to effective conservation.

- This creates economic incentives that can lead to unsustainable development.

Fisheries provide an excellent example of the way in which short-term economic forces promote unsustainability, even when the long-term economic benefits of sustainable fisheries are known to be high. For the fisher, the decision whether to catch fishes now or leave them in the sea will depend on their future value. If the value of a fish stock 5 years in the future is perceived to be less than the money that could be made after 5 years by catching the fish now, selling them, and investing the money in a bank, then there is an economic incentive to fish as hard as possible in the short-term. This is known as '**discounting the future**'.

- Economic discounting affects the decisions made by users of the marine environment.

Discount rates are used to measure the rate at which the perceived value of a resource, such as a fished stock, falls over time. Discount rates

● High discount rates have encouraged fishermen and whalers to fish some populations close to extinction.

reflect the cost of return on alternative investments. Thus if you 'invest' some money in fish by leaving them in the sea, you require that its value should grow at least as fast as the value of money in the bank. If the value of fish in the sea grows more slowly, or if the future value of the fish might be jeopardized by activities of competing fishers, then it is a good economic strategy to catch the fish sooner rather than later.

The present value (PV) of income V, t years into the future is:

$$PV(V_t) = \frac{V_t}{(1+\delta)^t}$$

where δ is the discount rate. The decline in perceived value of a unit of income at different discount rates is shown in Figure 15.1. High discount rates (typically 0.1 to 0.2) tend to be used by fishers because fishers are uncertain about reaping the benefits from fishes left in the sea, especially when their competitors might catch them and processes such as stock recruitment are highly unpredictable. Fishers' rates are typically higher than bank interest rates and thus it generally pays to catch fish and invest the profits. This explains why species such as whales, with very low growth rates, were 'mined' rather than fished sustainably. Market instruments that capture at a private level the social and global values of sustainable fishing through, for example, premium pricing for sustainably harvested fish, may be an important step towards sustainability.

Our arguments about the effects of discounting on fisheries also apply to other forward projections of value, for example in the context of climate change. Thus if global warming has costs in 25 years, the willingness to invest in mitigating them now would be very low, even if the discount rate were small (Hanley 1998).

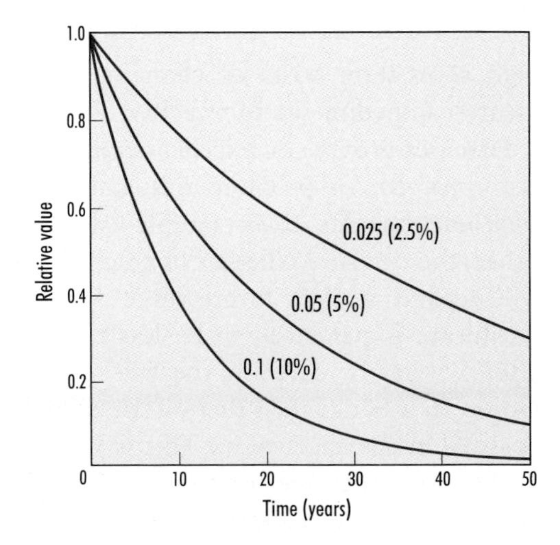

Fig. 15.1 The decline in the perceived value of a unit of income at different discount rates.

Where areas are being converted for human uses such as aquaculture, there are several key questions to ask about the costs of such conversion. For example, what is being lost and what gained, who are the beneficiaries and losers, and what is the economic rationale for conservation? Valuing the functions of intact ecosystems alone is inadequate because human-converted ecosystems also have value to society. Thus the intact and the modified ecosystem must be directly compared. Two marine case studies are instructive in this regard. Thus full economic valuation of Philippine mangrove as against the aquaculture to which it may be converted (assuming 6% discount rate over 30 years) shows that conversion makes sense in terms of short-term private benefits. However when external costs such as loss of long-term timber and charcoal supply, offshore fisheries, and storm protection, are incorporated, the total economic value (TEV) of mangrove is around 70% greater than that of shrimp farms. Similarly, on Philippine reefs (assuming 10% discount over 10 years) destructive fishing practices had high initial benefits to the users, but the benefits of sustainable reef fisheries and tourism were then lost. The TEV of intact reef was 75% higher than that of a destructively fished reef (Balmford et al. 2002).

- Comparing full economic values of ecosystems used in different ways can help to guide decisions on development.

15.5 Conservation Policy and Legislation

Human impacts on the marine environment are determined by the actions of individuals, villages, towns, cities, nations, and businesses steered by government regulations or incentives, social pressure, conscience, or market forces. These impacts can be mitigated, even eliminated, by a variety of means, but in this the actions of individuals are of paramount importance. Human behaviour can be influenced by many means including education, persuasion, economic incentives, legal pressure, and military force. Broadly speaking, conservation faces a choice between top-down and bottom-up approaches, or more usually some mixture of the two, to achieve its goals.

- Human impacts on the marine environment are determined by many societal factors.

The greater the proportion of society that is influenced by its own desire to conserve, the more effective conservation is likely to be. In practical terms, this means that humans impacting the environment should see and share in the benefits of conservation and sustainable development. This requires effective education. Two countries that exemplify this are Australia and New Zealand. These countries have adopted extremely 'green' policies towards use of the marine environment.

- Conservation is most likely to be effective if society wants to support it.

Education and experience in the early years of life have a major influence on attitudes and actions. Children who regard conservation as an important issue are likely to retain this opinion when they diverge into a range of careers or lifestyles. Education, at least for most children and young adults, is based on a syllabus that is mandated by regional or national government. The content of this syllabus can influence how people subsequently respond to, and deal with, sustainable development issues. In many countries, the general 'greening' of government policy is now reflected in the school syllabus and this has encouraged children to think about the role of conservation and sustainable development and has also increased awareness of issues such as climate change that may not be locally visible. However, children attending a small school on a Pacific island where 90% of animal protein comes from fish will have a very different perception of the marine environment from children at a landlocked inner city school where their exposure to marine species is through aquaria, supermarket counters, and restaurants.

● The education system has a major influence on society's view of the marine environment.

Nature reserves and public aquaria also play an increasing role in teaching parties of visiting schoolchildren and adults about the marine environment. For example, the Monterey Bay aquarium in California, USA opened in 1984 and is visited by 1.8 million people a year. This aquarium runs education programmes to raise awareness of over-fishing of ocean species such as the bluefin tuna (Fig. 15.2). Thus by displaying related tuna species in aquaria, visitors are able to see animals that would rarely be seen except in cans and photographs. This must have a major effect on perceptions of these animals, which will increasingly be seen as beautiful and endangered sea creatures rather than overpriced food in short supply.

Fig. 15.2 Monterey Bay Aquarium is one of many large aquariums that runs education programmes with a strong conservation message. Copyright Monterey Bay Aquarium.

For those parts of society that do not see the benefit of sustainable development or are unable to support it, economic incentives and legal measures are needed. Rigorous policing and heavy fines do change the behaviour of some groups of people but they are unlikely to be effective without the support of wider society, and this support is most likely to be gained through persuasion and bottom-up processes such as those fostered by education.

● Legal measures to promote sustainability do help, but they still need to be supported by most of society.

Policies on sustainable development are influenced by the electorate, non-governmental organizations (NGOs), the media, science, industry, and economic and military concerns. Governments are lobbied on many issues by people with many perspectives (**stakeholders**), and for every argument in favour of conservation and sustainable development there are likely to be opposing arguments from those sectors of society that fear losses of food, income, or access rights.

Nevertheless, there is a general 'greening' of government policy, driven in part by international gatherings and declarations such as the Rio Convention on Biological Diversity and the Johannesburg World Summit on Sustainable Development (WSSD) (Box 15.1). Several governments have expressed aspirations to move towards embedding the principles of sustainable development in all aspects of policy, rather than having an environment ministry that has to compete with, rather than work with, other ministries such as industry, transport, and fisheries for funding and influence. Indeed, the treaty of the European Community states 'that environmental protection requirements are to be integrated into the definition of Community policies and activities, in particular with a view to promoting sustainable development'. Of course, it often takes a long time for such high level aspirations to be converted into action.

● Governments are undoubtedly 'greening' their policies, but thus far this 'greening' is reflected in aspiration rather than action.

Many current systems for marine environmental management are confused by conflicting interests and shared responsibility. For example, in British estuaries human activities are regulated by at least 80 parliamentary acts, and a series of EU regulations. Even though several estuarine sites of important conservation status have been designated for protection as SSSIs (Sites of Special Scientific Interest), port and harbour authorities still regulate shipping activity in these areas and gravel or sand extraction can be licensed.

● Conflicting interests can be an impediment to effective conservation action.

Even in cases where the need for rapid conservation action is recognised, many groups often have to liaise and reach agreement before action can be taken. Thus the North Atlantic right whale (*Eubalaena glacialis*) was brought near to extinction by commercial whaling, and by the 1900s the population numbered around 300 individuals. Despite complete cessation of commercial whaling, the right whale was still at risk from entanglement in fishing gear and collision with ships. A population analysis in 1998 suggested that the rates of mortality affecting the whale population could drive the whale to extinction within 200 years (Caswell et al. 1999). The only way to save the whale was to reduce mortality, and this could best be achieved by reducing the probability of ship collisions, which accounted for 80% of all deaths. Such collisions often occurred in the Bay of Fundy, where a shipping route crosses one of the main feeding areas used by the whales. In late 2002, after several years of pressure from the WWF and other bodies, the International Maritime Organization gave the Canadian Government permission to reroute shipping in the Bay of Fundy. The new shipping lanes came into effect on 1 July 2003 and are expected to reduce the risk of ship collisions by 80% and should help to prevent the extinction of the population (Fig. 15.3). Clearly in such cases, good scientific advice and both national and international understanding and agreement are needed.

● International agencies play an important, but not always successful, role in trying to coordinate conservation action.

International agencies can play a key role in coordinating conservation actions across many areas of jurisdiction. Thus whales, turtles, fishes, and water currents do not respect national boundaries and local or national conservation will often be ineffective. The International Whaling Commission, for example, tries to regulate whaling on a global scale, though agreement is hard to reach in such an international agency when countries perceive conservation issues in different ways and some seek to continue exploitation while others strive to ban it.

The recognition that management of many human activities in the marine environment has become overly complex and heavily influenced by sectoral concerns has prompted the development of an ecosystem approach (Box 15.3). This approach builds on other attempts to improve

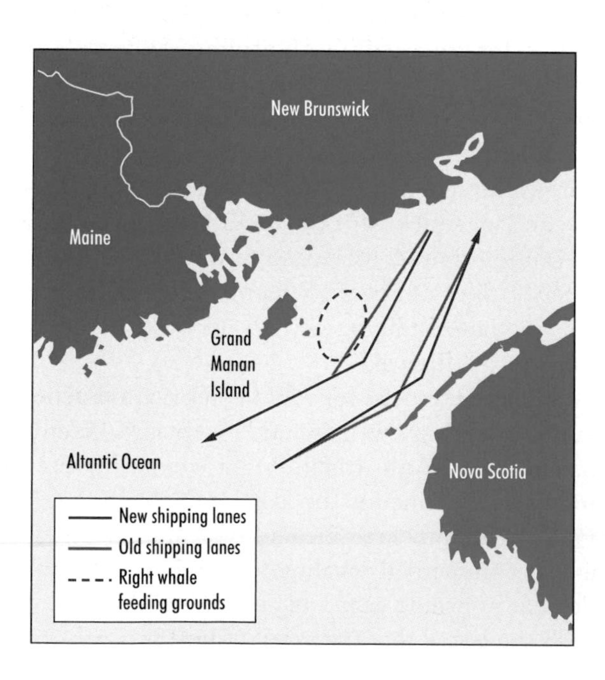

Fig. 15.3 Shipping lanes in the north-west Atlantic were moved to reduce the probability that endangered right whales would be struck by ships.

cooperation among sectors, through approaches such as 'integrated coastal zone management' (ICZM), which have already been widely used with various levels of success in some coastal regions. Integrated coastal zone management (ICZM) is a dynamic, multidisciplinary, and iterative process that promotes sustainable management of coastal waters and lands (coastal zone). ICZM brings together all those involved in the development, management and use of the coast in a framework that facilitates integration of their interests and responsibilities. Over the long-term, ICZM seeks to balance cultural, economic, environmental, recreational, and social objectives in order to achieve common goals. In Europe the environmental concerns that have led to the development of ICZM thinking include the impacts of fishing (Chapter 12), aquaculture (Chapter 13), pollution (Chapter 14), rapid human population growth, and poor employment opportunities in coastal communities.

● Integrated coastal zone management provides a framework for management of multiple human impacts and activities. The integrated approach of ICZM is consistent with the ecosystem approach.

A European Union (EU) ICZM Demonstration Programme consisting of 35 projects in a range of socio-economic, cultural, administrative, and physical conditions across Europe in 1996 to 1999 concluded that a sectoral approach (for example, dealing with issues impact by impact) to management did not meet the needs of managing complex issues in coastal areas. Management should therefore be inherently inter- and multidisciplinary; promote integration of the terrestrial (e.g. human settlements, watershed considerations) and marine (e.g. fisheries) components; highlight the need for integration of all relevant policy areas, sectors, and levels of administration; and use informed participation and cooperation of all interested and affected parties to assess the societal objectives. In short, ICZM implies a new style of governance that involves partnership with all of the segments of civil society, and solicits the collaboration of all coastal zone stakeholders in the conception and implementation of a development model that is in their mutual interest (European Commission 2000).

Eight principles have been drawn from successful coastal management initiatives and helped to form the basis for a call on EU member states to set up national strategies for ICZM by 2006 (European Commission 2002). These principles include (1) a broad overall perspective; (2) a long-term perspective taking into account the precautionary principle; (3) adaptive management; (4) local specificity and diversity; (5) working with natural processes and carrying capacity of ecosystems; (6) involving all parties in the management process; (7) support and involvement of relevant administration bodies at national, regional, and local levels; (8) a combination of instruments to facilitate coherence between sectoral policy objectives. Thus in Europe, the move towards ICZM recognizes the important role of coastal lands and waters for nature and ICZM is an ambitious strategy, or at least a wish list, for addressing some of the overarching environmental concerns for the coastal zone.

ICZM is considered a prerequisite for successful management of coastal and marine resources around the world and many international bodies such as the Food and Agriculture Organization of the United Nations, Intergovernmental Oceanographic Commission, United Nations Environment Programme, and the World Bank support and promote ICZM.

When it comes to environmental decision-making at a global level, the dichotomy in the sustainability, profitability, and capacity for species and habitat conservation in the developing and developed world needs to be recognized. Developing countries still have some of the greatest assets in wilderness areas and biological diversity, but they have the greatest human population growth rates and fewer resources for conservation. It is also worth remembering that opportunities for conservation in developed Western nations are often subsidized by poor people. The removal of capacity in developed countries can have harsh social and economic consequences in the short term but, in the longer term, these are often softened by government subsidies and job opportunities in other sectors of the economy. In the poorer countries of the world, fishing is often the occupation of last resort for families with no other opportunities for subsistence. The scale of poverty and reliance on fisheries in the developing world has been widely described elsewhere (Kent 1998). Perhaps 1 billion people in 40 developing countries may lose access to their primary source of protein as a result of over-fishing (UNDP 1999).

● The capacity for conservation differs in wealthy and poor countries.

The international community will have to intervene to provide funding for conservation in the poorer countries. This would, for example, include providing the means to generate alternatives sources of livelihood to alleviate the poverty that lies at the root, for example, of destructive fishing practices and unsustainable coastal development that threaten biological diversity.

● Opportunities for conservation in the developed world may be subsidized by poor countries.

15.6 Conservation in Action

Action to meet the objectives of marine conservation is being taken on many fronts, and this has involved many levels of ecological (species, habitats, ecosystems) and administrative (local and international, NGO and governmental) organization. To show how all of these levels are being incorporated is beyond the scope of this book; instead we give three examples that help to illustrate the range of actions involved. The first of these is the assessment of environmental risk in any project, in the process referred to as '**environmental impact assessment**' (EIA). The second is **ecolabelling**, which provides consumers with a basis for environmental choice and seeks voluntary participation of an industry in practices that meet sustainable development objectives. The third is the use of marine reserves (marine protected areas), from which fishing and other forms of extractive use are excluded.

As we consider these various forms of conservation action, it is worth considering the overall context in which conservation policy is implemented. Most conservation action is ultimately driven by high-level objectives. As we have seen, these objectives are often chosen internationally at fora such as the World Summit on Sustainable Development (Box 15.1). Then, sustainable development strategies have to be developed to help meet the objectives. Conservation actions consistent with the strategies are then taken to meet the objectives. Progress towards objectives is monitored, and strategies and actions (hopefully not objectives!) amended to ensure objectives are met.

- Conservation action involves many levels of ecological and societal organization.

- Conservation action is guided by high level objectives and strategies to meet the objectives.

15.6.1 EIA

The aim of EIA is to prevent, reduce, or offset any adverse impacts of each and every major development affecting the environment, including the sea (Barrow 1997). EIA is a formal process to identify and predict the environmental impacts of a project, with a view to mitigating adverse impacts or addressing them in revised plans. Such assessments are required for activities such as oil-rig and wind-farm construction and aggregate extraction. Interestingly, however, fisheries development is usually exempt from EIA. The omission of fishing has led to some internal argument among the industries concerned in Europe, since fishing has been identified as the greatest threat to the marine environment by the Oslo and Paris Commission, which upholds the Convention for the Protection of the Marine Environment of the North-East Atlantic from dumping and land-derived and offshore sources of pollution. The purpose of EIA is to support the objectives of conservation and sustainable development by integrating environmental protection requirements into the planning process at the earliest possible stage. The EIA process serves to predict the environmental, social, economic, and cultural consequences of a proposed activity and to assess plans to mitigate any adverse impacts resulting from that activity. Although there is much variation in practice, particularly among countries, EIA often provides a focus for the involvement of society in reviewing the potential impact of the activity, through submissions from the public and government bodies.

- Environmental impact assessment aims to predict the impacts of development and to mitigate or avoid those impacts that are unsustainable.

Once it is decided that an EIA should go ahead, a six-stage process begins. This involves scoping, quantification, report production, decision on project, and, if the development is allowed to go ahead, development and monitoring. During scoping, all the potential impacts of the projects are listed, and during the quantification stage, the magnitude of each impact is assessed in relation to the existing habitats, species, and activities at the proposed development site. Based on scoping and quantification, an Environmental Statement is produced and this is reviewed by government and associated agencies to determine whether development should go ahead.

● Environmental impact assessment is a six-stage process.

A developing and more comprehensive form of impact assessment is Strategic Environmental Assessment (SEA). SEA assesses the wider impacts of any development, taking account of interactions with other forms of development and the wider environment.

15.6.2 Ecolabelling

Consumer choice can have a powerful influence on the use of the marine environment as demonstrated by the boom in sales of '**dolphin friendly**' or '**dolphin safe**' tuna when the public saw the first pictures of dolphin kills in tuna purse seine net fisheries. Today, increasingly detailed information is available on the provenance of wild caught fish.

● The choices made by consumers may influence the behaviour of fishers and other users of the marine environment.

In Great Britain, the Marine Conservation Society publishes a 'Good Fish Guide' to help consumers who like to eat seafood but are concerned about the impacts of fishing on fish stocks, marine wildlife, and habitats to choose fish that come from sustainably managed fisheries that minimize damage to the marine environment and do not harm other wildlife.

Internationally, the Marine Stewardship Council (MSC) runs a scheme for assessing and certifying fisheries as sustainable. The MSC was founded in 1996 as a joint venture between the World Wide Fund for Nature and Unilever, became an independent charitable organization in 1999, and is now funded primarily by the Packard Foundation. Products bearing the MSC label have won the patronage of royalty and celebrity chefs alike, and retail giants are beginning to put MSC labelled products on their shelves, with c. 200 different products currently on offer. Fisheries certified under the MSC scheme include the Western Australian Rock Lobster, New Zealand Hoki, and Alaska Salmon. The certification process is conducted by a panel of independent scientists who assess the fishery against criteria for the status of the stock, ecosystem impacts due to fishing, and the management systems in place (Philipps et al. 2003).

● Several groups now certify or recommend fishery products from sustainable fisheries. See **www.mcsuk.org** and **www.msc.org**.

With commitment to improving fishing practices by the industry, qualification for the MSC label should be relatively straightforward for fisheries where the choices of how much fish to catch and how to catch it are under the control of a single group of fishers. It will be in their interest to invest in long-term sustainability and in labelling that will attract higher prices. However, as we saw in Chapter 12, the majority of the world's fisheries are not like this, and most fishers have little control over the setting of fishing quotas and share the fish resource with other fishers and nations who may not choose to fish responsibly and are caught up in the race to fish. Fishers who participate in these fisheries are effectively excluded from even considering MSC certification due to the actions of others beyond their control. Thus the MSC scheme focuses very much on consumer choice as the driver for change in fisheries. If fishers have little capacity to change their practices, then potential demand for MSC products may soon outstrip supply (Kaiser and Edwards-Jones 2006).

● If fishers have little capacity to change their practices, demand for ecolabelled products could outstrip supply.

All certification schemes rely heavily on value judgement, and the organizers make significant efforts to develop clear criteria for assessing sustainability and to find unbiased experts to do the assessments. Nevertheless, people differ widely in their views about acceptable human impacts in the marine environment, and one person's conservation disaster may be seen as sustainable and acceptable by others.

15.6.3 Marine reserves

Reserves, where human activity is spatially controlled or banned, have useful roles to play in marine conservation and there has been an upsurge of interest in them, often stimulated by growing appreciation of the effects of fishing on habitats and species. It is notable that $<1\%$ of the marine environment has reserve status as opposed to 6% of land, and this alone suggests that marine reserves need to be more extensive than they are. No reserve can be fully effective at protecting marine life if ships can spill oil and other toxic substances in the vicinity, if currents feed contaminants to the animals living there, if animals migrate from the reserve and can be affected by fishing or pollution elsewhere, and if climate change drives the protected species to other locations. The proper solution to these problems lies in the effective regulation of the human activities involved. Thus far, the greatest progress in tackling large-scale issues in marine conservation has involved tackling the impacts directly. For example, by stopping whalers killing whales, preventing yachts using TBT on their hulls, and controlling discharges of heavy metals and radioactivity into the marine environment (Chapter 14). With some of these activities controlled, and moves towards greater limitation of fishing effort, now is a good time to start using reserves more widely to support marine conservation.

● Marine reserves protect a very small proportion of the marine environment. In the open marine environment, the use of reserves for conservation must go hand in hand with control of human activity outside the reserve.

One of the success stories in marine reserve conservation is that of the large Great Barrier Reef Marine Park in north-eastern Australia (Box 15.4). However, many marine reserves are small and vulnerable to climatic events, land-derived pollution and run-off, and poaching.

Small reserves on and around coral reefs have provided most of the evidence for the potential conservation benefits of reserves. Most important has been the demonstration that when protection is effective, many site-attached reef fish species and invertebrates attain greater densities and sizes than in areas where they are fished. This may be important to fisheries because these species may produce larvae that

● Small reserves can provide conservation benefits, but are easily affected by external factors.

Box 15.4 The Great Barrier Reef Marine Park

The Great Barrier Reef (GBR) Marine Park is an example of one of the first attempts to conserve and manage at the ecosystem scale. The GBR Marine Park Act (1975) underpins the conservation of the GBR, providing for conservation of the GBR and sustainable use of the surrounding region. The main tool for protecting and preserving the GBR, as required by the Act, is zoning. Zoning separates conflicting human activities and protects the most vulnerable areas. The Park is managed by the Great Barrier Reef Marine Park Authority (GBRMPA), and the GBR was recognized as a World Heritage Site in 1981.

The GBR Marine Park is divided into four sections for management purposes (Figure a), and the challenge for managers is deciding who can do what and where in each of the sections. Zoning plans for each section (e.g. Figure b) provide for activities that are as-of-right, with permission or prohibited, and widely available maps ensure that all potential users are aware of restrictions on their movements and activities.

The least restrictive zones are 'general use' in which activities such as shipping and trawling may be permitted, but mining, oil drilling, commercial spear-fishing, and spear-fishing with breathing apparatus are not. Other zones provide for habitat protection, estuarine conservation, or general conservation, while permitting restricted commercial or recreational use. The most restrictive zones are 'marine national park zones', which are no-take 'areas'; 'scientific research zones', which only allow access for agreed scientific research; and 'preservation zones' where all entry is prohibited except in emergencies and to conduct scientific research that would be impossible elsewhere. In 2004, a new zoning plan was introduced to incorporate 28 new coastal protected areas and a Representative Areas Program, which ensures that examples of the entire range of habitats and biodiversity are represented within highly protected areas such as the Marine National Park zones.

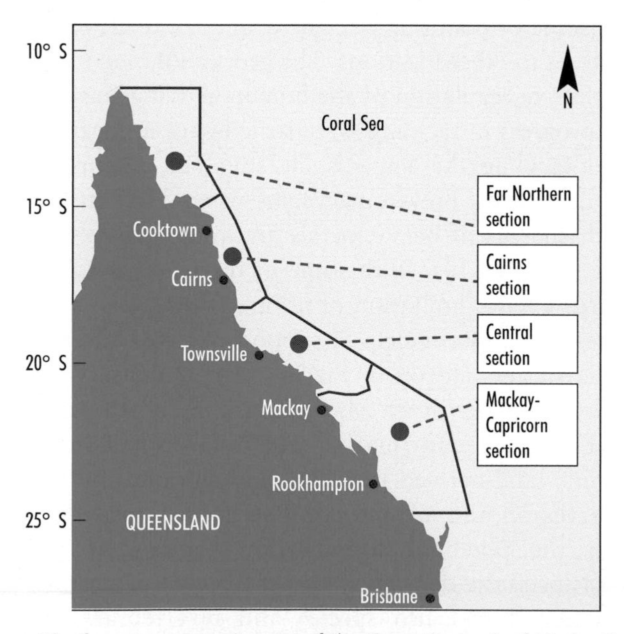

Fig. a The four management areas of the Great Barrier Reef Marine Park.

continues

BOX 15.4 continued

	Central use zone	Habitat protection zone	Conservation park zone	Buffer zone	Scientific research zone	Marine national park zone	Preservation zone	State zoning only	Estuarine conservation zone
Aquaculture	Permit	Permit	Permit					State zoning only	Permit
Bait netting									
Boating, diving, photography									
Crabbing (trapping)									
Harvest fishing for aquarium fish, coral and beachworm	Permit	Permit	Permit						
Harvest fishing for sea cucumber, Irochus, tropical rock lobster	Permit	Permit							
Limited collecting									
Limited spear fishing (snorkel only)									
Line fishing									
Netting (other than bait netting)									
Research (other than limited impact research)	Permit	Permit	Permit	Permit	Permit	Permit	Permit		Permit
Shipping (other than in a designated shipping area)		Permit	Permit	Permit	Permit	Permit			Permit
Tourism Program	Permit	Permit	Permit	Permit	Permit	Permit			Permit
Traditional use of marine resources									
Trawling									
Trolling									

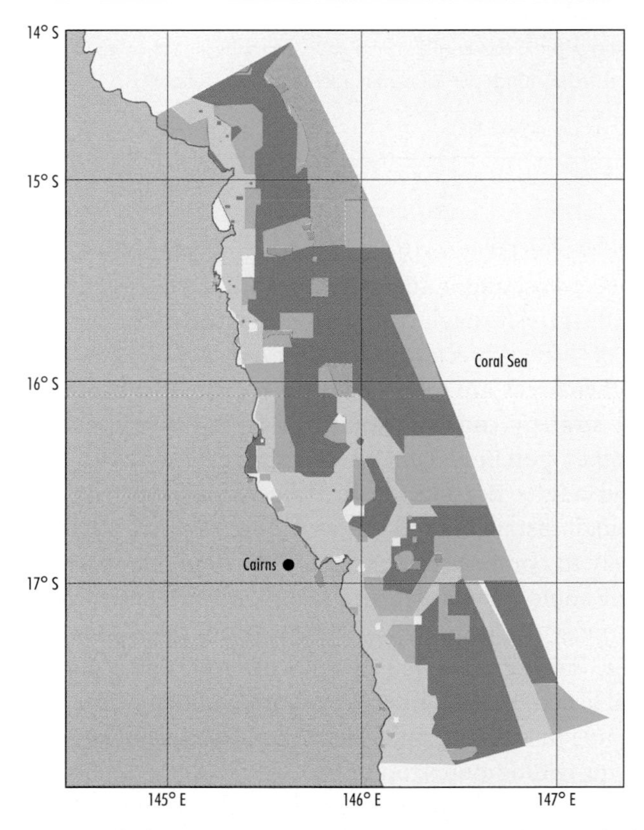

Fig. b Zoning of the Cairns/Cooktown Management Area of the Great Barrier Reef Marine Park with a key to permitted activities.

For further details see the Great Barrier Reef Marine Park Authority website: **www.gbrmpa.gov.au**.

Box 15.5 Paying for marine conservation

Policing Marine Protected Areas (MPAs) is an expensive business and governments rarely provide adequate funding. Given that the policing and management of a small MPA requires several staff and around €100 000 per year, income from recreational divers that use MPAs could make an important financial contribution to effective marine conservation.

Over 200 MPAs in the Caribbean and Central America contain coral reefs that are likely to attract recreational scuba-divers. Indeed, a survey of MPA use by dive operators showed that 46% conduct at least 80% of their diving in a MPA.

At present, only 25% of MPAs with coral reefs charge divers an entry or user fee, and this is typically €2–3 per dive or diver. However, divers have already shown willingness to pay up to €25 per day to access some MPAs, and this is only a fraction of their total expenditure on equipment, training, accommodation, and entertainment.

The wider Caribbean and Central America is a prime location for diving and attracts 57% of all international scuba-diving tourists. Provided the recreational use of MPAs can remain consistent with the goals of conservation and sustainable development, higher charges for diving access to all MPAs would help secure much needed funding for MPA protection and management. Indeed, if user fees were raised to around €25 per diver this would raise approximately €93 million per year, sufficient to cover some 78% of the predicted shortfall in funds for MPA management in this region.

Source: Green and Donnelly (2003)

will support recruitment to fished areas, may themselves migrate out into the fished areas, and in the case of rotational closure may ultimately be caught directly. Although there is some evidence for rapid build-up of abundance of fished species soon after protection, this is very soon dissipated when areas are reopened, thus rotational closure seems not to be a useful strategy, unless fishing effort is otherwise curbed. With respect to other potential benefits of marine reserves to fisheries, late juveniles and adults do migrate to fished areas, but the movement is often over small distances, and unless fishing effort is otherwise reduced this is unlikely to compensate for the loss of yield due to protection. The protection of adult fishes to ensure larval recruitment to fished areas is perhaps the most useful potential benefit of marine reserves to sustainable fisheries, but regrettably, it is the least well understood.

● Marine reserves often attract tourists and significant tourist revenue.

● Marine reserves provide recreational and educational opportunities.

Once established, marine reserves can undoubtedly have an important educational role in drawing public attention to areas of special ecological significance, providing opportunities to see relatively undisturbed marine habitats, and encouraging people to observe the benefits of conservation.

The Great Barrier Reef Marine Park and some other reserves around the world also show that marine reserves can be very successful economically (Box 15.5). This is important, as economic success is likely to foster community support and good enforcement, both of which are paramount to successful management. On many coral reefs, diving tourism is a key source of income to support reserve management. However, not all areas benefit from tourist expenditure and, as with other conservation tools, there are socio-economic constraints to the effective management of marine reserves. Some of the best studied marine reserves are at Sumilon and Apo Islands in the central Philippines, and Russ and Alcala (1996) have described 20 years of hopes and frustrations on Sumilon Island. A marine reserve was established there in 1974, but there have been several major breakdowns of management. Remarkably, this was at a site where the aggregate fisheries yield from the island was higher when the marine reserve was operating (Russ & Alcala 1996). One of the key problems with management on the island seemed to be that fishers were not convinced of the real benefits of the marine reserve and were not fully involved in its management. There are other examples of repeated breakdown of reserves on tropical reefs. Together they indicate that management imposed from outside the community and without the support of the community is unlikely to work.

● Reserves will tend to be more effective when local people support them.

Given the small areas that are designated as reserves in the marine environment and the key role that they can play in protecting species and habitats of conservation concern, they should be used more widely to support conservation objectives. However, the designation and effective protection of marine reserves must go hand-in-hand with direct control of overall human impacts (Fig. 15.4).

Fig. 15.4 This sign at Tai O fishing village in Hong Kong informs people that they can still fish, but only if they do it in an environmentally sustainable manner. The use of indiscriminate and destructive fishing practices is illegal and fishers are subject to heavy fines and a prison sentence if they ignore the warnings. (Photo copyright: Johanna Junback.)

In this section on conservation in action, our examples have shown that, because of the multiple concerns, the complexity of the issues and interdisciplinary demands of decision-makers, conservation action is as much art as science. The uncertainty of environmental decision-making was well illustrated by the concerns for deep-sea pollution raised by the disposal of the *Brent Spar* oil platform (Gage & Gordon 1995, Angel & Rice 1996). Shell abandoned plans to dispose of this redundant oil storage platform by sinking it in water 2200 m deep in the north-eastern Atlantic. Shell had been given scientific advice that deep-sea disposal was the best environmental option, but environmental groups protested that toxic residues would leak from the platform and contaminate the food-chain. Greenpeace boarded the platform in a bid to prevent its sinking and eventually the *Brent Spar* was towed to a fjord in Norway to be dismantled. It may well be that the environmental costs of disposing of the platform inshore have been greater than those of deep-sea disposal and the arrival of the platform inshore also caused controversy in Norway. Scientists have continued to debate the pros and cons of the various disposal options, and yet the desire of Greenpeace to make an issue of the disposal and the ensuing campaigns at fuel stations in Germany effectively forced the final disposal option.

● The environmental costs of disposing of the *Brent Spar* oil platform inshore may have been greater than those of disposal in the deep sea. This issue started a lively scientific debate.

15.7 The Future

Every contemporary conservation action takes place in a marine environment that has already been dramatically altered by human activities. Some of the biggest impediments to successful conservation are the massive changes to marine habitats and species that have already occurred and the emphasis on reactive rather than proactive management. However, awareness of marine conservation issues is growing and many governments have made international commitments to sustainable development and the ecosystem approach. Few of these international commitments have yet to be translated into successful conservation action, but there are some encouraging examples that suggest combined controls on human impacts and large-scale marine zoning schemes, which include marine reserves, could lead to long-term improvements in sustainability. The wealthy developed world currently has more capacity to implement conservation measures and yet many poorer nations are responsible for species-rich and productive marine environments. In many poorer nations, prospects for improved sustainability are not good, unless the international community commits to supporting and financing the ecosystem approach and subsidizing the very high short-term social and economic costs associated with moving towards sustainability.

● Current conservation actions take place in a marine environment that has already been dramatically altered by human activities.

CHAPTER SUMMARY

- Conservation is the sustainable management of the marine environment. Sustainable management requires human intervention to maintain or create an environment that ensures that the wellbeing of future generations is not compromised.
- National and international agreements express an aspiration to move towards sustainable development, and these underpin much of the conservation and environmental protection legislation that we see today.
- There are ethical, ecological and economic reasons to conserve but short-term economic forces often drive unsustainable development despite the long-term economic benefits of conservation.
- Recent conservation policy reflects a high-level aspiration to achieve sustainable development. However, converting high-level aspirations into practical, funded and effective conservation actions remains a challenge.
- Sustainable development includes not only biological and ecological considerations, but also involves a consideration of the consequences of actions (e.g. conservation) for local economies, and social and cultural structures.
- Conservation actions involve many levels of ecological (species, habitats, ecosystems) and administrative (local to international) organisation. Their success depends on the wealth and cultural values of society.

FURTHER READING

Kunin and Lawton (1996) provide a holistic analysis of the importance of biodiversity and the necessity for conservation. Norse (1993) provides a conservationist's point of view of the integration of conservation in decision making, while Sutherland (1998) provides an objective view of the role of conservation with case studies.

- Kunin, W. E. & Lawton, J. H. 1996. Does biodiversity matter? Evaluating the case for conserving species. In K. J. Gaston (ed.) *Biodiversity: a Biology of Numbers and Difference*. Blackwell Science, Oxford, pp. 283–308.
- Norse, E. A. (ed.) 1993. *Global Marine Biological Diversity: a Strategy for Building Conservation into Decision Making*. Center for Marine Conservation, Washington.
- Sutherland, W. J. (ed.) 1998. *Conservation Science and Action*. Blackwell Science, Oxford.

WEBLINKS

Below we have provided a selection of weblinks that will help you gain a more in–depth understanding of some of the issues we have covered in *Marine Ecology: Processes, Systems and Impacts*. More importantly these websites open the door to specific regional examples of marine ecology at work, current issues and data that you can download and use. Weblinks evolve all the time, so while we have tried our best to ensure all of these links are current some will inevitably change. If they have changed, typing some of the key words in the description text will probably take you relatively quickly to the new location. Alternatively, visit the book's online resource centre at www.oxfordtextbooks.co.uk/orc/kaiser/ to access up-to-date URLs for all links given here.

Antarctic & Arctic: There are many websites dealing with Artic and Antarctic issues. However, by going to the following websites it is possible to follow links to this wealth of information and direct links to the organizations at which Polar research is conducted:

http://www.antarctica.ac.uk
http://www.arcus.org/index.html
http://nsidc.org/index.html
http://www.scar.org

Antarctic living resources: This is the website of the Commission for the Conservation of Antarctic Marine Living Resources. http://www.ccamlr.org/pu/e/gen-intro.htm. It contains information on Southern Ocean ecosystems and efforts to take an ecosystem approach to management of fishery resources there.

Antarctic spring phytoplankton bloom: Part of the NASA Goddard Space Centre's web site, this provides images of the seasonal development of the spring phytoplankton bloom in the north Atlantic. It also provides information on patchiness and some of the problems this presents for sampling. http://daac.gsfc.nasa.gov/CAMPAIGN_DOCS/OCDST/nab.html

Aquaculture: The Food and Agriculture Organisation of the United Nations compiles and publishes downloadable data and figures on the cultivation of aquatic organisms around the world http://www.fao.org. To find out more about aquaculture research visit the University of Stirling, Institute of Aquaculture website http://www.aquaculture.stir.ac.uk.

Benthos: The virtual handbook written by Tom Brey of the Alfred Wegner Institute is a mine of information regarding benthic ecology and production processes: this can be cited as: T. Brey, 2001. Population dynamics in benthic invertebrates. A virtual handbook. Version 01.2. http://www.awi-bremerhaven.de/Benthic/Ecosystem/FoodWeb/Handbook/main.html Alfred Wegener Institute for Polar and Marine Research, Germany.

Biodiversity research: DIVERSITAS is a major international programme dealing with biodiversity and ecosystem processes, as well as the links between ecosystem services and society. Many marine and intertidal aspects are included in this programme http://www.diversitas.org/.

Biodiversity and ecosystem function: A major website holding up-to-date literature on marine biodiversity and ecosystem functioning can be found at http://www.abdn.ac.uk/ecosystem.

Bioluminescence webpage: http://www.lifesci.ucsb.edu/~biolum/. A site full of excellent information on the biology of bioluminescence with pictures and video clips.

Carbon cycle: For a general introduction to the marine carbon cycle and the links between chemistry and biology and global climate change see: http://calspace.ucsd.edu/virtualmuseum/climatechange1/06_3.shtml.

Climate change: A wealth of information about global climate change can be found at the website of the Intergovernmental Panel on Climate Change: http://www.ipcc.ch/. The North Atlantic Oscillation Website, http://www.ldeo.columbia.edu/NAO/, provides everything you wanted to know about this climate pattern, describing the underlying mechanism, plotting data and linking to other useful NAO web resources.

Coccolithophorids: For a detailed introduction into the biology, biogeochemistry and geology of this important group of phytoplankton go to: http://www.soes.soton.ac.uk/staff/tt/eh/index.html.

Consequences of climate change: A global analysis of the current and predicted effects of climate warming can be found at http://www.ipcc.ch. This site contains lots of informative reports and illustrations of the predicted effects of global warming.

Conservation: The website of the Society for Conservation Biology http://conbio.net/scb/ provides up-to-date access to key issues affecting the conservation of all natural resources. The Marine Conservation Society of the United Kingdom http://www.mscuk.org and Marine Conservation Biology Institute USA http://www.mcbi.org provide lots of useful links to other organizations, conservation work opportunities and projects and scientific information on current topical issues.

Coral reefs: For a global perspective on coral reefs see http://www.reefbase.org and for the International Society for Reef Studies webpage go to: http://www.fit.edu/isrs.

Deep-Sea: NOAA Vents programme. http://www.pmel.noaa.gov/vents/home.html. A site dedicated to information and research on the geology and biology of hydrothermal vents, with lots of pictures and video clips.

Detecting change in biological communities: For information on detecting change in communities and key considerations in experimental design, see the manual on offer at http://www.primer-e.com.

Diatoms: A key resource for anyone wanting to learn more about the diatoms is the International Society for Diatom Research, which includes links to many other diatom websites: http://www.isdr.org/.

Estuaries: The website of the Estuarine Research Federation http://www.erf.org/ provides information from the world's largest estuarine science organization, including publications, education and links.

European Network of Excellence: MarBEF is a major network of European marine ecologists that involves over 80 different institutes. This is an excellent starting point for finding out what is going on in Europe and an ideal starting point for finding placement work opportunities or employment http://www.marbef.org/outreach/whereis/intertidal.html.

Evolution & Diversity: The Tree of Life Web Project (ToL) is a collaborative effort of biologists from around the world. On more than 3000 world wide web pages, the project provides information about the diversity of organisms on Earth, their evolutionary history (phylogeny), and characteristics. http://tolweb.org/tree/.

Exploitation of non-biological resources and renewable energy: For exploitation of non-renewable offshore resources and alternative forms of energy generation such as windfarms see http://www.thecrownestate.co.uk. This website has links to reports and information regarding the amount of material removed from the seabed and the potential of wind and wave energy to meet future energy requirements.

Flagellates: A website dealing with flagellates that are important in microbial processes in marine systems, and well introduced at http://tolweb.org/notes/?note_id=50.

Fisheries: Up-to-date global fisheries statistics including biological and economic information is available through http://www.fao.org. Regional information for Europe is available at http://www.ices.dk where it is possible to download a database of European fisheries statistics. Current research in the U.S. can be accessed through http://www.nmfs.noaa.org which is a highly informative website with respect to current issues in fisheries and gives access to free to use photographic images. A particularly informative industry-run website can be found at http://www.fishingnj.org, which gives the fishers' angle on current issues. A more commercial perspective is given at http://www.seafish.co.uk where you will also find excellent seafood recipes. Other informative research websites include those of the Centre for Environment, Fisheries, and Aquaculture Science http://www.cefas.co.uk, Fisheries Research Services Aberdeen http://www.marlab.ac.uk, the Australian Institute of Marine Science http://www.aims.gov.au. Visit the Worldfish Center website for more information about Asian and African fisheries and projects http://www.worldfishcenter.org.

Habitat listing: The MarLIN database held at the Marine Biological Association of the United Kingdom gives access to a wealth of information about intertidal and subtidal coastal habitats with links to primary and 'grey' literature that is invaluable for learning and research http://www.marlin.ac.uk.

Harmful algal blooms: Information about algal blooms and harmful algal blooms, red tides and algal toxicity can be found at: http://www.whoi.edu/science/B/redtide/ as well as http://www.bigelow.org/hab/.

Longterm data sets: There are number of long-term data oceanographic data sets for which biological, chemical and physical data are collected to examine seasonal, interannual and even decadal variations:
Hawaii Ocean Time-series (HOTS) – http://hahana.soest.hawaii.edu/hot/hot_jgofs.html
Bermuda Atlantic Time-series (BATS) – http://bbsr.edu/cintoo/bats/bats.html
Monterey Bay time-series study – http://www.mbari.org/bog/Projects/centralcal/summary/ts_summary.htm
VOS Underway pCO2 Programm – http:www.pmel.noaa.gov/co2/uwpco2/
Continuous Plankton Recorder Survey – http: //192.171.163.165/cpr_survey.htm.

Mangroves: For mangroves, see http://www.ncl.ac.uk/tcmweb/tcm/mglinks.htm. A site listing most of the major websites dealing with mangroves and wetlands.

Maps of major marine environments:
http://www.oceansatlas.org/cds_static/en/coastal_marine_habitats__en_2384_all_1.html
http://www.teachers.ash.org.au/ jmresources/marine/environments.html

Marine Institutes: There are a number of websites that give extensive listings of places around the world where marine research takes place:
http://oceanlink.island.net/career/careerlinks.html
http://www.skio.peachnet.edu/resources/marinelinks.php
http://www.sams.ac.uk/activities/web%20links/links.htm
http://www.eurocean.org/categories.php?category_no = 13

Marine systems: GLOBEC http://www.pml.ac.uk/globec/ is the International Geosphere-Biosphere Programme (IGBP) core project responsible for understanding how Global Change will affect the abundance, diversity and productivity of marine populations. This web site gives summaries of recent GLOBEC-related work and access to data.

Meiofauna: The web site of the International Association of Meiobenthologists and all you wanted to know about meiofauna http://www.meiofauna.org/.

Microbial loop: Background information about the microbial loop and why it is important can be found at: http://www.bigelow.org/bacteria/ as well as http://www.uib.no/ums/magazine/updates/Loopmi/loopmi.htm.

Satellite imagery: The Sea-viewing Wide Field-of-view Sensor (SeaWiFS) Project http://seawifs.gsfc.nasa.gov/SEAWIFS.html uses satellite observations to provide quantitative data on global ocean bio-optical properties. This website provides an overview of the project, has some excellent summary maps and even enables you to produce mapped globes from a perspective of your choice.

Seagrasses: WCMC Global Seagrass Database. http://www.unep-wcmc.org/marine/seagrassatlas/. Based around the world atlas of seagrasses, the website gives some introductory information, but most interestingly has a series of on-line maps showing the distribution of seagrass species around the world.

Seaweeds or macroalgae: Probably the best starting point for any seaweed-related enquiries. http://www.seaweed.ie. Algaebase http://algaebase.org has details on 57000 algal species, 1500 images, 33000 bibliographic items, 104000 distributional algal records, and a 27000-word on-line glossary.

Viruses: This website gives a good general overview of aquatic viruses and current research topics in the field, as well as further weblinks: http://www.uib.no/ums/magazine/updates/Viruses/viruses.htm.

REFERENCES

Abram, N. J., Gagan, M. K., McCuloch, M. T., Chappell, J. & Hantoro, W. S. 2003. Coral reef death during the 1997 Indian Ocean Dipole linked to Indonesian wildfires. *Science* 301: 952–5.

Acevedo-Gutierrez, A., Croll, D. A. & Tershy, B. R. 2002. High feeding costs limit dive time in the largest whales. *Journal of Experimental Biology* 205: 1747–53.

Adam, P. 1993. *Saltmarsh Ecology*. Cambridge University Press, Cambridge.

Aebischer, N. J., Coolson, J. C. & Colebrook, J. M. 1990. Parallel long-term trends across four marine trophic levels and weather. *Nature* 347: 753–5.

Agardy, T. 2000. Effects of fisheries on marine ecosystems: a conservationist's perspective. *ICES Journal of Marine Science* 57: 761–5.

Aguilera, J., Bischof, K., Karsten, U., Hanelt, D. & Wiencke, C. 2002. Seasonal variation in ecophysiological patterns in macroalgae from an Arctic fjord: II. Pigment accumulation and biochemical defence systems. *Marine Biology*. 140: 1087–95.

Ainley, D. G., Tynan, C. T. & Stirling, I. 2003. Sea ice; A critical habitat for polar marine mammals. In D. N. Thomas & Dieckmann, G. S. (eds), *Sea Ice – An Introduction to its Physics, Chemistry, Biology and Geology*, Blackwell, Oxford.

Aioi, K., Komatsu, T. & Morita, K. 1998. The world's longest seagrass, *Zostera caulescens* from northern Japan. *Aquatic Botany* 61: 87–93.

Aksnes, D. L. & Giske, J. 1993. A theoretical-model of aquatic visual feeding. *Ecological Modelling* 67: 233–50.

Allen, R. L. 1985. Dolphins and the purse-seine fishery for yellowfin tuna. In J. R. Beddington, R. J. H. Beverton & D. M. Lavigne (eds), *Marine mammals and fisheries*, George Allen & Unwin, London, pp. 236–52.

Aller, J. Y. & Stupakoff, I. 1996. The distribution and seasonal characteristics of benthic communities on the Amazon shelf as indicators of physical processes. *Continental Shelf Research*, 16: 717–51.

Allsopp, W. H. L. 1997. World aquaculture review: performance and perspectives. In E. L. Pikitch, D. D. Huppert & M. P. Sissenwine (eds), *Global trends: fisheries management*, American Fisheries Society Symposium, Bethesda, Maryland, pp. 153–65.

Alongi, D. M. 1998. *Coastal Ecosystem Processes*. CRC Press, Cambridge MA.

Alverson, D. L., Freeberg, M. H., Pope, J. G. & Murawski, S. A. 1994. A global assessment of fisheries bycatch and discards. *FAO Fisheries Technical Paper* 339: 233 pp.

Anderson, P. & Sorensen, H. M. 1986. Population dynamics and trophic coupling in pelagic microorganisms in eutrophic coastal waters. *Marine Ecology Progress Series* 33: 99–109.

Angel, M. & Rice, A. 1996. The ecology of the deep ocean and its relevance to global waste management. *Journal of Applied Ecology* 33, 915–26.

Arcos, J. M., Ruiz, X., Bearhop, S. & Furness, R. W. 2002. Mercury levels in seabirds and their fish prey at the Ebro Delta (NW Mediterranean): the role of trawler discards as a source of contamination. *Marine Ecology Progress Series* 232: 281–90.

Arístegui, J., Agustí, S., Middelburg, J. J. & Duarte, C. M. 2005. Respiration in the Mesopelagic and Bathypelagic Zones of the Oceans. In P. A. del Giorgio & P. J. le B. Williams (eds), *Respiration in Aquatic Ecosystems*, Oxford University Press, Oxford, pp. 181–205.

Arntz, W. E., Brey, T. & Gallardo, V. A. 1994. Antarctic zoobenthos. *Oceanography and Marine Biology, An annual review* 32: 241–304.

Arntz, W. E., Gutt, J. & Klages, M. 1997. Antarctic marine biodiversity: an overview. In B. Battaglia (ed.), *Antarctic communities: species, structure and survival*, Cambridge University Press, Cambridge, pp. 3–14.

Aronson, R. B., Macintyre, I. G., Wapnick, C. M. & O'Neill, M. W. 2004. Phase shifts, alternative states, and the unprecedented convergence of tow reef systems. *Ecology* 85: 1876–91.

Atkinson, A., Seigel, V., Pakhomov, E. & Rothery, P. 2004. Long-term decline in krill stock and increase in salps within the Southern Ocean. *Nature* 432: 100–3.

Attrill, M. J. 2002. A testable linear model for diversity trends in estuaries. *Journal of Animal Ecology* 71: 262–9.

Attrill, M. J. & Power, M. 2002. Climatic influence on a marine fish assemblage. *Nature* 417: 275–8.

Attrill, M. J. & Power, M. 2000a. Modelling the effect of drought on estuarine water quality. *Water Research* 34: 1584–94.

Attrill, M. J. & Power, M. 2000b. Effect on invertebrate populations of drought induced changes in estuarine water quality. *Marine Ecology Progress Series* 203: 133–143.

Attrill, M. J. & Rundle, S. D. 2002. Ecotone or ecocline: ecological boundaries in estuaries. *Estuarine and Coastal Shelf Science* 55: 929–36.

Attrill M. J., Hartnoll R. G. & Thurston M. H. 1990. A depth-related distribution of the red crab, Geryon trispinosus (Herbst) [=G. tridens Kroyer]: indications of vertical migration. *Progress in Oceanography* 24: 197–206.

Attrill, M. J., Power, M. & Thomas, R. M. 1999. Modelling estuarine Crustacea population fluctuations in response to physico-chemical trends. *Marine Ecology Progress Series* 178: 89–99.

Attrill, M. J., Rundle, S. D. & Thomas, R. M. 1996. The influence of drought-induced low freshwater flow on an upper-estuarine macroinvertebrate community. *Water Research* 30: 261–8.

Attrill, M. J., Stafford, R. & Rowden, A. A. 2001. Latitudinal diversity patterns in estuarine tidal flats: indications of a global cline. *Ecography* 24: 318–24.

Attrill, M. J., Strong, J. A. & Rowden, A. A. 2000. Are macroinvertebrate communities influenced by seagrass structural complexity? *Ecography* 23: 114–21.

Auster, P. J. & Langton, R. W. 1999. The effects of fishing on fish habitat. In L. Benaka (ed.), *Fish habitat: essential fish habitat and restoration*, American Fisheries Society, Bethesda, Maryland, pp. 150–87.

Auster, P. J., Lindholm, J., Schaub, S., Funnell, G., Kaufman, L. S. & Valentine, P. C. 2003. Use of sand wave habitats by silver hake. *Journal of Fish Biology* 62: 143–52.

Auster, P. J., Malatesta, R. & Donaldson, C. 1997. Distributional responses to small-scale habitat variability by early juvenile silver hake, Merluccius bilinearis. *Environmental Biology of Fishes* 50: 195–200.

Azam, F. 1998. Microbial Control of Oceanic Carbon Flux: the Plot Thickens. *Science* 280: 694–6.

Azam, F. & Worden, A. Z. 2004. Microbes, molecules, and marine ecosystems. *Science* 303: 1622–4.

Babin, M., Morel, A., Fournier-Sicre, V., Fell, F. & Stramski, D. 2003. Light scattering properties of marine particles in coastal and open ocean waters as related to the particle mass concentration. *Limnology and Oceanography* 48: 843–59.

Backhaus, J. O., Hegseth, E. N., Wehde, H., Irigoien, X., Hatten, K. & Logemann, K. 2003. Convection and primary production in winter. *Marine Ecology Progress Series* 251: 1–14.

Baco, A. R. & Smith, C. R. 2003. High species richness in deep-sea chemoautotrophic whale skeleton communities. *Marine Ecology Progress Series* 260: 109–14.

Bagoien, E., Kaartvedt, S., Aksnes, D. L. & Eiane, K. 2001. Vertical distribution and mortality of overwintering Calanus. *Limnology and Oceanography* 46: 1494–510.

Baird, D., Evans, P. R., Milne, H. & Pienkowksi, M. W. 1985. Utilisation by shorebirds of benthic invertebrate production in intertidal areas. *Oceanography and Marine Biology: Annual Review* 23: 575–97.

Baker, S. M. 1909. On the causes of the zoning of brown seaweeds on the seashore, II. *New Phytologist* 9: 54–67.

Bakun, A. 1990. Global climate change and intensification of coastal upwelling. *Science* 247: 198–201.

Ball, M. C. & Pidsley, S. M. 1995. Growth responses to salinity in relation to distribution of two mangrove species, *Sonneratia alba* and *S. lanceolata*, in northern Australia. *Functional Ecology* 9: 77–85.

Balmford, A., Bruner, A., Cooper, P., Costanza, R., Farber, S., Green, R. E., Jenkins, M., Jefferiss, P., Jessamy, V., Madden, J., Munro, K., Myers, N., Naeem, S., Paavola, J., Rayment, M., Rosendo, S., Roughgarden, J., Trumper, K. & Turner, R. K. 2002. Economic reasons for conserving wild nature. *Science* 297: 950–3.

Baran, E. & Hambrey, J. 1998. Mangrove conservation and coastal management in Southeast Asia: what impact on fishery resources? *Marine Pollution Bulletin* 37: 431–40.

Barnes, D. K. A. 2002. Biodiversity – Invasions by marine life on plastic debris. *Nature* 416: 808–9.

Barnes, D. K. A. 2002. Polarisation of competition increases with latitude. *Proceedings of the Royal Society of London B* 1504: 2061–9.

Barnes, D. K. A. 2003. Competition asymmetry with taxon divergence. *Proceedings of the Royal Society of London B* 270: 557–62.

Barnes, D. K. A. & Brockington, S. 2003. Zoobenthic biodiversity, biomass and abundance at Adelaide Island, Antarctica. *Marine Ecology Progress Series* 249: 145–55.

Barnes, D. K. A. & Dick, M. H. 2000. Overgrowth competition between clades: Implications for interpretation of the fossil record and overgrowth indices. *Biological Bulletin* 199: 85–94.

Barnes, R. S. K. & Hughes, R. N. 1999. *An Introduction to Marine Ecology*. Blackwells Publishing, Oxford.

Barnes, R. S. K. 1989. What, if anything, is a brackish water fauna? *Transactions of the Royal Society of Edinburgh, Earth Sciences* 80: 235–40.

Barnes, R. S. K. 1994. *The Brackish-water Fauna of NW Europe*. Cambridge University Press, Cambridge.

Barrow, C. J. 1997. *Environmental and Social Impact Assessment: an Introduction*. Edward Arnold, London.

Bathmann, U., Scharek, R., Klass, C., Dubischer, C. D. & Smetacek, V. 1997. Spring development of phytoplankton biomass and composition in major water masses of the Atlantic sector of the Southern Ocean. *Deep-Sea Research* 44: 51–67.

Baum, J. K., Myers, R. A., Kehler, D. G., Worm, B., Harley, S. J. & Doherty, P. A. 2003. Collapse and conservation of shark populations in the northwest Atlantic. *Science* 299: 389–92.

Beardsley, R. C., Epstein, A. W., Chen, C. S., Wishner, K. F., Macaulay, M. C. & Kenney, R. D. 1996. Spatial variability in zooplankton abundance near feeding right whales in the Great South Channel. *Deep-Sea Research Part II – Topical Studies in Oceanography* 43: 1601–25.

Beaugrand, G., Brander, K. M., Lindley, J. A., Souissi, S. & Reid, P. C. 2003. Plankton effect on cod recruitment in the North Sea. *Nature* 426: 661–4.

Beaugrand, G., Reid, P. C., Ibanez, F., Lindley, J. A. & Edwards, M. 2002. Reorganization of North Atlantic marine copepod biodiversity and climate. *Science* 296: 1692–4.

Béjà, O., Suzuki, M. T., Heidelberg, J. F., Nelson, W. C., Preston, C. M., Hamada, T., Eisen, J. A., Fraser, C. M. & DeLong, E. F. 2002. Unsuspected diversity among marine aerobic anoxygenic phototrophs. *Nature* 415: 630–3.

Bellwood, D. R. & Hughes, T. P. 2001. Regional-scale assembly rules and biodiversity of coral reefs. *Science* 292: 1532–4.

Bellwood, D. R., Hughes, T. P., Folke, C. & Nyström, C. 2004. Confronting the coral reef crisis. *Nature* 429: 827–33.

Benfield, M. C., Davis, C. S., Wiebe, P. H., Gallager, S. M., Lough, R. G. & Copley, N. J. 1996. Video Plankton Recorder estimates of copepod, pteropod and larvacean distributions from a stratified region of Georges Bank with comparative measurements from a MOCNESS sampler. *Deep-Sea Research Part II – Topical Studies in Oceanography* 43: 1925–45.

Benner, R. 2002. Chemical Composition and Reactivity. In D. A. Hansell & C. A. Carlson (eds), *Biogeochemistry of Marine Dissolved Organic Matter*, Academic Press, Amsterdam, pp. 59–90.

Bentamy, A., Grima, N., Quilfen, Y., Harscoat, V., Maroni, C. & Pouliquen, S. 1996. *An Atlas of Surface Wind from ERS-1 Scatterometer Measurements*, IFREMER publication, pp. 229, IFREMER, DRO/OS, BP 70, 29280 Plouzane, France.

Berghahn, R. 1996. Episodic mass invasions of juvenile gadoids into the Wadden Sea and their consequences for the population dynamics of the brown shrimp (*Crangon crangon*). *Marine Ecology* 17: 251–60.

Berman-Frank, I., Lundgren, P., Chen, Y-B., Küpper, H., Kolber, Z., Bergman, B. & Falkowski, P. 2001. Segregation of nitrogen fixation and oxygenic photosynthesis in the marine cyanobacterium *Trichodesmium*. *Science* 294: 1534–7.

Beukema, J. J. 2002. Expected change in the benthic fauan of Wadden sea tidal flats as a result of sea-level rise or bottom subsidence. *Netherlands Journal of Sea Research* 47: 25–39.

Bibby, C. J. 1998. Selecting areas for conservation. In W. J. Sutherland (ed.), *Conservation Science and Action*, Blackwell, Oxford, pp. 176–201.

Bildstein, K. L. 1990. Status, conservation and management of the scarlet ibis *Eudocimus rubber* in the Caroni Swamp, Trinidad, West Indies. *Biological Conservation* 54: 61–78.

Biles, C. L., Solan, M., Isaksson, I., Paterson, D. M., Emes, C. & Raffaelli, D. G. 2003. Flow modifies the effect of biodiversity on ecosystem functioning: an in situ study of estuarine sediments. *Journal of Experimental Marine Biology and Ecology* 285–6: 165–78.

Billett, D. M. S., Lampitt, R. S., Rice, A. L. & Mantoura, R. F. 1983. Seasonal sedimentation of phytoplankton to the deep-sea benthos. *Nature* 302: 520–2.

Binder, B. J., Chisholm, S. W., Olson, R. J., Frankel, S. L. & Worden, A. Z. 1996. Dynamics of picophytoplankton, ultraphytoplankton and bacteria in the central equatorial Pacific. *Deep Sea Research* (part 2) 43: 907–31.

Bingham, B. L. & Young, C. M. 1995. Stochastic events and dynamics of a mangrove root epifaunal community. *PSZNI: Marine Ecology* 16: 145–63.

Birkeland, C. 1997. *Life and Death on Coral Reefs*. Chapman & Hall, New York.

Bjorndal, K. 1980. Nutrition and grazing behavior of the green turtle *Chelonia mydas*. *Marine Biology* 56: 147–54.

Black, E. A. & Truscott, J. 1994. Strategies for regulation of aquaculture site selection in coastal areas. *Journal of Applied Ichthyology* 10: 294–306.

Blome, D. & Riemann, F. 1999. Antarctic sea ice nematodes, with description of *Geomonhystera glaciei* sp. nov. (Monhysteridae). *Mitteilung des Hamburgischen Zoologischen Museum Instituts* 96: 15–20.

Bolam, S. G., Huxham, M. & Fernandes, T. F. 2002. Diversity, biomass and ecosystem processes in the marine benthos. *Ecological Monographs* 72: 599–615.

Bologna, P. A. X. & Heck, K. L. 1999. Differential predation and growth rates of bay scallops within a seagrass habitat. *Journal of Experimental Marine Biology and Ecology* 239: 299–314.

Borg, J. A., Attrill, M. J., Rowden, A. A., Schembri, P. J. & Jones, M. B. 2002. A quantitative technique for sampling motile macroinvertebrates in beds of the seagrass *Posidonia oceanica* (L.) Delile. *Scientia Marina* 66(1): 53–58.

Boschi, E. E. 2000. Species of decapod crustaceans and their distributions in the American marine zoogeographic provinces. *Revista de Investigacion y Desarrollo Pesquero* 13: 1–136.

Bowden, D. A., Rowden, A. A. & Attrill, M. J. 2001. Effect of patch size and in-patch location on the infaunal macro-invertebrate community of *Zostera marina* seagrass beds. *Journal of Experimental Marine Biology and Ecology* 259: 133–54.

Boyd, P. W. 2002. The role of iron in the biogeochemistry of the Southern Ocean and equatorial Pacific: a comparison of in situ iron enrichments. *Deep-Sea Research Part II – Topical Studies in Oceanography* 49: 1803–21.

Boyd, P. W. and 34 others 2000. A mesoscale phytoplankton bloom in the polar Southern Ocean stimulated by iron fertilization. *Nature* 407: 695–702.

Branch, G. M. & Pringle, A. 1987. The impact of the sand prawn *Callianassa kraussi* Stebbing on sediment turnover and on bacteria, meiofauna and benthic microfauna. *Journal of Experimental Marine Biology and Ecology* 107: 219–35.

Brentnall, S. J., Richards, K. J., Brindley, J. & Murphy, E. 2003. Plankton patchiness and its effect on larger-scale productivity. *Journal of Plankton Research* 25: 121–40.

Brey, T. 1998. Growth performance and mortality in aquatic macrobenthic invertebrates. *Advances in Marine Biology* 35: 153–243.

Brey, T. & Gerdes, D. 1997. Is Antarctic benthic biomass really higher than elsewhere? *Antarctic Science* 9: 266–7.

Bricaud, A., Morel, A. & Barale, V. 1999. MERIS potential for ocean colour studies in the open ocean. *International Journal of Remote Sensing* 20: 1757–69.

Brierley, A. S. & Thomas, D. N. 2002. Ecology of Southern Ocean pack ice. *Advances in Marine Biology* 43: 171–276.

Brierley, A. S., Ward, P., Watkins, J. L. & Goss, C. 1998. Acoustic discrimination of Southern Ocean zooplankton. *Deep-Sea Research Part II – Topical Studies in Oceanography* 45: 1155–73.

Brierley, A. S., Fernandes, P. G., Brandon, M. A., Armstromg, F., Millard, N. W., McPhail, S. D., Stevenson, P., Pebody, M., Perrett, J., Squires, M., Bone, D. G. & Griffiths, G. 2002. Antarctic krill under sea ice: elevated abundance in a narrow band just south of ice edge. *Science* 295: 1890–2.

Briggs, J. C. 2003. Marine centres of origin as evolutionary engines. *Journal of Biogeography* 30: 1–18.

Broecker, W. S. 1997. Thermohaline circulation, the Achilles heel of our climate system: Will man-made CO_2 upset the current balance? *Science* 278: 1582–8.

Broecker, W. S., Sutherland, S. & Peng, T.-H. 1999. A possible 20th-century slowdown of Southern Ocean deep-water formation. *Science* 286: 1132–5.

Brown, A. C. & McLachlan, A. 1990. *Ecology of Sandy Shores*. Elsevier, Amsterdam.

Brown, A. C. & McLachlan, A. 2002. Sandy shore ecosystems and the threats facing them: some predictions for the year 2025. *Environmental Conservation* 29: 62–77.

Brown, C. & Laland, K. N. 2001. Suboski and Templeton revisited: Social learning and life skills training for hatchery reared fish. *Journal of Fish Biology* 59: 471–93.

Brown, J. 1995. *Macroecology*. University of Chicago Press, Chicago.

Bryant, A., Heath, M., Broekhuizen, N., Ollason, J., Gurney, W. & Greenstreet, S. 1995. Modeling the predation, growth and population dynamics of fish within a spatially resolved shelf sea ecosystem model. *Netherlands Journal of Sea Research* 33: 407–21.

Buesseler, K. O. & Boyd, P. W. 2003. Will ocean fertilization work? *Science* 300: 67–8.

Buesseler, K. O., Andrews, J. E., Pike, S. M. & Charette, M. A. 2004. The Effects of Iron Fertilization on Carbon Sequestration in the Southern Ocean. *Science* 304: 414–17.

Buma, A. G. J., de Boer, M. K. & Boelen, P. 2001. Depth distributions of DNA damage in Antarctic marine phyto- and bacterioplankton exposed to summertime ultraviolet radiation. *Journal of Phycology* 37: 200–8.

Burke, W. T. 1994. *The New International Law of Fisheries: UNCLOS 1982 and Beyond*. Oxford: Clarendon Press.

Burkhardt, S. & Riebesell, U. 1997. CO_2 availability affects elemental composition (C : N : P) of the marine diatom *Skeletonema costatum*. *Marine Ecology Progress Series* 155: 67–76.

Burkhardt, S., Riebesell, U. & Zondervan, I. 1999a. Stable carbon isotope fractionation by marine phytoplankton in response to daylength, growth rate, and CO_2 availability. *Marine Ecology Progress Series* 184: 31–41.

Burkhardt, S., Riebesell, U. & Zondervan, I. 1999b. Effects of growth rate, CO_2 concentration, and cell size on the stable carbon isotope fractionation in marine phytoplankton. *Geochemica et Cosmochima Acta* 63: 3729–41.

Burkhardt, S., Amoroso, G., Riebesell, U. & Sultemeyer, D. 2001. CO_2 and HCO_3^- uptake in marine diatoms acclimated to different CO_2 concentrations. *Limnology and Oceanography* 46: 1378–91.

Caddy, J. 1998. A short review of precautionary reference points and some proposals for their use in data-poor situations. *FAO Fisheries Technical Paper* 379: 30 pp.

Cadée, G. C., González, H. & Schnack-Schiel, S. B. 1992. Krill diet affects faecal string settling. *Polar Biology* 12: 75–80.

Caley, M. J. & Schluter, D. 1997. The relationship between local and regional diversity. *Ecology* 78: 70–80.

Carr, M. E. & Kearns, E. J. 2003. Production regimes in four Eastern Boundary Current systems. *Deep-Sea Research Part II – Topical Studies in Oceanography* 50: 3199–221.

Cary, S. C., Shank, T. & Stein, J. 1998. Worms bask in extreme temperatures. *Nature* 391: 545–6.

Caswell, H., Fujiwara, M. & Brault, S. 1999. Declining survival probability threatens North Atlantic right whale. *Proceedings of the National Academy of Sciences* 96: 3308–13.

Cavicchioloi, R. 2002. Extremophiles and the search for extraterrestrial life. *Astrobiology* 2: 281–92.

Cavicchioli, R., Siddiqui, K. S., Andrews, D. & Sowers, K. R. 2002. Low-temperature extremophiles and their applications. *Current Opinion in Biotechnology* 13: 253–61.

Chamberlain, G. 1997. Sustainability of world shrimp farming. In E. L. Pikitch, D. D. Huppert & M. P. Sissenwine (eds), *Global Trends: Fisheries Management*, American Fisheries Society Symposium, Bethesda, Maryland, pp. 195–212.

Chan, T. U. & Hamilton, D. P. 2001. The effect of freshwater flow on the succession and biomass of phytoplankton in a seasonal estuary. *Marine and Freshwater Research* 52: 869–84.

Chapelle, G. & Peck, L. S. 1999. Polar gigantism dictated by oxygen availability. *Nature* 399: 114–15.

Chapelle, G. & Peck, L. S. 2004. Amphipod crustacean size spectra: new insights in the relationship between size and oxygen. *Oikos* 106: 167–75.

Chapman, P. M. & Brinkhurst, R. O. 1981. Seasonal changes in the interstitial salinities and seasonal movements of subtidal benthic invertebrates in the Fraser River estuary, B.C. Estuarine Coastal Shelf. *Science* 12: 49–66.

Charlier, R. & Charlier, C. 1992. Environmental, economic and social aspects of marine aggregates' exploitation. *Environmental Conservation* 19: 29–37.

Chase, J. M. 1998. Central-place forager effects on food web dynamics and spatial patterns in Northern California meadows. *Ecology* 79: 1236–45.

Chavez, F. P., Ryan, J., Lluch-Cota, S. E. & Niquen, M. 2003. From anchovies to sardines and back: Multidecadal change in the Pacific Ocean. *Science* 299: 217–21.

Chisholm, S. W., Falkowski, P. G. & Cullen, J. J. 2001. Dis-crediting ocean fertilization. *Science* 294: 309–10.

Chyba, F. F. & Phillips, C. B. 2002. Europa as an abode of life. *Origins of Life and Evolution of the Biosphere* 32: 47–68.

Clark, P. U., Pisias, N. G., Stocker, T. F. & Weaver, A. J. 2002. The role of the thermohaline circulation in abrupt climate change. *Nature* 415: 863–9.

Clarke, A. 1992. Is there a latitudinal diversity cline in the sea? *Trends in Ecology and Evolution* 7: 286–7.

Clarke, K. & Warwick, R. 1994. *Change in Marine Communities: An Approach to Statistical Analysis and Interpretation.* Natural Environmental Research Council, Plymouth Marine Laboratory, Plymouth.

Cliff, G. 1991. Shark Attacks on the South-African Coast between 1960 and 1990. *South African Journal of Science* 87: 513–18.

Coale, K. H. and 47 others 2004. Southern Ocean iron enrichment experiment: Carbon cycling in high- and low-Si Waters. *Science* 304: 408–14.

Cohen, J. E., Pimm, S. L., Yodzis, P. & Saldana, J. 1993. Body Sizes of Animal Predators and Animal Prey in Food Webs. *Journal of Animal Ecology* 62: 67–78.

Coles, S. L. & Brown, B. E. 2003. Coral bleaching – capacity for acclimatization and adaptation. *Advances in Marine Biology* 46: 183–223.

Collie, J. S., Hall, S. J., Kaiser, M. J. & Poiner, I. R. 2000. A quantitative analysis of fishing impacts on shelf-sea benthos. *Journal of Animal Ecology* 69, 785–799.

Colman, J. S. 1933. The nature of intertidal zonation of plants and animals. *Journal of the Marine Biological Association of the United Kingdom* 61: 71–93.

Comiso, J. C. 2002. A rapidly declining Arctic perennial ice cover. *Geophysical Research Letters* 29, doi: 10. 1029/2002GL015650.

Comiso, J. C. 2003a. Warming Trends in the Arctic. *Journal of Climate* 16: 3498–510.

Comiso, J. C. 2003b. Large-scale characteristics and variability of the global sea ice cover. In D. N. Thomas & G. S. Dieckmann (eds), *Sea Ice – An Introduction to its Physics, Chemistry, Biology and Geology*. Blackwells Publishing, Oxford.

Connell, J. H. 1961a. The influence of inter-specific competition and other factors on the distribution of the barnacle *Chthamalus stellatus. Ecology* 42: 710–23.

Connell, J. H. 1961b. Effects of competition, predation by *Thais lapillus* and other factors on natural populations of the barnacle *Balanus balanoides. Ecological Monographs* 31: 61–104.

Connell, J. H. 1978. Diversity in tropical rain forests and coral reefs. *Science* 199: 1302–10.

Connolly, R. M. 1997. Differences in composition of small, motile invertebrate assemblages from seagrass and unvegetated habitats in a southern Australian estuary. *Hydrobiologia* 346: 137–48.

Constable, A. J. & Nicol, S. 2002. Defining smaller-scale management units to further develop the ecosystem approach in managing large-scale pelagic krill fisheries in Antarctica. *CCAMLR Science* 9: 117–31.

Cooper, D. J., Watson, A. J. & Ling, R. D. 1998. Variation of PCO_2 along a North Atlantic shipping route (UK to Caribbean): a year of automated observations. *Marine Chemistry* 60: 147–64.

Costanza, R., d'Arge, R., de Groot, R., Farber, S., Grasso, M., Hannon, B., Limburg, K., Naeem, S., O'Neill, R. V., Paruelo, J., Raskin, R. G., Sutton, P. & van derBelt, M. 1997. The value of the world's ecosystem services and natural capital. *Nature* 387: 253–60.

Costello, M., Deady, S., Pike, A. & Fives, J., 1994. Parasites and diseases of wrasse (Labridae) being used as cleaner-fish on salmon farms in Ireland and Scotland. In M. D. J. Sayer, J. W. Treasurer & M. J. Costello (eds), *Wrasse – Biology and Use in Aquaculture*, Fishing News Books, Oxford, pp. 211–27.

Coutant, C. C. 1990. Temperature-oxygen habitat for freshwater and coastal striped bass in a changing climate. *Transactions of the American Fisheries Society* 119: 240–53.

Crame, J. A. 1999. An evolutionary perspective on marine faunal connections between southernmost South America and Antarctica. *Scientia Marina* 63: 1–14.

Crame, J. A. 2000. Evolution of taxonomic gradients in the marine realm: evidence from the composition of recent bivalve faunas. *Paleobiology* 26: 188–214.

Cranfield, H. J., Michael, H. J. & Doonan, I. J. 1999. Changes in the distribution of epifaunal reefs and oysters during 130 years of dredging for oysters in Foveaux Strait, southern New Zealand. *Aquatic Conservation: Marine and Freshwater Ecosystems* 9: 461–83.

Cranfield, H. J., Rowden, A. A., Smith, D. J., Gordon, D. P. & Michael, K. P. 2004. Macrofaunal assemblages of benthic habitat of different complexity and the proposition of a model of biogenic reef habitat regeneration in Foveaux Strait, New Zealand. *Journal of Sea Research* 52: 109–26.

Creutzberg, F. 1984. A persistent chlorophyll a maximum coinciding with an enriched benthic zone. In P. E. Gibbs (ed.), *Proceedings of the Nineteenth European Marine Biology Symposium*, Cambridge University Press, Cambridge, pp. 97–108.

Croll, D. A., Clark, C. W., Calambokidis, J., Ellison, W. T. & Tershy, B. R. 2001. Effect of anthropogenic low-frequency noise on the foraging ecology of Balaenoptera whales. *Animal Conservation* 4: 13–27.

Cury, P., Bakun, A., Crawford, R. J. M., Jarre, A., Quinones, R. A., Shannon, L. J. & Verheye, H. M. 2000. Small pelagics in upwelling systems: patterns of interaction and structural changes in 'wasp-waist' ecosystems. *ICES Journal of Marine Science* 57: 603–18.

Cushing, D. H. 1975. *Marine ecology and fisheries*. Cambridge University Press, Cambridge.

Cushing, D. H. 1988. *The Provident Sea*. Cambridge University Press, Cambridge.

Cushing, D. H. 1990. Plankton Production and Year-Class Strength in Fish Populations – An Update of the Match Mismatch Hypothesis. *Advances in Marine Biology* 26: 249–93.

Dahl, E. 1952. Some aspects of the ecology and zonation of the fauna on sandy beaches. *Oikos* 4: 1–27.

Dahle, L. A. 1995. Off-shore fish farming systems. *INFOFISH-International* 2: 24–30.

Dalzell, P., Adams, T. J. H. & Polunin, N. V. C. 1996. Coastal fisheries in the Pacific Islands. *Oceanography and Marine Biology: Annual Review* 34: 395–531.

Davidson, A. T. & van der Heijden, A. 2000. Exposure of natural Antarctic marine microbial assemblages to ambient UV radiation: effects on bacterioplankton. *Aquatic Microbial Ecology* 21: 257–264.

Davoren, G. K., Montevecchi, W. A. & Anderson, J. T. 2003. Search strategies of a pursuit-diving marine bird and the persistence of prey patches. *Ecological Monographs* 73: 463–81.

Dayton, P. K. 1994. Community landscape: scale and stability in hard bottom marine communities. In P. Giller, A. Hildrew & D. Raffaelli (eds), *Aquatic Ecology: Scale, Pattern and Process*, Blackwell Scientific Publications, Oxford, pp. 289–332.

Dayton, P. K. 1985. Ecology of kelp communities. *Annual Review of Ecology and Systematics* 16: 215–45.

Dayton, P. K., Robilliard, G. A. & DeVries, A. L. 1969. Anchor ice formation in McMurdo Sound, Antarctica, and its biological effects. *Science* 163: 273–4.

de Baar, H. J. W. and 6 others 1995. Importance of iron for plankton blooms and carbon dioxide drawdown in the Southern Ocean. *Nature* 373: 412–15.

de Beer, D. 2000. Potentiometric microsensors for *in situ* measurements in aquatic environments. In J. Buffle & G. Horvai (eds), *In situ monitoring of aquatic systems: chemical analysis and speciation*, Wiley & Sons, London, pp. 161–94.

De Forges, B. R., Koslow, J. A. & Poore, G. C. B. 2000. Diversity and endemism of the benthic seamount fauna in the southwest Pacific. *Nature* 405: 944–6.

Deegan, L. A. & Garritt, R. H. 1997. Evidence for spatial variability in estuarine food webs. *Marine Ecology Progress Series* 147 (1–3): 31–47.

del Giorgio, P. A. & Williams, P. J. le B. (eds) 2005. *Respiration in Aquatic Ecosystems*, Oxford University Press, Oxford.

del Giorgio, P. A. & Williams, P. J. le B. 2005. The Global Significance of Respiration in Aquatic Ecosystems: From Single Cells to the Biosphere. In P. A. del Giorgio & P. J. le B. Williams (eds), *Respiration in Aquatic Ecosystems*, Oxford University Press, Oxford, pp. 267–303.

Deming, J. W. 2002. Psychrophiles and polar regions. *Current Opinion in Microbiology* 5: 301–9.

Dernie, K. M., Kaiser, M. J. & Warwick, R. M. 2003. Recovery rates of benthic communities following physical disturbance. *Journal of Animal Ecology* 72: 1043–56.

Diehl, S. 2002. Phytoplankton, light, and nutrients in a gradient of mixing depths: Theory. *Ecology* 83: 386–98.

Dinmore, T. A., Duplisea, D. E., Rackham, B. D., Maxwell, D. L. & Jennings, S. 2003. Impact of a large-scale area closure on patterns of fishing disturbance and the consequences for benthic production. *ICES Journal of Marine Sciences* 60: 371–80.

Distel, D. L. & Roberts, S. J. 1997. Bacterial endosymbionts in the gills of the deep-sea wood-boring bivalves *Xylophaga atlantica* and *Xylophaga washingtona*. *Biological Bulletin* 192(2): 253–61.

Distel, D. L., Baco, A. R., Chuang, E., Morrill, W., Cavanaugh, C. M. & Smith, C. R. 2000. Do mussels take wooden steps to deep-sea vents? *Nature* 403: 725–6.

Dittmar, T. & Kattner, G. 2003. The biogeochemistry of the river and shelf ecosystem of the Arctic Ocean: a review. *Marine Chemistry* 83: 103–20.

Doherty, P. J. 2002. Variable replenishment and the dynamics of reef fish populations. In P. F. Sale (ed.), *Coral Reef Fishes: Dynamics and Diversity in a Complex Ecosystem*, Academic Press, Amsterdam, pp. 327–55.

Dolan, J. R. & Pérez, M. T. 2000. Costs, benefits and characteristics of mixotrophy in marine oligotrichs. *Freshwater Biology* 45: 227–38.

Doty, M. S. 1946. Critical tidal factors that are correlated with the vertical distribution of marine algae and other organisms along the Pacific coast. *Ecology* 27: 315–28.

Dring, M. J. & Brown, F. A. 1982. Photosynthesis of intertidal brown algae during and after periods of emersion; a renewed search for physiological causes of zonation. *Marine Ecology Progress Series* 8: 301–8.

Duarte, C. M. 1991. Allometric scaling of seagrass form and productivity. *Marine Ecology Progress Series* 77: 289–300.

Duarte, C. M. 2000. Marine biodiversity and ecosystem services: an elusive link. *Journal of Experimental Marine Biology and Ecology* 250: 117–32.

Duarte, C. M. 2001. Seagrass Ecosystems. In S. L. Levin (ed.), *Encyclopedia of Biodiversity*, Vol. 5. Academic Press, pp. 255–68.

Duarte, C. M., Chiscano, C. L. 1999. Seagrass biomass and production: a reassessment. *Aquatic Botany* 65: 159–74.

Ducklow, H. 2000. Bacterial Production and Biomass in the Oceans. In D. L. Kirchman (ed.), *Microbial Ecology of the Oceans*, John Wiley & Sons, Inc. New York, pp. 85–120.

Dulvy, N. K. & Reynolds, J. D. 1997. Evolutionary transitions among egg-laying, live-bearing and maternal inputs in sharks and rays. *Proceedings of the Royal Society of London B* 264: 1309–15.

Dulvy, N. K., Sadovy, Y. & Reynolds, J. D. 2003. Extinction vulnerability in marine populations. *Fish and Fisheries* 4: 25–64.

Dunne, J. A., Williams, R. J. & Martinez, N. D. 2004. Network structure and robustness of marine food webs. *Marine Ecology Progress Series* 273: 291–302.

Durako, M., Zieman, J. & Robblee, M. 2003. Seagrass ecology. In W. Nuttle, J. Hunt & M. Robblee (eds), *A Synthesis of Research on Florida Bay*, Florida Bay Science Programme/US Geological Survey, pp. 7.1–7.36.

Durazo, R., Harrison, N. M. & Hill, A. E. 1998. Seabird observations at a tidal mixing front in the Irish Sea. *Estuarine Coastal and Shelf Science* 47: 153–64.

Dyer, K. R. 1997. *Estuaries: A Physical Introduction*. John Wiley, New York.

Eadie, B. J., McKee, B. A., Lansing, M. B., Robbins, J. A., Metz, S. & Trefry, J. H. 1994. Records of nutrient-enhanced coastal ocean productivity in sediments from the Louisiana continental shelf. *Estuaries* 17: 754–65.

Eastman, J. T. 2000. Antarctic notothenioid fishes as subjects for research in evolutionary biology. *Antarctic Science* 12: 276–87.

Edgar, G. J. 1999. Experimental analysis of structural versus trophic importance of seagrass beds. I. Effects on macrofaunal and meiofaunal invertebrates. *Vie Melieu* 49: 239–48.

Edgar, G. J. & Robertson, A. I. 1992. The influence of seagrass structure on the distribution and abundance of mobile epifauna: pattern and process in a western Australian *Amphipholis* bed. *Journal of Experimental Marine Biology and Ecology* 160: 13–31.

Edgar, G. J., Barrett, N. S., Graddon, D. J. & Last, P. R. 2000. The conservation significance of estuaries: a classification of Tasmanian estuaries using ecological, physical and demographic attributes as a case study. *Biological Conservation* 92: 383–97.

Eicken, H. 1992. The role of sea ice in structuring Antarctic ecosystems. *Polar Biology* 12: 3–13.

Eicken, H. 2003. From the microscopic, to the macroscopic, to the regional scale: Growth microstructure and properties of sea ice. In D. N. Thomas & G. S. Dieckmann (eds), *Sea Ice – An Introduction to its Physics, Chemistry, Biology and Geology*. Blackwells Publishing, Oxford.

Elahimanesh, P. 2001. Sturgeon in Iran. *Fisheries* 26: 41–4.

Ellison, A. M., Farnsworth, E. J. & Merkt, R. E. 1999. Origins of mangrove ecosystems and the mangrove biodiversity anomaly. *Global Ecology and Biogeography* 8: 95–115.

Elzenga, J. T., Prins, H. B. A. & Stefels, J. 2000. The role of extracellular carbonic anhydrase activity in inorganic carbon utilization of *Phaeocystis globosa* (Prymnesiophyceae): A comparison with other marine algae using the isotopic disequilibrium technique. *Limnology and Oceanography* 45: 372–80.

Emerson, C. W. 1989. Wind Stress Limitation of Benthic Secondary Production in Shallow, Soft-Sediment Communities. *Marine Ecology Progress Series* 53: 65–77.

Emmerson, M. C., Solan, M., Emes, C., Paterson, D. M. & Raffaelli, D. G. 2001. Consistent patterns and the idiosyncratic effects of biodiversity in marine ecosystems. *Nature* 411(6833): 73–7.

Estes, J. A. & Duggins, D. O. 1995. Sea otters and kelp forests in Alaska: generality and variation in a community ecology paradigm. *Ecological Monographs* 65: 75–100.

Estes, J. A., Tinker, M. T., Williams, T. M. & Doak, D. F. 1998. Killer whale predation on sea otters linking oceanic and nearshore ecosystems. *Science* 282: 473–6.

European Commission 2000. Communication from the Commission to the Council and the European Parliament on Integrated Coastal Zone Management: a strategy for Europe. COM (2000) 547 final. European Commission, Brussels.

European Commission 2002. Recommendation of the European Parliament and of the Council 2002/413/EC of 30 May 2002. concerning the implementation of Integrated Coastal Zone Management in Europe. Official Journal (L148/24 of 6 June 2002).

Evans, S. M., Birchenough, A. C. & Brancato, M. S. 2000. The TBT ban: out of the frying pan into the fire? *Marine Pollution Bulletin* 40: 204–11.

Fairbridge, R. W. 1980. The estuary: its definition and geodynamic cycle. In E. Olavsson & I. Cato (eds), *Chemistry and Geochemistry of Estuaries*, John Wiley & Sons, New York, pp. 1–35.

Fairweather, P. G. & Underwood, A. J. 1991. Experimental removals of a rocky intertidal predator: variations within two habitats in the effects on prey. *Journal of Experimental Marine Biology and Ecology* 154: 29–75.

Falkowski, P. G., Barber, R. T. & Smetacek, V. 1998. Biogeochemical controls and feedbacks on ocean primary production. *Science* 281: 200–6.

Falkowski, P. and 16 others 2000. The Global Carbon Cycle: A Test of Our Knowledge of Earth as a System. *Science* 290: 291–6.

FAO 1995. *Code of conduct for sustainable fisheries*. Food and Agriculture Organization of the United Nations, Rome.

FAO 2002. *The state of world fisheries and aquaculture* (SOFIA). Food and Agriculture Organization of the United Nations, Rome. pp. 150.

Farnsworth, E. J. & Ellison, A. M. 1996. Scale-dependent spatial and temporal variability in biogeography of mangrove root epibiont communities. *Ecological Monographs* 66: 45–66.

Fauchald, P., Erikstad, K. E. & Skarsfjord, H. 2000. Scale-dependent predator-prey interactions: The hierarchical spatial distribution of seabirds and prey. *Ecology* 81: 773–83.

Feely, R. A. +6 others 2004. Impact of anthropogenic CO_2 on the $CaCO_3$ system in the oceans. *Science* 305: 362–6.

Fegan, D. F. 1996. Sustainable shrimp farming in Asia: vision or pipedream. *Asian Aquaculture* 2: 22–4.

Fenchel, T. 2005. Respiration in Aquatic Protists. In P. A. del Giorgio & P. J. le B. Williams (eds), *Respiration in Aquatic Ecosystems*, Oxford University Press, Oxford, pp. 47–56.

Field, C. B., Behrenfeld, M. J., Randerson, J. T. & Falkowski, P. 1998. Primary Production of the Biosphere: Integrating Terrestrial and Oceanic Components. *Science* 281: 237–40.

Flach, E. & Tamaki, A. 2001. Competitive bioturbators on intertidal sandflats in the European Wadden Sea and in Ariake Sound, Japan. In *Ecological comparisons of sedimentary shores* (ed. Reise K.), Springer, Berlin, pp. 149–172.

Fogg, G. E. 1998. *The Biology of Polar Regions*. Oxford University Press, Oxford.

Foote, A. D., Osborne, R. W. & Hoelzel, A. R. 2004. Environment – Whale-call response to masking boat noise. *Nature* 428: 910.

Fuhrman, J. A. 1999. Marine viruses and their biogeochemical and ecological effects. *Nature* 399: 541–8.

Fuhrman, J. A. & Capone, D. 2001. Nifty nanoplankton. *Nature* 412: 593–4.

Fujikura, K., Shigeaki, K., Tamaki, K., Maki, Y., Hunt, J. C. & Okutani, T. 1999. The deepest chemosynthesis-based community yet discovered from the hadal zone, 7326 m deep, in the Japan Trench. *Marine Ecology Progress Series* 190: 17–26.

Gage, J. D. & Gordon, J. 1995. Sound bites, science and the Brent Spar: environmental considerations relevant to the deep-sea disposal option. *Marine Pollution Bulletin* 30: 772–9.

Gage, J. D. & Tyler, P. A. 1991. *Deep-Sea Biology: a Natural History of Organisms at the Deep-Sea Floor*. Cambridge University Press, Cambridge.

Gaines, S. D. & Roughgraden, J. 1985. Larval settlement rate, a leading detreminat of structure in an ecological community of the intertidal zone. *Proceedings of the National Academy of Sciences USA* 82: 3707–11.

Gaines, S. D., Brown, S. & Roughgraden, J. 1985. Spatial variation in larval concentrations as a cause of spatial variation in settlement for the barnacle *Balanus glandula*. *Oecologia* 67: 267–72.

Garrison, D. L. 1991. Antarctic sea ice biota. *American Zoologist* 31: 17–33.

Garrison, D. L., Close, A. R. & Reimnitz, E. 1989. Algae concentrated by frazil ice: evidence from laboratory experiments and field measurements. *Antarctic Science* 1: 313–16.

Garzoli, S. L., Richardson, P. L., Rae, C. M. D., Fratantoni, D. M., Goni, G. J. & Roubicek, A. J. 1999. Three Agulhas rings observed during the Benguela Current experiment. *Journal of Geophysical Research – Oceans* 104: 20971–85.

Gaston, K. J. & Spicer, J. I. 2004. *Biodiversity: An Introduction*. 2nd edition. Blackwell Science, Oxford.

Geider, R. J. & la Roche, J. 2002. Redfield Revisited: variability of C:N:P in marine microalgae and its biochemical basis. *European Journal of Phycology* 37: 1–17.

Geider, R. J. and 20 others 2001. Primary productivity of planet earth: biological determinants and physical constraints in terrestrial and aquatic habitats. *Global Change Biology* 7: 849–82.

Gerdes, D., Hilbig, B. & Montiel, A. 2003. Impact of iceberg scouring on macrobenthic communities in the high-Antarctic Weddell Sea. *Polar Biology* 26: 295–301.

Gibbons, M. J. & Griffiths, C. L. 1986. A comparison of macrofaunal and meiofaunal distribution and standing stock across a rocky shore with an estimate of their productivities. *Marine Biology* 93: 181–8.

Gibson, D. R. & Sommerville, C. 1996. The potential for viral problems related to the use of wrasse (Labridae) in the farming of Atlantic salmon. In M. D. J. Sayer, J. W. Treasurer & M. J. Costello (eds), *Wrasse – Biology and Use in Aquaculture*, Fishing News Books, Oxford, pp. 240–6.

Gibson, R. N. & Robb, L. 1992. The relationship between body size, sediment grain size and the burying ability of juvenile plaice *Pleuronectes platessa* L. *Journal of Fish Biology* 40: 771–8.

Gill, J. A., Sutherland, W. J. & Norris, K. 2001. Depletion models can predict shorebird distribution at different spatial scales. *Proceedings of the Royal Society of London*. Series B, 268: 369–76.

Givernaud, T. J., Cosson, J. & Givernaud-Mouradi, A. 1991. Etudes des populations de *Sargassum muticum* sur les cots de Basse, Normandies, France. In M. Elliott & J. P. Ducrotoy (eds), *Estuaries and Coasts: Spatial and Temporal Intercomparison*, Olsen & Olsen, pp. 129–32.

Glasby, G. P. 2000. Lessons learned from deep-sea mining. *Science* 289: 511–53.

Glud, R. N., Rysgaard, S. & Kühl, M. 2002. A laboratory study on O_2 dynamics and photosynthesis in ice algal communities: quantification by microsensors, O_2 exchange rates, ^{14}C incubations and a PAM fluorometer. *Aquatic Microbial Ecology* 27: 301–11.

Glynn, P. W. 1997. Bioerosion and coral reef growth: a dynamic balance. In C. Birkeland (ed.), *Life and Death on Coral Reefs*, Academic Press, New York, pp. 68–95.

Gnanadesikan, A., Sarmiento, J. L. & Slater, R. D. 2003. Effects of patchy ocean fertilization on atmospheric carbon dioxide and biological production. *Global Biogeochemical Cycles*, 17, art. no. 050.

Gordon, M. S., Ng, W. W. S. & Yip, A. Y. W. 1978. Aspects of the physiology of terrestrial life in amphibious fishes. III. The Chinese mudskipper *Periophthalmus cantonensis*. *Journal of Experimental Biology* 72: 57–75.

Goss-Custard, J. D., McGrorty, S. & Kirby, R. 1990. Inshore birds of the soft coasts and sea-level rise. In J. J. Beukeema, W. J. Wolff & J. J. W. N. Brouns (eds), *Expected Effects of Climate Change on Marine Coastal Ecosystems*, Kluwer, Dordecht, pp. 189–93.

Gould, S. J. & Calloway, C. B. 1980. Clams and brachiopods – ships that pass in the night. *Paleobiology* 6: 383–96.

Gould, S. J. 1989. *Wonderful life*. W. W. Norton & Co, New York.

Gowen, R. J. & Bradbury, N. B. 1987. The ecological impact of salmonid farming in coastal waters: a review. *Oceanography and Marine Biology: An Annual Review* 25: 563–75.

Grantham, B. A., Chan, F., Nielsen, K. J., Fox, D. S., Barth, J. A., Huyer, A., Lubchenco, J. & Menge, B. A. 2004. Upwelling-driven nearshore hypoxia signals ecosystem and oceanographic changes in the northeast Pacific. *Nature* 429: 749–53.

Grassle, J. F. 1989. Species diversity in deep-sea communities. *Trends in Ecology and Evolution* 4: 12–15.

Grassle, J. F. & Maciolek, N. J. 1992. Deep-sea species richness: regional and local diversity estimate from quantitative bottom samples. *American Naturalist* 139: 313–34.

Grassle, J. F. & Saunders, H. L. 1973. Life histories and the role of disturbance. *Deep Sea Research* 20: 643–59.

Gray, J. S. 2001. Marine diversity: the paradigms in patterns of species richness examined. *Scientia Marina* 65: 41–56.

Green, E. P. & Short, F. T. 2003. *World Atlas of Seagrasses*. University of California Press.

Green, E. P. & Donnelly, R. 2003. Recreational Scuba diving in Caribbean marine protected areas: do the users pay? *Ambio* 32: 140–4.

Gresty, K. A. & Quarmby, C. 1991. The Trophic Level of *Mytilicola intestinalis* Steuer (Copepoda: Poecilostomatoida) in *Mytilus edulis* L., as Determined from Stable Isotope Analysis. Proceedings of the Fourth International Conference on Copepoda; *Bulletin of the Plankton Society of Japan*, Special Volume, pp. 363–71.

Griffiths, G. E. 2003. *Technology and Applications of Autonomous Underwater Vehicles*. Taylor & Francis, London.

Gross, M. R. 1998. One species with two biologies: Atlantic salmon (*Salmo salar*) in the wild and aquaculture. *Canadian Journal of Fisheries and Aquatic Sciences* 55: 131–44.

Gutt, J. & Piepenburg, D. 2003. Scale-dependent impacts of catastrophic disturbances by grounding icebergs on the diversity of Antarctic benthos. *Marine Ecology Progress Series* 253: 77–83.

Gutt, J. 2000. Some 'driving forces' structuring communities of the sublittoral Antarctic macrobenthos. *Antarctic Science* 12: 297–313.

Gutt, J. 2001. On the direct impact of ice on marine benthic communities, a review. *Polar Biology* 24: 553–64.

Haamer, J. 1996. Improving water quality in a eutrophied fjord system with mussel farming. *Ambio* 25: 356–62.

Haas, C. 2004. Late-summer sea ice thickness variability in the Arctic Transpolar Drift 1991–2001 derived from ground-based electromagnetic sounding, *Geophysical Research Letters* 31, L09402, doi: 10. 1029/2003GL019394.

Haddock, S. H. D., McDougall, C. M. & Case, J. F. 2004. The Bioluminescence Web Page, **http://lifesci.ucsb.edu/~biolum/**.

Haines, J. & Maurer, D. 1980. Quantitative faunal associates of the serpulid polychaete *Hydroides dianthus*. *Marine Biology* 56: 43–7.

Hall, M. A. 1996. On bycatches. *Reviews in Fish Biology and Fisheries* 6: 319–52.

Hall, M. A. 1998. An ecological view of the tuna-dolphin problem: impacts and trade-offs. *Reviews in Fish Biology and Fisheries* 8: 1–34.

Hall, S. J. 1994. Physical disturbance and marine benthic communities: life in unconsolidated sediments. *Oceanography and Marine Biology Annual Review* 32: 179–239.

Hall, S. J. 1999. *The Effects of Fishing on Marine Ecosystems and Communities*. Oxford: Blackwell Science.

Hall, S. J. 2002. The continental shelf benthic ecosystem: current status, agents for change and future prospects. *Environmental Conservation* 29: 350–74.

Hall, S. J., Raffaelli, D. G. & Thrush, S. F. 1994. Patchiness and disturbance in shallow water benthic assemblages. In A. G. Hildrew, D. Raffaelli & P. S. Giller (eds), *Aquatic Ecology: Scale, Pattern and Process*, Blackwell Scientific Publications, Oxford, pp. 333–73.

Hall, S. J., Robertson, M. R., Basford, D. J. & Fryer, R. 1993. Pit-digging by the crab Cancer pagurus: a test for long-term, large-scale effects on infaunal community structure. *Journal of Animal Ecology* 62: 59–66.

Hall-Spencer, J. M. & Moore, P. G. 2000. Impact of scallop dredging on maerl grounds. In M. J. Kaiser & S. J. De Groot (eds), *Effects of Fishing on Non-Target Species and Habitats: Biological, Conservation and Socio-Economic Issues*, Blackwell Science, Oxford, pp. 105–18.

Hall-Spencer, J. M. & Moore, P. G. 2000. *Limaria hians* (Mollusca: Limacea): a neglected reef-forming keystone species. *Aquatic Conservation: Marine and Freshwater Ecosystems* 10: 267–77.

Hall-Spencer, J., Allain, V. & Fosså, J. H. 2002. Trawling damage to Northeast Atlantic ancient coral reefs. *Proceedings of the Royal Society of London B* 269: 507–11.

Hamm, C. E., Merkel, R., Springer, O., Jurkojc, P., Maier, C., Prechtel, K. & Smetacek, V. 2003. Architecture and material properties of diatom shells provide effective mechanical protection. *Nature* 421: 841–43.

Hamner, W. M., Jones, M. S., Carleton, J. H., Hauri, I. R. & Williams, D. M. 1988. Zooplankton, planktivorous fish and water currents on a windward reef face, Great Barrier Reef, Australia. *Bulletin of Marine Science* 42: 459–79.

Hanley, N. 1998. Economics of nature conservation. In W. J. Sutherland (ed.), *Conservation Science and Action*, Blackwell, Oxford, pp. 220–36.

Hannach, G. & Siglo, A. C. 1998. Photoinduction of UV-absorbing compounds in six species of marine phytoplankton. *Marine Ecology Progress Series* 174: 207–22.

Hare, S. R. & Mantua, N. J. 2000. Empirical evidence for North Pacific regime shifts in 1977 and 1989. *Progress in Oceanography* 47: 103–45.

Hargrave, B. T., Prouse, N. J., Phillips, G. A. & Cranford, P. J. 1994. Meal size and sustenance time in the deep-sea amphipod *Eurythenes gryllus* collected from the arctic-ocean. *Deep-Sea Research Part 1* 41: 1489–1508.

Hartstein, N. D. & Rowden, A. A. 2004. Effect of biodeposits from mussel culture on macroinvertebrate assemblages at sites of different hydrodynamic regime. *Marine Environmental Research* 57: 339–57.

Harvell, C. D., Kim, K., Burkholder, J. M., Colwell, R. R., Epstein, P. R., Grimes, D. J., Hofmann, E. E., Lipp, E. K., Osterhaus, A. D. M. E., Overstreet, R. M., Porter, J. W., Smith, G. W. & Vasta, G. R. 1999. Emerging marine diseases-climate links and anthropogenic factors. *Science* 285: 1505–10.

Hatcher, B. G. 1988. The primary productivity of coral reefs: a beggar's banquet. *Trends in Ecology and Evolution* 3: 106–11.

Hatcher, B. G. 1997. Organic production and decomposition. In C. Birkeland (ed.), *Life and Death on Coral Reefs*, Chapman & Hall, New York, pp. 140–74.

Hauxwell, J., Cebrian, J., Furlong, C. & Valiela, I. 2001. Macroalgal canopies contribute to eelgrass (*Zostera marina*) decline in temperate estuarine ecosystems. *Ecology* 82: 1007–22.

Hauxwell, J., Cebrian, J. & Valiela, I. 2003. Eelgrass *Zostera marina* loss in temperate estuaries: relationship to land-derived nitrogen loads and effect of light limitation imposed by algae. *Marine Ecology Progress Series* 247: 59–73.

Hawkins, S. J. & Hartnoll, R. G. 1983. Grazing of intertidal algae by the marine invertebrates. *Oceanography and Marine Biology: Annual Review* 21: 195–282.

Hawkins, S. J. & Hartnoll, R. G. 1985. Factors determining the upper limits of intertidal canopy-forming algae. *Marine Ecology Progress Series* 20: 265–71.

Hays, G. C. 2003. A review of the adaptive significance and ecosystem consequences of zooplankton diel vertical migrations. *Hydrobiologia* 503: 163–70.

Hays, G. C., Houghton, J. D. R. & Myers, A. E. 2004. Endangered species – Pan-Atlantic leatherback turtle movements. *Nature* 429: 522.

Heath, M. R. 1992. Field Investigations of the early Life Stages of Marine Fish. In J. H. S. Blaxter & A. J. Southward (eds), *Advances in Marine Biology*, Academic Press, London, pp. 2–133.

Heath, M. R. 1995. Size spectrum dynamics and the planktonic ecosystem of Loch Linnhe. *ICES Journal of Marine Science* 52: 627–42.

Heath, M. R., Boyle, P. R., Gislason, A., Gurney, W. S. C., Hay, S. J., Head, E. J. H., Holmes, S., Ingvarsdóttir, A., Jónasdóttir, S. H., Lindique, P., Pollard, R. T., Rasmussen, J., Richards, K., Richardson, K., Smerdon, G. & Speirs, D. 2004. Comparative ecology of overwintering *Calanus finmarchicus* in the northern North Atlantic, and implications for life cycle patterns. *ICES Journal of Marine Science*, in press.

Heck, K. L. & Orth, R. J. 1980. Seagrass habitats: the roles of habitat complexity, competition and predation in structuring associated fish and motile macroinvertebrate assemblages. In: V. S. Kennedy (ed.), *Estuarine Perspectives*, Academic Press, pp. 449–64.

Hedgpeth, J. W. 1967. Ecological aspects of the Laguna Madre, a hypersaline estuary. In G. H. Lauff (ed), *Estuaries*, Washington, D. C., American Association for the Advancement of Science Publication 83, pp. 408–19.

Heip, C. & Craeymeersch, J. 1995. Benthic community structures in the North Sea. *Helgoländer Meeresunters* 49: 313–28.

Helbing, E. W., Villafañe, V. E., Ferrario, M. & Holm-Hansen, O. 1992. Impact of natural ultraviolet radiation on rates of photosynthesis and on specific marine phytoplankton species. *Marine Ecology Progress Series* 80: 89–100.

Helbling, E. W., Buma, A. G. J., de Boer, M. K. & Villafañe, V. E. 2001. In situ impact of solar ultraviolet radiation on photosynthesis and DNA in temperate marine phytoplankton. *Marine Ecology Progress Series* 211: 43–9.

Hemminga, M. A. & Duarte, C. M. 2000. *Seagrass Ecology*. Cambridge University Press, Cambridge.

Hemminga, M. A., Marba, N. & Stapel, J. 1999. Leaf nutrient resorption, leaf life span and the retention of nutrients in seagrass systems. *Aquatic Botany* 65: 141–58.

Herring, P. 2002. *The Biology of the Deep Ocean*. Oxford University Press, Oxford.

Hessler, R. R., Smithey, W. M., Boudrias, M. A., Keller, C. H., Lutz, R. A. & Childress, J. J. 1988. Temporal change in megafauna at the Rose Garden hydrothermal vent (Galapagos Rift; eastern tropical Pacific). *Deep-Sea Research* 35: 1681–709.

Hicks, G. R. F. 1985. Meiofauna associated with rocky shore algae. In *The Ecology of Rocky Coasts* (eds Moore P. G., Seed R.), Hodder and Stoughton, London, pp. 36–56.

Hilborn, R. 1996. Risk analysis in fisheries and natural resource management. *Human Ecological Risk Assessment* 12: 655–9.

Hilborn, R. & Walters, C. J. 1992. *Quantitative Fisheries Stock Assessment: Choice, Dynamics and Uncertainty*. New York: Chapman & Hall.

Hilborn, R. 1992. Hatcheries and the future of salmon in the Northwest. *Fisheries* 17: 5–8.

Hilborn, R. 1999. Confession of a reformed hatchery basher. *Fisheries* 24: 30–1.

Hill, A. E., James, I. D., Linden, P. F., Matthews, J. P., Prandle, D., Simpson, J. H., Gmitrowicz, E. M., Smeed, D. A., Lwiza, K. M. M., Durazo, R., Fox, A. D. & Bowers, D. G. 1993. Dynamics of Tidal Mixing Fronts in the North Sea. *Philosophical Transactions of the Royal Society of London Series A: Mathematical Physical and Engineering Sciences* 343: 431–46.

Hinz, H., Kaiser, M. J., Bergmann, M., Rogers, S. I. & Armstrong, M. 2003. Using habitat selection theory to identify potential 'Essential Fish Habitats'. *Journal of Fish Biology* 63: 1219–34.

Hiscock, K. 1983. Water movement. In R. Earll & D. G. Erwin (eds), *Sublittoral Ecology*, Oxford University Press, Oxford, pp. 58–96.

Hixon, M. A. 1991. Predation as a process structuring coral reef fish communities. In P. F. Sale (ed.) *The Ecology of Fishes on Coral Reefs*, Academic Press, San Diego, pp. 475–508.

Hogarth, P. J. 1999. *The Biology of Mangroves*. Oxford University Press, Oxford.

Holliday, D. V. & Pieper, R. E. 1995. Bioacoustical Oceanography at High-Frequencies. *ICES Journal of Marine Science* 52: 279–96.

Hooper, W. D. & Ash, H. B. 1934. Translation of *On Agriculture* by Marcus Terentius Varro. Harvard University Press, Cambridge, MA.

Hovel, K. A. & Lipcius, R. N. 2001. Habitat fragmentation in a seagrass landscape: patch size and complexity control blue crab survival. *Ecology* 82: 1814–29.

Hoyer, K., Karsten, U. & Wiencke, C. 2002. Induction of sunscreen compounds in Antarctic macroalgae by different radiation conditions. *Marine Biology* 41: 619–27.

Hubbard, D. K. 1997. Reefs as dynamic systems. In C. Birkeland (ed.), *Life and Death of Coral Reefs*, Chapman & Hall, New York, pp. 43–67.

Hughes, D. J., Atkinson, R. J. A. & Ansell, A. D. 1999. The annual cycle of sediment turnover by the echiuran worm *Maxmuelleria lankesteri* (Herdman) in a Scottish sea loch. *Journal of Experimental Marine Biology and Ecology* 238: 209–309.

Hunter, K. A. 1999. Direct disposal of liquified fossil fuel carbon dioxide in the ocean. *Marine and Freshwater Research* 50: 755–60.

Hurlbert, S. H. 1984. Pseudoreplication and the design of ecological field experiments. *Ecological Monographs* 54: 187–211.

Huryn, A. D. 1998. Ecosystem-level evidence for top-down and bottom-up control of preduction in a grassland stream system. *Oecologia* 115: 173–83.

Hutchings, J. A. 2001. Conservation biology of marine fishes: perceptions and caveats regarding assignment of extinction risk. *Canadian Journal of Fisheries and Aquatic Science* 58: 108–121.

Hutchings, J. A. & Myers, R. A. 1994. What can be learned from the collapse of a renewable resource- Atlantic Cod *Gadus morhua*, of Newfoundland and Canada. *Canadian Journal of Fisheries and Aquatic Sciences* 51: 2126–46.

Hutchinson, G. E. 1957. Concluding remarks. *Cold Spring Harbor Symposium on Quantitative Biology* 22: 415–27.

Hutchinson, G. E. 1961. The paradox of the plankton. *American Naturalist* 95: 137–45.

Ikebe, Y. & Oishi, T. 1996. Correlation between environmental parameters and behaviour during high tides in *Periophthalmus modestus*. *Journal of Fish Biology* 49: 139–47.

IPCC 2000. *Special Report on Emissions Scenarios*. Cambridge University Press, Cambridge.

IPCC 2001. *Climate Change 2001: Impacts, adaptation and vulnerability – Contribution of Working Group II to the Third Assessment Report of IPCC*. Cambridge University Press, Cambridge, UK.

Irlandi, E. A., Orlando, B. A. & Ambrose, W. G. 1999. The effect of habitat patch size on growth and survival of juvenile bay scallops (*Argopecten irradians*). *Journal of Experimental Marine Biology and Ecology* 235: 21–43.

Ishimatsu, A., Hishida, Y., Takita, T., Kanda, T., Oikawa, S., Takeda, T. & Huat, K. K. 1998. Mudskippers store air in their burrows. *Nature* 391: 237–8.

Iverson, R. L. 1990. Control of marine fish production. *Limnology and Oceanography* 35: 1593–604.

Jablonski, D. 1999. The future of the fossil record. *Science* 284: 2114–16.

Jackson, E. L., Rowden, A. A., Attrill, M. J., Bossey, S. F. & Jones, M. B. 2001. The importance of seagrass as a habitat for fishery species. *Oceanography and Marine Biology: An Annual Review* 39: 269–303.

Jackson, J. B. C. 1979. Overgrowth competition between encrusting cheilostome ectoprocts in a Jamaican cryptic reef environment. *Journal of Animal Ecology* 48: 805–23.

Janssen, H. H. & Gradinger, R. 1999. Turbellaria (Archoophora: Acoela) from Antarctic sea ice endofauna – examination of their micromorphology. *Polar Biology* 21: 410–16.

Jarman, S., Elliott, N., Nicol, S., McMinn, A. & Newman, S. 1999. The base composition of the krill genome and its potential susceptibility to damage by UV-B. *Antarctic Science* 11: 23–6.

Jenkins, G. & Hamer, P. 2001. Spatial variation in the use of seagrass and unvegetated habitats by post-settlement King George whiting in relation to meiofaunal distribution and macrophyte structure. *Marine Ecology Progress Series* 224: 219–29.

Jenness, M. I. & Duineveld, G. C. A. 1985. Effects of tidal currents on chlorophyll a content of sandy sediments in the southern North Sea. *Marine Ecology Progress Series* 21: 283–7.

Jennings, S. & Kaiser, M. J. 1998. The effects of fishing on marine ecosystems. *Advances in Marine Biology* 34: 201–352.

Jennings, S. & Warr, K. J. 2003. Smaller predator-prey body size ratios in longer food chains. *Proceedings of the Royal Society of London Series B: Biological Sciences* 270: 1413–17.

Jennings, S., Kaiser, M. J. & Reynolds J. D. 2001. *Marine Fisheries Ecology*. Oxford: Blackwell Science.

Jeppesen, E., Sondergaard, M., Jensen, J. P., Mortensen, E., Hansen, A. M. & Jorgensen, T. 1998. Cascading trophic interactions from fish to bacteria and nutrients after reduced sewage loading: an 18-year study of a shallow hypertrophic lake. *Ecosystems* 1: 250–67.

Jickells, T. D. 1998. Nutrient biogeochemistry of the coastal zone. *Science* 281: 217–222.

Jones, E. G., Collins, M. A., Bagley, P. M., Addison, S. & Priede, I. G. 1998. The fate of cetacean carcasses in the deep sea: Observations on consumption rates and scavenging species succession in the abyssal NE Atlantic Ocean. *Proceedings of the Royal Society of London B* 265: 1119–27.

Jones, G. P., McCormick, M. I., Srinivasan, M. & Eagle, J. V. 2004. Coral decline threatens fish biodiversity in marine reserves. *Proceedings of the National Academy of Sciences* USA 101: 8251–3.

Jones, G. P., Milicich, M. J., Emslie, M. J. & Lunow, C. 1999. Self-recruitment in a coral reef fish population. *Nature* 402: 802–804.

Jones, N. S. 1950. Marine bottom communities. *Biological Reviews* 25: 283–313.

Jones, N. S. 1951. The bottom fauna off the south of the Isle of Man. *Journal of Animal Ecology* 20.

Jones, S. E. & Jago, C. F. 1993. In situ assessment of modification of sediment properties by burrowing invertebrates. *Marine Biology* 115: 133–42.

Justić, D. 1991. Hypoxic conditions in the northern Adriatic Sea: historical development and ecological significance. In R. V. Tyson & T. H. Pearson (eds), *Modern and Ancient Continetal Shelf Anoxia*, Geological Society Special Publication, London, pp. 95–102.

Justić, D., Rabalais, N. N. & Turner, R. E. 2005. Coupling between climate variability and coastal eutrophication: evidence and outlook for the northern Gulf of Mexico. *Journal of Sea Research* 54: 25–35.

Kaiser, M. J. 2003. Detecting the effects of fishing on seabed community diversity: importance of scale and sample size. *Conservation Biology* 17: 512–20.

Kaiser, M. J. & de Groot, S. J. 2000. *The Effects of Fishing on Non-Target Species and Habitats: Biological, Conservation and Socio-Economic Issues*. Blackwell Science, Oxford.

Kaiser, M. J. & Edwards-Jones, G. 2006. The role of eco-labelling in fisheries management and conservation. *Conservation Biology*.

Kaiser, M. J., Laing, I., Utting, S. D. & Burnell, G. M. 1998. Environmental impacts of bivalve mariculture. *Journal of Shellfish Research* 17: 59–66.

Kaiser, M. J., Rogers, S. I. & Ellis, J. R. 1999a. Importance of benthic habitat complexity for demersal fish assemblages. In L. Benaka (ed.), *Fish Habitat: Essential Fish Habitat and Restoration*, American Fisheries Society, Bethesda, Maryland, pp. 212–23.

Kaiser, M. J., Cheney, K. L., Spence, F. E., Edwards, D. B. & Radford, K. J. 1999b. Fishing effects in northeast Atlantic shelf seas: patterns in fishing effort, diversity and community structure. VII. The effects of trawling disturbance on the fauna associated with the tubeheads of serpulid worms. *Fisheries Research* 40: 195–205.

Kaiser, M. J., Collie, J. S., Hall, S. J., Jennings, S. & Poiner, I. R. 2002. Modification of marine habitats by trawling activities: prognosis and solutions. *Fish and Fisheries* 3: 114–36.

Karentz, D., Cleaver, J. E. & Mitchell, D. L. 1991. Cell survival characteristics and molecular responses of Antarctic phytoplankton to Ultraviolet-B radiation. *Journal of Phycology* 27: 326–41.

Karl, D. M. 2002. Hidden in a sea of microbes. *Nature* 415: 590–1.

Keeling, C. D. & Whorf, T. P. 2000. The 1,800-year oceanic tidal cycle: A possible cause of rapid climate change. *Proceedings of the National Academy of Sciences* 97: 3814–19.

Kelley, D. S., Karson, J. A., Blackman, D. K., Fruh-Green, G. L., Butterfield, D. A., Lilley, M. D., Olson, E. J., Schrenk, M. O., Roe, K. K., Lebon, G. T. & Rivizzigno, P.; AT3-60 Shipboard Party. 2001. An off-axis hydrothermal vent field near the mid-Atlantic Ridge at 30 °N. *Nature* 412: 145–9.

Kennish, M. 2002. Environmental threats and environmental future of estuaries. *Environmental Conservation* 29: 78–107.

Kenny, A. J. & Rees, H. L. 1996. The effects of marine gravel extraction on the macrobenthos: results 2 years post-dredging. *Marine Pollution Bulletin* 32: 615–22.

Kent, G. 1998. Fisheries, food security and the poor. *Food Policy* 22: 393–404.

Kent, M., Gill, W. J., Weaver, R. E. & Armitage, R. 1997. Landscape and plant community boundaries in biogeography. *Progress in Physical Geography* 21(3): 315–54.

Keuskamp, D. 2004. Limited effects of grazer exclusion on the epiphytes of *Posidonia sinuosa* in South Australia. *Aquatic Botany* 78: 3–14.

Kideys, A. E. 2002. Fall and rise of the Black Sea ecosystem. *Science* 297: 1482–4.

King, D. A. 2004. Climate change science: adapt, mitigate, or ignore? *Science* 303: 176–7.

King, G. M. 2005. Ecophysiology of Microbial Respiration. In P. A. del Giorgio & P. J. le B. Williams (eds), *Respiration in Aquatic Ecosystems*, Oxford University Press, Oxford, pp. 18–35.

Kirchman, D. L. (ed.) 2000. *Microbial Ecology of the Oceans*. John Wiley & Sons, Inc., New York.

Kirk, J. T. O. 1994. *Light and Photosynthesis in Aquatic Ecosystems*. Cambridge University Press, Cambridge.

Kirst, G. O. & Wiencke, C. 1995. Ecophysiology of polar algae. *Journal of Phycology* 31: 181–99.

Klages, M., Vopel, K., Bluhm, H., Brey, T., Soltwedel, T. & Arntz, W. E. 2001. Deep-sea food falls: first observation of a natural event in the Arctic Ocean. *Polar Biology* 24: 292–5.

Klekowski, E. J., Temple, S. A., Siung-Chang, A. M., Kumarsingh, K. 1999. An association of mangrove mutation, scarlet ibis and mercury contamination in Trinidad, West Indies. *Environmental Pollution* 105: 185–9.

Koslow, J. A. 1997. Seamounts and the ecology of deep-sea fisheries. *American Scientist* 85: 168–76.

Koslow, J. A., Gowlett-Holmes, K., Lowry, J. K., O'Hara, T., Poore, G. C. B. & Williams, A. 2001. Seamount benthic macrofauna off southern Tasmania: community structure and impacts of trawling. *Marine Ecology Progress Series* 213: 111–25.

Krause, G. H. & Weis, E. 1991. Chlorophyll fluorescence and photosynthesis: the basics. *Annual Review of Plant Physiology and Plant Molecular Biology* 42: 313–49.

Krembs, C., Gradinger, R. & Spindler, M. 2000. Implications of brine channel geometry and surface area for the interaction of sympagic organisms in Arctic sea ice. *Journal of Experimental Marine Biology and Ecology* 243: 55–80.

Krembs, C., Eicken, H., Junge, K. & Deming, J. W. 2002. High concentrations of exopolymeric substances in Arctic winter sea ice: implications for the polar ocean carbon cycle and cryoprotection of diatoms. *Deep Sea Research (Part 1)* 49: 2163–81.

Kunin, W. E. & Lawton, J. H. 1996. Does biodiversity matter? Evaluating the case for conserving species. In K. J. Gaston (ed.), *Biodiversity: A Biology of Numbers and Difference*, Blackwell Science, Oxford, pp. 283–308.

Kunzig, R. 2000. *Mapping the Deep: the Extraordinary Story of Ocean Science*. Sort Of Books, London.

Kurlansky, M. 1997. *Cod: A Biography of the Fish That Changed the World*. New York: Walker & Company.

Lalli, C. M. & Parsons, T. R. 2004. *Biological Oceanography. An Introduction* (2nd Edition). Butterworth-Heinemann.

Lambshead, P. J. D. & Boucher, G. 2003. Marine nematode deep-sea biodiversity – hyperdiverse or hype? *Journal of Biogeography* 30: 475–85.

Lambshead, P. J. D., Platt, H. & Shaw, K. 1983. The detection of differences among assemblages of marine benthic species based on an assessment of dominance and diversity. *Journal of Natural History* 17: 859–74.

Lambshead, P. J. D., Tietjen, J., Moncreiff, C. B. & Ferrero, T. J. 2001. North Atlantic latitudinal diversity patterns in deep-sea marine nematode data. *Marine Ecology Progress Series* 210: 299–301.

Lampitt, R. S. 1985. Evidence for the seasonal deposition of detritus to the deep-sea floor and its subsequent resuspension. *Deep-Sea Research* 23A: 885–97.

Larkum, A. W. D. & den Hartog, C. 1989. Evolution and biogeography of seagrasses. In A. W. D. Larkum, A. J. McComb & S. A. Shepherd (eds), *Biology of Seagrasses*, Elsevier, pp. 112–56.

Lawton, J. H. & Jones, C. G. 1995. Linking species and ecosystems: organisms as ecosystem engineers. In J. H. Lawton & C. G. Jones (eds), *Linking Species and Ecosystems*, Kluwer Academic Publishers, pp. 141–150.

Lee, S. Y. 1998. Ecological role of grapsid crabs in mangrove ecosystems: a review. *Marine and Freshwater Research* 49: 335–43.

Lee, S. Y. 1999. The effect of mangrove leaf litter enrichment on macrobenthic colonization of defaunated sandy substrates. *Estuarine Coastal and Shelf Science* 49: 703–12.

Lee, S. Y., Fong, C. W. & We, R. S. S. 2001. The effects of seagrass (*Zostera japonica*) canopy structure on associated fauna: a study using artificial seagrass units and sampling of natural beds. *Journal of Experimental Marine Biology and Ecology* 259: 23–50.

Lefevre, N., Taylor, A. H., Gilbert, F. J. & Geider, R. J. 2003. Modeling carbon-to-nitrogen and carbon-to-chlorophyll-a ratios in the ocean at low latitudes: evaluation of the role of physiological plasticity. *Limnology and Oceanography* 48: 1796–807.

Legendre, L. +9 others 1992. Ecology of sea ice biota – 2. Global significance. *Polar Biology* 12: 429–44.

Lehody, P., Bertignac, M., Hampton, J., Lewis, A. & Picaut, J. 1997. El Nino Southern Oscillation and tuna in the western Pacific. *Nature* 389: 715–17.

Lenihan, H. S. & Peterson, C. H. 1998. How habitat degradation through fishery disturbance enhances impacts of hypoxia on oyster reefs. *Ecological Applications* 8: 128–40.

Lenton, T. M. & Watson, A. J. 2000. Redfield revisited, 1, Regulation of nitrate, phosphate, and oxygen in the ocean. *Global Biogeochemical Cycles* 14: 225–48.

Lenz, J. 2000. Introduction. In R. P. Harris, P. H. Wiebe, J. Lenz, H. R. Skjoldal & M. Huntley (eds), *ICES Zooplankton Methods Manual*, Academic Press, London.

Letourneau, K. K. & Dyer, L. A. 1998. Experimental test in a lowland tropical forest shows top-down effects through four trophic levels. *Ecology* 79: 1678–87.

Leventer, A. 1998. The fate of Antarctic 'Sea ice diatoms' and their use as paleoenvironmental indicators. *Antarctic Research Series* 73: 121–37.

Levin, L. A. 1994. Paleoecology and ecology of xenophyophores. *Palaios* 9: 32–41.

Lewin, R. 1986. Supply-side ecology. *Science* 234: 25–7.

Little, C. & Kitching, J. A. 2001. *Biology of Rocky Shores*. Oxford University Press, Oxford.

Lizotte, M. P. 2001. The contributions of sea ice algae to Antarctic marine primary production. *American Zoologist* 41: 57–73.

Lodge, S. M. 1948. Algal growth in the absence of Patella on an experimental strip of foreshore, Port St Mary, Isle of Man. *Proceedings and Transactions of the Liverpool Biological Society* 56: 78–83.

Loneragan, N. R., Bunn, S. E. & Kellaway, D. M. 1997. Are mangroves and seagrasses sources of organic carbon for penaeid prawns in a tropical Australian estuary? A multiple stable-isotope study. *Marine Biology* 130(2): 289–300.

Longhurst, A. R. 1995. Seasonal cycles of production and consumption. *Progress in Oceanography* 36: 77–167.

Longhurst, A. R. 1998. *Ecological Geography of the Sea*. Academic Press, San Diego.

Luther, G. W., Rozan, T. F., Taillefert, M., Nuzzio, D. B., Di Meo, C., Shank, T. M., Lutz, R. A. & Cary, S. C. 2001. Chemical speciation drives hydrothermal vent ecology. *Nature* 410: 813–16.

Lynam, C. P., Hay, S. J. & Brierley, A. S. 2004. Interannual variability in abundance of North Sea jellyfish and links to the North Atlantic Oscillation. *Limnology and Oceanography* 49, in press.

MacArthur, R. H. & Wilson, E. O. 1967. *The Theory of Island Biogeography*. Princeton University Press, Princeton.

MacIntyre, H. L., Kana, T. M., Anning, T. & Geider, R. J. 2002. Photoacclimation of photosynthesis irradiance response curves and photosynthetic pigments in microalgae and cyanobacteria. *Journal of Phycology* 38: 17–38.

Macintyre, I. G. & Glynn, P. W. 1976. Evolution of modern Caribbean fringing reef, Galeta Point, Panama. *American Association of Petroleum Geologists Bulletin* 72: 1054–72.

Mackas, D. L. & Tsuda, A. 1999. Mesozooplankton in the eastern and western subarctic Pacific: community structure, seasonal life histories, and interannual variability. *Progress in Oceanography* 43: 335–63.

Malin, G. & Kirst, G. O. 1997. Algal production of dimethyl sulfide and its atmospheric role. *Journal of Phycology* 33: 889–96.

Marion, G. M., Fritsen, C. H., Eicken, H. & Payne, M. C. 2003. The search for life on Europa: Limiting environmental factors, potential habitats, and earth analogues. *Astrobiology* 3: 785–811.

Martin, A. P. 2003. Phytoplankton patchiness: the role of lateral stirring and mixing. *Progress in Oceanography* 57: 125–74.

Martin, J. H. +43 others 2002. Testing the iron hypothesis in ecosystems of the equatorial Pacific Ocean. *Nature* 371: 123–9.

Matthiessen, P., Waldock, R., Thain, J., Waite, M. & Scrope-Howe, S. 1995. Changes in periwinkle (*Littorina littorea*) populations following the ban on TBT-based antifoulings on small boats in the United Kingdom. *Ecotoxicology and Environmental Safety* 30: 180–94.

McCarthy, I. D., Fuiman, L. A. & Alvarez, M. C. 2003. Aroclor 1254 affects growth and survival skills of Atlantic croaker Micropogonias undulates larvae. *Marine Ecology Progress Series*.

McClanahan, T., Polunin, N. & Done, T. 2002. Ecological states and the resilience of coral reefs. *Conservation Ecology* 16(2): 18. [online] **http://www.consecol.org/v016/iss2/art18.**

McClanahan, T. R. 1992. Resource utilization, competition and predation: a model and example from coral reef grazers. *Ecological Modelling* 61: 195–215.

McClanahan, T. R. 1995. A coral-reef ecosystem-fisheries model-impacts of fishing intensity and catch selection on reef structure and processes. *Ecological Modelling* 80: 1–19.

McClanahan, T. R. 2002. The near future of coral reefs. *Environmental Conservation* 29: 460–83.

McGlade, J. M. & Metuzals, K. 2000. Options for reductions of by-catches of harbour porpoises (*Phocoena phocoena*) in the North Sea. In M. J. Kaiser & S. J. de Groot (eds), *The Effects of Fishing on Non-Target Species and Habitats*, Blackwell Science, Oxford, pp. 332–53.

McIvor, C. C. & Smith, T. J. 1995. Differences in the crab fauna of mangrove areas at a Southwest Florida and a Northeast Australia location: implications for leaf litter processing. *Estuaries* 18: 591–7.

McKinney, F. K. 1995. One hundred million years of competitive interactions between bryozoan clades: asymmetrical but not escalating. *Biological Journal of the Linnean Society* 56: 465–81.

McKinney, F., Lidgard, S., Sepkoski, J. J. & Taylor, P. D. 1998. Decoupled temporal patterns of evolution and ecology in two post-paleozoic clades. *Science* 281: 807–9.

McLaren, B. E. & Peterson, R. O. 1994. Wolves, moose, and tree rings on Isle Royale. *Science* 266: 1555–8.

McManus, M. A., Alldredge, A. L., Barnard, A. H., Boss, E., Case, J. F., Cowles, T. J., Donaghay, P. L., Eisner, L. B., Gifford, D. J., Greenlaw, C. F., Herren, C. M., Holliday, D. V., Johnson, D., MacIntyre, S., McGehee, D. M., Osborn, T. R., Perry, M. J., Pieper, R. E., Rines, J. E. B., Smith, D. C., Sullivan, J. M., Talbot, M. K., Twardowski, M. S., Weidemann, A. & Zaneveld, J. R. 2003. Characteristics, distribution and persistence of thin layers over a 48 hour period. *Marine Ecology Progress Series* 261: 1–19.

McNeill, S. E. & Fairweather, P. G. 1993. Single large or several small marine reserves? An experimental approach with seagrass fauna. *Journal of Biogeography* 20: 429–40.

McShea, D. W. 1996. Metazoan complexity and evolution: Is there a trend? *Evolution* 50: 477–92.

Meinesz, M., Belsher, T., Thibaut, T., Antolic, B., Ben Mustapha, K., Boudouresque, C.-F., Chiaverini, D., Cinelli, F., Cottalorda, J.-M., Djellouli, A., El Abed, A., Orestano, C., Grau, A. M., Ivešа, L., Jaklin, A., Langar, H., Massuti-Pascual, E., Peirano, A., Tunesi, L., Vaugelas, J. de, Zavodnik, N. & Zuljevic, A. 2001. The introduced green alga *Caulerpa taxifolia* continues to spread in the Mediterranean. *Biological Invasions* 3: 201–10.

Menge, B. A. & Lubchenko, J. 1981. Community organisation in temperate and tropical rocky intertidal habitats. Prey refuges in relation to consumer pressure gradients. *Ecological Monographs* 51: 429–50.

Menge, B. A., Lubchenko, J. & Ashkenas, L. R. 1985. Diversity, heterogeneity and consumer pressure in a tropical intertidal community. *Oecologia* 65: 394–405.

Metcalfe, J. D. & Arnold, G. P. 1997. Tracking fish with electronic tags. *Nature* 387: 665–6.

Michel, C., Nielsen, T. G., Nozais, C. & Gosselin, M. 2002. Significance of sedimentation and grazing by ice micro- and meiofauna for carbon cycling in annual sea ice (northern Baffin Bay). *Aquatic Microbial Ecology* 30: 57–68.

Micheli, F. & Peterson, C. H. 1999. Estuarine vegetated habitats as corridors for predator movements. *Conservation Biology* 13: 869–81.

Middelburg, J. J., Duarte, C. M. & Gattuso, J.-P. 2004. Respiration in coastal benthic communities. In P. A. del Giorgio & Williams, P. J. le B. (eds), *Respiration in Aquatic Ecosystems*, Oxford University Press, Oxford, pp. 206–24.

Miller, P. 2004. Multi-spectral front maps for automatic detection of ocean colour features from SeaWiFS. *International Journal of Remote Sensing* 25: 1437–42.

Milliman, J. D. 1991. Flux and fate of fluvial sediment and water in coastal seas. In F. C. Mantoura, J.-C. Martin & R. Wollast (eds), *Ocean Margin Processes in Global Change* (eds), John Wiley & Sons, Chichester, UK, pp. 69–89.

Mills, C. E. 2001. Jellyfish blooms: are populations increasing globally in response to changing ocean conditions? *Hydrobiologia* 451: 55–68.

Montagna, P. & Harper, D. E. J. 1996. Benthic infaunal long-term response to offshore production platforms in the Gulf of Mexico. *Canadian Journal of Fisheries and Aquatic Sciences* 53: 2567–88.

Montagna, P. A. & Kalke, R. D. 1992. The effect of fresh-water inflow on meiofaunal and macrofaunal populations in the Guadalupe and Nueces estuaries, Texas. *Estuaries* 15(3): 307–26.

Montresor, M., Procaccini, G. & Stoecker, D. K. 1999. *Polarella glacialis*, gen. nov., sp. nov. (Dinophyceae): Suessiaceae are still alive. *Journal of Phycology* 35: 186–97.

Moran P. J. 1986. The *Acanthaster* phenomenon. *Oceanography and Marine Biology: Annual Review* 24: 379–48.

Mukai, H. 1993. Biogeography of the tropical seagrasses in the Western Pacific. *Australian Journal of Marine and Freshwater Research* 44: 1–17.

Muller-Parker, G. & D'Elia, C. F. 1997. Interactions between corals and their symbiotic algae. In C. F. Birkeland (ed.), *Life and Death of Coral Reefs*, Chapman & Hall, New York, pp. 96–113.

Mulsow, S., Landrum, P. F. & Robbins, J. A. 2002. Biological mixing responses to sublethal concentrations of DDT in sediments by *Heteromastus filiformis* using a 137Cs marker layer technique. *Marine Ecology Progress Series* 239: 181–91.

Mumby, P. J., Edwards, A. J., Arias-González, J. E., Lindeman, K. C., Blackwell, P. G., Gall, A., Gorczynska, M. I., Harborne, A. R., Pescod, C. L., Renken, H., Wabnitz, C. C. C. & Llewellyn, G. 2004. Mangroves enhance the biomass of coral reef fish communities in the Caribbean. *Nature* 427: 533–6.

Murphy, E. J., Boyd, P. W., Leakey, R. J. G., Atkinson, A., Edwards, E. S., Robinson, C., Priddle, J., Bury, S. J., Robins, D. B., Burkill, P. H., Savidge, G., Owens, N. J. P. & Turner, D. 1998. Carbon flux in ice-ocean-plankton systems of the Bellingshausen Sea during a period of ice retreat. *Journal of Marine Systems* 17: 207–27.

Myers, A. A. 1997. Biogeographic barriers and the development of marine biodiversity. *Estuarine and Coastal Shelf Science* 44: 241–8.

Nagata, T. 2000. Production mechanisms of dissolved organic matter. In D. L. Kirchman (ed.), *Microbial Ecology of the Oceans*, John Wiley & Sons, Inc. New York, Chapter 5, pp. 121–52.

Nakano, H., Koube, H., Umezawa, S., Momoyama, K., Hiraoka, M., Inouye, K. & Oseko, N. 1994. Mass mortalities of cultured Kuruma shrimp, *Penaeus japonicus*, in Japan in 1993: Epizoological survey and infection trials. *Fish Pathology* 29: 135–9.

Naylor, R. L., Goldburg, R., Primavera, J., Kautsky, N., Beveridge, M. C. M., Clay, J., Folke, C., Lubchenco, J., Mooney, H. A. & Troell, M. 2000. Effect of aquaculture on world fish supply. *Nature* 405: 1017–24.

Nelson, T. A. & Waaland, J. R. 1996. Seasonality of eelgrass, epiphyte, and grazer biomass and productivity in subtidal eelgrass meadows subject to moderate tidal amplitude. *Aquatic Botany* 56: 51–74.

New, M. B., Tacon, A. G. J. & Csavas, I. 1993. Farm-made aquafeeds, FAO (Food and Agriculture Organisation of the United Nations) RAPA (Regional Office for Asia and the Pacific) ASEAN (Association of Southeast Asian Nations) Commission of European Communities, Rome.

Nicol, S. & de la Mare, W. K. 1993. Ecosystem management and the Antarctic krill. *American Scientist* 81: 36–47.

Nixon, S. W. 2004. The artificial Nile. *American Scientist* 92: 158–65.

Norse, E. A. (ed.) 1993. *Global Marine Biological Diversity: A Strategy for Building Conservation into Decision Making*. Center for Marine Conservation, Washington.

Noske, R. A. 1995. The ecology of mangrove forest birds in Peninsular Malaysia. *Ibis* 137: 250–63.

Odor, R. K. 1992. Big squid in big currents. *South African Journal of Marine Science-Suid-Afrikaanse Tydskrif Vir Seewetenskap* 12: 225–35.

Ojeda, F. F. & Santilices, B. 1984. Ecological dominance of *Lessonia nigriscens* (Phaeophyta) in Central Chile. *Marine Ecology Progress Series* 19: 83–91.

Olsgard, F. & Gray, J. F. 1995. A comprehensive analysis of the effects of offshore oil and gas exploration and production on the benthic communities of the Norwegian continental shelf. *Marine Ecology Progress Series* 122: 277–306.

Orth, R. J., Luckenback, M. & Moore, K. A. 1994. Seed dispersal in a marine macrophyte: implications for colonization and restoration. *Ecology* 75: 1927–39.

Packard, T. T. 1985. Measurement of electron transport activity of marine microplankton. In P. J. L. Williams and H. W. Jannasch (eds.), *Advances in Aquatic Microbiology*, Academic Press, London. pp. 207–61.

Pace, M. L., Cole, J. J., Carpenter, S. R. & Kitchell, J. F. 1999. Trophic cascades revealed in diverse ecosystems. *Trends in Ecology and Evolution* 14: 483–8.

Paerl, H. W. 1995. Coastal eutrophication in relation to atmospheric nitrogen deposition: current perspectives. *Ophelia* 41: 237–59.

Paine, R. T. 1974. Intertidal community structure, experimental studies on the relationship between a dominant competitor and its principal predator. *Oecologia* 15: 93–120.

Paine, R. T. 1980. Food webs: linkage, interaction strength and community infrastructure. *Journal of Animal Ecology* 49: 667–85.

Palmisano, A. C. & Garrison, D. L. 1993. Microorganisms in Antarctic sea ice. In E. I. Friedman (ed.), *Antarctic Microbiology*, Wiley-Liss Inc., New York, pp. 167–218.

Pandolfi, J. M. 1999. Responses of Pleistocene coral reefs to environmental change over long temporal scales. *American Zoologist* 39: 113–30.

Parkinson, C. L. 2002. Trends in the length of the southern Ocean sea-ice season, 1979–99. *Annals of Glaciology* 34: 435–40.

Paterson, D. M. & Black, K. S. 1999. Water flow, sediment dynamics and benthic ecology. *Advances in Ecological Research* 29: 155–93.

Paulay, G. 1997. Diversity and distribution of reef organisms. In C. Birkeland (ed.), *Life and Death of Coral Reefs*, Chapman & Hall, New York, pp. 298–353.

Pauly, D. & Christensen, V. 1995. Primary production required to sustain global fisheries. *Nature* 374: 255–7.

Pauly, D. & MacLean, J. 2003. *In a Perfect Ocean. The State of Fisheries and Ecosystems in the North Atlantic Ocean.* Island Press, Washington.

Pauly, D., Christensen, V., Dalsgaard, J., Froese, R. & Torres, F. 1998. Fishing down marine food webs. *Science* 279: 860–3.

Pauly, D., Watson, R. & Alder, J. 2005. Global trends in world fisheries: impacts on marine ecosystems and food security. Phil. Trans. Roy. Soc. (in press).

Pearre, S. 2003. Eat and run? The hunger/satiation hypothesis in vertical migration: history, evidence and consequences. *Biological Reviews* 78: 1–79.

Pearson, T. & Rosenberg, R. 1978. Macrobenthic succession in relation to organic enrichment and pollution of the marine environment. *Oceanography and Marine Biology: An Annual Review* 16: 229–311.

Pechenik, J. A. 1999. On the advantages and disadvantages of larval stages in benthic marine invertebrate life cycles. *Marine Ecology Progress Series* 177: 269–97.

Pershing, A. J., Wiebe, P. H., Manning, J. P. & Copley, N. J. 2001. Evidence for vertical circulation cells in the well-mixed area of Georges Bank and their biological implications. *Deep-Sea Research Part II – Topical Studies in Oceanography* 48: 283–310.

Petersen, C. J. 1914. Valuation of the sea. II. The animal communities of the sea bottom and their importance for marine zoogeography. *Report of the Danish Biological Station* 21: 44pp. + appendix 68pp.

Peterson, C. H. & Black, R. 1987. Resource depletion by active suspension feeders on tidal flats, influence of local density and tidal elevation. *Limnology and Oceanography* 32: 143–66.

Peterson, C. H. & Black, R. 1988. Density-dependent mortality caused by physical stress interacting with biotic history. *American Naturalist* 131: 257–70.

Petz, W., Song, W. & Wilbert, N. 1995. Taxonomy and ecology of the ciliate fauna (Protozoa, Ciliophora) in the endopagial and pelagial of the Weddell Sea, Antarctica. *Stapfia* 40: 1–223.

Phillips, B., Ward, T. & Chafee, C. 2003. *Eco-Labelling in Fisheries: What Is It All About?* Blackwell Publishing, Oxford.

Phillips, M. J., Lin, C. K. & Beveridge, M. C. M. 1993. Shrimp culture and the environment: lessons from the world's most rapidly expanding warm water aquaculture sector. In Anon. (ed.), *Environment and Aquaculture in Developing Countries.* ICLARM Conference Proceedings, Manila, Philippines.

Pickett, S. T. A. & White, P. S. 1985. Patch dynamics: a synthesis. In S. T. A. Pickett & P. S. White (eds), *The Ecology of Natural Disturbance and Patch Dynamics*, Academic Press, New York, pp. 371–84.

Pikitch, E. K., Santora, C., Babcock, E. A., Bakun, A., Bonfil, R., Conover, D. O., Dayton, P., Doukakis, P., Fluharty, D., Heneman, B., Houde, E. D., Link, J., Livingston, P. A., Mangel, M., McAllister, M. K., Pope, J. & Sainsbury, K. J. 2004. Ecosystem-based fishery management. *Science* 305: 346–7.

Pitcher, T. J. & Hart, P. J. B. 1982. *Fisheries Ecology*. Croom Helm, Beckenham.

Platt, T., Fuentes-Yaco, C. & Frank, K. T. 2003. Spring algal bloom and larval fish survival. *Nature* 423: 398–9.

Polunin, N. V. C. 1996. Trophodynamics of reef fisheries productivity. In N. V. C. Polunin & C. M. Roberts (eds), *Reef Fisheries*, Chapman & Hall, London, pp. 113–35.

Polunin, N. V. C. (ed.) 2005. *Aquatic Ecosystems in 2025*. In press. Cambridge University Press, Cambridge.

Pomeroy, L. R. 1974. The ocean's food web, a changing paradigm. *Bioscience* 24: 499–504.

Pongthanapanich, T. 1996. Economic study suggests management guidelines for mangroves to derive optimal economic and social benefits. *Aquaculture Asia* 1: 16–17.

Posey, M. H. 1986. Changes in a benthic community associated with dense beds of a burrowing deposit feeder *Callianassa californiensis*. *Marine Ecology Progress Series* 31: 15–22.

Post, D. M., Taylor, J. P., Kitchell, J. F., Olson, M. H., Schindler, D. E. & Herwig, B. R. 1998. The role of migratory waterfowl as nutrient vectors in managed wetlands. *Conservation Biology* 12: 910–20.

Preen, A. R. 1995. Impacts of dugong foraging on seagrass habitats: observational and experimental evidence of cultivation grazing. *Marine Ecology Progress Series* 124: 201–13.

Priede, I. G., Bagley, P., Armstrong, J. D., Smith, K. L., Jr. & Merrrett, N. R. 1991. Direct measurement of active dispersal of food-falls by abyssal demersal fishes. *Nature* 351: 647–9.

Purcell, E. M. 1977. Life at low Reynolds number. *American Journal of Physics* 45: 3–11.

Raffaelli, D. G. & Hall, S. J. 1992. Compartments and predation in an estuarine food web. *Journal of Animal Ecology* 61: 551–60.

Raffaelli, D. G. & Hawkins, S. J. 1996. *Intertidal Ecology*. Chapman & Hall, London.

Raffaelli, D. G. & Moller, H. 2000. Manipulative experiments in animal ecology – do they promise more than they can deliver? *Advances in Ecological Research* 30: 299–330.

Raffaelli, D. G., Karakassis, I. & Galloway, A. 1991. Zonation schemes on sandy shores: a multivariate approach. *Journal of Experimental Marine Biology and Ecology* 148: 241–53.

Rahmstorf, S. 2002. Ocean circulation and climate during the past 120,000 years. *Nature* 419: 207–14.

Ramage, D. L. & Schiel, D. R. 1998. Reproduction in the seagrass *Zostera novazelandica* on intertidal platforms in southern New Zealand. *Marine Biology* 130: 479–89.

Ramsay, K., Kaiser, M. J. & Hughes, R. N. 1997. A field study of intraspecific competition for food in hermit crabs (*Pagurus bernhardus*). *Estuarine Coastal and Shelf Science* 44: 213–20.

Randall, D. J. & Farrell, A. P. 1997. *Deep-sea Fishes*, Fish Physiology Series No. 16. Academic Press.

Raven, J. A. 1997. CO_2 concentrating mechanisms: a direct role for thylakoid lumen acidification? *Plant, Cell and Environment* 20: 147–54.

Raven, J. A. & Beardall, J. 2005. Respiration in Aquatic Photolithotrophs. In P. A. del Giorgio & P. J. le B. Williams (eds), *Respiration in Aquatic Ecosystems*, Oxford University Press, Oxford, pp. 36–46.

Raven, J. A. & Geider, R. J. 2003. Adaptation, acclimation and regulation in algal photosynthesis. In A. W. D. Larkum, S. Douglas & J. A. Raven (eds), *Photosynthesis of Algae*, Kluwer Academic Publishers, pp. 385–412.

Ray, G. C. 1997. Do the metapopulation dynamics of estuarine fishes influence the stability of shelf ecosystems? *Bulletin of Marine Science* 60(3): 1040–49.

Reid, P. C., Borges, M. D. & Svendsen, E. 2001. A regime shift in the North Sea circa 1988 linked to changes in the North Sea horse mackerel fishery. *Fisheries Research* 50: 163–71.

Reimnitz, E., Kempema, E. W. & Barnes, P. W. 1987. Anchor ice, seabed freezing, and sediment dynamics in shallow Arctic seas. *Journal of Geophysical Research* 92: 14671–8.

Reise, K. 1985. *Tidal Flat Ecology. An Experimental Approach to Species Interactions*. Springer Verlag, Berlin.

Reise, K., Gollasch, S. & Wolff, W. J. 1999. Introduced marine species of the North Sea coasts. *Helgolander Meeresuntersuchungen* 52: 219–34.

Reusch, T. B. H., Stam, W. T. & Olsen, J. L. 1999a. Microsatellite loci in eelgrass *Zostera marina* reveal marked polymorphism within and among populations. *Molecular Ecology* 8: 317–21.

Reusch, T. B. H., Boström, C., Stam, W. T., Olsen, J. L. 1999b. An ancient eelgrass clone in the Baltic. *Marine Ecology Progress Series* 183: 301–4.

Rex, M. A. 1981. Community structure in the deep-sea benthos. *Annual Review of Ecology and Systematics* 12: 331–53.

Rex, M. A., Stuart, C. T., Hessler, R. R., Allen, J. A., Sanders H. L. & Wilson, G. D. F. 1993. Global scale latitudinal patterns of species diversity in the deep-sea benthos. *Nature* 365: 636–9.

Rhoads, D. C. 1974. Organism-sediment relations on the muddy sea floor. *Oceanography and Marine Biology Annual Review* 12: 263–300.

Rhoads, D. C., Boesch, D. F., Tang, Z.-C., Xu, F.-S., Huang, L.-Q. & Nilsen, K. J. 1985. Macrobenthos and sedimentary facies on the Changjiang delta platform and adjacent continental shelf, East China Sea. *Continental Shelf Research* 4: 189–213.

Richardson, L. L. 1998. Coral diseases: what is really known? *Trends in Ecology and Evolution* 13: 438–43.

Richmond, R. H. 1997. Reproduction and recruitment in corals: critical links in the persistence of reefs. In C. Birkeland (ed.), *Life and death of coral reefs*, Chapman & Hall, New York, pp. 175–97.

Riebesell, U., Schloss, I. & Smetacek, V. 1991. Aggregation of algae released from melting sea ice: implications for seeding and sedimentation. *Polar Biology* 11: 239–48.

Riebesell, U., Burkhardt, S., Dauelsberg, A. & Kroon, B. 2000. Carbon isotope fractionation by a marine diatom: dependence on the growth-rate-limiting resource. *Marine Ecology Progress Series* 193: 295–303.

Riemann, F. & Sime-Ngando, T. 1997. Note on sea ice nematodes (Monhysteroidea) from Resolute Passage, Canadian High Arctic. *Polar Biology* 18: 70–5.

Robertson, A. I. 1986. Leaf-burying crabs: their influence on energy flow and export from mixed mangrove forests (*Rhizophora* spp.) in northeastern Australia. *Journal of Experimental Marine Biology and Ecology* 102: 237–48.

Robertson, A. I., Alongi, D. M. & Boto, K. G. 1992. Food chains and carbon fluxes. In A. I. Robertson & D. M. Alongi (eds), *Tropical Mangrove Ecosystems*, American Geophysical Union, NY, pp. 293–326.

Robinson, C. & Williams, P. J. le B. 2005. Respiration and its measurement in surface marine waters. In P. A. del Giorgio & P. J. le B. Williams (eds), *Respiration in Aquatic Ecosystems*, Oxford University Press, Oxford, pp. 147–80.

Rochette, R., Tetreault, F. & Himmelman, J. H. 2001. Aggregation of whelks, *Buccinum undatum*, near feeding predators: the role of reproductive requirements. *Animal Behaviour* 61: 31–41.

Rodhouse, P. G., Prince, P. A., Trathan, P. N., Hatfield, E. M. C., Watkins, J. L., Bone, D. G., Murphy, E. J. & White, M. G. 1996. Cephalopods and mesoscale oceanography at the Antarctic Polar Front: Satellite tracked predators locate pelagic trophic interactions. *Marine Ecology Progress Series* 136: 37–50.

Rodhouse, P. G., Elvidge, C. D. & Trathan, P. N. 2001. Remote sensing of the global light-fishing fleet: An analysis of interactions with oceanography, other fisheries and predators. *Advances in Marine Biology* 39: 261–303.

Roelke, D. L. & Buyukates, Y. 2002. Dynamics of phytoplankton succession couplesd to species diversity as a system-level tool for study of *Microcystis* population dynamics in eutrophic lakes. *Limnology and Oceanography* 47: 1109–18.

Rogers, A. D. 2000. The role of the oceanic oxygen minima in generating biodiversity in the deep sea. *Deep-Sea Research Part II – Topical Studies in Oceanography* 47: 119–48.

Rose, C. D., Sharp, W. C., Kenworthy, W. J., Hunt, J. H., Lyons, W. G., Prager, E. J., Valentine, J. F., Hall, M. O., Whitfield, P. E. & Fourqurean, J. W. 1999. Overgrazing of a large seagrass bed by the sea urchin *Lytechinus variegatus* in outer Florida bay. *Marine Ecology Progress Series* 190: 211–22.

Rosenberry, R. 1994. World shrimp farming 1994. Annual Report. *Shrimp News International*, San Diego, California, USA.

Rowden, A. A., Jones, M. B. & Morris, A. W. 1998. The role of *Callianassa subterranea* (Montagu) (Thalassinidia) in sediment resuspension in the North Sea. *Continental Shelf Research* 18: 1365–80.

Rowe, G. T., Polloni, P. T. & Haedrich, R. L. 1982. The deep-sea macrobenthos on the continental margin of the NW Atlantic Ocean. *Deep-Sea Research* 21: 641–50.

Roy, K., Jablonski, D., Valentine, J. W. & Rosenberg, G. 1998. Marine latitudinal diversity gradients: tests of causal hypotheses. *Proceedings of the National Academy of Sciences USA* 95: 3699–702.

Rudnick, D. L. & Davis, R. E. 2003. Red noise and regime shifts. *Deep-Sea Research Part I – Oceanographic Research Papers* 50: 691–9.

Russ, G. R. 1994. Sumilon Island Reserve: 20 years of hopes and frustrations. *Naga* 7: 8–12.

Russ, G. R. & Alcala, A. C. 1996. Do marine reserves export adult fish biomass- evidence from Apo Island, Central Philippines. *Marine Ecology Progress Series* 132: 1–9.

Russell, G. & Veltkamp, C. J. 1984. Epiphyte survival on skin-shedding macrophytes. *Marine Ecology Progress Series* 18: 149–53.

Russell, G., Hawkins, S. J., Evans, L. C., Jones, H. D. & Holmes, G. D. 1983. Restoration of a disused dock basin as a habitat for marine benthos and fish. *Journal of Applied Ecology* 20: 43–58.

Ruxton, G. D. & Houston, D. C. 2004. Energetic feasibility of an obligate marine scavenger. *Marine Ecology Progress Series* 266: 59–63.

Ryther, J. H. 1969. Potential productivity of the sea. *Science* 130: 602–8.

Sabine, C. L. +14 others 2004. The oceanic sink for anthropogenic CO_2. *Science* 305: 367–71.

Sainsbury, J. C. 1996. *Commercial fishing methods: an introduction to vessels and gears*. Fishing News Books, Oxford.

Saintilan, N. 1997. Above- and below-ground biomasses of two species of mangrove on the Hawkesbury River estuary, New South Wales. *Marine and Freshwater Research* 48: 147–52.

Sale, P. F. (ed.) 2002. *Coral Reef Fishes: Dynamics and Diversity in a Complex Ecosystem*. Academic Press, San Diego.

Salvat, B. 1964. Les conditions hydrodynamiques interstitielles des sediments meubles intertidaux et la repartition verticale de la faune endognee. *C. R. Acad. Sci. Paris* 259: 1576–9.

Sarmiento, J. L., Gruber, N., Brzezinski, M. & Dunne, J. P. 2004. High-latitude controls of thermocline nutrients and low latitude biological productivity. *Nature*, 427: 56–60.

Scanlan, D. J. & West, N. J. 2002. Molecular ecology of the marine cyanobacterial genera *Prochlorococcus* and *Synechococcus*. *FEMS Microbiology Ecology* 40: 1–12.

Schafer, W. 1972. *Ecology and Palaeoecology of Marine Environments*. University of Chicago Press, Chicago.

Scheffer, M., Carpenter, S., Foley, J. A., Folkes, C. & Walker, B. 2001. Catastrophic shifts in ecosystems. *Nature* 413: 591–6.

Schnack-Schiel, S. 2003. The macrobiology of sea ice. In D. N. Thomas & G. S. Dieckmann (eds), *Sea Ice – An Introduction to its Physics, Chemistry, Biology and Geology*, Blackwells Publishing, Oxford.

Schnack-Schiel, S. B. +5 others 2001. Meiofauna in sea ice of the Weddell Sea (Antarctica). *Polar Biology* 24: 724–8.

Schofield, O., Bergmann, T., Bissett, P., Grassle, J. F., Haidvogel, D. B., Kohut, J., Moline, M. & Glenn, S. M. 2002. The long-term ecosystem observatory: An integrated coastal observatory. *IEEE Journal of Oceanic Engineering* 27: 146–54.

Schwamborn, R., Ekau, W., Voss, M. & Saint-Paul, U. 2002. How important are mangroves as a carbon source for decapod crustacean larvae in a tropical estuary? *Marine Ecology Progress Series* 229: 195–205.

Seibel, B. A., Thuesen, E. V. & Childress, J. J. 2000. Light-limitation on predator-prey interactions: Consequences for metabolism and locomotion of deep-sea cephalopods. *Biological Bulletin* 198: 284–98.

Sepkoski, J. J., McKinney, F. K. & Lidgard, S. 2000. Competitive displacement among post-Paleozoic cyclostome and cheilostome bryozoans. *Paleobiology* 26: 7–18.

Sheldon, R. W., Prakash, A. & Sutcliffe, W. H. 1972. The size distribution of particles in the ocean. *Limnology and Oceanography* 17: 327–40.

Sheppard, C. R. C., Price, A. & Roberts, C. M. 1992. *Marine Ecology of the Arabian Region*. San Diego: Academic Press.

Sherr, E. & Sherr, B. 2000. Marine microbes: an overview. In D. L. Kirchman (ed.), *Microbial Ecology of the Oceans*, John Wiley & Sons, Inc. New York, pp. 13–46.

Shiomoto, A., Tadokoro, K. & Nagasawa, K. 1997. Trophic relations in the subarctic North Pacific ecosystem: possible feeding effects from pink salmon. *Marine Ecology Progress Series* 150: 75–85.

Short, F. T., Burdick, D. M. & Kaldy, J. E. 1995. Mesocosm experiments quantify the effects of eutrophication on eelgrass, *Zostera marina*. *Limnology and Oceanography* 40: 740–9.

Shpigel, M., Neori, A., Popper, D. M. & Gordin, H. 1993. A proposed model for 'environmentally clean' land-based culture of fish, bivalves and seaweeds. *Aquaculture* 117: 115–28.

Siegenthaler, U. & Sarmiento, J. L. 1993. Atmospheric carbon dioxide and the ocean. *Nature* 365: 119–25.

Skei, J., Larsson, P., Rosenberg, R., Jonsson, P., Olsson, M. & Broman, D. 2000. Eutrophication and contaminants in aquatic ecosystems. *Ambio* 29: 184–94.

Smart, C. W. & Gooday, A. J. 1997. Recent benthic foraminifera in the Abyssal northeast Atlantic Ocean: Relation to phytodetrital inputs. *Journal of Foriminiferal Research* 27: 85–92.

Smayda, T. J. 1990. Novel and nuisance phytoplankton bolloms in the sea: evidence for a global epidemic. In E. Graneli, B. Sundstom, L. Edler & D. M. Anderson (eds), *Toxic Marine Phytoplankton*, Elsevier, Amsterdam, pp. 29–40.

Smetacek, V. & Passow, U. 1990. Spring bloom initiation and Sverdrup's critical-depth model. *Limnology and Oceanography* 35: 228–34.

Smith, C. R. & Baco, A. R. 2003. Ecology of whale falls at the deep-sea floor. *Oceanography and Marine Biology* 41: 311–54.

Smith, I. P., Jensen, A. C., Collins K. J. & Mattey, E. L. 2001. Movement of wild European lobsters *Homarus gammarus* in natural habitat. *Marine Ecology Progress Series* 222: 177–86.

Smith, T. D. 1994. *Scaling fisheries: the science of measuring the effects of fishing, 1855–1955*. Cambridge: Cambridge University Press.

Smith, T. J. 1992. Forest Structure. In A. I. Robertson & D. M. Alongi (eds), *Tropical Mangrove Ecosystems*, American Geophysical Union, New York, pp. 101–36.

Smith, T. J., Boto, K. G., Frusher, S. D. & Giddins, R. L. 1991. Keystone species and mangrove forest dynamics: the influence of burrowing by crabs on soil nutrient status and forest productivity. *Estuarine Coastal and Shelf Science* 33: 419–32.

Snelgrove, P. & Butman, C. 1994. Animal-sediment relationships revistied: cause versus effect. *Oceanography and Marine Biology: An Annual Review* 32: 111–77.

Snowdon, P. et al. 2000. Synthesis of allometrics, review of root biomass and design of future woody biomass sampling strategies. Australian Greenhouse Office, Report No. 17.

Solan, M. & Kennedy, R. 2002. Observation and quantification of in situ animal-sediment relationships using time-lapse sediment profile imagery (t-SPI). *Marine Ecology Progress Series* 228: 179–91.

Somerfield, P. J., Olsgard, F. & Carr, M. 1997. A further examination of two new taxonomic distinctness measures. *Marine Ecology Progress Series* 154: 303–6.

Sommer, U. (ed.) 1989. *Plankton Ecology*. Springer Verlag, Berlin.

Sommer, U., Stibor, H., Katechakis, A., Sommer, F. & Hansen, T. 2002. Pelagic food web configurations at different levels of nutrient richness and their implications for the ratio fish production: primary production. *Hydrobiologia* 484: 11–20.

Sousa, W. P., Quek, S. P. & Mitchell, B. J. 2003. Regeneration of *Rhizophora mangle* in a Caribbean mangrove forest: interacting effects of canopy disturbance and a stem-boring beetle. *Oecologia* 137: 436–45.

Southward, A. & Southward, E. 1978. Recolonization of rockey shores in Cornwall after use of toxic dispersants to clean up the Torrey Canyon spill. *Journal of the Fisheries Research Board of Canada* 35: 682–706.

Spalding, M. D., Ravilious, C. & Green, E. P. 2001. *World Atlas of Coral Reefs*. University of California Press, Berkeley.

Spindler, M. 1994. Notes on the biology of the sea ice zones in the Arctic and Antarctic. *Polar Biology* 14: 319–24.

Spurgeon, J. P. G. 1992. The economic valuation of coral reefs. *Marine Pollution Bulletin* 24: 529–36.

Squire, V. A., Dugan, J. P., Wadhams, P., Rottier, P. J. & Liu, A. K. 1995. Of ocean waves and sea-ice. *Annual Reviews of Fluid Mechanics* 27: 115–68.

Stanhope, M. J., Hartwick, B. & Baillie, D. 1993. Molecular phylogeographic evidence for multiple shifts in habitat preference in the diversification of an amphipod species. *Molecular Ecology* 2: 99–112.

Stehli, F. G. & Wells, J. W. 1971. Diversity and age patterns in hermatypic corals. *Systematic Zoology* 20: 115–26.

Steneck, R. S., Graham, M. H., Bourque, B. J., Corbett, D., Erlandson, J. M., Estes, J. A. & Tegner, M. J. 2002. Kelp forest ecosystems: biodiversity, stability, resilience and future. *Environmental Conservation* 29: 436–59.

Stephenson, T. A. & Stephenson, A. 1949. The universal features of zonation between tidemarks on rocky coasts. *Journal of Ecology* 38: 289–305.

Stephenson, T. A. & Stephenson, A. 1972. *Life Between Tidemarks on Rocky Shores*. W. H. Freeman, San Francisco.

Stoddart, D. R. 1984. Coral reefs of Seychelles and adjacent regions. In D. R. Stoddart (ed.), *Biogeography and Ecology of the Seychelles Islands*, Dr W Junk, The Hague, pp. 63–81.

Stoecker, D. K. 1999. Mixotrophy among dinoflagellates. *Journal of Eukaryotic Microbiology* 46: 397–401.

Strass, V. H. & Nöthig, E.-M. 1996. Seasonal shifts in ice edge phytoplankton blooms in the Barents Sea related to the water column stability. *Polar Biology* 16: 409–22.

Strom, S. L. 2000. Bactivory: interactions between Bacteria and their Grazers. In D. L. Kirchman (ed.), *Microbial Ecology of the Oceans*, John Wiley & Sons, Inc. New York, Chapter 12 pp. 351–86.

Struder-Kypke, M. C. & Montagnes, D. J. S. 2002. Development of web-based guides to planktonic protists. *Aquatic Microbial Ecology* 27: 203–7.

Suchanek, A. 1992. Extreme biodiversity in the marine environment: mussel bed communities of *Mytilus californianus*. *Northwest Environ. J.* 8: 150–2.

Suggett, D. J., Oxborough, K., Baker, N. R., MacIntyre, H. L., Kana, T. M. & Geider, R. J. 2003. Fast repetition rate and pulse amplitude modulation chlorophyll-a fluorescence measurements for assessment of photosynthetic electron transport in marine phytoplankton. *European Journal of Phycology* 38: 371–84.

Sutherland, W. J. 1996. *From Individual Behaviour to Population Ecology*. Oxford University Press, Oxford.

Sutherland, W. J. (ed.) 1998. *Conservation Science and Action*. Blackwell, Oxford.

Svånstad, T. & Kristiansen, T. S. 1990. Enhancement studies of coastal cod in western Norway. Part II. Migration of reared coastal cod. *Journal du Conseil International pour l'Exploration de la Mer* 47: 13–22.

Sverdrup, H. U. 1953. On conditions for the vernal blooming of phytoplankton. *J. Cons. Perm. Int. Exp.* 18: 237–95.

Takahashi, T. 2004. Enhanced: The fate of industrial carbon dioxide. *Science* 305: 352–53.

Takahashi, T., Olafsson, J., Goddard, J. G., Chipman, D. W. & Sutherland, S. C. 1993. Seasonal variations of CO_2 and nutrients in the high-latitude surface oceans: a comparative study. *Global Biogeochemical Cycles* 7: 843–78.

Tanaka, M. & Nandakumar, K. 1994. Measurement of the degree of intransitivity in a community of sessile organisms. *Journal of Experimental Marine Biology and Ecology* 182: 85–95.

Tarling, G. A. 2003. Sex-dependent diel vertical migration in northern krill *Meganyctiphanes norvegica* and its consequences for population dynamics. *Marine Ecology Progress Series* 260: 173–88.

Taylor, A. H. & Gangopadhyay, A. 2001. A simple model of interannual displacements of the Gulf Stream. *Journal of Geophysical Research – Oceans* 106: 13849–60.

Tedengren, M., André, C., Johannesson, K. & Kautsky, N. 1990. Genotypic and phenotypic differences between Baltic and North Sea populations of *Mytilus edulis* evaluated through reciprocal transplantations. III. Physiology. *Marine Ecology Progress Series* 59: 221–9.

Tegner, M. J. & Dayton, P. K. 2000. Ecosystem effects of fishing in kelp forest communities. *ICES Journal of Marine Sciences*.

Thayer, G. W., Bjorndal, K. A., Ogden, J. C., Williams, J. L. & Zieman, J. C. 1984. Role of larger herbivores in seagrass communities. *Estuaries* 7: 351–76.

Theil, H. & Schriever, G. 1990. Deep-sea mining, environmental impact and the DISCOL project. *Ambio* 19: 245–50.

Thomas, D. N. & Dieckmann, G. S. 2002. Antarctic sea ice – a habitat for extremophiles. *Science* 295: 641–4.

Thompson, D., Moss, S. E. W. & Lovell, P. 2003. Foraging behaviour of South American fur seals *Arctocephalus australis*: extracting fine scale foraging behaviour from satellite tracks. *Marine Ecology Progress Series* 260: 285–96.

Thompson, R. C., Crowe, T. P. & Hawkins, S. J. 2002. Rocky intertidal communities: past environmental changes, present status and predictions for the next 25 years. *Environmental Conservation* 29: 168–91.

Thorpe, S. E., Heywood, K. J., Stevens, D. P. & Brandon, M. A. 2004. Tracking passive drifters in a high resolution ocean model: implications for interannual variability of larval krill transport to South Georgia. *Deep-Sea Research Part I – Oceanographic Research Papers* 51: 909–20.

Thorpe, S., Landeghem, K., Hogan, L. & Holland, P. 1997. *Economic Effects on Australian Southern Bluefin Tuna Farming of a Quarantine Ban on Imported Pilchards*. Australian Bureau of Agricultural and Resource Economics, Canberra, Australia.

Thorson, G. 1957. Bottom communities. In G. W. Hedgpeth (ed.) *Treatise on Marine Ecology and Paleoecology, 1 Ecology*. Geological Society of America, New York, pp. 461–534.

Thorson, G. 1971. *Life in the Sea*. Weidenfield & Nicholson, London.

Thrush, S. F., Pridmore, R. D., Hewitt, J. E. & Cummings, V. J. 1991. Impact of ray feeding disturbances on sandflat macrobenthos: do communities dominated by polychaetes or shellfish respond differently? *Marine Ecology Progress Series* 69: 245–52.

Tomlinson, P. B. 1986. *The Botany of Mangroves*. Cambridge University Press: Cambridge.

Tuck, G. N., Polacheck, T., Croxall, J. P. & Weimerskirch, H. 2001. Modelling the impact of fishery by-catches on albatross populations. *Journal of Applied Ecology* 38: 1182–96.

Tudhope, A. W., Chilcot, C. P., McCulloch, M. T., Cook, E. R., Chappell, J., Ellam, R. M., Lea, D. W., Lough, J. M & Shimmield, G. B. 2001. Variability in the El Niño-Southern Oscillation through a Glacial-Interglacial Cycle. *Science* 291: 1511–17.

Turner, R. E. & Rabalais, N. N. 1994. Coastal eutrophication near the Mississippi River delta. *Nature* 368: 619.

Uchino, O., Bojkov, R. D., Balis, D. S., Akagi, K., Hayashi M. & Kajihara, R. 1999. Essential characteristics of the Antarctic-spring ozone decline: Update to 1998. *Geophysical Research Letters* 26: 1377–80.

Underwood, A. J. 1978. A refutation of critical tidal levels as determinants of the structure of intertidal communities on British shores. *Journal of Experimental Marine Biology and Ecology* 33: 261–76.

Underwood, A. J., Jernakoff, P. 1981. Effects of interactions between algae and grazing gastropods on the structure of a low intertidal algal community. *Oecologia* 48: 221–33.

Underwood, A. J. 1981. Techniques of analysis of variance in experimental marine biology and ecology. *Oceanography and Marine Biology: Annual Review* 19: 513–605.

Underwood, A. J. 1991. Beyond BACI: Experimental designs for detecting human environmental impacts on temporal variations in natural populations. *Australian Journal of Marine and Freshwater Research* 42: 569–88.

Underwood, A. J. 1997. *Experiments in Ecology: Their Design and Interpretation Using Analysis of Variance*. Cambridge University Press, Cambridge.

UNDP 2003. Human Development Report 2003. Oxford University Press, New York.

Van Dover, C. L. 2000. *The Ecology of Deep-Sea Hydrothermal Vents*. Princeton University Press, Princeton, NJ.

van Duren, L. A. & Videler, J. J. 2003. Escape from viscosity: the kinematics and hydrodynamics of copepod foraging and escape swimming. *Journal of Experimental Biology* 206: 269–79.

Van Katwijk, M. M. 2003. Reintroduction of eelgrass (*Zostera marina* L.) in the Dutch Wadden Sea: a research overview and management vision. In W. J. Wolff, K. Essink, A. Kellerman & M. A. van Leeuwe (eds), *Challenges to the Wadden Sea*, Ministry of Agriculture, Nature Management and Fisheries, The Netherlands, pp. 173–95.

Vannini, M., Cannicci, S. & Ruwa, K. 1995. Effect of light intensity on vertical migrations of the tree crab, *Sesarma leptosoma*. *Journal of Experimental Marine Biology and Ecology* 185: 181–9.

Vaughan, D. & Doake, C. 1996. Recent atmospheric warming and retreat of ice shelves on the Antarctic Peninsula. *Nature* 379: 328–31.

Verity, P. G. & Smetacek, V. 1996. Organism life cycles, predation, and the structure of marine pelagic ecosystems. *Marine Ecology Progress Series* 130: 277–93.

Veron, J. E. N. 2000. *Corals of the World*. Australian Institute of Marine Science, Townsville, Queensland, Australia.

Vincent, W. F. & Roy, S. 1993. Solar ultraviolet-B radiation and aquatic primary production: damage, protection and recovery. *Environment Review* 1: 1–12.

Vrijenhoek, R. C., Shank, T. & Lutz, R. A. 1998. Gene flow and dispersal in deep-sea hydrothermal vent animals. *Cahiers de Biologie Marine* 39: 363–6.

Wadhams, P. 2000. *Ice in the Ocean*. Gordon & Breach Science. Reading, UK.

Walther, G.-R., Post, E., Convey, P., Menzel, A., Parmesan, C., Beebee, T. J. C., Fromentin, J.-M., Hoegh-Guldberg, O. & Bairlein, A. 2002. Ecological response to recent climate change. *Nature* 416: 389–95.

Waluda, C. M., Rodhouse, P. G., Podesta, G. P., Trathan, P. N. & Pierce, G. J. 2001. Surface oceanography of the inferred hatching grounds of *Illex argentinus* (Cephalopoda : Ommastrephidae) and influences on recruitment variability. *Marine Biology* 139: 671–9.

Wapnick, C. M., Precht, W. F. & Aronson, R. B. 2004. Millennial-scale dynamics of staghorn coral in Discovery Bay, Jamaica. *Ecology Letters* 7: 354–61.

Warwick, R. M. 1986. A new method for detecting pollution effects on marine macrobhenthic communities. *Marine Biology* 92: 557–62.

Warwick, R. M. & Uncles, R. J. 1980. Distribution of benthic macrofauna associations in the Bristol Channel in relation to tidal stress. *Marine Ecology Progress Series* 3: 97–103.

Warwick, R. M., Clarke, K. R. & Gee, J. M. 1990. The effect of disturbance by soldier crabs *Mictyris platycheles* H. Milne Edwards on meiobenthic community structure. *Journal of Experimental Marine Biology and Ecology* 135: 19–33.

Watanabe, T. & Nomura, M. 1990. Current status of aquaculture in Japan. In M. Mohan Joseph (ed.), *Aquaculture in Asia*, Asian Fisheries Society, Indian Branch, pp. 223–53.

Watson, R. & Pauly, D. 2001. Systematic distortions in world fisheries catch trends. *Nature* 414: 534–36.

Weaver, A. J. & Hillaire-Marcel, C. 2004. Global warming and the next ice age. *Science* 304: 400–2.

Webster, P. J., Rowden, A. A. & Attrill, M. J. 1998. Effect of shoot density on the infaunal macroinvertebrate community within a *Zostera marina* seagrass bed. *Estuarine Coastal and Shelf Science* 47: 351–8.

Weimerskirch, H., Brothers, N. & Jouventin, P. 1997. Population dynamics of wandering albatross *Diomedea exulans* and Amsterdam albatross *D. amsterdamensis* in the Indian Ocean and their relationships with long-line fisheries: conservation implications. *Biological Conservation* 79: 257–70.

Weissenberger, J., Dieckmann, G. S., Gradinger, R. & Spindler, M. 1992. Sea ice: A cast technique to examine and analyse brine pockets and channel structure. *Limnology and Oceanography* 37: 179–83.

White, W. B. & Annis, J. L. 2004. The influence of the Antarctic circumpolar wave on El Niño from 1950 through 2001. *Journal of Geophysical Research* 109, C06019, doi: 10. 1029/2002JC001666.

White, W. B. & Peterson, R. 1996. An Antarctic Circumpolar Wave in surface pressure, wind, temperature, and sea ice extent. *Nature* 380: 699–702.

Whittaker, R. H. 1974. *Communities and Ecosystems*. MacMillan, New York.

Widdicombe, S., Austen, M. C., Kendall, M. A., Warwick, R. M. & Jones, M. B. 2000. Bioturbation as a mechanism for setting and maintaining levels of diversity in subtidal macrobenthic communities. *Hydrobiologia* 440: 367–75.

Wiebe, P. H. & Benfield, M. C. 2003. From the Hensen net toward four-dimensional biological oceanography. *Progress in Oceanography* 56: 7–136.

Wildish, D. J. & Peer, D. 1983. Tidal current speed and production of benthic macrofauna in the lower Bay of Fundy. *Canadian Journal of Fisheries and Aquatic Sciences* 40 (Suppl. 1): 309–21.

Williams, P. J. le B. 1998. The balance of plankton respiration and photosynthesis in the open oceans. *Nature* 394: 55–7.

Williams, P. J. le B. & del Giorgio, P. A. 2005. Respiration in aquatic ecosystems: history and background. In P. A. del Giorgio & P. J. le B. Williams (eds), *Respiration in Aquatic Ecosystems*, Oxford University Press, Oxford, pp. 1–17.

Williams, S. L. & Ruckelshaus, M. H. 1993. Effects of nitrogen availability and herbivory on eelgrass (*Zostera marina*) and epiphytes. *Ecology* 74: 904–18.

Winter, C., Moeseneder, M. M. & Herndl, G. J. 2001. Impact of UV radiation on bacterioplankton community composition. *Applied and Environmental Microbiology* 67: 665–72.

Wolfe, G. V. 2000. The chemical defense ecology of marine unicellular plankton: Constraints, mechanisms, and impacts. *Biological Bulletin* 198: 225–44.

Worm, B., Lotze, H. K., Hillebrand, H. & Sommer, U. 2002. Consumer versus resource control of species diversity and ecosystem functioning. *Nature* 417: 848–51.

Wotton, R. S. 2004. The Essential Role of Exopolymers (EPS) in Aquatic Ecosystems. *Oceanography and Marine Biology: An Annual Review* 42: 57–94.

Yamaguchi, A., Watanabe, Y., Ishida, H., Harimoto, T., Furusawa, K., Suzuki, S., Ishizaka, J., Ikeda, T. & Masayuki, M. T. 2002. Community and trophic structures of pelagic copepods down to greater depths in the western subarctic Pacific (WEST-COSMIC). *Deep-Sea Research Part I – Oceanographic Research Papers* 49: 1007–25.

Yates, M. G., Stillman, R. A. & Goss-Custard, J. D. 2000. Contrasting interference functions and foraging dispersion in two species of shorebird (Charadrii). *Journal of Animal Ecology* 69(2): 314–22.

Yodzis, P. 1998. Local trophodynamics and the interaction of marine mammals and fisheries in the Benguela ecosystem. *Journal of Animal Ecology* 67: 635–58.

Yodzis, P. 2000. Diffuse effects in food webs. *Ecology* 81: 261–6.

Young, J. R. & Ziveri, P. 2000. Calculation of coccolith volume and its use in calibration of carbonate flux estimates. *Deep-Sea Research* (Part II), 47: 1679–1700.

INDEX

Page numbers for figures/tables are shown in **bold** type. Page numbers for main entries which have subheadings refer to general/introductory aspects of that topic. Species appear under their common/family names.